Values of Selected Physical Constants

Quantity	Symbol	SI Units	cgs Units
Avogadro's number	N_A	6.023×10^{23} molecules/mol	6.023×10^{23} molecules/mol
Boltzmann's constant	k	1.38×10^{-23} J/atom-K	1.38×10^{-16} erg/atom-K 8.62×10^{-5} eV/atom-K
Bohr magneton	μ_B	9.27×10^{-24} A-m^2	9.27×10^{-21} erg/gauss[a]
Electron charge	e	1.602×10^{-19} C	4.8×10^{-10} statcoul[b]
Electron mass	—	9.11×10^{-31} kg	9.11×10^{-28} g
Gas constant	R	8.31 J/mol-K	1.987 cal/mol-K
Permeability of a vacuum	μ_0	1.257×10^{-6} henry/m	unity[a]
Permittivity of a vacuum	ϵ_0	8.85×10^{-12} farad/m	unity[b]
Planck's constant	h	6.63×10^{-34} J-s	6.63×10^{-27} erg-s 4.13×10^{-15} eV-s
Velocity of light in a vacuum	c	3×10^8 m/s	3×10^{10} cm/s

[a] In cgs-emu units.
[b] In cgs-esu units.

Unit Abbreviations

A = ampere
Å = angstrom
Btu = British thermal unit
C = Coulomb
°C = degrees Celsius
cal = calorie (gram)
cm = centimeter
eV = electron volt
°F = degrees Fahrenheit
ft = foot
g = gram

in. = inch
J = joule
K = degrees Kelvin
kg = kilogram
lb_f = pound force
lb_m = pound mass
m = meter
Mg = megagram
mm = millimeter
mol = mole
MPa = megapascal

N = newton
nm = nanometer
P = poise
Pa = Pascal
s = second
T = temperature
μm = micrometer (micron)
W = watt
psi = pounds per square inch

SI Multiple and Submultiple Prefixes

Factor by Which Multiplied	Prefix	Symbol
10^9	giga	G
10^6	mega	M
10^3	kilo	k
10^{-2}	centi[a]	c
10^{-3}	milli	m
10^{-6}	micro	μ
10^{-9}	nano	n
10^{-12}	pico	p

[a] Avoided when possible.

Fundamentals of Materials Science and Engineering
An Interactive e · Text

FIFTH EDITION

Fundamentals of Materials Science and Engineering

An Interactive  . Text

William D. Callister, Jr.

Department of Metallurgical Engineering
The University of Utah

John Wiley & Sons, Inc.

New York   Chichester   Weinheim   Brisbane   Singapore   Toronto

Front Cover: The object that appears on the front cover depicts a monomer unit for polycarbonate (or PC, the plastic that is used in many eyeglass lenses and safety helmets). Red, blue, and yellow spheres represent carbon, hydrogen, and oxygen atoms, respectively.

Back Cover: Depiction of a monomer unit for polyethylene terephthalate (or PET, the plastic used for beverage containers). Red, blue, and yellow spheres represent carbon, hydrogen, and oxygen atoms, respectively.

Editor *Wayne Anderson*
Marketing Manager *Katherine Hepburn*
Associate Production Director *Lucille Buonocore*
Senior Production Editor *Monique Calello*
Cover and Text Designer *Karin Gerdes Kincheloe*
Cover Illustration *Roy Wiemann*
Illustration Studio *Wellington Studio*

This book was set in 10/12 Times Roman by Bi-Comp, Inc., and printed and bound by Von Hoffmann Press. The cover was printed by Phoenix Color Corporation.

This book is printed on acid-free paper. ⊚

The paper in this book was manufactured by a mill whose forest management programs include sustained yield harvesting of its timberlands. Sustained yield harvesting principles ensure that the number of trees cut each year does not exceed the amount of new growth.

To order books or for customer service call 1-800-CALL-WILEY (225-5945).

ISBN 0-471-39551-X

Printed in the United States of America

10 9 8 7 6 5 4 3 2 1

Preface

Fundamentals of Materials Science and Engineering is an alternate version of my text, **Materials Science and Engineering: An Introduction, Fifth Edition.** The contents of both are the same, but the order of presentation differs and **Fundamentals** utilizes newer technologies to enhance teaching and learning.

With regard to the order of presentation, there are two common approaches to teaching materials science and engineering—one that I call the "traditional" approach, the other which most refer to as the "integrated" approach. With the traditional approach, structures/characteristics/properties of metals are presented first, followed by an analogous discussion of ceramic materials and polymers. **Introduction, Fifth Edition** is organized in this manner, which is preferred by many materials science and engineering instructors. With the integrated approach, one particular structure, characteristic, or property for all three material types is presented before moving on to the discussion of another structure/characteristic/property. This is the order of presentation in **Fundamentals.**

Probably the most common criticism of college textbooks is that they are too long. With most popular texts, the number of pages often increases with each new edition. This leads instructors and students to complain that it is impossible to cover all the topics in the text in a single term. After struggling with this concern (trying to decide what to delete without limiting the value of the text), we decided to divide the text into two components. The first is a set of "core" topics—sections of the text that are most commonly covered in an introductory materials course, and second, "supplementary" topics—sections of the text covered less frequently. Furthermore, we chose to provide only the core topics in print, but the entire text (both core and supplementary topics) is available on the CD-ROM that is included with the print component of **Fundamentals.** Decisions as to which topics to include in print and which to include only on the CD-ROM were based on the results of a recent survey of instructors and confirmed in developmental reviews. The result is a printed text of approximately 525 pages and an *Interactive eText* on the CD-ROM, which consists of, in addition to the complete text, a wealth of additional resources including interactive software modules, as discussed below.

The text on the CD-ROM with all its various links is navigated using Adobe Acrobat™. These links within the *Interactive eText* include the following: (1) from the Table of Contents to selected *eText* sections; (2) from the index to selected topics within the *eText;* (3) from reference to a figure, table, or equation in one section to the actual figure/table/equation in another section (all figures can be enlarged and printed); (4) from end-of-chapter Important Terms and Concepts to their definitions within the chapter; (5) from in-text boldfaced terms to their corresponding glossary definitions/explanations; (6) from in-text references to the corresponding appendices; (7) from some end-of-chapter problems to their answers; (8) from some answers to their solutions; (9) from software icons to the corresponding interactive modules; and (10) from the opening splash screen to the supporting web site.

The interactive software included on the CD-ROM and noted above is the same that accompanies **Introduction, Fifth Edition.** This software, *Interactive Materials Science and Engineering, Third Edition* consists of interactive simulations and animations that enhance the learning of key concepts in materials science and engineering, a materials selection database, and *E-Z Solve: The Engineer's Equation Solving and Analysis Tool.* Software components are executed when the user clicks on the icons in the margins of the *Interactive eText;* icons for these several components are as follows:

Crystallography and Unit Cells   Tensile Tests

Ceramic Structures   Diffusion and Design Problem

Polymer Structures   Solid Solution Strengthening

Dislocations   Phase Diagrams

E-Z Solve   Database

My primary objective in **Fundamentals** as in **Introduction, Fifth Edition** is to present the basic fundamentals of materials science and engineering on a level appropriate for university/college students who are well grounded in the fundamentals of calculus, chemistry, and physics. In order to achieve this goal, I have endeavored to use terminology that is familiar to the student who is encountering the discipline of materials science and engineering for the first time, and also to define and explain all unfamiliar terms.

The second objective is to present the subject matter in a logical order, from the simple to the more complex. Each chapter builds on the content of previous ones.

The third objective, or philosophy, that I strive to maintain throughout the text is that if a topic or concept is worth treating, then it is worth treating in sufficient detail and to the extent that students have the opportunity to fully understand it without having to consult other sources. In most cases, some practical relevance is provided. Discussions are intended to be clear and concise and to begin at appropriate levels of understanding.

The fourth objective is to include features in the book that will expedite the learning process. These learning aids include numerous illustrations and photographs to help visualize what is being presented, learning objectives, "Why Study . . ." items that provide relevance to topic discussions, end-of-chapter questions and problems, answers to selected problems, and some problem solutions to help in self-assessment, a glossary, list of symbols, and references to facilitate understanding the subject matter.

The fifth objective, specific to **Fundamentals,** is to enhance the teaching and learning process using the newer technologies that are available to most instructors and students of engineering today.

Most of the problems in **Fundamentals** require computations leading to numerical solutions; in some cases, the student is required to render a judgment on the basis of the solution. Furthermore, many of the concepts within the discipline of materials science and engineering are descriptive in nature. Thus, questions have also been included that require written, descriptive answers; having to provide a written answer helps the student to better comprehend the associated concept. The questions are of two types: with one type, the student needs only to restate in his/her own words an explanation provided in the text material; other questions require the student to reason through and/or synthesize before coming to a conclusion or solution.

The same engineering design instructional components found in **Introduction, Fifth Edition** are incorporated in **Fundamentals.** Many of these are in Chapter 20, "Materials Selection and Design Considerations," that is on the CD-ROM. This chapter includes five different case studies (a cantilever beam, an automobile valve spring, the artificial hip, the thermal protection system for the Space Shuttle, and packaging for integrated circuits) relative to the materials employed and the rationale behind their use. In addition, a number of design-type (i.e., open-ended) questions/problems are found at the end of this chapter.

Other important materials selection/design features are Appendix B, "Properties of Selected Engineering Materials," and Appendix C, "Costs and Relative Costs for Selected Engineering Materials." The former contains values of eleven properties (e.g., density, strength, electrical resistivity, etc.) for a set of approximately one hundred materials. Appendix C contains prices for this same set of materials. The materials selection database on the CD-ROM is comprised of these data.

SUPPORTING WEB SITE

The web site that supports **Fundamentals** can be found at *www.wiley.com/college/callister.* It contains student and instructor's resources which consist of a more extensive set of learning objectives for all chapters, an index of learning styles (an electronic questionnaire that accesses preferences on ways to learn), a glossary (identical to the one in the text), and links to other web resources. Also included with the Instructor's Resources are suggested classroom demonstrations and lab experiments. Visit the web site often for new resources that we will make available to help teachers teach and students learn materials science and engineering.

INSTRUCTORS' RESOURCES

Resources are available on another CD-ROM specifically for instructors who have adopted **Fundamentals.** These include the following: 1) detailed solutions of all end-of-chapter questions and problems; 2) a list (with brief descriptions) of possible classroom demonstrations and laboratory experiments that portray phenomena and/or illustrate principles that are discussed in the book (also found on the web site); references are also provided that give more detailed accounts of these demonstrations; and 3) suggested course syllabi for several engineering disciplines.

Also available for instructors who have adopted **Fundamentals** as well as **Introduction, Fifth Edition** is an online assessment program entitled *eGrade.* It is a browser-based program that contains a large bank of materials science/engineering problems/questions and their solutions. Each instructor has the ability to construct homework assignments, quizzes, and tests that will be automatically scored, recorded in a gradebook, and calculated into the class statistics. These self-scoring problems/questions can also be made available to students for independent study or pre-class review. Students work online and receive immediate grading and feedback.

Tutorial and Mastery modes provide the student with hints integrated within each problem/question or a tailored study session that recognizes the student's demonstrated learning needs. For more information, visit *www.wiley.com/college/egrade.*

ACKNOWLEDGMENTS

Appreciation is expressed to those who have reviewed and/or made contributions to this alternate version of my text. I am especially indebted to the following individuals: Carl Wood of Utah State University, Rishikesh K. Bharadwaj of Systran Federal Corporation, Martin Searcy of the Agilent Technologies, John H. Weaver of The University of Minnesota, John B. Hudson of Rensselaer Polytechnic Institute, Alan Wolfenden of Texas A & M University, and T. W. Coyle of the University of Toronto.

I am also indebted to Wayne Anderson, Sponsoring Editor, to Monique Calello, Senior Production Editor, Justin Nisbet, Electronic Publishing Analyst at Wiley, and Lilian N. Brady, my proofreader, for their assistance and guidance in developing and producing this work. In addition, I thank Professor Saskia Duyvesteyn, Department of Metallurgical Engineering, University of Utah, for generating the *e-Grade* bank of questions/problems/solutions.

Since I undertook the task of writing my first text on this subject in the early 1980's, instructors and students, too numerous to mention, have shared their input and contributions on how to make this work more effective as a teaching and learning tool. To all those who have helped, I express my sincere thanks!

Last, but certainly not least, the continual encouragement and support of my family and friends is deeply and sincerely appreciated.

WILLIAM D. CALLISTER, JR.
Salt Lake City, Utah
August 2000

Contents

Chapters 14 through 21 discuss just supplementary topics, and are found only on the CD-ROM (and not in print)

List of Symbols

The number of the section in which a symbol is introduced or explained is given in parentheses.

A = area

$\mathring{A}$ = angstrom unit

A_i = atomic weight of element i (2.2)

APF = atomic packing factor (3.4)

%RA = ductility, in percent reduction in area (7.6)

a = lattice parameter: unit cell x-axial length (3.4)

a = crack length of a surface crack (9.5a, 9.5b)

at% = atom percent (5.6)

B = magnetic flux density (induction) (18.2)

B_r = magnetic remanence (18.7)

BCC = body-centered cubic crystal structure (3.4)

b = lattice parameter: unit cell y-axial length (3.11)

$\mathbf{b}$ = Burgers vector (5.7)

C = capacitance (12.17)

C_i = concentration (composition) of component i in wt% (5.6)

C_i' = concentration (composition) of component i in at% (5.6)

C_v, C_p = heat capacity at constant volume, pressure (17.2)

CPR = corrosion penetration rate (16.3)

CVN = Charpy V-notch (9.8)

%CW = percent cold work (8.11)

c = lattice parameter: unit cell z-axial length (3.11)

c = velocity of electromagnetic radiation in a vacuum (19.2)

D = diffusion coefficient (6.3)

D = dielectric displacement (12.18)

d = diameter

d = average grain diameter (8.9)

d_{hkl} = interplanar spacing for planes of Miller indices h, k, and l (3.19)

E = energy (2.5)

E = modulus of elasticity or Young's modulus (7.3)

$\mathscr{E}$ = electric field intensity (12.3)

E_f = Fermi energy (12.5)

E_g = band gap energy (12.6)

$E_r(t)$ = relaxation modulus (7.15)

%EL = ductility, in percent elongation (7.6)

e = electric charge per electron (12.7)

e^- = electron (16.2)

erf = Gaussian error function (6.4)

exp = e, the base for natural logarithms

F = force, interatomic or mechanical (2.5, 7.2)

$\mathscr{F}$ = Faraday constant (16.2)

FCC = face-centered cubic crystal structure (3.4)

G = shear modulus (7.3)

H = magnetic field strength (18.2)

H_c = magnetic coercivity (18.7)

HB = Brinell hardness (7.16)

HCP = hexagonal close-packed crystal structure (3.4)

HK = Knoop hardness (7.16)

HRB, HRF = Rockwell hardness: B and F scales (7.16)

HR15N, HR45W = superficial Rockwell hardness: 15N and 45W scales (7.16)

HV = Vickers hardness (7.16)

h = Planck's constant (19.2)

(hkl) = Miller indices for a crystallographic plane (3.13)

I = electric current (12.2)

I = intensity of electromagnetic radiation (19.3)

i = current density (16.3)

i_C = corrosion current density (16.4)

J = diffusion flux (6.3)

J = electric current density (12.3)

K = stress intensity factor (9.5a)

K_c = fracture toughness (9.5a, 9.5b)

K_{Ic} = plane strain fracture toughness for mode I crack surface displacement (9.5a, 9.5b)

k = Boltzmann's constant (5.2)

k = thermal conductivity (17.4)

l = length

l_c = critical fiber length (15.4)

ln = natural logarithm

log = logarithm taken to base 10

M = magnetization (18.2)

$\overline{M}_n$ = polymer number-average molecular weight (4.5)

$\overline{M}_w$ = polymer weight-average molecular weight (4.5)

mol% = mole percent

N = number of fatigue cycles (9.10)

N_A = Avogadro's number (3.5)

N_f = fatigue life (9.10)

n = principal quantum number (2.3)

n = number of atoms per unit cell (3.5)

n = strain-hardening exponent (7.7)

n = number of electrons in an electrochemical reaction (16.2)

n = number of conducting electrons per cubic meter (12.7)

n = index of refraction (19.5)

n' = for ceramics, the number of formula units per unit cell (3.7)

n_n = number-average degree of polymerization (4.5)

n_w = weight-average degree of polymerization (4.5)

P = dielectric polarization (12.18)

P–B ratio = Pilling–Bedworth ratio (16.10)

p = number of holes per cubic meter (12.10)

Q = activation energy

Q = magnitude of charge stored (12.17)

R = atomic radius (3.4)

R = gas constant

r = interatomic distance (2.5)

r = reaction rate (11.3, 16.3)

r_A, r_C = anion and cation ionic radii (3.6)

S = fatigue stress amplitude (9.10)

SEM = scanning electron microscopy or microscope

T = temperature

T_c = Curie temperature (18.6)

T_C = superconducting critical temperature (18.11)

T_g = glass transition temperature (11.15)

T_m = melting temperature

TEM = transmission electron microscopy or microscope

TS = tensile strength (7.6)

t = time

t_r = rupture lifetime (9.16)

U_r = modulus of resilience (7.6)

$[uvw]$ = indices for a crystallographic direction (3.12)

V = electrical potential difference (voltage) (12.2)

V_C = unit cell volume (3.4)

V_C = corrosion potential (16.4)

V_H = Hall voltage (12.13)

V_i = volume fraction of phase i (10.7)

v = velocity

vol% = volume percent

W_i = mass fraction of phase i (10.7)

wt% = weight percent (5.6)

x = length

x = space coordinate

Y = dimensionless parameter or function in fracture toughness expression (9.5a, 9.5b)

y = space coordinate

z = space coordinate

α = lattice parameter: unit cell $y-z$ interaxial angle (3.11)

α, β, γ = phase designations

α_l = linear coefficient of thermal expansion (17.3)

β = lattice parameter: unit cell $x-z$ interaxial angle (3.11)

γ = lattice parameter: unit cell $x-y$ interaxial angle (3.11)

γ = shear strain (7.2)

Δ = finite change in a parameter the symbol of which it precedes

ϵ = engineering strain (7.2)

ϵ = dielectric permittivity (12.17)

ϵ_r = dielectric constant or relative permittivity (12.17)

$\dot{\epsilon}_s$ = steady-state creep rate (9.16)

ϵ_T = true strain (7.7)

η = viscosity (8.16)

η = overvoltage (16.4)

θ = Bragg diffraction angle (3.19)

θ_D = Debye temperature (17.2)

λ = wavelength of electromagnetic radiation (3.19)

μ = magnetic permeability (18.2)

μ_B = Bohr magneton (18.2)

μ_r = relative magnetic permeability (18.2)

μ_e = electron mobility (12.7)

μ_h = hole mobility (12.10)

ν = Poisson's ratio (7.5)

ν = frequency of electromagnetic radiation (19.2)

ρ = density (3.5)

ρ = electrical resistivity (12.2)

ρ_t = radius of curvature at the tip of a crack (9.5a, 9.5b)

σ = engineering stress, tensile or compressive (7.2)

σ = electrical conductivity (12.3)

σ^* = longitudinal strength (composite) (15.5)

σ_c = critical stress for crack propagation (9.5a, 9.5b)

σ_{fs} = flexural strength (7.10)

σ_m = maximum stress (9.5a, 9.5b)

σ_m = mean stress (9.9)

σ'_m = stress in matrix at composite failure (15.5)

σ_T = true stress (7.7)

σ_w = safe or working stress (7.20)

σ_y = yield strength (7.6)

τ = shear stress (7.2)

τ_c = fiber–matrix bond strength/ matrix shear yield strength (15.4)

τ_{crss} = critical resolved shear stress (8.6)

χ_m = magnetic susceptibility (18.2)

SUBSCRIPTS

c = composite

cd = discontinuous fibrous composite

cl = longitudinal direction (aligned fibrous composite)

ct = transverse direction (aligned fibrous composite)

f = final

f = at fracture

f = fiber

i = instantaneous

m = matrix

m, max = maximum

min = minimum

0 = original

0 = at equilibrium

0 = in a vacuum

A familiar item that is fabricated from three different material types is the beverage container. Beverages are marketed in aluminum (metal) cans (top), glass (ceramic) bottles (center), and plastic (polymer) bottles (bottom). (Permission to use these photographs was granted by the Coca-Cola Company.)

After careful study of this chapter you should be able to do the following:

1. List six different property classifications of materials that determine their applicability.
2. Cite the four components that are involved in the design, production, and utilization of materials, and briefly describe the interrelationships between these components.
3. Cite three criteria that are important in the materials selection process.
4. (a) List the three primary classifications of solid materials, and then cite the distinctive chemical feature of each.
   (b) Note the other three types of materials and, for each, its distinctive feature(s).

1.1 HISTORICAL PERSPECTIVE

Materials are probably more deep-seated in our culture than most of us realize. Transportation, housing, clothing, communication, recreation, and food production—virtually every segment of our everyday lives is influenced to one degree or another by materials. Historically, the development and advancement of societies have been intimately tied to the members' ability to produce and manipulate materials to fill their needs. In fact, early civilizations have been designated by the level of their materials development (i.e., Stone Age, Bronze Age).

The earliest humans had access to only a very limited number of materials, those that occur naturally: stone, wood, clay, skins, and so on. With time they discovered techniques for producing materials that had properties superior to those of the natural ones; these new materials included pottery and various metals. Furthermore, it was discovered that the properties of a material could be altered by heat treatments and by the addition of other substances. At this point, materials utilization was totally a selection process, that is, deciding from a given, rather limited set of materials the one that was best suited for an application by virtue of its characteristics. It was not until relatively recent times that scientists came to understand the relationships between the structural elements of materials and their properties. This knowledge, acquired in the past 60 years or so, has empowered them to fashion, to a large degree, the characteristics of materials. Thus, tens of thousands of different materials have evolved with rather specialized characteristics that meet the needs of our modern and complex society; these include metals, plastics, glasses, and fibers.

The development of many technologies that make our existence so comfortable has been intimately associated with the accessibility of suitable materials. An advancement in the understanding of a material type is often the forerunner to the stepwise progression of a technology. For example, automobiles would not have been possible without the availability of inexpensive steel or some other comparable substitute. In our contemporary era, sophisticated electronic devices rely on components that are made from what are called semiconducting materials.

1.2 MATERIALS SCIENCE AND ENGINEERING

The discipline of *materials science* involves investigating the relationships that exist between the structures and properties of materials. In contrast, *materials engineering* is, on the basis of these structure–property correlations, designing or engineering the structure of a material to produce a predetermined set of properties. Throughout this text we draw attention to the relationships between material properties and structural elements.

"Structure" is at this point a nebulous term that deserves some explanation. In brief, the structure of a material usually relates to the arrangement of its internal components. Subatomic structure involves electrons within the individual atoms and interactions with their nuclei. On an atomic level, structure encompasses the organization of atoms or molecules relative to one another. The next larger structural realm, which contains large groups of atoms that are normally agglomerated together, is termed "microscopic," meaning that which is subject to direct observation using some type of microscope. Finally, structural elements that may be viewed with the naked eye are termed "macroscopic."

The notion of "property" deserves elaboration. While in service use, all materials are exposed to external stimuli that evoke some type of response. For example, a specimen subjected to forces will experience deformation; or a polished metal surface will reflect light. Property is a material trait in terms of the kind and magnitude of response to a specific imposed stimulus. Generally, definitions of properties are made independent of material shape and size.

Virtually all important properties of solid materials may be grouped into six different categories: mechanical, electrical, thermal, magnetic, optical, and deteriorative. For each there is a characteristic type of stimulus capable of provoking different responses. Mechanical properties relate deformation to an applied load or force; examples include elastic modulus and strength. For electrical properties, such as electrical conductivity and dielectric constant, the stimulus is an electric field. The thermal behavior of solids can be represented in terms of heat capacity and thermal conductivity. Magnetic properties demonstrate the response of a material to the application of a magnetic field. For optical properties, the stimulus is electromagnetic or light radiation; index of refraction and reflectivity are representative optical properties. Finally, deteriorative characteristics indicate the chemical reactivity of materials. The chapters that follow discuss properties that fall within each of these six classifications.

In addition to structure and properties, two other important components are involved in the science and engineering of materials, viz. "processing" and "performance." With regard to the relationships of these four components, the structure of a material will depend on how it is processed. Furthermore, a material's performance will be a function of its properties. Thus, the interrelationship between processing, structure, properties, and performance is linear, as depicted in the schematic illustration shown in Figure 1.1. Throughout this text we draw attention to the relationships among these four components in terms of the design, production, and utilization of materials.

We now present an example of these processing-structure-properties-performance principles with Figure 1.2, a photograph showing three thin disk specimens placed over some printed matter. It is obvious that the optical properties (i.e., the light transmittance) of each of the three materials are different; the one on the left is transparent (i.e., virtually all of the reflected light passes through it), whereas the disks in the center and on the right are, respectively, translucent and opaque. All of these specimens are of the same material, aluminum oxide, but the leftmost one is what we call a single crystal—that is, it is highly perfect—which gives rise to its transparency. The center one is composed of numerous and very small single

Processing ——→ Structure ——→ Properties ——→ Performance

FIGURE 1.1 The four components of the discipline of materials science and engineering and their linear interrelationship.

FIGURE 1.2
Photograph showing the light transmittance of three aluminum oxide specimens. From left to right: single-crystal material (sapphire), which is transparent; a polycrystalline and fully dense (nonporous) material, which is translucent; and a polycrystalline material that contains approximately 5% porosity, which is opaque. (Specimen preparation, P. A. Lessing; photography by J. Telford.)

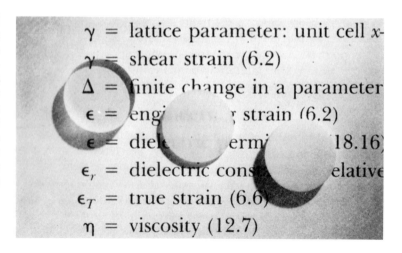

γ = lattice parameter: unit cell x-

γ = shear strain (6.2)

Δ = finite change in a parameter

ϵ = engineering strain (6.2)

ϵ = dielectric permittivity (18.16)

ϵ_r = dielectric constant, relative

ϵ_T = true strain (6.6)

η = viscosity (12.7)

crystals that are all connected; the boundaries between these small crystals scatter a portion of the light reflected from the printed page, which makes this material optically translucent. And finally, the specimen on the right is composed not only of many small, interconnected crystals, but also of a large number of very small pores or void spaces. These pores also effectively scatter the reflected light and render this material opaque.

Thus, the structures of these three specimens are different in terms of crystal boundaries and pores, which affect the optical transmittance properties. Furthermore, each material was produced using a different processing technique. And, of course, if optical transmittance is an important parameter relative to the ultimate in-service application, the performance of each material will be different.

1.3 WHY STUDY MATERIALS SCIENCE AND ENGINEERING?

Why do we study materials? Many an applied scientist or engineer, whether mechanical, civil, chemical, or electrical, will at one time or another be exposed to a design problem involving materials. Examples might include a transmission gear, the superstructure for a building, an oil refinery component, or an integrated circuit chip. Of course, materials scientists and engineers are specialists who are totally involved in the investigation and design of materials.

Many times, a materials problem is one of selecting the right material from the many thousands that are available. There are several criteria on which the final decision is normally based. First of all, the in-service conditions must be characterized, for these will dictate the properties required of the material. On only rare occasions does a material possess the maximum or ideal combination of properties. Thus, it may be necessary to trade off one characteristic for another. The classic example involves strength and ductility; normally, a material having a high strength will have only a limited ductility. In such cases a reasonable compromise between two or more properties may be necessary.

A second selection consideration is any deterioration of material properties that may occur during service operation. For example, significant reductions in mechanical strength may result from exposure to elevated temperatures or corrosive environments.

Finally, probably the overriding consideration is that of economics: What will the finished product cost? A material may be found that has the ideal set of properties but is prohibitively expensive. Here again, some compromise is inevitable. The cost of a finished piece also includes any expense incurred during fabrication to produce the desired shape.

The more familiar an engineer or scientist is with the various characteristics and structure–property relationships, as well as processing techniques of materials, the more proficient and confident he or she will be to make judicious materials choices based on these criteria.

1.4 CLASSIFICATION OF MATERIALS

Solid materials have been conveniently grouped into three basic classifications: metals, ceramics, and polymers. This scheme is based primarily on chemical makeup and atomic structure, and most materials fall into one distinct grouping or another, although there are some intermediates. In addition, there are three other groups of important engineering materials—composites, semiconductors, and biomaterials. Composites consist of combinations of two or more different materials, whereas semiconductors are utilized because of their unusual electrical characteristics; biomaterials are implanted into the human body. A brief explanation of the material types and representative characteristics is offered next.

METALS

Metallic materials are normally combinations of metallic elements. They have large numbers of nonlocalized electrons; that is, these electrons are not bound to particular atoms. Many properties of metals are directly attributable to these electrons. Metals are extremely good conductors of electricity and heat and are not transparent to visible light; a polished metal surface has a lustrous appearance. Furthermore, metals are quite strong, yet deformable, which accounts for their extensive use in structural applications.

CERAMICS

Ceramics are compounds between metallic and nonmetallic elements; they are most frequently oxides, nitrides, and carbides. The wide range of materials that falls within this classification includes ceramics that are composed of clay minerals, cement, and glass. These materials are typically insulative to the passage of electricity and heat, and are more resistant to high temperatures and harsh environments than metals and polymers. With regard to mechanical behavior, ceramics are hard but very brittle.

POLYMERS

Polymers include the familiar plastic and rubber materials. Many of them are organic compounds that are chemically based on carbon, hydrogen, and other nonmetallic elements; furthermore, they have very large molecular structures. These materials typically have low densities and may be extremely flexible.

COMPOSITES

A number of composite materials have been engineered that consist of more than one material type. Fiberglass is a familiar example, in which glass fibers are embedded within a polymeric material. A composite is designed to display a combination of the best characteristics of each of the component materials. Fiberglass acquires strength from the glass and flexibility from the polymer. Many of the recent material developments have involved composite materials.

SEMICONDUCTORS

Semiconductors have electrical properties that are intermediate between the electrical conductors and insulators. Furthermore, the electrical characteristics of these materials are extremely sensitive to the presence of minute concentrations of impurity atoms, which concentrations may be controlled over very small spatial regions. The semiconductors have made possible the advent of integrated circuitry that has totally revolutionized the electronics and computer industries (not to mention our lives) over the past two decades.

BIOMATERIALS

Biomaterials are employed in components implanted into the human body for replacement of diseased or damaged body parts. These materials must not produce toxic substances and must be compatible with body tissues (i.e., must not cause adverse biological reactions). All of the above materials—metals, ceramics, polymers, composites, and semiconductors—may be used as biomaterials. {For example, in Section 20.8 are discussed some of the biomaterials that are utilized in artificial hip replacements.}

1.5 ADVANCED MATERIALS

Materials that are utilized in high-technology (or high-tech) applications are sometimes termed *advanced materials.* By high technology we mean a device or product that operates or functions using relatively intricate and sophisticated principles; examples include electronic equipment (VCRs, CD players, etc.), computers, fiber-optic systems, spacecraft, aircraft, and military rocketry. These advanced materials are typically either traditional materials whose properties have been enhanced or newly developed, high-performance materials. Furthermore, they may be of all material types (e.g., metals, ceramics, polymers), and are normally relatively expensive. In subsequent chapters are discussed the properties and applications of a number of advanced materials—for example, materials that are used for lasers, integrated circuits, magnetic information storage, liquid crystal displays (LCDs), fiber optics, and the thermal protection system for the Space Shuttle Orbiter.

1.6 MODERN MATERIALS' NEEDS

In spite of the tremendous progress that has been made in the discipline of materials science and engineering within the past few years, there still remain technological challenges, including the development of even more sophisticated and specialized materials, as well as consideration of the environmental impact of materials production. Some comment is appropriate relative to these issues so as to round out this perspective.

Nuclear energy holds some promise, but the solutions to the many problems that remain will necessarily involve materials, from fuels to containment structures to facilities for the disposal of radioactive waste.

Significant quantities of energy are involved in transportation. Reducing the weight of transportation vehicles (automobiles, aircraft, trains, etc.), as well as increasing engine operating temperatures, will enhance fuel efficiency. New high-strength, low-density structural materials remain to be developed, as well as materials that have higher-temperature capabilities, for use in engine components.

Furthermore, there is a recognized need to find new, economical sources of energy, and to use the present resources more efficiently. Materials will undoubtedly play a significant role in these developments. For example, the direct conversion of solar into electrical energy has been demonstrated. Solar cells employ some rather complex and expensive materials. To ensure a viable technology, materials that are highly efficient in this conversion process yet less costly must be developed.

Furthermore, environmental quality depends on our ability to control air and water pollution. Pollution control techniques employ various materials. In addition, materials processing and refinement methods need to be improved so that they produce less environmental degradation, that is, less pollution and less despoilage of the landscape from the mining of raw materials. Also, in some materials manufacturing processes, toxic substances are produced, and the ecological impact of their disposal must be considered.

Many materials that we use are derived from resources that are nonrenewable, that is, not capable of being regenerated. These include polymers, for which the prime raw material is oil, and some metals. These nonrenewable resources are gradually becoming depleted, which necessitates: 1) the discovery of additional reserves, 2) the development of new materials having comparable properties with less adverse environmental impact, and/or 3) increased recycling efforts and the development of new recycling technologies. As a consequence of the economics of not only production but also environmental impact and ecological factors, it is becoming increasingly important to consider the "cradle-to-grave" life cycle of materials relative to the overall manufacturing process.

{The roles that materials scientists and engineers play relative to these, as well as other environmental and societal issues, are discussed in more detail in Chapter 21.}

REFERENCES

The October 1986 issue of *Scientific American,* Vol. 255, No. 4, is devoted entirely to various advanced materials and their uses. Other references for Chapter 1 are textbooks that cover the basic fundamentals of the field of materials science and engineering.

Ashby, M. F. and D. R. H. Jones, *Engineering Materials 1, An Introduction to Their Properties and Applications,* 2nd edition, Pergamon Press, Oxford, 1996.

Ashby, M. F. and D. R. H. Jones, *Engineering Materials 2, An Introduction to Microstructures, Processing and Design,* Pergamon Press, Oxford, 1986.

Askeland, D. R., *The Science and Engineering of Materials,* 3rd edition, Brooks/Cole Publishing Co., Pacific Grove, CA, 1994.

Barrett, C. R., W. D. Nix, and A. S. Tetelman, *The Principles of Engineering Materials,* Prentice Hall, Inc., Englewood Cliffs, NJ, 1973.

Flinn, R. A. and P. K. Trojan, *Engineering Materials and Their Applications,* 4th edition, John Wiley & Sons, New York, 1990.

Jacobs, J. A. and T. F. Kilduff, *Engineering Materials Technology,* 3rd edition, Prentice Hall, Upper Saddle River, NJ, 1996.

McMahon, C. J., Jr. and C. D. Graham, Jr., *Introduction to Engineering Materials: The Bicycle and the Walkman,* Merion Books, Philadelphia, 1992.

Murray, G. T., *Introduction to Engineering Materials—Behavior, Properties, and Selection,* Marcel Dekker, Inc., New York, 1993.

Ohring, M., *Engineering Materials Science,* Academic Press, San Diego, CA, 1995.

Ralls, K. M., T. H. Courtney, and J. Wulff, *Introduction to Materials Science and Engineering,* John Wiley & Sons, New York, 1976.

Schaffer, J. P., A. Saxena, S. D. Antolovich, T. H. Sanders, Jr., and S. B. Warner, *The Science and*

Design of Engineering Materials, 2nd edition, WCB/McGraw-Hill, New York, 1999.

Shackelford, J. F., *Introduction to Materials Science for Engineers,* 5th edition, Prentice Hall, Inc., Upper Saddle River, NJ, 2000.

Smith, W. F., *Principles of Materials Science and Engineering,* 3rd edition, McGraw-Hill Book Company, New York, 1995.

Van Vlack, L. H., *Elements of Materials Science and Engineering,* 6th edition, Addison-Wesley Publishing Co., Reading, MA, 1989.

This micrograph, which represents the surface of a gold specimen, was taken with a sophisticated atomic force microscope (AFM). Individual atoms for this (111) crystallographic surface plane are resolved. Also note the dimensional scale (in the nanometer range) below the micrograph. (Image courtesy of Dr. Michael Green, TopoMetrix Corporation.)

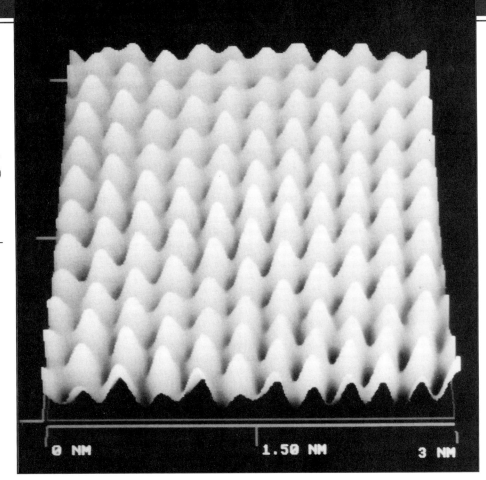

0 NM   1.50 NM   3 NM

Why Study *Atomic Structure and Interatomic Bonding?*

An important reason to have an understanding of interatomic bonding in solids is that, in some instances, the type of bond allows us to explain a material's properties. For example, consider carbon, which may exist as both graphite and diamond. Whereas graphite is relatively soft and has a "greasy" feel to it, diamond is the hardest known material. This dramatic disparity in properties is directly attributable to a type of interatomic bonding found in graphite that does not exist in diamond (see Section 3.9).

After careful study of this chapter you should be able to do the following:

1. Name the two atomic models cited, and note the differences between them.
2. Describe the important quantum-mechanical principle that relates to electron energies.
3. (a) Schematically plot attractive, repulsive, and net energies versus interatomic separation for two atoms or ions.
   (b) Note on this plot the equilibrium separation and the bonding energy.
4. (a) Briefly describe ionic, covalent, metallic, hydrogen, and van der Waals bonds.
   (b) Note what materials exhibit each of these bonding types.

2.1 INTRODUCTION

Some of the important properties of solid materials depend on geometrical atomic arrangements, and also the interactions that exist among constituent atoms or molecules. This chapter, by way of preparation for subsequent discussions, considers several fundamental and important concepts, namely: atomic structure, electron configurations in atoms and the periodic table, and the various types of primary and secondary interatomic bonds that hold together the atoms comprising a solid. These topics are reviewed briefly, under the assumption that some of the material is familiar to the reader.

ATOMIC STRUCTURE

2.2 FUNDAMENTAL CONCEPTS

Each atom consists of a very small nucleus composed of protons and neutrons, which is encircled by moving electrons. Both electrons and protons are electrically charged, the charge magnitude being 1.60×10^{-19} C, which is negative in sign for electrons and positive for protons; neutrons are electrically neutral. Masses for these subatomic particles are infinitesimally small; protons and neutrons have approximately the same mass, 1.67×10^{-27} kg, which is significantly larger than that of an electron, 9.11×10^{-31} kg.

Each chemical element is characterized by the number of protons in the nucleus, or the **atomic number** (Z).[1] For an electrically neutral or complete atom, the atomic number also equals the number of electrons. This atomic number ranges in integral units from 1 for hydrogen to 92 for uranium, the highest of the naturally occurring elements.

The *atomic mass* (A) of a specific atom may be expressed as the sum of the masses of protons and neutrons within the nucleus. Although the number of protons is the same for all atoms of a given element, the number of neutrons (N) may be variable. Thus atoms of some elements have two or more different atomic masses, which are called **isotopes.** The **atomic weight** of an element corresponds to the weighted average of the atomic masses of the atom's naturally occurring isotopes.[2] The **atomic mass unit (amu)** may be used for computations of atomic weight. A scale has been established whereby 1 amu is defined as $\frac{1}{12}$ of the atomic mass of

---

[1] Terms appearing in boldface type are defined in the Glossary, which follows Appendix E.

[2] The term "atomic mass" is really more accurate than "atomic weight" inasmuch as, in this context, we are dealing with masses and not weights. However, atomic weight is, by convention, the preferred terminology, and will be used throughout this book. The reader should note that it is *not* necessary to divide molecular weight by the gravitational constant.

the most common isotope of carbon, carbon 12 (^{12}C) ($A = 12.00000$). Within this scheme, the masses of protons and neutrons are slightly greater than unity, and

$$A \cong Z + N \tag{2.1}$$

The atomic weight of an element or the molecular weight of a compound may be specified on the basis of amu per atom (molecule) or mass per mole of material. In one **mole** of a substance there are 6.023×10^{23} (Avogadro's number) atoms or molecules. These two atomic weight schemes are related through the following equation:

1 amu/atom (or molecule) = 1 g/mol

For example, the atomic weight of iron is 55.85 amu/atom, or 55.85 g/mol. Sometimes use of amu per atom or molecule is convenient; on other occasions g (or kg)/mol is preferred; the latter is used in this book.

2.3 ELECTRONS IN ATOMS

ATOMIC MODELS

During the latter part of the nineteenth century it was realized that many phenomena involving electrons in solids could not be explained in terms of classical mechanics. What followed was the establishment of a set of principles and laws that govern systems of atomic and subatomic entities that came to be known as **quantum mechanics.** An understanding of the behavior of electrons in atoms and crystalline solids necessarily involves the discussion of quantum-mechanical concepts. However, a detailed exploration of these principles is beyond the scope of this book, and only a very superficial and simplified treatment is given.

One early outgrowth of quantum mechanics was the simplified **Bohr atomic model,** in which electrons are assumed to revolve around the atomic nucleus in discrete orbitals, and the position of any particular electron is more or less well defined in terms of its orbital. This model of the atom is represented in Figure 2.1.

Another important quantum-mechanical principle stipulates that the energies of electrons are quantized; that is, electrons are permitted to have only specific values of energy. An electron may change energy, but in doing so it must make a quantum jump either to an allowed higher energy (with absorption of energy) or to a lower energy (with emission of energy). Often, it is convenient to think of these allowed electron energies as being associated with *energy levels* or *states.*

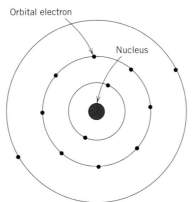

FIGURE 2.1 Schematic representation of the Bohr atom.

FIGURE 2.2 (*a*) The first three electron energy states for the Bohr hydrogen atom. (*b*) Electron energy states for the first three shells of the wave-mechanical hydrogen atom. (Adapted from W. G. Moffatt, G. W. Pearsall, and J. Wulff, *The Structure and Properties of Materials,* Vol. I, *Structure,* p. 10. Copyright © 1964 by John Wiley & Sons, New York. Reprinted by permission of John Wiley & Sons, Inc.)

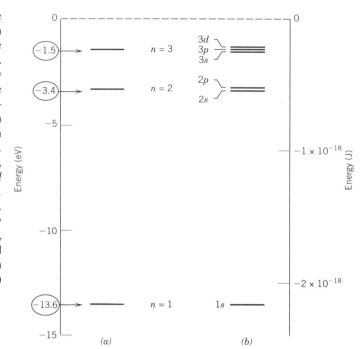

These states do not vary continuously with energy; that is, adjacent states are separated by finite energies. For example, allowed states for the Bohr hydrogen atom are represented in Figure 2.2*a*. These energies are taken to be negative, whereas the zero reference is the unbound or free electron. Of course, the single electron associated with the hydrogen atom will fill only one of these states.

Thus, the Bohr model represents an early attempt to describe electrons in atoms, in terms of both position (electron orbitals) and energy (quantized energy levels).

This Bohr model was eventually found to have some significant limitations because of its inability to explain several phenomena involving electrons. A resolution was reached with a **wave-mechanical model,** in which the electron is considered to exhibit both wavelike and particle-like characteristics. With this model, an electron is no longer treated as a particle moving in a discrete orbital; but rather, position is considered to be the probability of an electron's being at various locations around the nucleus. In other words, position is described by a probability distribution or electron cloud. Figure 2.3 compares Bohr and wave-mechanical models for the hydrogen atom. Both these models are used throughout the course of this book; the choice depends on which model allows the more simple explanation.

QUANTUM NUMBERS

Using wave mechanics, every electron in an atom is characterized by four parameters called **quantum numbers.** The size, shape, and spatial orientation of an electron's probability density are specified by three of these quantum numbers. Furthermore, Bohr energy levels separate into electron subshells, and quantum numbers dictate the number of states within each subshell. Shells are specified by a *principal quantum number n,* which may take on integral values beginning with unity; sometimes these shells are designated by the letters *K, L, M, N, O,* and so on, which correspond, respectively, to $n = 1, 2, 3, 4, 5, \ldots$, as indicated in Table 2.1. It should also be

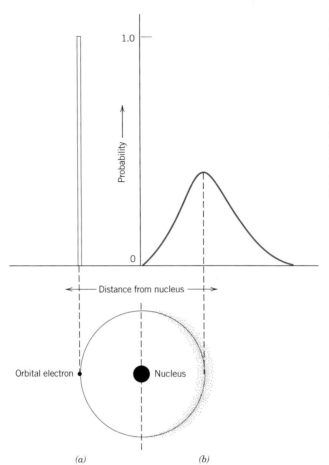

FIGURE 2.3 Comparison of the (*a*) Bohr and (*b*) wave-mechanical atom models in terms of electron distribution. (Adapted from Z. D. Jastrzebski, *The Nature and Properties of Engineering Materials,* 3rd edition, p. 4. Copyright © 1987 by John Wiley & Sons, New York. Reprinted by permission of John Wiley & Sons, Inc.)

Table 2.1 The Number of Available Electron States in Some of the Electron Shells and Subshells

Principal Quantum Number n	Shell Designation	Subshells	Number of States	Number of Electrons Per Subshell	Per Shell
1	K	s	1	2	2
2	L	s	1	2	8
		p	3	6	
3	M	s	1	2	18
		p	3	6	
		d	5	10	
4	N	s	1	2	32
		p	3	6	
		d	5	10	
		f	7	14	

noted that this quantum number, and it only, is also associated with the Bohr model. This quantum number is related to the distance of an electron from the nucleus, or its position.

The second quantum number, l, signifies the subshell, which is denoted by a lowercase letter—an s, p, d, or f; it is related to the shape of the electron subshell. In addition, the number of these subshells is restricted by the magnitude of n. Allowable subshells for the several n values are also presented in Table 2.1. The number of energy states for each subshell is determined by the third quantum number, m_l. For an s subshell, there is a single energy state, whereas for p, d, and f subshells, three, five, and seven states exist, respectively (Table 2.1). In the absence of an external magnetic field, the states within each subshell are identical. However, when a magnetic field is applied these subshell states split, each state assuming a slightly different energy.

Associated with each electron is a *spin moment,* which must be oriented either up or down. Related to this spin moment is the fourth quantum number, m_s, for which two values are possible ($+\frac{1}{2}$ and $-\frac{1}{2}$), one for each of the spin orientations.

Thus, the Bohr model was further refined by wave mechanics, in which the introduction of three new quantum numbers gives rise to electron subshells within each shell. A comparison of these two models on this basis is illustrated, for the hydrogen atom, in Figures 2.2a and 2.2b.

A complete energy level diagram for the various shells and subshells using the wave-mechanical model is shown in Figure 2.4. Several features of the diagram are worth noting. First, the smaller the principal quantum number, the lower the energy level; for example, the energy of a $1s$ state is less than that of a $2s$ state, which in turn is lower than the $3s$. Second, within each shell, the energy of a subshell level increases with the value of the l quantum number. For example, the energy of a $3d$ state is greater than a $3p$, which is larger than $3s$. Finally, there may be overlap in energy of a state in one shell with states in an adjacent shell, which is especially true of d and f states; for example, the energy of a $3d$ state is greater than that for a $4s$.

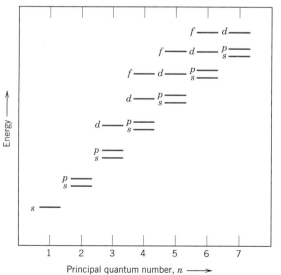

FIGURE 2.4 Schematic representation of the relative energies of the electrons for the various shells and subshells. (From K. M. Ralls, T. H. Courtney, and J. Wulff, *Introduction to Materials Science and Engineering,* p. 22. Copyright © 1976 by John Wiley & Sons, New York. Reprinted by permission of John Wiley & Sons, Inc.)

ELECTRON CONFIGURATIONS

The preceding discussion has dealt primarily with **electron states**—values of energy that are permitted for electrons. To determine the manner in which these states are filled with electrons, we use the **Pauli exclusion principle,** another quantum-mechanical concept. This principle stipulates that each electron state can hold no more than two electrons, which must have opposite spins. Thus, s, p, d, and f subshells may each accommodate, respectively, a total of 2, 6, 10, and 14 electrons; Table 2.1 summarizes the maximum number of electrons that may occupy each of the first four shells.

Of course, not all possible states in an atom are filled with electrons. For most atoms, the electrons fill up the lowest possible energy states in the electron shells and subshells, two electrons (having opposite spins) per state. The energy structure for a sodium atom is represented schematically in Figure 2.5. When all the electrons occupy the lowest possible energies in accord with the foregoing restrictions, an atom is said to be in its **ground state.** However, electron transitions to higher energy states are possible, as discussed in Chapters 12 {and 19.} The **electron configuration** or structure of an atom represents the manner in which these states are occupied. In the conventional notation the number of electrons in each subshell is indicated by a superscript after the shell–subshell designation. For example, the electron configurations for hydrogen, helium, and sodium are, respectively, $1s^1$, $1s^2$, and $1s^2 2s^2 2p^6 3s^1$. Electron configurations for some of the more common elements are listed in Table 2.2.

At this point, comments regarding these electron configurations are necessary. First, the **valence electrons** are those that occupy the outermost filled shell. These electrons are extremely important; as will be seen, they participate in the bonding between atoms to form atomic and molecular aggregates. Furthermore, many of the physical and chemical properties of solids are based on these valence electrons.

In addition, some atoms have what are termed "stable electron configurations"; that is, the states within the outermost or valence electron shell are completely filled. Normally this corresponds to the occupation of just the s and p states for the outermost shell by a total of eight electrons, as in neon, argon, and krypton; one exception is helium, which contains only two $1s$ electrons. These elements (Ne, Ar, Kr, and He) are the inert, or noble, gases, which are virtually unreactive chemically. Some atoms of the elements that have unfilled valence shells assume

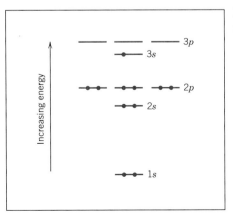

FIGURE 2.5 Schematic representation of the filled energy states for a sodium atom.

Table 2.2 A Listing of the Expected Electron Configurations for Some of the Common Elements[a]

Element	Symbol	Atomic Number	Electron Configuration
Hydrogen	H	1	$1s^1$
Helium	He	2	$1s^2$
Lithium	Li	3	$1s^22s^1$
Beryllium	Be	4	$1s^22s^2$
Boron	B	5	$1s^22s^22p^1$
Carbon	C	6	$1s^22s^22p^2$
Nitrogen	N	7	$1s^22s^22p^3$
Oxygen	O	8	$1s^22s^22p^4$
Fluorine	F	9	$1s^22s^22p^5$
Neon	Ne	10	$1s^22s^22p^6$
Sodium	Na	11	$1s^22s^22p^63s^1$
Magnesium	Mg	12	$1s^22s^22p^63s^2$
Aluminum	Al	13	$1s^22s^22p^63s^23p^1$
Silicon	Si	14	$1s^22s^22p^63s^23p^2$
Phosphorus	P	15	$1s^22s^22p^63s^23p^3$
Sulfur	S	16	$1s^22s^22p^63s^23p^4$
Chlorine	Cl	17	$1s^22s^22p^63s^23p^5$
Argon	Ar	18	$1s^22s^22p^63s^23p^6$
Potassium	K	19	$1s^22s^22p^63s^23p^64s^1$
Calcium	Ca	20	$1s^22s^22p^63s^23p^64s^2$
Scandium	Sc	21	$1s^22s^22p^63s^23p^63d^14s^2$
Titanium	Ti	22	$1s^22s^22p^63s^23p^63d^24s^2$
Vanadium	V	23	$1s^22s^22p^63s^23p^63d^34s^2$
Chromium	Cr	24	$1s^22s^22p^63s^23p^63d^54s^1$
Manganese	Mn	25	$1s^22s^22p^63s^23p^63d^54s^2$
Iron	Fe	26	$1s^22s^22p^63s^23p^63d^64s^2$
Cobalt	Co	27	$1s^22s^22p^63s^23p^63d^74s^2$
Nickel	Ni	28	$1s^22s^22p^63s^23p^63d^84s^2$
Copper	Cu	29	$1s^22s^22p^63s^23p^63d^{10}4s^1$
Zinc	Zn	30	$1s^22s^22p^63s^23p^63d^{10}4s^2$
Gallium	Ga	31	$1s^22s^22p^63s^23p^63d^{10}4s^24p^1$
Germanium	Ge	32	$1s^22s^22p^63s^23p^63d^{10}4s^24p^2$
Arsenic	As	33	$1s^22s^22p^63s^23p^63d^{10}4s^24p^3$
Selenium	Se	34	$1s^22s^22p^63s^23p^63d^{10}4s^24p^4$
Bromine	Br	35	$1s^22s^22p^63s^23p^63d^{10}4s^24p^5$
Krypton	Kr	36	$1s^22s^22p^63s^23p^63d^{10}4s^24p^6$

[a] When some elements covalently bond, they form sp hybrid bonds. This is especially true for C, Si, and Ge.

stable electron configurations by gaining or losing electrons to form charged ions, or by sharing electrons with other atoms. This is the basis for some chemical reactions, and also for atomic bonding in solids, as explained in Section 2.6.

Under special circumstances, the s and p orbitals combine to form hybrid sp^n orbitals, where n indicates the number of p orbitals involved, which may have a value of 1, 2, or 3. The 3A, 4A, and 5A group elements of the periodic table (Figure 2.6) are those which most often form these hybrids. The driving force for the formation of hybrid orbitals is a lower energy state for the valence electrons. For carbon the sp^3 hybrid is of primary importance in organic and polymer chemistries.

The shape of the sp^3 hybrid is what determines the 109° (or tetrahedral) angle found in polymer chains (Chapter 4).

2.4 THE PERIODIC TABLE

All the elements have been classified according to electron configuration in the **periodic table** (Figure 2.6). Here, the elements are situated, with increasing atomic number, in seven horizontal rows called periods. The arrangement is such that all elements that are arrayed in a given column or group have similar valence electron structures, as well as chemical and physical properties. These properties change gradually and systematically, moving horizontally across each period.

The elements positioned in Group 0, the rightmost group, are the inert gases, which have filled electron shells and stable electron configurations. Group VIIA and VIA elements are one and two electrons deficient, respectively, from having stable structures. The Group VIIA elements (F, Cl, Br, I, and At) are sometimes termed the halogens. The alkali and the alkaline earth metals (Li, Na, K, Be, Mg, Ca, etc.) are labeled as Groups IA and IIA, having, respectively, one and two electrons in excess of stable structures. The elements in the three long periods, Groups IIIB through IIB, are termed the transition metals, which have partially filled d electron states and in some cases one or two electrons in the next higher energy shell. Groups IIIA, IVA, and VA (B, Si, Ge, As, etc.) display characteristics that are intermediate between the metals and nonmetals by virtue of their valence electron structures.

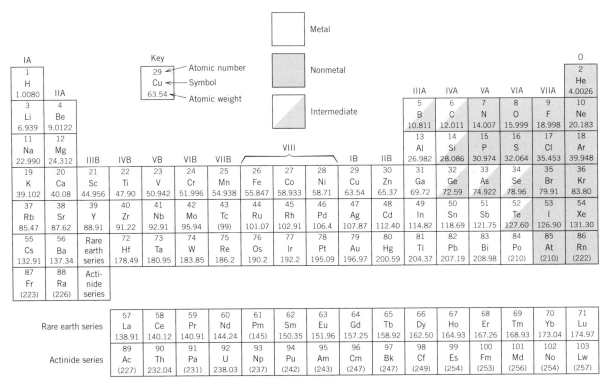

FIGURE 2.6 The periodic table of the elements. The numbers in parentheses are the atomic weights of the most stable or common isotopes.

IA	IIA	IIIB	IVB	VB	VIB	VIIB	VIII	VIII	VIII	IB	IIB	IIIA	IVA	VA	VIA	VIIA	0
1 H 2.1																	2 He –
3 Li 1.0	4 Be 1.5											5 B 2.0	6 C 2.5	7 N 3.0	8 O 3.5	9 F 4.0	10 Ne –
11 Na 0.9	12 Mg 1.2											13 Al 1.5	14 Si 1.8	15 P 2.1	16 S 2.5	17 Cl 3.0	18 Ar –
19 K 0.8	20 Ca 1.0	21 Sc 1.3	22 Ti 1.5	23 V 1.6	24 Cr 1.6	25 Mn 1.5	26 Fe 1.8	27 Co 1.8	28 Ni 1.8	29 Cu 1.9	30 Zn 1.6	31 Ga 1.6	32 Ge 1.8	33 As 2.0	34 Se 2.4	35 Br 2.8	36 Kr –
37 Rb 0.8	38 Sr 1.0	39 Y 1.2	40 Zr 1.4	41 Nb 1.6	42 Mo 1.8	43 Tc 1.9	44 Ru 2.2	45 Rh 2.2	46 Pd 2.2	47 Ag 1.9	48 Cd 1.7	49 In 1.7	50 Sn 1.8	51 Sb 1.9	52 Te 2.1	53 I 2.5	54 Xe –
55 Cs 0.7	56 Ba 0.9	57–71 La–Lu 1.1–1.2	72 Hf 1.3	73 Ta 1.5	74 W 1.7	75 Re 1.9	76 Os 2.2	77 Ir 2.2	78 Pt 2.2	79 Au 2.4	80 Hg 1.9	81 Tl 1.8	82 Pb 1.8	83 Bi 1.9	84 Po 2.0	85 At 2.2	86 Rn –
87 Fr 0.7	88 Ra 0.9	89–102 Ac–No 1.1–1.7															

FIGURE 2.7   The electronegativity values for the elements. (Adapted from Linus Pauling, *The Nature of the Chemical Bond,* 3rd edition. Copyright 1939 and 1940, 3rd edition copyright © 1960, by Cornell University. Used by permission of the publisher, Cornell University Press.)

As may be noted from the periodic table, most of the elements really come under the metal classification. These are sometimes termed **electropositive** elements, indicating that they are capable of giving up their few valence electrons to become positively charged ions. Furthermore, the elements situated on the right-hand side of the table are **electronegative;** that is, they readily accept electrons to form negatively charged ions, or sometimes they share electrons with other atoms. Figure 2.7 displays electronegativity values that have been assigned to the various elements arranged in the periodic table. As a general rule, electronegativity increases in moving from left to right and from bottom to top. Atoms are more likely to accept electrons if their outer shells are almost full, and if they are less "shielded" from (i.e., closer to) the nucleus.

ATOMIC BONDING IN SOLIDS

2.5 BONDING FORCES AND ENERGIES

An understanding of many of the physical properties of materials is predicated on a knowledge of the interatomic forces that bind the atoms together. Perhaps the principles of atomic bonding are best illustrated by considering the interaction between two isolated atoms as they are brought into close proximity from an infinite separation. At large distances, the interactions are negligible; but as the atoms approach, each exerts forces on the other. These forces are of two types, attractive and repulsive, and the magnitude of each is a function of the separation or interatomic distance. The origin of an attractive force F_A depends on the particular type of bonding that exists between the two atoms. Its magnitude varies with the distance, as represented schematically in Figure 2.8a. Ultimately, the outer electron shells of the two atoms begin to overlap, and a strong repulsive force F_R comes into play. The net force F_N between the two atoms is just the sum of both attractive and repulsive components; that is,

$$F_N = F_A + F_R \tag{2.2}$$

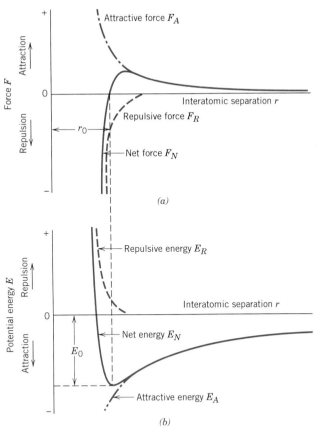

FIGURE 2.8 (a) The dependence of repulsive, attractive, and net forces on interatomic separation for two isolated atoms. (b) The dependence of repulsive, attractive, and net potential energies on interatomic separation for two isolated atoms.

which is also a function of the interatomic separation, as also plotted in Figure 2.8a. When F_A and F_R balance, or become equal, there is no net force; that is,

$$F_A + F_R = 0 \tag{2.3}$$

Then a state of equilibrium exists. The centers of the two atoms will remain separated by the equilibrium spacing r_0, as indicated in Figure 2.8a. For many atoms, r_0 is approximately 0.3 nm (3 Å). Once in this position, the two atoms will counteract any attempt to separate them by an attractive force, or to push them together by a repulsive action.

Sometimes it is more convenient to work with the potential energies between two atoms instead of forces. Mathematically, energy (E) and force (F) are related as

$$E = \int F \, dr \tag{2.4}$$

Or, for atomic systems,

$$E_N = \int_{\infty}^{r} F_N \, dr \tag{2.5}$$

$$= \int_{\infty}^{r} F_A \, dr + \int_{\infty}^{r} F_R \, dr \tag{2.6}$$

$$= E_A + E_R \tag{2.7}$$

in which E_N, E_A, and E_R are respectively the net, attractive, and repulsive energies for two isolated and adjacent atoms.

Figure 2.8*b* plots attractive, repulsive, and net potential energies as a function of interatomic separation for two atoms. The net curve, which is again the sum of the other two, has a potential energy trough or well around its minimum. Here, the same equilibrium spacing, r_0, corresponds to the separation distance at the minimum of the potential energy curve. The **bonding energy** for these two atoms, E_0, corresponds to the energy at this minimum point (also shown in Figure 2.8*b*); it represents the energy that would be required to separate these two atoms to an infinite separation.

Although the preceding treatment has dealt with an ideal situation involving only two atoms, a similar yet more complex condition exists for solid materials because force and energy interactions among many atoms must be considered. Nevertheless, a bonding energy, analogous to E_0 above, may be associated with each atom. The magnitude of this bonding energy and the shape of the energy-versus-interatomic separation curve vary from material to material, and they both depend on the type of atomic bonding. Furthermore, a number of material properties depend on E_0, the curve shape, and bonding type. For example, materials having large bonding energies typically also have high melting temperatures; at room temperature, solid substances are formed for large bonding energies, whereas for small energies the gaseous state is favored; liquids prevail when the energies are of intermediate magnitude. In addition, as discussed in Section 7.3, the mechanical stiffness (or modulus of elasticity) of a material is dependent on the shape of its force-versus-interatomic separation curve (Figure 7.7). The slope for a relatively stiff material at the $r = r_0$ position on the curve will be quite steep; slopes are shallower for more flexible materials. Furthermore, how much a material expands upon heating or contracts upon cooling (that is, its linear coefficient of thermal expansion) is related to the shape of its E_0-versus-r_0 curve {(see Section 17.3).} A deep and narrow "trough," which typically occurs for materials having large bonding energies, normally correlates with a low coefficient of thermal expansion and relatively small dimensional alterations for changes in temperature.

Three different types of primary or chemical bond are found in solids—ionic, covalent, and metallic. For each type, the bonding necessarily involves the valence electrons; furthermore, the nature of the bond depends on the electron structures of the constituent atoms. In general, each of these three types of bonding arises from the tendency of the atoms to assume stable electron structures, like those of the inert gases, by completely filling the outermost electron shell.

Secondary or physical forces and energies are also found in many solid materials; they are weaker than the primary ones, but nonetheless influence the physical properties of some materials. The sections that follow explain the several kinds of primary and secondary interatomic bonds.

2.6 PRIMARY INTERATOMIC BONDS

IONIC BONDING

Perhaps **ionic bonding** is the easiest to describe and visualize. It is always found in compounds that are composed of both metallic and nonmetallic elements, elements that are situated at the horizontal extremities of the periodic table. Atoms of a metallic element easily give up their valence electrons to the nonmetallic atoms. In the process all the atoms acquire stable or inert gas configurations and, in addition, an electrical charge; that is, they become ions. Sodium chloride (NaCl) is the classical ionic material. A sodium atom can assume the electron structure of neon (and a net single positive charge) by a transfer of its one valence 3*s* electron

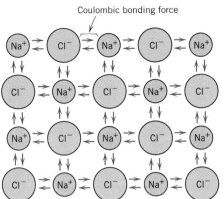

Coulombic bonding force

FIGURE 2.9   Schematic representation of ionic bonding in sodium chloride (NaCl).

to a chlorine atom. After such a transfer, the chlorine ion has a net negative charge and an electron configuration identical to that of argon. In sodium chloride, all the sodium and chlorine exist as ions. This type of bonding is illustrated schematically in Figure 2.9.

The attractive bonding forces are **coulombic;** that is, positive and negative ions, by virtue of their net electrical charge, attract one another. For two isolated ions, the attractive energy E_A is a function of the interatomic distance according to[3]

$$E_A = -\frac{A}{r} \tag{2.8}$$

An analogous equation for the repulsive energy is

$$E_R = \frac{B}{r^n} \tag{2.9}$$

In these expressions, A, B, and n are constants whose values depend on the particular ionic system. The value of n is approximately 8.

Ionic bonding is termed nondirectional, that is, the magnitude of the bond is equal in all directions around an ion. It follows that for ionic materials to be stable, all positive ions must have as nearest neighbors negatively charged ions in a three-dimensional scheme, and vice versa. The predominant bonding in ceramic materials is ionic. Some of the ion arrangements for these materials are discussed in Chapter 3.

Bonding energies, which generally range between 600 and 1500 kJ/mol (3 and 8 eV/atom), are relatively large, as reflected in high melting temperatures.[4] Table

---

[3] The constant A in Equation 2.8 is equal to

$$\frac{1}{4\pi\epsilon_0}(Z_1 e)(Z_2 e)$$

where ϵ_0 is the permittivity of a vacuum (8.85×10^{-12} F/m), Z_1 and Z_2 are the valences of the two ion types, and e is the electronic charge (1.602×10^{-19} C).

[4] Sometimes bonding energies are expressed per atom or per ion. Under these circumstances the electron volt (eV) is a conveniently small unit of energy. It is, by definition, the energy imparted to an electron as it falls through an electric potential of one volt. The joule equivalent of the electron volt is as follows: 1.602×10^{-19} J $= 1$ eV.

Table 2.3 Bonding Energies and Melting Temperatures for Various Substances

Bonding Type	Substance	Bonding Energy		Melting Temperature (°C)
		kJ/mol (kcal/mol)	eV/Atom, Ion, Molecule	
Ionic	NaCl	640 (153)	3.3	801
	MgO	1000 (239)	5.2	2800
Covalent	Si	450 (108)	4.7	1410
	C (diamond)	713 (170)	7.4	>3550
Metallic	Hg	68 (16)	0.7	−39
	Al	324 (77)	3.4	660
	Fe	406 (97)	4.2	1538
	W	849 (203)	8.8	3410
van der Waals	Ar	7.7 (1.8)	0.08	−189
	Cl_2	31 (7.4)	0.32	−101
Hydrogen	NH_3	35 (8.4)	0.36	−78
	H_2O	51 (12.2)	0.52	0

2.3 contains bonding energies and melting temperatures for several ionic materials. Ionic materials are characteristically hard and brittle and, furthermore, electrically and thermally insulative. As discussed in subsequent chapters, these properties are a direct consequence of electron configurations and/or the nature of the ionic bond.

COVALENT BONDING

In **covalent bonding** stable electron configurations are assumed by the sharing of electrons between adjacent atoms. Two atoms that are covalently bonded will each contribute at least one electron to the bond, and the shared electrons may be considered to belong to both atoms. Covalent bonding is schematically illustrated in Figure 2.10 for a molecule of methane (CH_4). The carbon atom has four valence electrons, whereas each of the four hydrogen atoms has a single valence electron. Each hydrogen atom can acquire a helium electron configuration (two 1s valence electrons) when the carbon atom shares with it one electron. The carbon now has four additional shared electrons, one from each hydrogen, for a total of eight valence

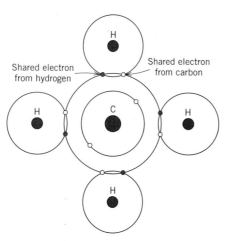

FIGURE 2.10 Schematic representation of covalent bonding in a molecule of methane (CH_4).

Shared electron from hydrogen

Shared electron from carbon

electrons, and the electron structure of neon. The covalent bond is directional; that is, it is between specific atoms and may exist only in the direction between one atom and another that participates in the electron sharing.

Many nonmetallic elemental molecules (H_2, Cl_2, F_2, etc.) as well as molecules containing dissimilar atoms, such as CH_4, H_2O, HNO_3, and HF, are covalently bonded. Furthermore, this type of bonding is found in elemental solids such as diamond (carbon), silicon, and germanium and other solid compounds composed of elements that are located on the right-hand side of the periodic table, such as gallium arsenide (GaAs), indium antimonide (InSb), and silicon carbide (SiC).

The number of covalent bonds that is possible for a particular atom is determined by the number of valence electrons. For N' valence electrons, an atom can covalently bond with at most $8 - N'$ other atoms. For example, $N' = 7$ for chlorine, and $8 - N' = 1$, which means that one Cl atom can bond to only one other atom, as in Cl_2. Similarly, for carbon, $N' = 4$, and each carbon atom has $8 - 4$, or four, electrons to share. Diamond is simply the three-dimensional interconnecting structure wherein each carbon atom covalently bonds with four other carbon atoms. This arrangement is represented in Figure 3.16.

Covalent bonds may be very strong, as in diamond, which is very hard and has a very high melting temperature, >3550°C (6400°F), or they may be very weak, as with bismuth, which melts at about 270°C (518°F). Bonding energies and melting temperatures for a few covalently bonded materials are presented in Table 2.3. Polymeric materials typify this bond, the basic molecular structure being a long chain of carbon atoms that are covalently bonded together with two of their available four bonds per atom. The remaining two bonds normally are shared with other atoms, which also covalently bond. Polymeric molecular structures are discussed in detail in Chapter 4.

It is possible to have interatomic bonds that are partially ionic and partially covalent, and, in fact, very few compounds exhibit pure ionic or covalent bonding. For a compound, the degree of either bond type depends on the relative positions of the constituent atoms in the periodic table (Figure 2.6) or the difference in their electronegativities (Figure 2.7). The wider the separation (both horizontally—relative to Group IVA—and vertically) from the lower left to the upper-right-hand corner (i.e., the greater the difference in electronegativity), the more ionic the bond. Conversely, the closer the atoms are together (i.e., the smaller the difference in electronegativity), the greater the degree of covalency. The percent ionic character of a bond between elements A and B (A being the most electronegative) may be approximated by the expression

$$\% \text{ ionic character} = \{1 - \exp[-(0.25)(X_A - X_B)^2]\} \times 100 \qquad (2.10)$$

where X_A and X_B are the electronegativities for the respective elements.

METALLIC BONDING

Metallic bonding, the final primary bonding type, is found in metals and their alloys. A relatively simple model has been proposed that very nearly approximates the bonding scheme. Metallic materials have one, two, or at most, three valence electrons. With this model, these valence electrons are not bound to any particular atom in the solid and are more or less free to drift throughout the entire metal. They may be thought of as belonging to the metal as a whole, or forming a "sea of electrons" or an "electron cloud." The remaining nonvalence electrons and atomic nuclei form what are called *ion cores,* which possess a net positive charge equal in magnitude to the total valence electron charge per atom. Figure 2.11 is a

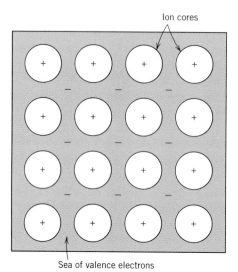

Ion cores

Sea of valence electrons

FIGURE 2.11 Schematic illustration of metallic bonding.

schematic illustration of metallic bonding. The free electrons shield the positively charged ion cores from mutually repulsive electrostatic forces, which they would otherwise exert upon one another; consequently the metallic bond is nondirectional in character. In addition, these free electrons act as a "glue" to hold the ion cores together. Bonding energies and melting temperatures for several metals are listed in Table 2.3. Bonding may be weak or strong; energies range from 68 kJ/mol (0.7 eV/atom) for mercury to 850 kJ/mol (8.8 eV/atom) for tungsten. Their respective melting temperatures are −39 and 3410°C (−38 and 6170°F).

Metallic bonding is found for Group IA and IIA elements in the periodic table, and, in fact, for all elemental metals.

Some general behaviors of the various material types (i.e., metals, ceramics, polymers) may be explained by bonding type. For example, metals are good conductors of both electricity and heat, as a consequence of their free electrons (see Sections 12.5, 12.6, {and 17.4}). By way of contrast, ionically and covalently bonded materials are typically electrical and thermal insulators, due to the absence of large numbers of free electrons.

Furthermore, in Section 8.5 we note that at room temperature, most metals and their alloys fail in a ductile manner; that is, fracture occurs after the materials have experienced significant degrees of permanent deformation. This behavior is explained in terms of deformation mechanism (Section 8.3), which is implicitly related to the characteristics of the metallic bond. Conversely, at room temperature ionically bonded materials are intrinsically brittle as a consequence of the electrically charged nature of their component ions (see Section 8.15).

2.7 SECONDARY BONDING OR VAN DER WAALS BONDING

Secondary, van der Waals, or physical bonds are weak in comparison to the primary or chemical ones; bonding energies are typically on the order of only 10 kJ/mol (0.1 eV/atom). Secondary bonding exists between virtually all atoms or molecules, but its presence may be obscured if any of the three primary bonding types is present. Secondary bonding is evidenced for the inert gases, which have stable

FIGURE 2.12 Schematic illustration of van der Waals bonding between two dipoles.

Atomic or molecular dipoles

electron structures, and, in addition, between molecules in molecular structures that are covalently bonded.

Secondary bonding forces arise from atomic or molecular **dipoles.** In essence, an electric dipole exists whenever there is some separation of positive and negative portions of an atom or molecule. The bonding results from the coulombic attraction between the positive end of one dipole and the negative region of an adjacent one, as indicated in Figure 2.12. Dipole interactions occur between induced dipoles, between induced dipoles and polar molecules (which have permanent dipoles), and between polar molecules. **Hydrogen bonding,** a special type of secondary bonding, is found to exist between some molecules that have hydrogen as one of the constituents. These bonding mechanisms are now discussed briefly.

FLUCTUATING INDUCED DIPOLE BONDS

A dipole may be created or induced in an atom or molecule that is normally electrically symmetric; that is, the overall spatial distribution of the electrons is symmetric with respect to the positively charged nucleus, as shown in Figure 2.13a. All atoms are experiencing constant vibrational motion that can cause instantaneous and short-lived distortions of this electrical symmetry for some of the atoms or molecules, and the creation of small electric dipoles, as represented in Figure 2.13b. One of these dipoles can in turn produce a displacement of the electron distribution of an adjacent molecule or atom, which induces the second one also to become a dipole that is then weakly attracted or bonded to the first; this is one type of van der Waals bonding. These attractive forces may exist between large numbers of atoms or molecules, which forces are temporary and fluctuate with time.

The liquefaction and, in some cases, the solidification of the inert gases and other electrically neutral and symmetric molecules such as H_2 and Cl_2 are realized because of this type of bonding. Melting and boiling temperatures are extremely low in materials for which induced dipole bonding predominates; of all possible intermolecular bonds, these are the weakest. Bonding energies and melting temperatures for argon and chlorine are also tabulated in Table 2.3.

POLAR MOLECULE-INDUCED DIPOLE BONDS

Permanent dipole moments exist in some molecules by virtue of an asymmetrical arrangement of positively and negatively charged regions; such molecules are termed **polar molecules.** Figure 2.14 is a schematic representation of a hydrogen

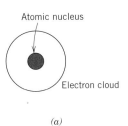

Atomic nucleus

Electron cloud

(a)

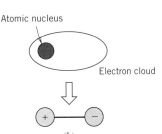

Atomic nucleus

Electron cloud

(b)

FIGURE 2.13 Schematic representations of (a) an electrically symmetric atom and (b) an induced atomic dipole.

FIGURE 2.14   Schematic representation of a polar hydrogen chloride (HCl) molecule.

chloride molecule; a permanent dipole moment arises from net positive and negative charges that are respectively associated with the hydrogen and chlorine ends of the HCl molecule.

Polar molecules can also induce dipoles in adjacent nonpolar molecules, and a bond will form as a result of attractive forces between the two molecules. Furthermore, the magnitude of this bond will be greater than for fluctuating induced dipoles.

PERMANENT DIPOLE BONDS

Van der Waals forces will also exist between adjacent polar molecules. The associated bonding energies are significantly greater than for bonds involving induced dipoles.

The strongest secondary bonding type, the hydrogen bond, is a special case of polar molecule bonding. It occurs between molecules in which hydrogen is covalently bonded to fluorine (as in HF), oxygen (as in H_2O), and nitrogen (as in NH_3). For each H—F, H—O, or H—N bond, the single hydrogen electron is shared with the other atom. Thus, the hydrogen end of the bond is essentially a positively charged bare proton that is unscreened by any electrons. This highly positively charged end of the molecule is capable of a strong attractive force with the negative end of an adjacent molecule, as demonstrated in Figure 2.15 for HF. In essence, this single proton forms a bridge between two negatively charged atoms. The magnitude of the hydrogen bond is generally greater than that of the other types of secondary bonds, and may be as high as 51 kJ/mol (0.52 eV/molecule), as shown in Table 2.3. Melting and boiling temperatures for hydrogen fluoride and water are abnormally high in light of their low molecular weights, as a consequence of hydrogen bonding.

2.8 MOLECULES

At the conclusion of this chapter, let us take a moment to discuss the concept of a **molecule** in terms of solid materials. A molecule may be defined as a group of atoms that are bonded together by strong primary bonds. Within this context, the entirety of ionic and metallically bonded solid specimens may be considered as a single molecule. However, this is not the case for many substances in which covalent bonding predominates; these include elemental diatomic molecules (F_2, O_2, H_2, etc.) as well as a host of compounds (H_2O, CO_2, HNO_3, C_6H_6, CH_4, etc.). In the

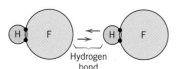

FIGURE 2.15   Schematic representation of hydrogen bonding in hydrogen fluoride (HF).

condensed liquid and solid states, bonds between molecules are weak secondary ones. Consequently, molecular materials have relatively low melting and boiling temperatures. Most of those that have small molecules composed of a few atoms are gases at ordinary, or ambient, temperatures and pressures. On the other hand, many of the modern polymers, being molecular materials composed of extremely large molecules, exist as solids; some of their properties are strongly dependent on the presence of van der Waals and hydrogen secondary bonds.

S U M M A R Y

This chapter began with a survey of the fundamentals of atomic structure, presenting the Bohr and wave-mechanical models of electrons in atoms. Whereas the Bohr model assumes electrons to be particles orbiting the nucleus in discrete paths, in wave mechanics we consider them to be wavelike and treat electron position in terms of a probability distribution.

Electron energy states are specified in terms of quantum numbers that give rise to electron shells and subshells. The electron configuration of an atom corresponds to the manner in which these shells and subshells are filled with electrons in compliance with the Pauli exclusion principle. The periodic table of the elements is generated by arrangement of the various elements according to valence electron configuration.

Atomic bonding in solids may be considered in terms of attractive and repulsive forces and energies. The three types of primary bond in solids are ionic, covalent, and metallic. For ionic bonds, electrically charged ions are formed by the transference of valence electrons from one atom type to another; forces are coulombic. There is a sharing of valence electrons between adjacent atoms when bonding is covalent. With metallic bonding, the valence electrons form a "sea of electrons" that is uniformly dispersed around the metal ion cores and acts as a form of glue for them.

Both van der Waals and hydrogen bonds are termed secondary, being weak in comparison to the primary ones. They result from attractive forces between electric dipoles, of which there are two types—induced and permanent. For the hydrogen bond, highly polar molecules form when hydrogen covalently bonds to a nonmetallic element such as fluorine.

I M P O R T A N T   T E R M S   A N D   C O N C E P T S

Atomic mass unit (amu)	Electronegative	Periodic table
Atomic number	Electropositive	Polar molecule
Atomic weight	Ground state	Primary bonding
Bohr atomic model	Hydrogen bond	Quantum mechanics
Bonding energy	Ionic bond	Quantum number
Coulombic force	Isotope	Secondary bonding
Covalent bond	Metallic bond	Valence electron
Dipole (electric)	Mole	van der Waals bond
Electron configuration	Molecule	Wave-mechanical model
Electron state	Pauli exclusion principle	

Note: In each chapter, most of the terms listed in the "Important Terms and Concepts" section are defined in the Glossary, which follows Appendix E. The others are important enough to warrant treatment in a full section of the text and can be referenced from the table of contents or the index.

REFERENCES

Most of the material in this chapter is covered in college-level chemistry textbooks. Below, two are listed as references.

Kotz, J. C. and P. Treichel, Jr., *Chemistry and Chemical Reactivity,* 4th edition, Saunders College Publishing, Fort Worth, TX, 1999.

Masterton, W. L. and C. N. Hurley, *Chemistry, Principles and Reactions,* 3rd edition, Saunders College Publishing, Philadelphia, 1996.

QUESTIONS AND PROBLEMS

2.1 **(a)** What is an isotope?

(b) Why are the atomic weights of the elements not integers? Cite two reasons.

2.2 Cite the difference between atomic mass and atomic weight.

2.3 **(a)** How many grams are there in 1 amu of a material?

(b) Mole, in the context of this book, is taken in units of gram-mole. On this basis, how many atoms are there in a pound-mole of a substance?

2.4 **(a)** Cite two important quantum-mechanical concepts associated with the Bohr model of the atom.

(b) Cite two important additional refinements that resulted from the wave-mechanical atomic model.

2.5 Relative to electrons and electron states, what does each of the four quantum numbers specify?

2.6 Allowed values for the quantum numbers of electrons are as follows:

$$n = 1, 2, 3, \ldots$$
$$l = 0, 1, 2, 3, \ldots, n - 1$$
$$m_l = 0, \pm 1, \pm 2, \pm 3, \ldots, \pm l$$
$$m_s = \pm \tfrac{1}{2}$$

The relationships between n and the shell designations are noted in Table 2.1. Relative to the subshells,

$l = 0$ corresponds to an s subshell

$l = 1$ corresponds to a p subshell

$l = 2$ corresponds to a d subshell

$l = 3$ corresponds to an f subshell

For the K shell, the four quantum numbers for each of the two electrons in the $1s$ state, in the order of nlm_lm_s, are $100(\tfrac{1}{2})$ and $100(-\tfrac{1}{2})$.

Write the four quantum numbers for all of the electrons in the L and M shells, and note which correspond to the s, p, and d subshells.

2.7 Give the electron configurations for the following ions: Fe^{2+}, Fe^{3+}, Cu^{+}, Ba^{2+}, Br^{-}, and S^{2-}.

2.8 Cesium bromide (CsBr) exhibits predominantly ionic bonding. The Cs^+ and Br^- ions have electron structures that are identical to which two inert gases?

2.9 With regard to electron configuration, what do all the elements in Group VIIA of the periodic table have in common?

2.10 Without consulting Figure 2.6 or Table 2.2, determine whether each of the electron configurations given below is an inert gas, a halogen, an alkali metal, an alkaline earth metal, or a transition metal. Justify your choices.

(a) $1s^2 2s^2 2p^6 3s^2 3p^6 3d^7 4s^2$.

(b) $1s^2 2s^2 2p^6 3s^2 3p^6$.

(c) $1s^2 2s^2 2p^5$.

(d) $1s^2 2s^2 2p^6 3s^2$.

(e) $1s^2 2s^2 2p^6 3s^2 3p^6 3d^2 4s^2$.

(f) $1s^2 2s^2 2p^6 3s^2 3p^6 4s^1$.

2.11 **(a)** What electron subshell is being filled for the rare earth series of elements on the periodic table?

(b) What electron subshell is being filled for the actinide series?

2.12 Calculate the force of attraction between a K^+ and an O^{2-} ion the centers of which are separated by a distance of 1.5 nm.

2.13 The net potential energy between two adjacent ions, E_N, may be represented by the sum of Equations 2.8 and 2.9, that is,

$$E_N = -\frac{A}{r} + \frac{B}{r^n} \qquad (2.11)$$

Calculate the bonding energy E_0 in terms of the parameters A, B, and n using the following procedure:

1. Differentiate E_N with respect to r, and then set the resulting expression equal to zero, since the curve of E_N versus r is a minimum at E_0.

2. Solve for r in terms of A, B, and n, which yields r_0, the equilibrium interionic spacing.

3. Determine the expression for E_0 by substitution of r_0 into Equation 2.11.

2.14 For a K^+–Cl^- ion pair, attractive and repulsive energies E_A and E_R, respectively, depend on the distance between the ions r, according to

$$E_A = -\frac{1.436}{r}$$

$$E_R = \frac{5.86 \times 10^{-6}}{r^9}$$

For these expressions, energies are expressed in electron volts per K^+–Cl^- pair, and r is the distance in nanometers. The net energy E_N is just the sum of the two expressions above.

(a) Superimpose on a single plot E_N, E_R, and E_A versus r up to 1.0 nm.

(b) On the basis of this plot, determine (i) the equilibrium spacing r_0 between the K^+ and Cl^- ions, and (ii) the magnitude of the bonding energy E_0 between the two ions.

(c) Mathematically determine the r_0 and E_0 values using the solutions to Problem 2.13 and compare these with the graphical results from part b.

2.15 Consider some hypothetical $X^+ - Y^-$ ion pair for which the equilibrium interionic spacing and bonding energy values are 0.35 nm and -6.13 eV, respectively. If it is known that n in Equation 2.11 has a value of 10, using the results of Problem 2.13, determine explicit expressions for attractive and repulsive energies, E_A and E_R of Equations 2.8 and 2.9.

2.16 The net potential energy E_N between two adjacent ions is sometimes represented by the expression

$$E_N = -\frac{C}{r} + D \exp\left(-\frac{r}{\rho}\right) \quad (2.12)$$

in which r is the interionic separation and C, D, and ρ are constants whose values depend on the specific material.

(a) Derive an expression for the bonding energy E_0 in terms of the equilibrium interionic separation r_0 and the constants D and ρ using the following procedure:

1. Differentiate E_N with respect to r and set the resulting expression equal to zero.
2. Solve for C in terms of D, ρ, and r_0.
3. Determine the expression for E_0 by substitution for C in Equation 2.12.

(b) Derive another expression for E_0 in terms of r_0, C, and ρ using a procedure analogous to the one outlined in part a.

2.17 (a) Briefly cite the main differences between ionic, covalent, and metallic bonding.

(b) State the Pauli exclusion principle.

2.18 Offer an explanation as to why covalently bonded materials are generally less dense than ionically or metallically bonded ones.

2.19 Compute the percents ionic character of the interatomic bonds for the following compounds: TiO_2, $ZnTe$, $CsCl$, $InSb$, and $MgCl_2$.

2.20 Make a plot of bonding energy versus melting temperature for the metals listed in Table 2.3. Using this plot, approximate the bonding energy for copper, which has a melting temperature of 1084°C.

2.21 Using Table 2.2, determine the number of covalent bonds that are possible for atoms of the following elements: germanium, phosphorus, selenium, and chlorine.

2.22 What type(s) of bonding would be expected for each of the following materials: brass (a copper-zinc alloy), rubber, barium sulfide (BaS), solid xenon, bronze, nylon, and aluminum phosphide (AlP)?

2.23 Explain why hydrogen fluoride (HF) has a higher boiling temperature than hydrogen chloride (HCl) (19.4 vs. −85°C), even though HF has a lower molecular weight.

2.24 On the basis of the hydrogen bond, explain the anomalous behavior of water when it freezes. That is, why is there volume expansion upon solidification?

Chapter 3 / Structures of Metals and Ceramics

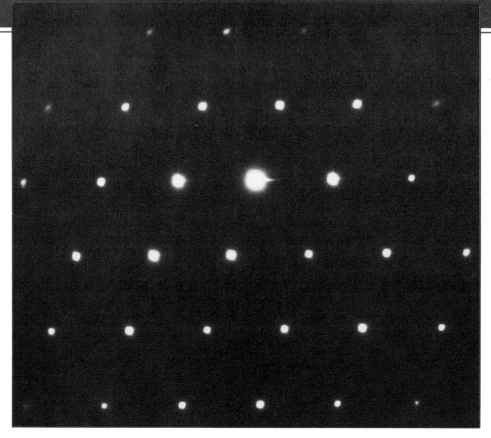

*H*igh velocity electron beams that are produced when electrons are accelerated across large voltages become wavelike in character. Their wavelengths are shorter than interatomic spacings, and thus these beams may be diffracted by atomic planes in crystalline materials, in the same manner as x-rays experience diffraction.

This photograph shows a diffraction pattern produced for a single crystal of gallium arsenide using a transmission electron microscope. The brightest spot near the center is produced by the incident electron beam, which is parallel to a $\langle 110 \rangle$ crystallographic direction. Each of the other white spots results from an electron beam that is diffracted by a specific set of crystallographic planes. (Photograph courtesy of Dr. Raghaw S. Rai, Motorola, Inc., Austin, Texas.)

Why Study *Structures of Metals and Ceramics?*

The properties of some materials are directly related to their crystal structures. For example, pure and undeformed magnesium and beryllium, having one crystal structure, are much more brittle (i.e., fracture at lower degrees of deformation) than are pure and undeformed metals such as gold and silver that have yet another crystal structure (see Section 8.5). {Also, the permanent magnetic and ferroelectric behaviors of some ceramic materials are explained by their crystal structures (Sections 18.4 and 12.23).}

Furthermore, significant property differences exist between crystalline and noncrystalline materials having the same composition. For example, noncrystalline ceramics and polymers normally are optically transparent; the same materials in crystalline (or semicrystalline) form tend to be opaque or, at best, translucent.

After studying this chapter you should be able to do the following:

1. Describe the difference in atomic/molecular structure between crystalline and noncrystalline materials.
2. Draw unit cells for face-centered cubic, body-centered cubic, and hexagonal close-packed crystal structures.
3. Derive the relationships between unit cell edge length and atomic radius for face-centered cubic and body-centered cubic crystal structures.
4. Compute the densities for metals having face-centered cubic and body-centered cubic crystal structures given their unit cell dimensions.
5. Sketch/describe unit cells for sodium chloride, cesium chloride, zinc blende, diamond cubic, fluorite, and perovskite crystal structures. Do likewise for the atomic structures of graphite and a silica glass.
6. Given the chemical formula for a ceramic compound, the ionic radii of its component ions, determine the crystal structure.
7. Given three direction index integers, sketch the direction corresponding to these indices within a unit cell.
8. Specify the Miller indices for a plane that has been drawn within a unit cell.
9. Describe how face-centered cubic and hexagonal close-packed crystal structures may be generated by the stacking of close-packed planes of atoms. Do the same for the sodium chloride crystal structure in terms of close-packed planes of anions.
10. Distinguish between single crystals and polycrystalline materials.
11. Define *isotropy* and *anisotropy* with respect to material properties.

3.1 INTRODUCTION

Chapter 2 was concerned primarily with the various types of atomic bonding, which are determined by the electron structure of the individual atoms. The present discussion is devoted to the next level of the structure of materials, specifically, to some of the arrangements that may be assumed by atoms in the solid state. Within this framework, concepts of crystallinity and noncrystallinity are introduced. For crystalline solids the notion of crystal structure is presented, specified in terms of a unit cell. Crystal structures found in both metals and ceramics are then detailed, along with the scheme by which crystallographic directions and planes are expressed. Single crystals, polycrystalline, and noncrystalline materials are considered.

CRYSTAL STRUCTURES

3.2 FUNDAMENTAL CONCEPTS

Solid materials may be classified according to the regularity with which atoms or ions are arranged with respect to one another. A **crystalline** material is one in which the atoms are situated in a repeating or periodic array over large atomic distances; that is, long-range order exists, such that upon solidification, the atoms will position themselves in a repetitive three-dimensional pattern, in which each atom is bonded to its nearest-neighbor atoms. All metals, many ceramic materials, and certain polymers form crystalline structures under normal solidification conditions. For those that do not crystallize, this long-range atomic order is absent; these *noncrystalline* or *amorphous* materials are discussed briefly at the end of this chapter.

Some of the properties of crystalline solids depend on the **crystal structure** of the material, the manner in which atoms, ions, or molecules are spatially arranged. There is an extremely large number of different crystal structures all having long-range atomic order; these vary from relatively simple structures for metals, to exceedingly complex ones, as displayed by some of the ceramic and polymeric

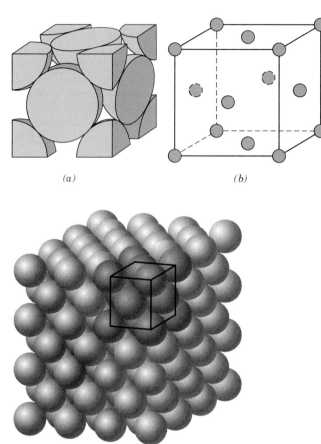

(a)   (b)

(c)

FIGURE 3.1 For the face-centered cubic crystal structure: (*a*) a hard sphere unit cell representation, (*b*) a reduced-sphere unit cell, and (*c*) an aggregate of many atoms. (Figure *c* adapted from W. G. Moffatt, G. W. Pearsall, and J. Wulff, *The Structure and Properties of Materials,* Vol. I, *Structure,* p. 51. Copyright © 1964 by John Wiley & Sons, New York. Reprinted by permission of John Wiley & Sons, Inc.)

materials. The present discussion deals with several common metallic and ceramic crystal structures. The next chapter is devoted to structures for polymers.

When describing crystalline structures, atoms (or ions) are thought of as being solid spheres having well-defined diameters. This is termed the *atomic hard sphere model* in which spheres representing nearest-neighbor atoms touch one another. An example of the hard sphere model for the atomic arrangement found in some of the common elemental metals is displayed in Figure 3.1*c*. In this particular case all the atoms are identical. Sometimes the term **lattice** is used in the context of crystal structures; in this sense "lattice" means a three-dimensional array of points coinciding with atom positions (or sphere centers).

3.3 UNIT CELLS

The atomic order in crystalline solids indicates that small groups of atoms form a repetitive pattern. Thus, in describing crystal structures, it is often convenient to subdivide the structure into small repeat entities called **unit cells.** Unit cells for most crystal structures are parallelepipeds or prisms having three sets of parallel faces; one is drawn within the aggregate of spheres (Figure 3.1*c*), which in this case happens to be a cube. A unit cell is chosen to represent the symmetry of the crystal structure, wherein all the atom positions in the crystal may be generated by translations of the unit cell integral distances along each of its edges. Thus, the unit cell is the basic structural unit or building block of the crystal structure and defines the crystal structure by virtue of its geometry and the atom positions within. Convenience usually dictates that parallelepiped corners coincide with centers of the hard sphere atoms. Furthermore, more than a single unit cell may be chosen for a particular crystal structure; however, we generally use the unit cell having the highest level of geometrical symmetry.

3.4 METALLIC CRYSTAL STRUCTURES

The atomic bonding in this group of materials is metallic, and thus nondirectional in nature. Consequently, there are no restrictions as to the number and position of nearest-neighbor atoms; this leads to relatively large numbers of nearest neighbors and dense atomic packings for most metallic crystal structures. Also, for metals, using the hard sphere model for the crystal structure, each sphere represents an ion core. Table 3.1 presents the atomic radii for a number of metals. Three relatively simple crystal structures are found for most of the common metals: face-centered cubic, body-centered cubic, and hexagonal close-packed.

THE FACE-CENTERED CUBIC CRYSTAL STRUCTURE

The crystal structure found for many metals has a unit cell of cubic geometry, with atoms located at each of the corners and the centers of all the cube faces. It is aptly called the **face-centered cubic (FCC)** crystal structure. Some of the familiar metals having this crystal structure are copper, aluminum, silver, and gold (see also Table 3.1). Figure 3.1a shows a hard sphere model for the FCC unit cell, whereas in Figure 3.1b the atom centers are represented by small circles to provide a better perspective of atom positions. The aggregate of atoms in Figure 3.1c represents a section of crystal consisting of many FCC unit cells. These spheres or ion cores touch one another across a face diagonal; the cube edge length a and the atomic radius R are related through

$$a = 2R \sqrt{2} \tag{3.1}$$

This result is obtained as an example problem.

For the FCC crystal structure, each corner atom is shared among eight unit cells, whereas a face-centered atom belongs to only two. Therefore, one eighth of

Table 3.1 Atomic Radii and Crystal Structures for 16 Metals

Metal	Crystal Structure[a]	Atomic Radius[b] (nm)	Metal	Crystal Structure	Atomic Radius (nm)
Aluminum	FCC	0.1431	Molybdenum	BCC	0.1363
Cadmium	HCP	0.1490	Nickel	FCC	0.1246
Chromium	BCC	0.1249	Platinum	FCC	0.1387
Cobalt	HCP	0.1253	Silver	FCC	0.1445
Copper	FCC	0.1278	Tantalum	BCC	0.1430
Gold	FCC	0.1442	Titanium (α)	HCP	0.1445
Iron (α)	BCC	0.1241	Tungsten	BCC	0.1371
Lead	FCC	0.1750	Zinc	HCP	0.1332

[a] FCC = face-centered cubic; HCP = hexagonal close-packed; BCC = body-centered cubic.

[b] A nanometer (nm) equals 10^{-9} m; to convert from nanometers to angstrom units (Å), multiply the nanometer value by 10.

each of the eight corner atoms and one half of each of the six face atoms, or a total of four whole atoms, may be assigned to a given unit cell. This is depicted in Figure 3.1a, where only sphere portions are represented within the confines of the cube. The cell comprises the volume of the cube, which is generated from the centers of the corner atoms as shown in the figure.

Corner and face positions are really equivalent; that is, translation of the cube corner from an original corner atom to the center of a face atom will not alter the cell structure.

Two other important characteristics of a crystal structure are the **coordination number** and the **atomic packing factor (APF).** For metals, each atom has the same number of nearest-neighbor or touching atoms, which is the coordination number. For face-centered cubics, the coordination number is 12. This may be confirmed by examination of Figure 3.1a; the front face atom has four corner nearest-neighbor atoms surrounding it, four face atoms that are in contact from behind, and four other equivalent face atoms residing in the next unit cell to the front, which is not shown.

The APF is the fraction of solid sphere volume in a unit cell, assuming the atomic hard sphere model, or

$$\text{APF} = \frac{\text{volume of atoms in a unit cell}}{\text{total unit cell volume}} \tag{3.2}$$

For the FCC structure, the atomic packing factor is 0.74, which is the maximum packing possible for spheres all having the same diameter. Computation of this APF is also included as an example problem. Metals typically have relatively large atomic packing factors to maximize the shielding provided by the free electron cloud.

THE BODY-CENTERED CUBIC CRYSTAL STRUCTURE

Another common metallic crystal structure also has a cubic unit cell with atoms located at all eight corners and a single atom at the cube center. This is called a **body-centered cubic (BCC)** crystal structure. A collection of spheres depicting this crystal structure is shown in Figure 3.2c, whereas Figures 3.2a and 3.2b are diagrams

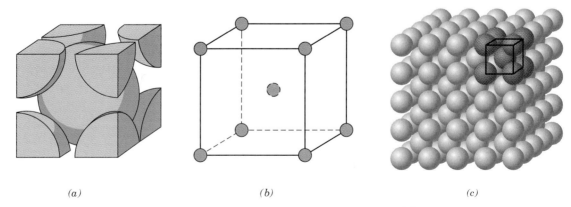

(a)   *(b)*   *(c)*

FIGURE 3.2 For the body-centered cubic crystal structure, (a) a hard sphere unit cell representation, (b) a reduced-sphere unit cell, and (c) an aggregate of many atoms. (Figure (c) from W. G. Moffatt, G. W. Pearsall, and J. Wulff, *The Structure and Properties of Materials,* Vol. I, *Structure,* p. 51. Copyright © 1964 by John Wiley & Sons, New York. Reprinted by permission of John Wiley & Sons, Inc.)

of BCC unit cells with the atoms represented by hard sphere and reduced-sphere models, respectively. Center and corner atoms touch one another along cube diagonals, and unit cell length a and atomic radius R are related through

$$a = \frac{4R}{\sqrt{3}}$$

(3.3)

Chromium, iron, tungsten, as well as several other metals listed in Table 3.1 exhibit a BCC structure.

Two atoms are associated with each BCC unit cell: the equivalent of one atom from the eight corners, each of which is shared among eight unit cells, and the single center atom, which is wholly contained within its cell. In addition, corner and center atom positions are equivalent. The coordination number for the BCC crystal structure is 8; each center atom has as nearest neighbors its eight corner atoms. Since the coordination number is less for BCC than FCC, so also is the atomic packing factor for BCC lower—0.68 versus 0.74.

THE HEXAGONAL CLOSE-PACKED CRYSTAL STRUCTURE

Not all metals have unit cells with cubic symmetry; the final common metallic crystal structure to be discussed has a unit cell that is hexagonal. Figure 3.3a shows a reduced-sphere unit cell for this structure, which is termed **hexagonal close-packed (HCP)**; an assemblage of several HCP unit cells is presented in Figure 3.3b. The top and bottom faces of the unit cell consist of six atoms that form regular hexagons and surround a single atom in the center. Another plane that provides three additional atoms to the unit cell is situated between the top and bottom planes. The atoms in this midplane have as nearest neighbors atoms in both of the adjacent two planes. The equivalent of six atoms is contained in each unit cell; one-sixth of each of the 12 top and bottom face corner atoms, one-half of each of the 2 center face atoms, and all the 3 midplane interior atoms. If a and c represent, respectively,

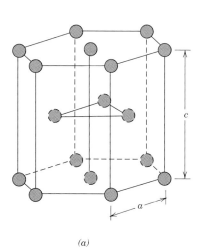

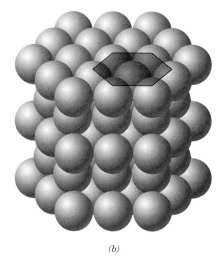

(a)

(b)

FIGURE 3.3   For the hexagonal close-packed crystal structure, (a) a reduced-sphere unit cell (a and c represent the short and long edge lengths, respectively), and (b) an aggregate of many atoms. (Figure (b) from W. G. Moffatt, G. W. Pearsall, and J. Wulff, *The Structure and Properties of Materials*, Vol. I, *Structure*, p. 51. Copyright © 1964 by John Wiley & Sons, New York. Reprinted by permission of John Wiley & Sons, Inc.)

the short and long unit cell dimensions of Figure 3.3a, the c/a ratio should be 1.633; however, for some HCP metals this ratio deviates from the ideal value.

The coordination number and the atomic packing factor for the HCP crystal structure are the same as for FCC: 12 and 0.74, respectively. The HCP metals include cadmium, magnesium, titanium, and zinc; some of these are listed in Table 3.1.

EXAMPLE PROBLEM 3.1

Calculate the volume of an FCC unit cell in terms of the atomic radius R.

SOLUTION

In the FCC unit cell illustrated,

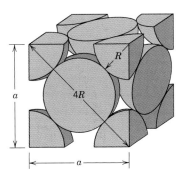

the atoms touch one another across a face-diagonal the length of which is $4R$. Since the unit cell is a cube, its volume is a^3, where a is the cell edge length. From the right triangle on the face,

$$a^2 + a^2 = (4R)^2$$

or, solving for a,

$$a = 2R\sqrt{2} \tag{3.1}$$

The FCC unit cell volume V_C may be computed from

$$V_C = a^3 = (2R\sqrt{2})^3 = 16R^3\sqrt{2} \tag{3.4}$$

EXAMPLE PROBLEM 3.2

Show that the atomic packing factor for the FCC crystal structure is 0.74.

SOLUTION

The APF is defined as the fraction of solid sphere volume in a unit cell, or

$$\text{APF} = \frac{\text{total sphere volume}}{\text{total unit cell volume}} = \frac{V_S}{V_C}$$

Both the total sphere and unit cell volumes may be calculated in terms of the atomic radius R. The volume for a sphere is $\frac{4}{3}\pi R^3$, and since there are four atoms per FCC unit cell, the total FCC sphere volume is

$$V_S = (4)\frac{4}{3}\pi R^3 = \frac{16}{3}\pi R^3$$

From Example Problem 3.1, the total unit cell volume is

$$V_C = 16R^3\sqrt{2}$$

Therefore, the atomic packing factor is

$$\text{APF} = \frac{V_S}{V_C} = \frac{(\frac{16}{3})\pi R^3}{16R^3\sqrt{2}} = 0.74$$

3.5 DENSITY COMPUTATIONS—METALS

A knowledge of the crystal structure of a metallic solid permits computation of its theoretical density ρ through the relationship

$$\rho = \frac{nA}{V_C N_A} \qquad (3.5)$$

where

n = number of atoms associated with each unit cell

A = atomic weight

V_C = volume of the unit cell

N_A = Avogadro's number (6.023×10^{23} atoms/mol)

EXAMPLE PROBLEM 3.3

Copper has an atomic radius of 0.128 nm (1.28 Å), an FCC crystal structure, and an atomic weight of 63.5 g/mol. Compute its theoretical density and compare the answer with its measured density.

SOLUTION

Equation 3.5 is employed in the solution of this problem. Since the crystal structure is FCC, n, the number of atoms per unit cell, is 4. Furthermore, the atomic weight A_{Cu} is given as 63.5 g/mol. The unit cell volume V_C for FCC was determined in Example Problem 3.1 as $16R^3\sqrt{2}$, where R, the atomic radius, is 0.128 nm.

Substitution for the various parameters into Equation 3.5 yields

$$\rho = \frac{nA_{Cu}}{V_C N_A} = \frac{nA_{Cu}}{(16R^3\sqrt{2})N_A}$$

$$= \frac{(4\ \text{atoms/unit cell})(63.5\ \text{g/mol})}{[16\sqrt{2}(1.28 \times 10^{-8}\ \text{cm})^3/\text{unit cell}](6.023 \times 10^{23}\ \text{atoms/mol})}$$

$$= 8.89\ \text{g/cm}^3$$

The literature value for the density of copper is 8.94 g/cm³, which is in very close agreement with the foregoing result.

3.6 CERAMIC CRYSTAL STRUCTURES

Because ceramics are composed of at least two elements, and often more, their crystal structures are generally more complex than those for metals. The atomic bonding in these materials ranges from purely ionic to totally covalent; many ceramics exhibit a combination of these two bonding types, the degree of ionic character being dependent on the electronegativities of the atoms. Table 3.2 presents the percent ionic character for several common ceramic materials; these values were determined using Equation 2.10 and the electronegativities in Figure 2.7.

For those ceramic materials for which the atomic bonding is predominantly ionic, the crystal structures may be thought of as being composed of electrically charged ions instead of atoms. The metallic ions, or **cations,** are positively charged, because they have given up their valence electrons to the nonmetallic ions, or **anions,** which are negatively charged. Two characteristics of the component ions in crystalline ceramic materials influence the crystal structure: the magnitude of the electrical charge on each of the component ions, and the relative sizes of the cations and anions. With regard to the first characteristic, the crystal must be electrically neutral; that is, all the cation positive charges must be balanced by an equal number of anion negative charges. The chemical formula of a compound indicates the ratio of cations to anions, or the composition that achieves this charge balance. For example, in calcium fluoride, each calcium ion has a +2 charge (Ca^{2+}), and associated with each fluorine ion is a single negative charge (F^-). Thus, there must be twice as many F^- as Ca^{2+} ions, which is reflected in the chemical formula CaF_2.

The second criterion involves the sizes or ionic radii of the cations and anions, r_C and r_A, respectively. Because the metallic elements give up electrons when ionized, cations are ordinarily smaller than anions, and, consequently, the ratio r_C/r_A is less than unity. Each cation prefers to have as many nearest-neighbor anions as possible. The anions also desire a maximum number of cation nearest neighbors.

Stable ceramic crystal structures form when those anions surrounding a cation are all in contact with that cation, as illustrated in Figure 3.4. The coordination number (i.e., number of anion nearest neighbors for a cation) is related to the cation–anion radius ratio. For a specific coordination number, there is a critical or minimum r_C/r_A ratio for which this cation–anion contact is established (Figure 3.4), which ratio may be determined from pure geometrical considerations (see Example Problem 3.4).

Table 3.2 For Several Ceramic Materials, Percent Ionic Character of the Interatomic Bonds

Material	Percent Ionic Character
CaF_2	89
MgO	73
NaCl	67
Al_2O_3	63
SiO_2	51
Si_3N_4	30
ZnS	18
SiC	12

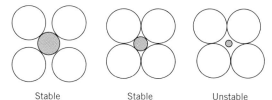

FIGURE 3.4 Stable and unstable anion–cation coordination configurations. Open circles represent anions; colored circles denote cations.

Stable   Stable   Unstable

The coordination numbers and nearest-neighbor geometries for various r_C/r_A ratios are presented in Table 3.3. For r_C/r_A ratios less than 0.155, the very small cation is bonded to two anions in a linear manner. If r_C/r_A has a value between 0.155 and 0.225, the coordination number for the cation is 3. This means each cation is surrounded by three anions in the form of a planar equilateral triangle, with the cation located in the center. The coordination number is 4 for r_C/r_A between 0.225 and 0.414; the cation is located at the center of a tetrahedron, with anions at each of the four corners. For r_C/r_A between 0.414 and 0.732, the cation may be thought of as being situated at the center of an octahedron surrounded by six anions, one at each corner, as also shown in the table. The coordination number is 8 for r_C/r_A between 0.732 and 1.0, with anions at all corners of a cube and a cation positioned at the center. For a radius ratio greater than unity, the coordination number is 12. The most common coordination numbers for ceramic materials are 4, 6, and 8. Table 3.4 gives the ionic radii for several anions and cations that are common in ceramic materials.

EXAMPLE PROBLEM 3.4

Show that the minimum cation-to-anion radius ratio for the coordination number 3 is 0.155.

SOLUTION

For this coordination, the small cation is surrounded by three anions to form an equilateral triangle as shown below—triangle *ABC;* the centers of all four ions are coplanar.

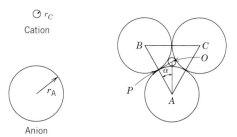

This boils down to a relatively simple plane trigonometry problem. Consideration of the right triangle *APO* makes it clear that the side lengths are related to the anion and cation radii r_A and r_C as

$$\overline{AP} = r_A$$

Table 3.3 Coordination Numbers and Geometries for Various Cation–Anion Radius Ratios (r_C/r_A)

Coordination Number	Cation–Anion Radius Ratio	Coordination Geometry
2	<0.155	
3	0.155–0.225	
4	0.225–0.414	
6	0.414–0.732	
8	0.732–1.0	

Source: W. D. Kingery, H. K. Bowen, and D. R. Uhlmann, *Introduction to Ceramics,* 2nd edition. Copyright © 1976 by John Wiley & Sons, New York. Reprinted by permission of John Wiley & Sons, Inc.

and

$$\overline{AO} = r_A + r_C$$

Furthermore, the side length ratio $\overline{AP}/\overline{AO}$ is a function of the angle α as

$$\frac{\overline{AP}}{\overline{AO}} = \cos \alpha$$

Table 3.4   Ionic Radii for Several Cations and Anions (for a Coordination Number of 6)

Cation	Ionic Radius (nm)	Anion	Ionic Radius (nm)
Al^{3+}	0.053	Br^-	0.196
Ba^{2+}	0.136	Cl^-	0.181
Ca^{2+}	0.100	F^-	0.133
Cs^+	0.170	I^-	0.220
Fe^{2+}	0.077	O^{2-}	0.140
Fe^{3+}	0.069	S^{2-}	0.184
K^+	0.138		
Mg^{2+}	0.072		
Mn^{2+}	0.067		
Na^+	0.102		
Ni^{2+}	0.069		
Si^{4+}	0.040		
Ti^{4+}	0.061		

The magnitude of α is 30°, since line $\overline{AO}$ bisects the 60° angle BAC. Thus,

$$\frac{\overline{AP}}{\overline{AO}} = \frac{r_A}{r_A + r_C} = \cos 30° = \frac{\sqrt{3}}{2}$$

Or, solving for the cation–anion radius ratio,

$$\frac{r_C}{r_A} = \frac{1 - \sqrt{3}/2}{\sqrt{3}/2} = 0.155$$

AX-TYPE CRYSTAL STRUCTURES

Some of the common ceramic materials are those in which there are equal numbers of cations and anions. These are often referred to as AX compounds, where A denotes the cation and X the anion. There are several different crystal structures for AX compounds; each is normally named after a common material that assumes the particular structure.

Rock Salt Structure

Perhaps the most common AX crystal structure is the *sodium chloride* (NaCl), or *rock salt,* type. The coordination number for both cations and anions is 6, and therefore the cation–anion radius ratio is between approximately 0.414 and 0.732. A unit cell for this crystal structure (Figure 3.5) is generated from an FCC arrangement of anions with one cation situated at the cube center and one at the center of each of the 12 cube edges. An equivalent crystal structure results from a face-centered arrangement of cations. Thus, the rock salt crystal structure may be thought of as two interpenetrating FCC lattices, one composed of the cations, the other of anions. Some of the common ceramic materials that form with this crystal structure are NaCl, MgO, MnS, LiF, and FeO.

Cesium Chloride Structure

Figure 3.6 shows a unit cell for the *cesium chloride* (CsCl) crystal structure; the coordination number is 8 for both ion types. The anions are located at each of the corners of a cube, whereas the cube center is a single cation. Interchange of anions

FIGURE 3.5 A unit cell for the rock salt, or sodium chloride (NaCl), crystal structure.

● Na⁺   ◯ Cl⁻

with cations, and vice versa, produces the same crystal structure. This is *not* a BCC crystal structure because ions of two different kinds are involved.

Zinc Blende Structure

A third AX structure is one in which the coordination number is 4; that is, all ions are tetrahedrally coordinated. This is called the *zinc blende*, or *sphalerite*, structure, after the mineralogical term for zinc sulfide (ZnS). A unit cell is presented in Figure 3.7; all corner and face positions of the cubic cell are occupied by S atoms, while the Zn atoms fill interior tetrahedral positions. An equivalent structure results if Zn and S atom positions are reversed. Thus, each Zn atom is bonded to four S atoms, and vice versa. Most often the atomic bonding is highly covalent in compounds exhibiting this crystal structure (Table 3.2), which include ZnS, ZnTe, and SiC.

A_mX_p-TYPE CRYSTAL STRUCTURES

If the charges on the cations and anions are not the same, a compound can exist with the chemical formula A_mX_p, where m and/or $p \neq 1$. An example would be AX_2, for which a common crystal structure is found in *fluorite* (CaF_2). The ionic radii ratio r_C/r_A for CaF_2 is about 0.8 which, according to Table 3.3, gives a coordination number of 8. Calcium ions are positioned at the centers of cubes, with fluorine

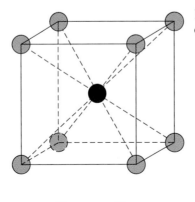

FIGURE 3.6 A unit cell for the cesium chloride (CsCl) crystal structure.

● Cs⁺   ◯ Cl⁻

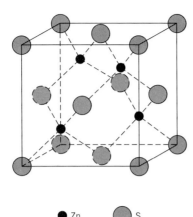

FIGURE 3.7   A unit cell for the zinc blende (ZnS) crystal structure.

● Zn   ◉ S

ions at the corners. The chemical formula shows that there are only half as many Ca^{2+} ions as F^- ions, and therefore the crystal structure would be similar to CsCl (Figure 3.6), except that only half the center cube positions are occupied by Ca^{2+} ions. One unit cell consists of eight cubes, as indicated in Figure 3.8. Other compounds that have this crystal structure include UO_2, PuO_2, and ThO_2.

$A_mB_nX_p$-TYPE CRYSTAL STRUCTURES

It is also possible for ceramic compounds to have more than one type of cation; for two types of cations (represented by A and B), their chemical formula may be designated as $A_mB_nX_p$. Barium titanate ($BaTiO_3$), having both Ba^{2+} and Ti^{4+} cations, falls into this classification. This material has a *perovskite crystal structure* and rather interesting electromechanical properties to be discussed later. At temperatures above 120°C (248°F), the crystal structure is cubic. A unit cell of this structure is shown in Figure 3.9; Ba^{2+} ions are situated at all eight corners of the cube and a single Ti^{4+} is at the cube center, with O^{2-} ions located at the center of each of the six faces.

Table 3.5 summarizes the rock salt, cesium chloride, zinc blende, fluorite, and perovskite crystal structures in terms of cation–anion ratios and coordination numbers, and gives examples for each. Of course, many other ceramic crystal structures are possible.

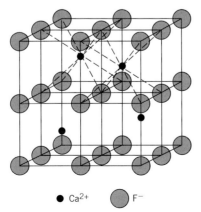

FIGURE 3.8   A unit cell for the fluorite (CaF_2) crystal structure.

● Ca^{2+} ◉ F^-

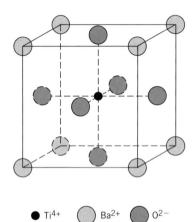

FIGURE 3.9 A unit cell for the perovskite crystal structure.

● Ti^{4+} ◯ Ba^{2+} ◗ O^{2-}

EXAMPLE PROBLEM 3.5

On the basis of ionic radii, what crystal structure would you predict for FeO?

SOLUTION

First, note that FeO is an AX-type compound. Next, determine the cation–anion radius ratio, which from Table 3.4 is

$$\frac{r_{Fe^{2+}}}{r_{O^{2-}}} = \frac{0.077 \text{ nm}}{0.140 \text{ nm}} = 0.550$$

This value lies between 0.414 and 0.732, and, therefore, from Table 3.3 the coordination number for the Fe^{2+} ion is 6; this is also the coordination number of O^{2-}, since there are equal numbers of cations and anions. The predicted crystal structure will be rock salt, which is the AX crystal structure having a coordination number of 6, as given in Table 3.5.

Table 3.5   Summary of Some Common Ceramic Crystal Structures

Structure Name	Structure Type	Anion Packing	Coordination Numbers		Examples
			Cation	*Anion*	
Rock salt (sodium chloride)	AX	FCC	6	6	NaCl, MgO, FeO
Cesium chloride	AX	Simple cubic	8	8	CsCl
Zinc blende (sphalerite)	AX	FCC	4	4	ZnS, SiC
Fluorite	AX_2	Simple cubic	8	4	CaF_2, UO_2, ThO_2
Perovskite	ABX_3	FCC	12(A) 6(B)	6	$BaTiO_3$, $SrZrO_3$, $SrSnO_3$
Spinel	AB_2X_4	FCC	4(A) 6(B)	4	$MgAl_2O_4$, $FeAl_2O_4$

Source: W. D. Kingery, H. K. Bowen, and D. R. Uhlmann, *Introduction to Ceramics,* 2nd edition. Copyright © 1976 by John Wiley & Sons, New York. Reprinted by permission of John Wiley & Sons, Inc.

3.7 DENSITY COMPUTATIONS—CERAMICS

It is possible to compute the theoretical density of a crystalline ceramic material from unit cell data in a manner similar to that described in Section 3.5 for metals. In this case the density ρ may be determined using a modified form of Equation 3.5, as follows:

$$\rho = \frac{n'(\Sigma A_C + \Sigma A_A)}{V_C N_A} \tag{3.6}$$

where

n' = the number of formula units[1] within the unit cell

ΣA_C = the sum of the atomic weights of all cations in the formula unit

ΣA_A = the sum of the atomic weights of all anions in the formula unit

V_C = the unit cell volume

N_A = Avogadro's number, 6.023×10^{23} formula units/mol

EXAMPLE PROBLEM 3.6

On the basis of crystal structure, compute the theoretical density for sodium chloride. How does this compare with its measured density?

SOLUTION

The density may be determined using Equation 3.6, where n', the number of NaCl units per unit cell, is 4 because both sodium and chloride ions form FCC lattices. Furthermore,

$$\Sigma A_C = A_{Na} = 22.99 \text{ g/mol}$$

$$\Sigma A_A = A_{Cl} = 35.45 \text{ g/mol}$$

Since the unit cell is cubic, $V_C = a^3$, a being the unit cell edge length. For the face of the cubic unit cell shown below,

$$a = 2r_{Na^+} + 2r_{Cl^-}$$

r_{Na^+} and r_{Cl^-} being the sodium and chlorine ionic radii, given in Table 3.4 as 0.102 and 0.181 nm, respectively.
Thus,

$$V_C = a^3 = (2r_{Na^+} + 2r_{Cl^-})^3$$

---

[1] By "formula unit" we mean all the ions that are included in the chemical formula unit. For example, for $BaTiO_3$, a formula unit consists of one barium ion, a titanium ion, and three oxygen ions.

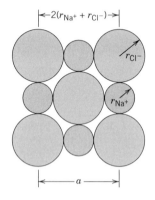

And finally,

$$\rho = \frac{n'(A_{Na} + A_{Cl})}{(2r_{Na^+} + 2r_{Cl^-})^3 N_A}$$

$$= \frac{4(22.99 + 35.45)}{[2(0.102 \times 10^{-7}) + 2(0.181 \times 10^{-7})]^3(6.023 \times 10^{23})}$$

$$= 2.14 \text{ g/cm}^3$$

This compares very favorably with the experimental value of 2.16 g/cm^3.

3.8 SILICATE CERAMICS

Silicates are materials composed primarily of silicon and oxygen, the two most abundant elements in the earth's crust; consequently, the bulk of soils, rocks, clays, and sand come under the silicate classification. Rather than characterizing the crystal structures of these materials in terms of unit cells, it is more convenient to use various arrangements of an SiO_4^{4-} tetrahedron (Figure 3.10). Each atom of silicon is bonded to four oxygen atoms, which are situated at the corners of the tetrahedron; the silicon atom is positioned at the center. Since this is the basic unit of the silicates, it is often treated as a negatively charged entity.

Often the silicates are not considered to be ionic because there is a significant covalent character to the interatomic Si–O bonds (Table 3.2), which bonds are directional and relatively strong. Regardless of the character of the Si–O bond, there is a -4 charge associated with every SiO_4^{4-} tetrahedron, since each of the four oxygen atoms requires an extra electron to achieve a stable electronic structure. Various silicate structures arise from the different ways in which the SiO_4^{4-} units can be combined into one-, two-, and three-dimensional arrangements.

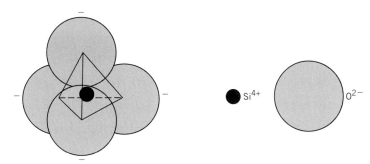

FIGURE 3.10 A silicon–oxygen (SiO_4^{4-}) tetrahedron.

SILICA

Chemically, the most simple silicate material is silicon dioxide, or silica (SiO_2). Structurally, it is a three-dimensional network that is generated when every corner oxygen atom in each tetrahedron is shared by adjacent tetrahedra. Thus, the material is electrically neutral and all atoms have stable electronic structures. Under these circumstances the ratio of Si to O atoms is 1 : 2, as indicated by the chemical formula.

If these tetrahedra are arrayed in a regular and ordered manner, a crystalline structure is formed. There are three primary polymorphic crystalline forms of silica: quartz, cristobalite (Figure 3.11), and tridymite. Their structures are relatively complicated, and comparatively open; that is, the atoms are not closely packed together. As a consequence, these crystalline silicas have relatively low densities; for example, at room temperature quartz has a density of only 2.65 g/cm³. The strength of the Si–O interatomic bonds is reflected in a relatively high melting temperature, 1710°C (3110°F).

Silica can also be made to exist as a noncrystalline solid or glass; its structure is discussed in Section 3.20.

THE SILICATES (CD-ROM)

3.9 CARBON

Carbon is an element that exists in various polymorphic forms, as well as in the amorphous state. This group of materials does not really fall within any one of the traditional metal, ceramic, polymer classification schemes. However, it has been decided to discuss these materials in this chapter since graphite, one of the polymorphic forms, is sometimes classified as a ceramic. This treatment focuses on the

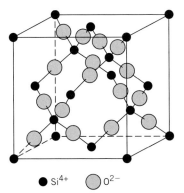

FIGURE 3.11 The arrangement of silicon and oxygen atoms in a unit cell of cristobalite, a polymorph of SiO_2.

FIGURE 3.16 A unit cell for the diamond cubic crystal structure.

⬡ C

structures of graphite and diamond {and the new fullerenes.} The characteristics and current and potential uses of these materials are discussed in Section 13.11.

DIAMOND

Diamond is a metastable carbon polymorph at room temperature and atmospheric pressure. Its crystal structure is a variant of the zinc blende, in which carbon atoms occupy all positions (both Zn and S), as indicated in the unit cell shown in Figure 3.16. Thus, each carbon bonds to four other carbons, and these bonds are totally covalent. This is appropriately called the *diamond cubic* crystal structure, which is also found for other Group IVA elements in the periodic table [e.g., germanium, silicon, and gray tin, below 13°C (55°F)].

GRAPHITE

Graphite has a crystal structure (Figure 3.17) distinctly different from that of diamond and is also more stable than diamond at ambient temperature and pressure. The graphite structure is composed of layers of hexagonally arranged carbon atoms; within the layers, each carbon atom is bonded to three coplanar neighbor atoms by strong covalent bonds. The fourth bonding electron participates in a weak van der Waals type of bond between the layers.

FULLERENES (CD-ROM)

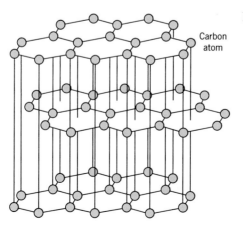

Carbon atom

FIGURE 3.17 The structure of graphite.

3.10 POLYMORPHISM AND ALLOTROPY

Some metals, as well as nonmetals, may have more than one crystal structure, a phenomenon known as **polymorphism.** When found in elemental solids, the condition is often termed **allotropy.** The prevailing crystal structure depends on both the temperature and the external pressure. One familiar example is found in carbon as discussed in the previous section: graphite is the stable polymorph at ambient conditions, whereas diamond is formed at extremely high pressures. Also, pure iron has a BCC crystal structure at room temperature, which changes to FCC iron at 912°C (1674°F). Most often a modification of the density and other physical properties accompanies a polymorphic transformation.

3.11 CRYSTAL SYSTEMS

Since there are many different possible crystal structures, it is sometimes convenient to divide them into groups according to unit cell configurations and/or atomic arrangements. One such scheme is based on the unit cell geometry, that is, the shape of the appropriate unit cell parallelepiped without regard to the atomic positions in the cell. Within this framework, an x, y, z coordinate system is established with its origin at one of the unit cell corners; each of the x, y, and z axes coincides with one of the three parallelepiped edges that extend from this corner, as illustrated in Figure 3.19. The unit cell geometry is completely defined in terms of six parameters: the three edge lengths a, b, and c, and the three interaxial angles α, β, and γ. These are indicated in Figure 3.19, and are sometimes termed the **lattice parameters** of a crystal structure.

On this basis there are found crystals having seven different possible combinations of a, b, and c, and α, β, and γ, each of which represents a distinct **crystal system.** These seven crystal systems are cubic, tetragonal, hexagonal, orthorhombic, rhombohedral, monoclinic, and triclinic. The lattice parameter relationships and unit cell sketches for each are represented in Table 3.6. The cubic system, for which $a = b = c$ and $\alpha = \beta = \gamma = 90°$, has the greatest degree of symmetry. Least symmetry is displayed by the triclinic system, since $a \neq b \neq c$ and $\alpha \neq \beta \neq \gamma$.

From the discussion of metallic crystal structures, it should be apparent that both FCC and BCC structures belong to the cubic crystal system, whereas HCP falls within hexagonal. The conventional hexagonal unit cell really consists of three parallelepipeds situated as shown in Table 3.6.

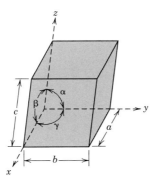

FIGURE 3.19 A unit cell with x, y, and z coordinate axes, showing axial lengths (a, b, and c) and interaxial angles (α, β, and γ).

Table 3.6 Lattice Parameter Relationships and Figures Showing Unit Cell Geometries for the Seven Crystal Systems

Crystal System	Axial Relationships	Interaxial Angles	Unit Cell Geometry
Cubic	$a = b = c$	$\alpha = \beta = \gamma = 90°$	
Hexagonal	$a = b \neq c$	$\alpha = \beta = 90°, \gamma = 120°$	
Tetragonal	$a = b \neq c$	$\alpha = \beta = \gamma = 90°$	
Rhombohedral	$a = b = c$	$\alpha = \beta = \gamma \neq 90°$	
Orthorhombic	$a \neq b \neq c$	$\alpha = \beta = \gamma = 90°$	
Monoclinic	$a \neq b \neq c$	$\alpha = \gamma = 90° \neq \beta$	
Triclinic	$a \neq b \neq c$	$\alpha \neq \beta \neq \gamma \neq 90°$	

CRYSTALLOGRAPHIC DIRECTIONS AND PLANES

When dealing with crystalline materials, it often becomes necessary to specify some particular crystallographic plane of atoms or a crystallographic direction. Labeling conventions have been established in which three integers or indices are used to designate directions and planes. The basis for determining index values is the unit cell, with a coordinate system consisting of three (x, y, and z) axes situated at one of the corners and coinciding with the unit cell edges, as shown in Figure 3.19. For some crystal systems—namely, hexagonal, rhombohedral, monoclinic, and triclinic—the three axes are *not* mutually perpendicular, as in the familiar Cartesian coordinate scheme.

3.12 CRYSTALLOGRAPHIC DIRECTIONS

A crystallographic direction is defined as a line between two points, or a vector. The following steps are utilized in the determination of the three directional indices:

1. A vector of convenient length is positioned such that it passes through the origin of the coordinate system. Any vector may be translated throughout the crystal lattice without alteration, if parallelism is maintained.

2. The length of the vector projection on each of the three axes is determined; *these are measured in terms of the unit cell dimensions a, b, and c.*

3. These three numbers are multiplied or divided by a common factor to reduce them to the smallest integer values.

4. The three indices, not separated by commas, are enclosed in square brackets, thus: [uvw]. The u, v, and w integers correspond to the reduced projections along the x, y, and z axes, respectively.

For each of the three axes, there will exist both positive and negative coordinates. Thus negative indices are also possible, which are represented by a bar over the appropriate index. For example, the [$1\bar{1}1$] direction would have a component in the $-y$ direction. Also, changing the signs of all indices produces an antiparallel

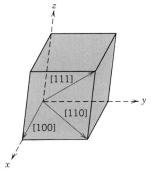

FIGURE 3.20 The [100], [110], and [111] directions within a unit cell.

direction; that is, $[\bar{1}1\bar{1}]$ is directly opposite to $[1\bar{1}1]$. If more than one direction or plane is to be specified for a particular crystal structure, it is imperative for the maintaining of consistency that a positive–negative convention, once established, not be changed.

The [100], [110], and [111] directions are common ones; they are drawn in the unit cell shown in Figure 3.20.

EXAMPLE PROBLEM 3.7

Determine the indices for the direction shown in the accompanying figure.

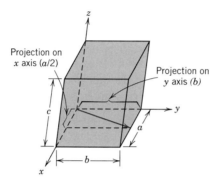

SOLUTION

The vector, as drawn, passes through the origin of the coordinate system, and therefore no translation is necessary. Projections of this vector onto the x, y, and z axes are, respectively, $a/2$, b, and $0c$, which become $\frac{1}{2}$, 1, and 0 in terms of the unit cell parameters (i.e., when the a, b, and c are dropped). Reduction of these numbers to the lowest set of integers is accompanied by multiplication of each by the factor 2. This yields the integers 1, 2, and 0, which are then enclosed in brackets as [120].

This procedure may be summarized as follows:

	x	y	z
Projections	$a/2$	b	$0c$
Projections (in terms of a, b, and c)	$\frac{1}{2}$	1	0
Reduction	1	2	0
Enclosure		[120]	

EXAMPLE PROBLEM 3.8

Draw a $[1\bar{1}0]$ direction within a cubic unit cell.

SOLUTION

First construct an appropriate unit cell and coordinate axes system. In the accompanying figure the unit cell is cubic, and the origin of the coordinate system, point O, is located at one of the cube corners.

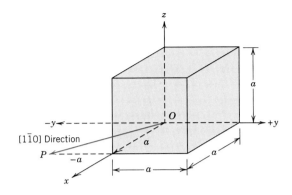

This problem is solved by reversing the procedure of the preceding example. For this [1$\bar{1}$0] direction, the projections along the x, y, z axes are a, $-a$, and $0a$, respectively. This direction is defined by a vector passing from the origin to point P, which is located by first moving along the x axis a units, and from this position, parallel to the y axis $-a$ units, as indicated in the figure. There is no z component to the vector, since the z projection is zero.

For some crystal structures, several nonparallel directions with different indices are actually equivalent; this means that the spacing of atoms along each direction is the same. For example, in cubic crystals, all the directions represented by the following indices are equivalent: [100], [$\bar{1}$00], [010], [0$\bar{1}$0], [001], and [00$\bar{1}$]. As a convenience, equivalent directions are grouped together into a *family*, which are enclosed in angle brackets, thus: ⟨100⟩. Furthermore, directions in cubic crystals having the same indices without regard to order or sign, for example, [123] and [$\bar{2}$1$\bar{3}$], are equivalent. This is, in general, not true for other crystal systems. For example, for crystals of tetragonal symmetry, [100] and [010] directions are equivalent, whereas [100] and [001] are not.

HEXAGONAL CRYSTALS

A problem arises for crystals having hexagonal symmetry in that some crystallographic equivalent directions will not have the same set of indices. This is circumvented by utilizing a four-axis, or *Miller–Bravais*, coordinate system as shown in Figure 3.21. The three a_1, a_2, and a_3 axes are all contained within a single plane

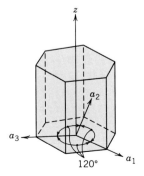

FIGURE 3.21 Coordinate axis system for a hexagonal unit cell (Miller–Bravais scheme).

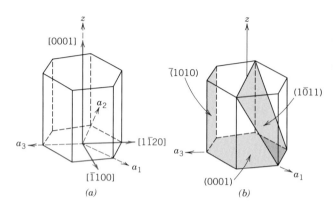

FIGURE 3.22 For the hexagonal crystal system, (a) [0001], [$1\bar{1}00$], and [$11\bar{2}0$] directions, and (b) the (0001), ($10\bar{1}1$), and ($\bar{1}010$) planes.

(called the basal plane), and at 120° angles to one another. The z axis is perpendicular to this basal plane. Directional indices, which are obtained as described above, will be denoted by four indices, as [$uvtw$]; by convention, the first three indices pertain to projections along the respective a_1, a_2, and a_3 axes in the basal plane.

Conversion from the three-index system to the four-index system,

$$[u'v'w'] \longrightarrow [uvtw]$$

is accomplished by the following formulas:

$$u = \frac{n}{3}(2u' - v')$$ (3.7a)

$$v = \frac{n}{3}(2v' - u')$$ (3.7b)

$$t = -(u + v)$$ (3.7c)

$$w = nw'$$ (3.7d)

where primed indices are associated with the three-index scheme and unprimed, with the new Miller–Bravais four-index system; n is a factor that may be required to reduce u, v, t, and w to the smallest integers. For example, using this conversion, the [010] direction becomes [$\bar{1}2\bar{1}0$]. Several different directions are indicated in the hexagonal unit cell (Figure 3.22a).

3.13 CRYSTALLOGRAPHIC PLANES

The orientations of planes for a crystal structure are represented in a similar manner. Again, the unit cell is the basis, with the three-axis coordinate system as represented in Figure 3.19. In all but the hexagonal crystal system, crystallographic planes are specified by three **Miller indices** as (hkl). Any two planes parallel to each other are equivalent and have identical indices. The procedure employed in determination of the h, k, and l index numbers is as follows:

1. If the plane passes through the selected origin, either another parallel plane must be constructed within the unit cell by an appropriate translation, or a new origin must be established at the corner of another unit cell.

2. At this point the crystallographic plane either intersects or parallels each of the three axes; the length of the planar intercept for each axis is determined in terms of the lattice parameters a, b, and c.

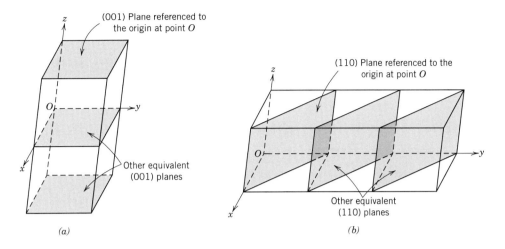

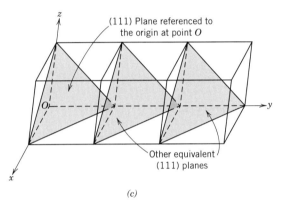

FIGURE 3.23 Representations of a series each of (a) (001), (b) (110), and (c) (111) crystallographic planes.

3. The reciprocals of these numbers are taken. A plane that parallels an axis may be considered to have an infinite intercept, and, therefore, a zero index.

4. If necessary, these three numbers are changed to the set of smallest integers by multiplication or division by a common factor.[2]

5. Finally, the integer indices, not separated by commas, are enclosed within parentheses, thus: (hkl).

An intercept on the negative side of the origin is indicated by a bar or minus sign positioned over the appropriate index. Furthermore, reversing the directions of all indices specifies another plane parallel to, on the opposite side of and equidistant from, the origin. Several low-index planes are represented in Figure 3.23.

One interesting and unique characteristic of cubic crystals is that planes and directions having the same indices are perpendicular to one another; however, for other crystal systems there are no simple geometrical relationships between planes and directions having the same indices.

---

[2] On occasion, index reduction is not carried out {(e.g., for x-ray diffraction studies that are described in Section 3.19);} for example, (002) is not reduced to (001). In addition, for ceramic materials, the ionic arrangement for a reduced-index plane may be different from that for a nonreduced one.

EXAMPLE PROBLEM 3.9

Determine the Miller indices for the plane shown in the accompanying sketch (a).

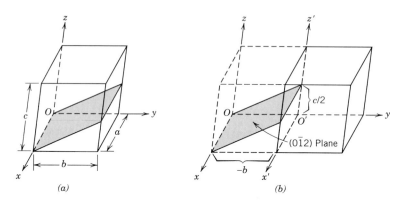

(a)   (b)

SOLUTION

Since the plane passes through the selected origin O, a new origin must be chosen at the corner of an adjacent unit cell, taken as O' and shown in sketch (b). This plane is parallel to the x axis, and the intercept may be taken as ∞a. The y and z axes intersections, referenced to the new origin O', are $-b$ and $c/2$, respectively. Thus, in terms of the lattice parameters a, b, and c, these intersections are ∞, -1, and $\frac{1}{2}$. The reciprocals of these numbers are 0, -1, and 2; and since all are integers, no further reduction is necessary. Finally, enclosure in parentheses yields $(0\bar{1}2)$.

These steps are briefly summarized below:

	x	y	z
Intercepts	∞a	$-b$	$c/2$
Intercepts (in terms of lattice parameters)	∞	-1	$\frac{1}{2}$
Reciprocals	0	-1	2
Reductions (unnecessary)			
Enclosure		$(0\bar{1}2)$	

EXAMPLE PROBLEM 3.10

Construct a $(0\bar{1}1)$ plane within a cubic unit cell.

SOLUTION

To solve this problem, carry out the procedure used in the preceding example in reverse order. To begin, the indices are removed from the parentheses, and reciprocals are taken, which yields ∞, -1, and 1. This means that the particular plane parallels the x axis while intersecting the y and z axes at $-b$ and c, respectively, as indicated in the accompanying sketch (a). This plane has been drawn in sketch (b). A plane is indicated by lines representing its intersections with the planes that constitute the faces of the unit cell or their extensions. For example, in this figure, line ef is the intersection between the $(0\bar{1}1)$ plane and

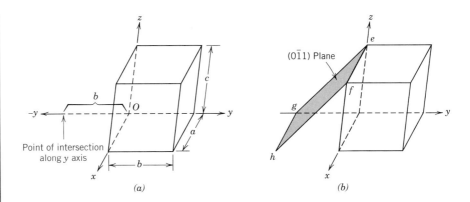

(a)   (b)

the top face of the unit cell; also, line *gh* represents the intersection between this same (0$\bar{1}$1) plane and the plane of the bottom unit cell face extended. Similarly, lines *eg* and *fh* are the intersections between (0$\bar{1}$1) and back and front cell faces, respectively.

ATOMIC ARRANGEMENTS

The atomic arrangement for a crystallographic plane, which is often of interest, depends on the crystal structure. The (110) atomic planes for FCC and BCC crystal structures are represented in Figures 3.24 and 3.25; reduced-sphere unit cells are also included. Note that the atomic packing is different for each case. The circles represent atoms lying in the crystallographic planes as would be obtained from a slice taken through the centers of the full-sized hard spheres.

A "family" of planes contains all those planes that are crystallographically equivalent—that is, having the same atomic packing; and a family is designated by indices that are enclosed in braces—e.g., {100}. For example, in cubic crystals the (111), ($\bar{1}\bar{1}\bar{1}$), ($\bar{1}$11), (1$\bar{1}$1), (11$\bar{1}$), ($\bar{1}\bar{1}$1), ($\bar{1}$1$\bar{1}$), and (1$\bar{1}\bar{1}$) planes all belong to the {111} family. On the other hand, for tetragonal crystal structures, the {100} family would contain only the (100), ($\bar{1}$00), (010), and (0$\bar{1}$0) since the (001) and (00$\bar{1}$) planes are not crystallographically equivalent. Also, in the cubic system only, planes having the same indices, irrespective of order and sign, are equivalent. For example, both (1$\bar{2}$3) and (3$\bar{1}$2) belong to the {123} family.

HEXAGONAL CRYSTALS

For crystals having hexagonal symmetry, it is desirable that equivalent planes have the same indices; as with directions, this is accomplished by the Miller–Bravais

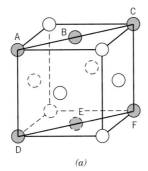

(a)

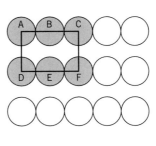

(b)

FIGURE 3.24 (a) Reduced-sphere FCC unit cell with (110) plane. (b) Atomic packing of an FCC (110) plane. Corresponding atom positions from (a) are indicated.

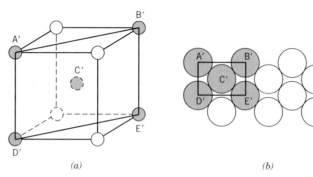

FIGURE 3.25 (a) Reduced-sphere BCC unit cell with (110) plane. (b) Atomic packing of a BCC (110) plane. Corresponding atom positions from (a) are indicated.

system shown in Figure 3.21. This convention leads to the four-index (*hkil*) scheme, which is favored in most instances, since it more clearly identifies the orientation of a plane in a hexagonal crystal. There is some redundancy in that *i* is determined by the sum of *h* and *k* through

$$i = -(h + k) \tag{3.8}$$

Otherwise the three *h*, *k*, and *l* indices are identical for both indexing systems. Figure 3.22*b* presents several of the common planes that are found for crystals having hexagonal symmetry.

3.14 LINEAR AND PLANAR ATOMIC DENSITIES (CD-ROM)

3.15 CLOSE-PACKED CRYSTAL STRUCTURES

METALS

It may be remembered from the discussion on metallic crystal structures (Section 3.4) that both face-centered cubic and hexagonal close-packed crystal structures have atomic packing factors of 0.74, which is the most efficient packing of equal-sized spheres or atoms. In addition to unit cell representations, these two crystal structures may be described in terms of close-packed planes of atoms (i.e., planes having a maximum atom or sphere-packing density); a portion of one such plane is illustrated in Figure 3.27*a*. Both crystal structures may be generated by the stacking of these close-packed planes on top of one another; the difference between the two structures lies in the stacking sequence.

Let the centers of all the atoms in one close-packed plane be labeled *A*. Associated with this plane are two sets of equivalent triangular depressions formed by three adjacent atoms, into which the next close-packed plane of atoms may rest. Those having the triangle vertex pointing up are arbitrarily designated as *B* positions, while the remaining depressions are those with the down vertices, which are marked *C* in Figure 3.27*a*.

A second close-packed plane may be positioned with the centers of its atoms over either *B* or *C* sites; at this point both are equivalent. Suppose that the *B* positions are arbitrarily chosen; the stacking sequence is termed *AB*, which is illustrated in Figure 3.27*b*. The real distinction between FCC and HCP lies in where

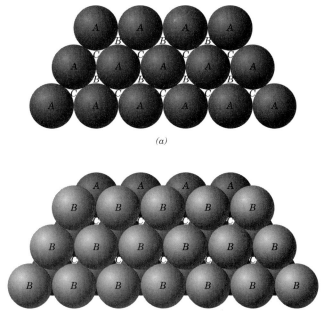

(a)

(b)

FIGURE 3.27 (a) A portion of a close-packed plane of atoms; *A*, *B*, and *C* positions are indicated. (b) The *AB* stacking sequence for close-packed atomic planes. (Adapted from W. G. Moffatt, G. W. Pearsall, and J. Wulff, *The Structure and Properties of Materials,* Vol. I, *Structure,* p. 50. Copyright © 1964 by John Wiley & Sons, New York. Reprinted by permission of John Wiley & Sons, Inc.)

the third close-packed layer is positioned. For HCP, the centers of this layer are aligned directly above the original *A* positions. This stacking sequence, *ABABAB* . . . , is repeated over and over. Of course, the *ACACAC* . . . arrangement would be equivalent. These close-packed planes for HCP are (0001)-type planes, and the correspondence between this and the unit cell representation is shown in Figure 3.28.

For the face-centered crystal structure, the centers of the third plane are situated over the *C* sites of the first plane (Figure 3.29*a*). This yields an *ABCABCABC* . . . stacking sequence; that is, the atomic alignment repeats every third plane. It is more difficult to correlate the stacking of close-packed planes to the FCC unit cell. However, this relationship is demonstrated in Figure 3.29*b*; these planes are of the (111) type. The significance of these FCC and HCP close-packed planes will become apparent in Chapter 8.

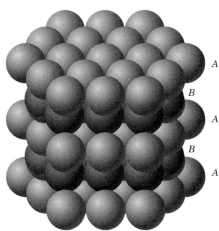

A
B
A
B
A

FIGURE 3.28 Close-packed plane stacking sequence for hexagonal close-packed. (Adapted from W. G. Moffatt, G. W. Pearsall, and J. Wulff, *The Structure and Properties of Materials,* Vol. I, *Structure,* p. 51. Copyright © 1964 by John Wiley & Sons, New York. Reprinted by permission of John Wiley & Sons, Inc.)

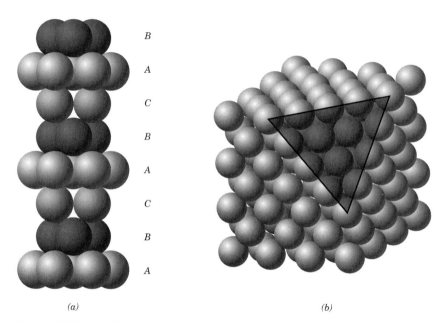

B
A
C
B
A
C
B
A

(a)

(b)

FIGURE 3.29   (*a*) Close-packed stacking sequence for face-centered cubic. (*b*) A corner has been removed to show the relation between the stacking of close-packed planes of atoms and the FCC crystal structure; the heavy triangle outlines a (111) plane. (Figure (*b*) from W. G. Moffatt, G. W. Pearsall, and J. Wulff, *The Structure and Properties of Materials,* Vol. I, *Structure,* p. 51. Copyright © 1964 by John Wiley & Sons, New York. Reprinted by permission of John Wiley & Sons, Inc.)

CERAMICS

A number of ceramic crystal structures may also be considered in terms of close-packed planes of ions (as opposed to *atoms* for metals), as well as unit cells. Ordinarily, the close-packed planes are composed of the large anions. As these planes are stacked atop each other, small interstitial sites are created between them in which the cations may reside.

These interstitial positions exist in two different types, as illustrated in Figure 3.30. Four atoms (three in one plane, and a single one in the adjacent plane) surround one type, labeled T in the figure; this is termed a **tetrahedral position,** since straight lines drawn from the centers of the surrounding spheres form a four-sided tetrahedron. The other site type, denoted as O in Figure 3.30, involves six ion spheres, three in each of the two planes. Because an octahedron is produced by joining these six sphere centers, this site is called an **octahedral position.** Thus, the coordination numbers for cations filling tetrahedral and octahedral positions are 4 and 6, respectively. Furthermore, for each of these anion spheres, one octahedral and two tetrahedral positions will exist.

Ceramic crystal structures of this type depend on two factors: (1) the stacking of the close-packed anion layers (both FCC and HCP arrangements are possible, which correspond to $ABCABC$ . . . and $ABABAB$ . . . sequences, respectively), and (2) the manner in which the interstitial sites are filled with cations. For example, consider the rock salt crystal structure discussed above. The unit cell has cubic symmetry, and each cation (Na^+ ion) has six Cl^- ion nearest neighbors, as may be verified from Figure 3.5. That is, the Na^+ ion at the center has as nearest neighbors

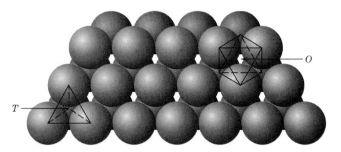

FIGURE 3.30 The stacking of one plane of close-packed spheres (anions) on top of another; tetrahedral and octahedral positions between the planes are designated by *T* and *O*, respectively. (From W. G. Moffatt, G. W. Pearsall, and J. Wulff, *The Structure and Properties of Materials,* Vol. 1, *Structure.* Copyright © 1964 by John Wiley & Sons, New York. Reprinted by permission of John Wiley & Sons, Inc.)

the six Cl^- ions that reside at the centers of each of the cube faces. The crystal structure, having cubic symmetry, may be considered in terms of an FCC array of close-packed planes of anions, and all planes are of the {111} type. The cations reside in octahedral positions because they have as nearest neighbors six anions. Furthermore, all octahedral positions are filled, since there is a single octahedral site per anion, and the ratio of anions to cations is 1:1. For this crystal structure, the relationship between the unit cell and close-packed anion plane stacking schemes is illustrated in Figure 3.31.

Other, but not all, ceramic crystal structures may be treated in a similar manner; included are the zinc blende and perovskite structures. The *spinel structure* is one of the $A_mB_nX_p$ types, which is found for magnesium aluminate or spinel ($MgAl_2O_4$). With this structure, the O^{2-} ions form an FCC lattice, whereas Mg^{2+} ions fill tetrahedral sites and Al^{3+} reside in octahedral positions. Magnetic ceramics, or ferrites, have a crystal structure that is a slight variant of this spinel structure; and the magnetic characteristics are affected by the occupancy of tetrahedral and octahedral positions {(see Section 18.5).}

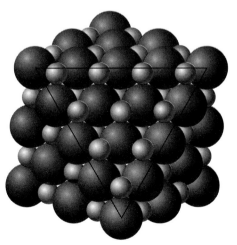

FIGURE 3.31 A section of the rock salt crystal structure from which a corner has been removed. The exposed plane of anions (light spheres inside the triangle) is a {111}-type plane; the cations (dark spheres) occupy the interstitial octahedral positions.

CRYSTALLINE AND NONCRYSTALLINE MATERIALS

3.16 SINGLE CRYSTALS

For a crystalline solid, when the periodic and repeated arrangement of atoms is perfect or extends throughout the entirety of the specimen without interruption, the result is a **single crystal.** All unit cells interlock in the same way and have the same orientation. Single crystals exist in nature, but they may also be produced artificially. They are ordinarily difficult to grow, because the environment must be carefully controlled.

If the extremities of a single crystal are permitted to grow without any external constraint, the crystal will assume a regular geometric shape having flat faces, as with some of the gem stones; the shape is indicative of the crystal structure. A photograph of several single crystals is shown in Figure 3.32. Within the past few years, single crystals have become extremely important in many of our modern technologies, in particular electronic microcircuits, which employ single crystals of silicon and other semiconductors.

3.17 POLYCRYSTALLINE MATERIALS

Most crystalline solids are composed of a collection of many small crystals or **grains;** such materials are termed **polycrystalline.** Various stages in the solidification of a polycrystalline specimen are represented schematically in Figure 3.33. Initially, small crystals or nuclei form at various positions. These have random crystallographic orientations, as indicated by the square grids. The small grains grow by the successive addition from the surrounding liquid of atoms to the structure of each. The extremities of adjacent grains impinge on one another as the solidification process approaches completion. As indicated in Figure 3.33, the crystallographic orientation varies from grain to grain. Also, there exists some atomic mismatch within the region where two grains meet; this area, called a **grain boundary,** is discussed in more detail in Section 5.8.

FIGURE 3.32 Photograph showing several single crystals of fluorite, CaF_2. (Smithsonian Institution photograph number 38181P.)

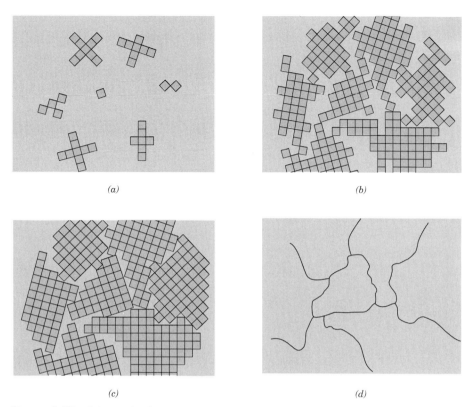

FIGURE 3.33 Schematic diagrams of the various stages in the solidification of a polycrystalline material; the square grids depict unit cells. (*a*) Small crystallite nuclei. (*b*) Growth of the crystallites; the obstruction of some grains that are adjacent to one another is also shown. (*c*) Upon completion of solidification, grains having irregular shapes have formed. (*d*) The grain structure as it would appear under the microscope; dark lines are the grain boundaries. (Adapted from W. Rosenhain, *An Introduction to the Study of Physical Metallurgy,* 2nd edition, Constable & Company Ltd., London, 1915.)

3.18 ANISOTROPY

The physical properties of single crystals of some substances depend on the crystallographic direction in which measurements are taken. For example, the elastic modulus, the electrical conductivity, and the index of refraction may have different values in the [100] and [111] directions. This directionality of properties is termed **anisotropy,** and it is associated with the variance of atomic or ionic spacing with crystallographic direction. Substances in which measured properties are independent of the direction of measurement are **isotropic.** The extent and magnitude of anisotropic effects in crystalline materials are functions of the symmetry of the crystal structure; the degree of anisotropy increases with decreasing structural symmetry—triclinic structures normally are highly anisotropic. The modulus of elasticity values at [100], [110], and [111] orientations for several materials are presented in Table 3.7.

For many polycrystalline materials, the crystallographic orientations of the individual grains are totally random. Under these circumstances, even though each

Table 3.7 **Modulus of Elasticity Values for Several Metals at Various Crystallographic Orientations**

Metal	Modulus of Elasticity (GPa)		
	[100]	[110]	[111]
Aluminum	63.7	72.6	76.1
Copper	66.7	130.3	191.1
Iron	125.0	210.5	272.7
Tungsten	384.6	384.6	384.6

Source: R. W. Hertzberg, *Deformation and Fracture Mechanics of Engineering Materials,* 3rd edition. Copyright © 1989 by John Wiley & Sons, New York. Reprinted by permission of John Wiley & Sons, Inc.

grain may be anisotropic, a specimen composed of the grain aggregate behaves isotropically. Also, the magnitude of a measured property represents some average of the directional values. Sometimes the grains in polycrystalline materials have a preferential crystallographic orientation, in which case the material is said to have a "texture."

3.19 X-RAY DIFFRACTION: DETERMINATION OF CRYSTAL STRUCTURES (CD-ROM)

3.20 NONCRYSTALLINE SOLIDS

It has been mentioned that **noncrystalline** solids lack a systematic and regular arrangement of atoms over relatively large atomic distances. Sometimes such materials are also called **amorphous** (meaning literally without form), or supercooled liquids, inasmuch as their atomic structure resembles that of a liquid.

An amorphous condition may be illustrated by comparison of the crystalline and noncrystalline structures of the ceramic compound silicon dioxide (SiO_2), which may exist in both states. Figures 3.38a and 3.38b present two-dimensional schematic diagrams for both structures of SiO_2, in which the SiO_4^{4-} tetrahedron is the basic unit (Figure 3.10). Even though each silicon ion bonds to four oxygen ions for both states, beyond this, the structure is much more disordered and irregular for the noncrystalline structure.

Whether a crystalline or amorphous solid forms depends on the ease with which a random atomic structure in the liquid can transform to an ordered state during solidification. Amorphous materials, therefore, are characterized by atomic or molecular structures that are relatively complex and become ordered only with some difficulty. Furthermore, rapidly cooling through the freezing temperature favors the formation of a noncrystalline solid, since little time is allowed for the ordering process.

Metals normally form crystalline solids; but some ceramic materials are crystalline, whereas others (i.e., the silica glasses) are amorphous. Polymers may be completely noncrystalline and semicrystalline consisting of varying degrees of crystallin-

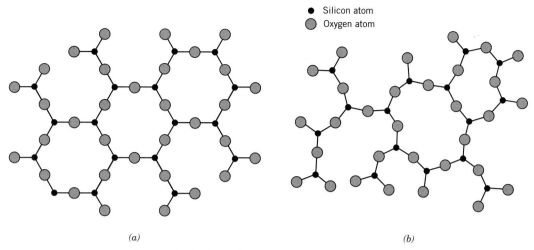

FIGURE 3.38 Two-dimensional schemes of the structure of (*a*) crystalline silicon dioxide and (*b*) noncrystalline silicon dioxide.

ity. More about the structure and properties of these amorphous materials is discussed below and in subsequent chapters.

SILICA GLASSES

Silicon dioxide (or silica, SiO_2) in the noncrystalline state is called *fused silica,* or *vitreous silica;* again, a schematic representation of its structure is shown in Figure 3.38*b*. Other oxides (e.g., B_2O_3 and GeO_2) may also form glassy structures (and polyhedral oxide structures {similar to those shown in Figure 3.12}); these materials, as well as SiO_2, are *network formers.*

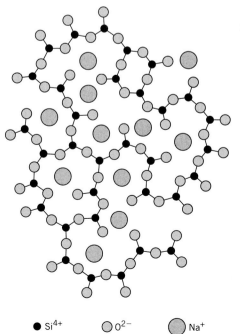

FIGURE 3.39 Schematic representation of ion positions in a sodium–silicate glass.

● Si^{4+} ○ O^{2-} ○ Na^+

The common inorganic glasses that are used for containers, windows, and so on are silica glasses to which have been added other oxides such as CaO and Na_2O. These oxides do not form polyhedral networks. Rather, their cations are incorporated within and modify the SiO_4^{4-} network; for this reason, these oxide additives are termed *network modifiers*. For example, Figure 3.39 is a schematic representation of the structure of a sodium–silicate glass. Still other oxides, such as TiO_2 and Al_2O_3, while not network formers, substitute for silicon and become part of and stabilize the network; these are called *intermediates*. From a practical perspective, the addition of these modifiers and intermediates lowers the melting point and viscosity of a glass, and makes it easier to form at lower temperatures {(Section 14.7).}

S U M M A R Y

Atoms in crystalline solids are positioned in an orderly and repeated pattern that is in contrast to the random and disordered atomic distribution found in noncrystalline or amorphous materials. Atoms may be represented as solid spheres, and, for crystalline solids, crystal structure is just the spatial arrangement of these spheres. The various crystal structures are specified in terms of parallelepiped unit cells, which are characterized by geometry and atom positions within.

Most common metals exist in at least one of three relatively simple crystal structures: face-centered cubic (FCC), body-centered cubic (BCC), and hexagonal close-packed (HCP). Two features of a crystal structure are coordination number (or number of nearest-neighbor atoms) and atomic packing factor (the fraction of solid sphere volume in the unit cell). Coordination number and atomic packing factor are the same for both FCC and HCP crystal structures.

For ceramics both crystalline and noncrystalline states are possible. The crystal structures of those materials for which the atomic bonding is predominantly ionic are determined by the charge magnitude and the radius of each kind of ion. Some of the simpler crystal structures are described in terms of unit cells; several of these were discussed (rock salt, cesium chloride, zinc blende, diamond cubic, graphite, fluorite, perovskite, and spinel structures).

Theoretical densities of metallic and crystalline ceramic materials may be computed from unit cell and atomic weight data.

Generation of face-centered cubic and hexagonal close-packed crystal structures is possible by the stacking of close-packed planes of atoms. For some ceramic crystal structures, cations fit into interstitial positions that exist between two adjacent close-packed planes of anions.

For the silicates, structure is more conveniently represented by means of interconnecting SiO_4^{4-} tetrahedra. Relatively complex structures may result when other cations (e.g., Ca^{2+}, Mg^{2+}, Al^{3+}) and anions (e.g., OH^-) are added. The structures of silica (SiO_2), silica glass, {and several of the simple and layered silicates} were presented.

Structures for the various forms of carbon—diamond, graphite, {and the fullerenes}—were also discussed.

Crystallographic planes and directions are specified in terms of an indexing scheme. The basis for the determination of each index is a coordinate axis system defined by the unit cell for the particular crystal structure. Directional indices are computed in terms of vector projections on each of the coordinate axes, whereas planar indices are determined from the reciprocals of axial intercepts. For hexagonal unit cells, a four-index scheme for both directions and planes is found to be more convenient.

{Crystallographic directional and planar equivalencies are related to atomic linear and planar densities, respectively.} The atomic packing (i.e., planar density) of spheres in a crystallographic plane depends on the indices of the plane as well as the crystal structure. For a given crystal structure, planes having identical atomic packing yet different Miller indices belong to the same family.

Single crystals are materials in which the atomic order extends uninterrupted over the entirety of the specimen; under some circumstances, they may have flat faces and regular geometric shapes. The vast majority of crystalline solids, however, are polycrystalline, being composed of many small crystals or grains having different crystallographic orientations.

Other concepts introduced in this chapter were: crystal system (a classification scheme for crystal structures on the basis of unit cell geometry); polymorphism (or allotropy) (when a specific material can have more than one crystal structure); and anisotropy (the directionality dependence of properties).

{X-ray diffractometry is used for crystal structure and interplanar spacing determinations. A beam of x-rays directed on a crystalline material may experience diffraction (constructive interference) as a result of its interaction with a series of parallel atomic planes according to Bragg's law. Interplanar spacing is a function of the Miller indices and lattice parameter(s) as well as the crystal structure.}

IMPORTANT TERMS AND CONCEPTS

Allotropy
Amorphous
Anion
Anisotropy
Atomic packing factor (APF)
Body-centered cubic (BCC)
Bragg's law
Cation
Coordination number
Crystal structure

Crystal system
Crystalline
Diffraction
Face-centered cubic (FCC)
Grain
Grain boundary
Hexagonal close-packed (HCP)
Isotropic
Lattice

Lattice parameters
Miller indices
Noncrystalline
Octahedral position
Polycrystalline
Polymorphism
Single crystal
Tetrahedral position
Unit cell

REFERENCES

Azaroff, L. F., *Elements of X-Ray Crystallography*, McGraw-Hill Book Company, New York, 1968. Reprinted by TechBooks, Marietta, OH, 1990.

Barrett, C. S. and T. B. Massalski, *Structure of Metals*, 3rd edition, Pergamon Press, Oxford, 1980.

Barsoum, M. W., *Fundamentals of Ceramics*, The McGraw-Hill Companies, Inc., New York, 1997.

Budworth, D. W., *An Introduction to Ceramic Science*, Pergamon Press, Oxford, 1970.

Buerger, M. J., *Elementary Crystallography*, John Wiley & Sons, New York, 1956.

Charles, R. J., "The Nature of Glasses," *Scientific American*, Vol. 217, No. 3, September 1967, pp. 126–136.

Chiang, Y. M., D. P. Birnie, III, and W. D. Kingery, *Physical Ceramics: Principles for Ceramic Science and Engineering*, John Wiley & Sons, Inc., New York, 1997.

Cullity, B. D., *Elements of X-Ray Diffraction*, 3rd edition, Addison-Wesley Publishing Co., Reading, MA, 1998.

Curl, R. F. and R. E. Smalley, "Fullerenes," *Scientific American*, Vol. 265, No. 4, October 1991, pp. 54–63.

Gilman, J. J., "The Nature of Ceramics," *Scientific American*, Vol. 217, No. 3, September 1967, pp. 112–124.

Hauth, W. E., "Crystal Chemistry in Ceramics," *American Ceramic Society Bulletin,* Vol. 30, 1951: No. 1, pp. 5–7; No. 2, pp. 47–49; No. 3, pp. 76–77; No. 4, pp. 137–142; No. 5, pp. 165–167; No. 6, pp. 203–205. A good overview of silicate structures.

Kingery, W. D., H. K. Bowen, and D. R. Uhlmann, *Introduction to Ceramics,* 2nd edition, John Wiley & Sons, New York, 1976. Chapters 1–4.

Richerson, D. W., *Modern Ceramic Engineering,* 2nd edition, Marcel Dekker, New York, 1992.

Schwartz, L. H. and J. B. Cohen, *Diffraction from Materials,* 2nd edition, Springer-Verlag, New York, 1987.

Van Vlack, L. H., *Physical Ceramics for Engineers,* Addison-Wesley Publishing Company, Reading, MA, 1964. Chapters 1–4 and 6–8.

Wyckoff, R. W. G., *Crystal Structures,* 2nd edition, Interscience Publishers, 1963. Reprinted by Krieger Publishing Company, Melbourne, FL, 1986.

QUESTIONS AND PROBLEMS

Note: To solve those problems having an asterisk (*) by their numbers, consultation of supplementary topics [appearing only on the CD-ROM (and not in print)] will probably be necessary.

3.1 What is the difference between atomic structure and crystal structure?

3.2 What is the difference between a crystal structure and a crystal system?

3.3 If the atomic radius of aluminum is 0.143 nm, calculate the volume of its unit cell in cubic meters.

3.4 Show for the body-centered cubic crystal structure that the unit cell edge length a and the atomic radius R are related through $a = 4R/\sqrt{3}$.

3.5 For the HCP crystal structure, show that the ideal c/a ratio is 1.633.

3.6 Show that the atomic packing factor for BCC is 0.68.

3.7 Show that the atomic packing factor for HCP is 0.74.

3.8 Iron has a BCC crystal structure, an atomic radius of 0.124 nm, and an atomic weight of 55.85 g/mol. Compute and compare its density with the experimental value found inside the front cover.

3.9 Calculate the radius of an iridium atom given that Ir has an FCC crystal structure, a density of 22.4 g/cm³, and an atomic weight of 192.2 g/mol.

3.10 Calculate the radius of a vanadium atom, given that V has a BCC crystal structure, a density of 5.96 g/cm³, and an atomic weight of 50.9 g/mol.

3.11 Some hypothetical metal has the simple cubic crystal structure shown in Figure 3.40. If its atomic weight is 70.4 g/mol and the atomic radius is 0.126 nm, compute its density.

3.12 Zirconium has an HCP crystal structure and a density of 6.51 g/cm³.

(a) What is the volume of its unit cell in cubic meters?

(b) If the c/a ratio is 1.593, compute the values of c and a.

3.13 Using atomic weight, crystal structure, and atomic radius data tabulated inside the front cover, compute the theoretical densities of lead, chromium, copper, and cobalt, and then

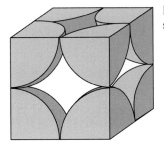

FIGURE 3.40 Hard-sphere unit cell representation of the simple cubic crystal structure.

compare these values with the measured densities listed in this same table. The c/a ratio for cobalt is 1.623.

3.14 Rhodium has an atomic radius of 0.1345 nm (1.345 Å) and a density of 12.41 g/cm³. Determine whether it has an FCC or BCC crystal structure.

3.15 Below are listed the atomic weight, density, and atomic radius for three hypothetical alloys. For each determine whether its crystal structure is FCC, BCC, or simple cubic and then justify your determination. A simple cubic unit cell is shown in Figure 3.40.

Alloy	Atomic Weight (g/mol)	Density (g/cm³)	Atomic Radius (nm)
A	77.4	8.22	0.125
B	107.6	13.42	0.133
C	127.3	9.23	0.142

3.16 The unit cell for tin has tetragonal symmetry, with a and b lattice parameters of 0.583 and 0.318 nm, respectively. If its density, atomic weight, and atomic radius are 7.30 g/cm³, 118.69 g/mol, and 0.151 nm, respectively, compute the atomic packing factor.

3.17 Iodine has an orthorhombic unit cell for which the a, b, and c lattice parameters are 0.479, 0.725, and 0.978 nm, respectively.

(a) If the atomic packing factor and atomic radius are 0.547 and 0.177 nm, respectively, determine the number of atoms in each unit cell.

(b) The atomic weight of iodine is 126.91 g/mol; compute its density.

3.18 Titanium has an HCP unit cell for which the ratio of the lattice parameters c/a is 1.58. If the radius of the Ti atom is 0.1445 nm, **(a)** determine the unit cell volume, and **(b)** calculate the density of Ti and compare it with the literature value.

3.19 Zinc has an HCP crystal structure, a c/a ratio of 1.856, and a density of 7.13 g/cm³. Calculate the atomic radius for Zn.

3.20 Rhenium has an HCP crystal structure, an atomic radius of 0.137 nm, and a c/a ratio of 1.615. Compute the volume of the unit cell for Re.

3.21 This is a unit cell for a hypothetical metal:

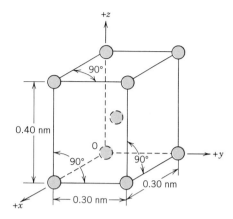

(a) To which crystal system does this unit cell belong?

(b) What would this crystal structure be called?

(c) Calculate the density of the material, given that its atomic weight is 141 g/mol.

3.22 Using the Molecule Definition File (MDF) on the CD-ROM that accompanies this book, generate a three-dimensional unit cell for the intermetallic compound $AuCu_3$ given the following: 1) the unit cell is cubic with an edge length of 0.374 nm, 2) gold atoms are situated at all cube corners, and 3) copper atoms are positioned at the centers of all unit cell faces.

3.23 Using the Molecule Definition File (MDF) on the CD-ROM that accompanies this book, generate a three-dimensional unit cell for the intermetallic compound AuCu given the following: 1) the unit cell is tetragonal with $a = 0.289$ nm and $c = 0.367$ nm (see Table 3.6), 2) gold atoms are situated at all unit cell corners, and 3) a copper atom is positioned at the center of the unit cell.

3.24 Sketch a unit cell for the body-centered orthorhombic crystal structure.

3.25 For a ceramic compound, what are the two characteristics of the component ions that determine the crystal structure?

3.26 Show that the minimum cation-to-anion radius ratio for a coordination number of 4 is 0.225.

3.27 Show that the minimum cation-to-anion radius ratio for a coordination number of 6 is 0.414. *Hint:* Use the NaCl crystal structure (Figure 3.5), and assume that anions and cations are just touching along cube edges and across face diagonals.

3.28 Demonstrate that the minimum cation-to-anion radius ratio for a coordination number of 8 is 0.732.

3.29 On the basis of ionic charge and ionic radii, predict the crystal structures for the following materials: **(a)** CsI, **(b)** NiO, **(c)** KI, and **(d)** NiS. Justify your selections.

3.30 Which of the cations in Table 3.4 would you predict to form iodides having the cesium chloride crystal structure? Justify your choices.

3.31 Compute the atomic packing factor for the cesium chloride crystal structure in which $r_C/r_A = 0.732$.

3.32 Table 3.4 gives the ionic radii for K^+ and O^{2-} as 0.138 and 0.140 nm, respectively. What would be the coordination number for each O^{2-} ion? Briefly describe the resulting crystal structure for K_2O. Explain why this is called the antifluorite structure.

3.33 Using the Molecule Definition File (MDF) on the CD-ROM that accompanies this book, generate a three-dimensional unit cell for lead oxide, PbO, given the following: (1) the unit cell is tetragonal with $a = 0.397$ nm and $c = 0.502$ nm, (2) oxygen ions are situated at all cube corners, and, in addition, at the centers of the two square faces, (3) one oxygen ion is positioned on each of two of the other opposing faces (rectangular) at the $0.5a$-$0.237c$ coordinate, and (4) for the other two rectangular and opposing faces, oxygen ions are located at the $0.5a$-$0.763c$ coordinate.

3.34 Calculate the density of FeO, given that it has the rock salt crystal structure.

3.35 Magnesium oxide has the rock salt crystal structure and a density of 3.58 g/cm³.

  (a) Determine the unit cell edge length.

  (b) How does this result compare with the edge length as determined from the radii in Table 3.4, assuming that the Mg^{2+} and O^{2-} ions just touch each other along the edges?

3.36 Compute the theoretical density of diamond given that the C—C distance and bond angle are 0.154 nm and 109.5°, respectively. How does this value compare with the measured density?

3.37 Compute the theoretical density of ZnS given that the Zn—S distance and bond angle are 0.234 nm and 109.5°, respectively. How does this value compare with the measured density?

3.38 Cadmium sulfide (CdS) has a cubic unit cell, and from x-ray diffraction data it is known that the cell edge length is 0.582 nm. If the measured density is 4.82 g/cm³, how many Cd^{2+} and S^{2-} ions are there per unit cell?

3.39 **(a)** Using the ionic radii in Table 3.4, compute the density of CsCl. *Hint:* Use a modification of the result of Problem 3.4.

  (b) The measured density is 3.99 g/cm³. How do you explain the slight discrepancy between your calculated value and the measured one?

3.40 From the data in Table 3.4, compute the density of CaF_2, which has the fluorite structure.

3.41 A hypothetical AX type of ceramic material is known to have a density of 2.65 g/cm³ and a unit cell of cubic symmetry with a cell edge length of 0.43 nm. The atomic weights of the A and X elements are 86.6 and 40.3 g/mol, respectively. On the basis of this information, which of the following crystal structures is (are) possible for this material: rock salt, cesium chloride, or zinc blende? Justify your choice(s).

3.42 The unit cell for $MgFe_2O_4$ (MgO-Fe_2O_3) has cubic symmetry with a unit cell edge length of 0.836 nm. If the density of this material is 4.52 g/cm³, compute its atomic packing factor. For this computation, you will need to use ionic radii listed in Table 3.4.

3.43 The unit cell for Al_2O_3 has hexagonal symmetry with lattice parameters $a = 0.4759$ nm and $c = 1.2989$ nm. If the density of this material is 3.99 g/cm³, calculate its atomic packing factor. For this computation use ionic radii listed in Table 3.4.

3.44 Compute the atomic packing factor for the diamond cubic crystal structure (Figure 3.16).

Assume that bonding atoms touch one another, that the angle between adjacent bonds is 109.5°, and that each atom internal to the unit cell is positioned $a/4$ of the distance away from the two nearest cell faces (a is the unit cell edge length).

3.45 Compute the atomic packing factor for cesium chloride using the ionic radii in Table 3.4 and assuming that the ions touch along the cube diagonals.

3.46 In terms of bonding, explain why silicate materials have relatively low densities.

3.47 Determine the angle between covalent bonds in an SiO_4^{4-} tetrahedron.

3.48 Draw an orthorhombic unit cell, and within that cell a $[12\bar{1}]$ direction and a (210) plane.

3.49 Sketch a monoclinic unit cell, and within that cell a $[0\bar{1}1]$ direction and a (002) plane.

3.50 Here are unit cells for two hypothetical metals:

(a) What are the indices for the directions indicated by the two vectors in sketch (*a*)?

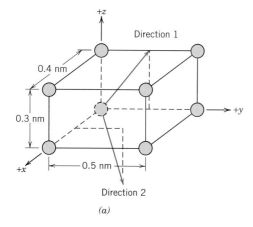

(a)

(b) What are the indices for the two planes drawn in sketch (*b*)?

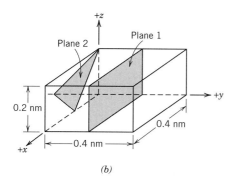

(b)

3.51 Within a cubic unit cell, sketch the following directions:

(a) $[\bar{1}10]$;   **(e)** $[\bar{1}\bar{1}1]$;
(b) $[\bar{1}2\bar{1}]$;   **(f)** $[\bar{1}22]$;
(c) $[0\bar{1}2]$;   **(g)** $[1\bar{2}3]$;
(d) $[1\bar{3}3]$;   **(h)** $[\bar{1}03]$.

3.52 Determine the indices for the directions shown in the following cubic unit cell:

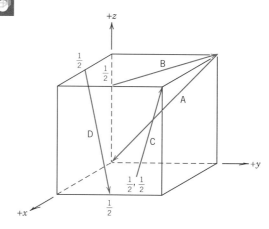

3.53 Determine the indices for the directions shown in the following cubic unit cell:

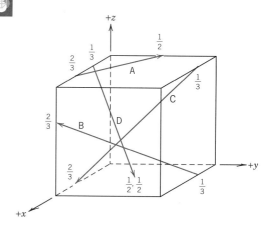

3.54 For tetragonal crystals, cite the indices of directions that are equivalent to each of the following directions:

(a) [101];
(b) [110];
(c) [010].

3.55 **(a)** Convert the [100] and [111] directions into the four-index Miller–Bravais scheme for hexagonal unit cells.

(b) Make the same conversion for the (010) and (101) planes.

3.56 Determine the Miller indices for the planes shown in the following unit cell:

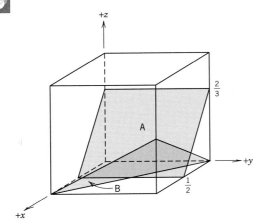

3.57 Determine the Miller indices for the planes shown in the following unit cell:

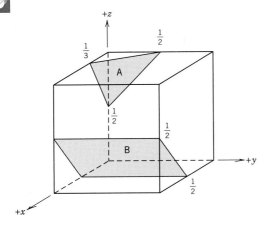

3.58 Determine the Miller indices for the planes shown in the following unit cell:

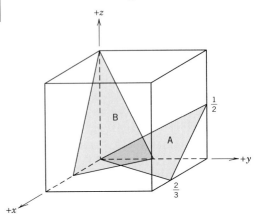

3.59 Sketch the $(1\bar{1}01)$ and $(11\bar{2}0)$ planes in a hexagonal unit cell.

3.60 Determine the indices for the planes shown in the hexagonal unit cells shown below.

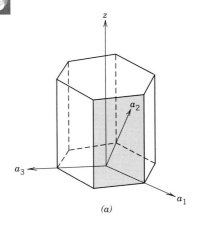

(a)

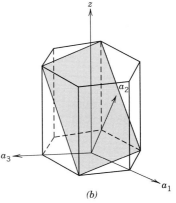

(b)

3.61 Sketch within a cubic unit cell the following planes:

(a) $(0\bar{1}\bar{1})$; (e) $(\bar{1}1\bar{1})$;
(b) $(11\bar{2})$; (f) $(1\bar{2}2)$;
(c) $(10\bar{2})$; (g) $(\bar{1}2\bar{3})$;
(d) $(1\bar{3}1)$; (h) $(0\bar{1}3)$.

3.62 Sketch the atomic packing of (a) the (100) plane for the FCC crystal structure, and (b) the (111) plane for the BCC crystal structure (similar to Figures 3.24b and 3.25b).

3.63 For each of the following crystal structures, represent the indicated plane in the manner of Figures 3.24 and 3.25, showing both anions and cations: (a) (100) plane for the rock salt crystal structure, (b) (110) plane for the cesium chloride crystal structure, (c) (111) plane for the zinc blende crystal structure, and (d) (110) plane for the perovskite crystal structure.

3.64 Consider the reduced-sphere unit cell shown in Problem 3.21, having an origin of the coordinate system positioned at the atom labeled with an O. For the following sets of planes, determine which are equivalent:

(a) (100), $(0\bar{1}0)$, and (001).
(b) (110), (101), (011), and $(\bar{1}10)$.
(c) (111), $(1\bar{1}1)$, $(11\bar{1})$, and $(\bar{1}1\bar{1})$.

3.65 Cite the indices of the direction that results from the intersection of each of the following pair of planes within a cubic crystal: (a) (110) and (111) planes; (b) (110) and $(1\bar{1}0)$ planes; and (c) $(10\bar{1})$ and (001) planes.

3.66 The zinc blende crystal structure is one that may be generated from close-packed planes of anions.

(a) Will the stacking sequence for this structure be FCC or HCP? Why?

(b) Will cations fill tetrahedral or octahedral positions? Why?

(c) What fraction of the positions will be occupied?

3.67 The corundum crystal structure, found for Al_2O_3, consists of an HCP arrangement of O^{2-} ions; the Al^{3+} ions occupy octahedral positions.

(a) What fraction of the available octahedral positions are filled with Al^{3+} ions?

(b) Sketch two close-packed O^{2-} planes stacked in an AB sequence, and note octahedral positions that will be filled with the Al^{3+} ions.

3.68 Iron sulfide (FeS) may form a crystal structure that consists of an HCP arrangement of S^{2-} ions.

(a) Which type of interstitial site will the Fe^{2+} ions occupy?

(b) What fraction of these available interstitial sites will be occupied by Fe^{2+} ions?

3.69 Magnesium silicate, Mg_2SiO_4, forms in the olivine crystal structure which consists of an HCP arrangement of O^{2-} ions.

(a) Which type of interstitial site will the Mg^{2+} ions occupy? Why?

(b) Which type of interstitial site will the Si^{4+} ions occupy? Why?

(c) What fraction of the total tetrahedral sites will be occupied?

(d) What fraction of the total octahedral sites will be occupied?

3.70* Compute and compare the linear densities of the [100], [110], and [111] directions for FCC.

3.71* Compute and compare the linear densities of the [110] and [111] directions for BCC.

3.72* Calculate and compare the planar densities of the (100) and (111) planes for FCC.

3.73* Calculate and compare the planar densities of the (100) and (110) planes for BCC.

3.74* Calculate the planar density of the (0001) plane for HCP.

3.75 Here are shown the atomic packing schemes for several different crystallographic directions for some hypothetical metal. For each direction the circles represent only those atoms contained within a unit cell, which circles are reduced from their actual size.

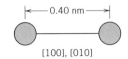

[100], [010]

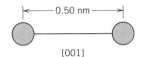

[001]

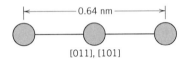

[011], [101]

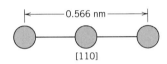

[110]

(a) To what crystal system does the unit cell belong?

(b) What would this crystal structure be called?

3.76 Below are shown three different crystallographic planes for a unit cell of some hypothetical metal; the circles represent atoms:

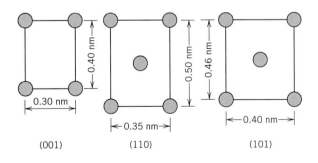

(001) (110) (101)

(a) To what crystal system does the unit cell belong?

(b) What would this crystal structure be called?

(c) If the density of this metal is 8.95 g/cm³, determine its atomic weight.

3.77 Explain why the properties of polycrystalline materials are most often isotropic.

3.78* Using the data for molybdenum in Table 3.1, compute the interplanar spacing for the (111) set of planes.

3.79* Determine the expected diffraction angle for the first-order reflection from the (113) set of planes for FCC platinum when monochromatic radiation of wavelength 0.1542 nm is used.

3.80* Using the data for aluminum in Table 3.1, compute the interplanar spacings for the (110) and (221) sets of planes.

3.81* The metal iridium has an FCC crystal structure. If the angle of diffraction for the (220) set of planes occurs at 69.22° (first-order reflection) when monochromatic x-radiation having a wavelength of 0.1542 nm is used, compute **(a)** the interplanar spacing for this set of planes, and **(b)** the atomic radius for an iridium atom.

3.82* The metal rubidium has a BCC crystal structure. If the angle of diffraction for the (321) set of planes occurs at 27.00° (first-order reflection) when monochromatic x-radiation having a wavelength of 0.0711 nm is used, compute **(a)** the interplanar spacing for this set of planes, and **(b)** the atomic radius for the rubidium atom.

3.83* For which set of crystallographic planes will a first-order diffraction peak occur at a diffraction angle of 46.21° for BCC iron when monochromatic radiation having a wavelength of 0.0711 nm is used?

3.84* Figure 3.37 shows an x-ray diffraction pattern for α-iron taken using a diffractometer and monochromatic x-radiation having a wavelength of 0.1542 nm; each diffraction peak on the pattern has been indexed. Compute the interplanar spacing for each set of planes indexed; also determine the lattice parameter of Fe for each of the peaks.

3.85* The diffraction peaks shown in Figure 3.37 are indexed according to the reflection rules for BCC (i.e., the sum $h + k + l$ must be even). Cite the h, k, and l indices for the first four diffraction peaks for FCC crystals consistent with h, k, and l all being either odd or even.

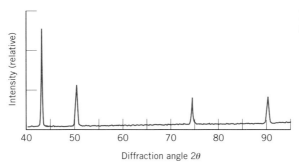

FIGURE 3.41 Diffraction pattern for polycrystalline copper.

3.86* Figure 3.41 shows the first four peaks of the x-ray diffraction pattern for copper, which has an FCC crystal structure; monochromatic x-radiation having a wavelength of 0.1542 nm was used.

(a) Index (i.e., give h, k, and l indices for) each of these peaks.

(b) Determine the interplanar spacing for each of the peaks.

(c) For each peak, determine the atomic radius for Cu and compare these with the value presented in Table 3.1.

3.87 Would you expect a material in which the atomic bonding is predominantly ionic in nature to be more or less likely to form a noncrystalline solid upon solidification than a covalent material? Why? (See Section 2.6.)

Design Problem

3.D1* Gallium arsenide (GaAs) and gallium phosphide (GaP) both have the zinc blende crystal structure and are soluble in one another at all concentrations. Determine the concentration in weight percent of GaP that must be added to GaAs to yield a unit cell edge length of 0.5570 nm. The densities of GaAs and GaP are 5.307 and 4.130 g/cm^3, respectively.

Chapter 4 / Polymer Structures

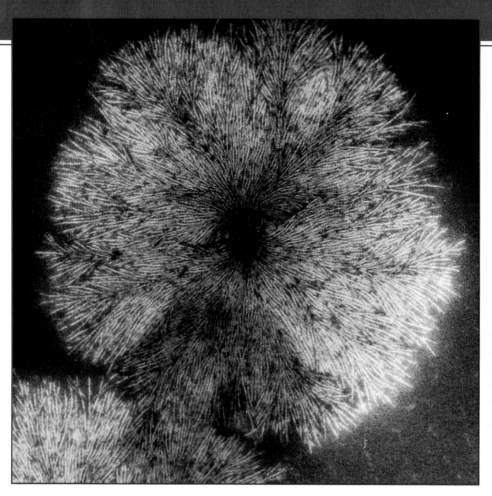

Transmission electron micrograph showing the spherulite structure in a natural rubber specimen. Chain-folded lamellar crystallites approximately 10 nm thick extend in radial directions from the center; they appear as white lines in the micrograph. 30,000×. (Photograph supplied by P. J. Phillips. First published in R. Bartnikas and R. M. Eichhorn, *Engineering Dielectrics*, Vol. IIA, *Electrical Properties of Solid Insulating Materials: Molecular Structure and Electrical Behavior.* Copyright ASTM. Reprinted with permission.)

Why Study *Polymer Structures?*

A relatively large number of chemical and structural characteristics affect the properties and behaviors of polymeric materials. Some of these influences are as follows:

1. Degree of crystallinity of semicrystalline polymers—on density, stiffness, strength, and ductility (Sections 4.11 and 8.18).

2. Degree of crosslinking—on the stiffness of rubber-like materials (Section 8.19).

{3. Polymer chemistry—on melting and glass-transition temperatures (Section 11.17).}

After careful study of this chapter you should be able to do the following:

1. Describe a typical polymer molecule in terms of its chain structure, and, in addition, how the molecule may be generated by repeating mer units.
2. Draw mer structures for polyethylene, polyvinyl chloride, polytetrafluoroethylene, polypropylene, and polystyrene.
3. Calculate number-average and weight-average molecular weights, and number-average and weight-average degrees of polymerization for a specified polymer.
4. Name and briefly describe:
   (a) the four general types of polymer molecular structures;
   {(b) the three types of stereoisomers;}
   {(c) the two kinds of geometrical isomers;}
   (d) the four types of copolymers.
5. Cite the differences in behavior and molecular structure for thermoplastic and thermosetting polymers.
6. Briefly describe the crystalline state in polymeric materials.
7. Briefly describe/diagram the spherulitic structure for a semicrystalline polymer.

4.1 INTRODUCTION

Naturally occurring polymers—those derived from plants and animals—have been used for many centuries; these materials include wood, rubber, cotton, wool, leather, and silk. Other natural polymers such as proteins, enzymes, starches, and cellulose are important in biological and physiological processes in plants and animals. Modern scientific research tools have made possible the determination of the molecular structures of this group of materials, and the development of numerous polymers, which are synthesized from small organic molecules. Many of our useful plastics, rubbers, and fiber materials are synthetic polymers. In fact, since the conclusion of World War II, the field of materials has been virtually revolutionized by the advent of synthetic polymers. The synthetics can be produced inexpensively, and their properties may be managed to the degree that many are superior to their natural counterparts. In some applications metal and wood parts have been replaced by plastics, which have satisfactory properties and may be produced at a lower cost.

As with metals and ceramics, the properties of polymers are intricately related to the structural elements of the material. This chapter explores molecular and crystal structures of polymers; Chapter 8 discusses the relationships between structure and some of the mechanical properties.

4.2 HYDROCARBON MOLECULES

Since most polymers are organic in origin, we briefly review some of the basic concepts relating to the structure of their molecules. First, many organic materials are *hydrocarbons;* that is, they are composed of hydrogen and carbon. Furthermore, the intramolecular bonds are covalent. Each carbon atom has four electrons that may participate in covalent bonding, whereas every hydrogen atom has only one bonding electron. A single covalent bond exists when each of the two bonding atoms contributes one electron, as represented schematically in Figure 2.10 for a molecule of methane (CH_4). Double and triple bonds between two carbon atoms involve the sharing of two and three pairs of electrons, respectively. For example, in ethylene, which has the chemical formula C_2H_4, the two carbon atoms are doubly bonded together, and each is also singly bonded to two hydrogen atoms, as represented by the structural formula

$$
\begin{array}{cc}
H & H \\
| & | \\
C & = C \\
| & | \\
H & H
\end{array}
$$

where $-$ and $=$ denote single and double covalent bonds, respectively. An example of a triple bond is found in acetylene, C_2H_2:

$$H - C \equiv C - H$$

Molecules that have double and triple covalent bonds are termed **unsaturated.** That is, each carbon atom is not bonded to the maximum (or four) other atoms; as such, it is possible for another atom or group of atoms to become attached to the original molecule. Furthermore, for a **saturated** hydrocarbon, all bonds are single ones (and saturated), and no new atoms may be joined without the removal of others that are already bonded.

Some of the simple hydrocarbons belong to the paraffin family; the chainlike paraffin molecules include methane (CH_4), ethane (C_2H_6), propane (C_3H_8), and butane (C_4H_{10}). Compositions and molecular structures for paraffin molecules are contained in Table 4.1. The covalent bonds in each molecule are strong, but only weak hydrogen and van der Waals bonds exist between molecules, and thus these hydrocarbons have relatively low melting and boiling points. However, boiling temperatures rise with increasing molecular weight (Table 4.1).

Hydrocarbon compounds with the same composition may have different atomic arrangements, a phenomenon termed **isomerism.** For example, there are two iso-

Table 4.1 Compositions and Molecular Structures for Some of the Paraffin Compounds: C_nH_{2n+2}

Name	Composition	Structure	Boiling Point (°C)
Methane	CH_4	$H-\underset{\underset{H}{|}}{\overset{\overset{H}{|}}{C}}-H$	−164
Ethane	C_2H_6	$H-\underset{\underset{H}{|}}{\overset{\overset{H}{|}}{C}}-\underset{\underset{H}{|}}{\overset{\overset{H}{|}}{C}}-H$	−88.6
Propane	C_3H_8	$H-\underset{\underset{H}{|}}{\overset{\overset{H}{|}}{C}}-\underset{\underset{H}{|}}{\overset{\overset{H}{|}}{C}}-\underset{\underset{H}{|}}{\overset{\overset{H}{|}}{C}}-H$	−42.1
Butane	C_4H_{10}	·	−0.5
Pentane	C_5H_{12}	·	36.1
Hexane	C_6H_{14}	·	69.0

mers for butane; normal butane has the structure

$$
\begin{array}{ccccccc}
& H & & H & & H & & H \\
& | & & | & & | & & | \\
H- & C & - & C & - & C & - & C & -H \\
& | & & | & & | & & | \\
& H & & H & & H & & H
\end{array}
$$

whereas a molecule of isobutane is represented as follows:

$$
\begin{array}{ccccc}
& & H & & \\
& & | & & \\
& H- & C & -H & \\
& H & | & H & \\
& | & | & | & \\
H- & C & -C- & C & -H \\
& | & | & | & \\
& H & H & H &
\end{array}
$$

Some of the physical properties of hydrocarbons will depend on the isomeric state; for example, the boiling temperatures for normal butane and isobutane are -0.5 and $-12.3°C$ (31.1 and 9.9°F), respectively.

There are numerous other organic groups, many of which are involved in polymer structures. Several of the more common groups are presented in Table 4.2, where R and R′ represent organic radicals—groups of atoms that remain as a single unit and maintain their identity during chemical reactions. Examples of singly bonded hydrocarbon radicals include the CH_3, C_2H_5, and C_6H_5 (methyl, ethyl, and phenyl) groups.

4.3 POLYMER MOLECULES

The molecules in polymers are gigantic in comparison to the hydrocarbon molecules heretofore discussed; because of their size they are often referred to as **macromolecules.** Within each molecule, the atoms are bound together by covalent interatomic bonds. For most polymers, these molecules are in the form of long and flexible chains, the backbone of which is a string of carbon atoms; many times each carbon atom singly bonds to two adjacent carbons atoms on either side, represented schematically in two dimensions as follows:

$$
\begin{array}{ccccccccccccc}
| & & | & & | & & | & & | & & | & & | \\
-C & - & C & - & C & - & C & - & C & - & C & - & C- \\
| & & | & & | & & | & & | & & | & & |
\end{array}
$$

Each of the two remaining valence electrons for every carbon atom may be involved in side-bonding with atoms or radicals that are positioned adjacent to the chain. Of course, both chain and side double bonds are also possible.

These long molecules are composed of structural entities called **mer** units, which are successively repeated along the chain. "Mer" originates from the Greek word *meros,* which means part; the term **polymer** was coined to mean many mers. We sometimes use the term **monomer,** which refers to a stable molecule from which a polymer is synthesized.

Table 4.2 Some Common Hydrocarbon Groups

Family	Characteristic Unit	Representative Compound	
Alcohols	R—OH	H—C—OH (with H above and H below)	Methyl alcohol
Ethers	R—O—R′	H—C—O—C—H (with H above and H below each C)	Dimethyl ether
Acids	R—C with OH and O	H—C—C with H, OH and O	Acetic acid
Aldehydes	R, C=O, H	H, C=O, H	Formaldehyde
Aromatic hydrocarbons	R (phenyl ring) a	OH (phenyl ring)	Phenol

a The simplified structure (phenyl ring) denotes a phenyl group,

4.4 THE CHEMISTRY OF POLYMER MOLECULES

Consider again the hydrocarbon ethylene (C_2H_4), which is a gas at ambient temperature and pressure, and has the following molecular structure:

$$
\begin{array}{c}
\text{H} \quad \text{H} \\
| \quad\; | \\
\text{C} = \text{C} \\
| \quad\; | \\
\text{H} \quad \text{H}
\end{array}
$$

If the ethylene gas is subjected catalytically to appropriate conditions of temperature and pressure, it will transform to polyethylene (PE), which is a solid polymeric material. This process begins when an active mer is formed by the reaction between an initiator or catalyst species (R·) and the ethylene mer unit, as follows:

$$
\text{R·} + \begin{matrix} H & H \\ | & | \\ C = C \\ | & | \\ H & H \end{matrix} \longrightarrow \text{R} - \begin{matrix} H & H \\ | & | \\ C - C· \\ | & | \\ H & H \end{matrix} \tag{4.1}
$$

The polymer chain then forms by the sequential addition of polyethylene monomer units to this active initiator-mer center. The active site, or unpaired electron (denoted by ·), is transferred to each successive end monomer as it is linked to the chain. This may be represented schematically as follows:

$$
\text{R} - \begin{matrix} H & H \\ | & | \\ C - C· \\ | & | \\ H & H \end{matrix} + \begin{matrix} H & H \\ | & | \\ C = C \\ | & | \\ H & H \end{matrix} \longrightarrow \text{R} - \begin{matrix} H & H & H & H \\ | & | & | & | \\ C - C - C - C· \\ | & | & | & | \\ H & H & H & H \end{matrix} \tag{4.2}
$$

 The final result, after the addition of many ethylene monomer units, is the polyethylene molecule, a portion of which is shown in Figure 4.1*a*. This representation is not strictly correct in that the angle between the singly bonded carbon atoms is not 180° as shown, but rather close to 109°. A more accurate three-dimensional model is one in which the carbon atoms form a zigzag pattern (Figure 4.1*b*), the C—C bond length being 0.154 nm. In this discussion, depiction of polymer molecules is frequently simplified using the linear chain model.

 If all the hydrogen atoms in polyethylene are replaced by fluorine, the resulting polymer is *polytetrafluoroethylene* (PTFE); its mer and chain structures are shown in Figure 4.2*a*. Polytetrafluoroethylene (having the trade name Teflon) belongs to a family of polymers called the fluorocarbons.

 Polyvinyl chloride (PVC), another common polymer, has a structure that is a slight variant of that for polyethylene, in which every fourth hydrogen is replaced with a Cl atom. Furthermore, substitution of the CH₃ methyl group

$$
\text{H} - \overset{\displaystyle \cdot}{\underset{\displaystyle |}{\overset{\displaystyle |}{C}}} - \text{H}
$$
$$
\text{H}
$$

FIGURE 4.1 For polyethylene, (*a*) a schematic representation of mer and chain structures, and (*b*) a perspective of the molecule, indicating the zigzag backbone structure.

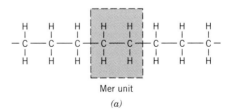

Mer unit

(a)

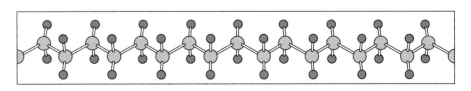

○ C ● H

(b)

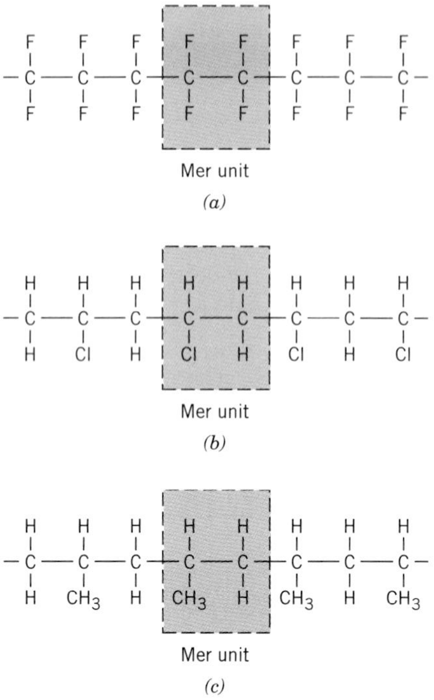

FIGURE 4.2 Mer and chain structures for (a) polytetrafluoroethylene, (b) polyvinyl chloride, and (c) polypropylene.

for each Cl atom in PVC yields *polypropylene* (PP). Polyvinyl chloride and polypropylene chain structures are also represented in Figure 4.2. Table 4.3 lists mer structures for some of the more common polymers; as may be noted, some of them, for example, nylon, polyester, and polycarbonate, are relatively complex. Mer structures for a large number of relatively common polymers are given in Appendix D.

When all the repeating units along a chain are of the same type, the resulting polymer is called a **homopolymer.** There is no restriction in polymer synthesis that prevents the formation of compounds other than homopolymers; and, in fact, chains may be composed of two or more different mer units, in what are termed **copolymers** (see Section 4.10).

The monomers discussed thus far have an active bond that may react to covalently bond with other monomers, as indicated above for ethylene; such a monomer is termed **bifunctional;** that is, it may bond with two other units in forming the two-dimensional chainlike molecular structure. However, other monomers, such as phenol-formaldehyde (Table 4.3), are **trifunctional;** they have three active bonds, from which a three-dimensional molecular network structure results.

4.5 MOLECULAR WEIGHT

Extremely large molecular weights[1] are to be found in polymers with very long chains. During the polymerization process in which these large macromolecules are

---

[1] "Molecular mass," "molar mass," and "relative molecular mass" are sometimes used and are really more appropriate terms than "molecular weight" in the context of the present discussion—in actual fact, we are dealing with masses and not weights. However, molecular weight is most commonly found in the polymer literature, and thus will be used throughout this book.

Table 4.3 A Listing of Mer Structures for 10 of the More Common Polymeric Materials

Polymer	Repeating (Mer) Structure
Polyethylene (PE)	
Polyvinyl chloride (PVC)	
Polytetrafluoroethylene (PTFE)	
Polypropylene (PP)	
Polystyrene (PS)	
Polymethyl methacrylate (PMMA)	
Phenol-formaldehyde (Bakelite)	
Polyhexamethylene adipamide (nylon 6,6)	

Table 4.3 (*Continued*)

Polymer	Repeating (Mer) Structure

Polyethylene terephthalate (PET, a polyester)

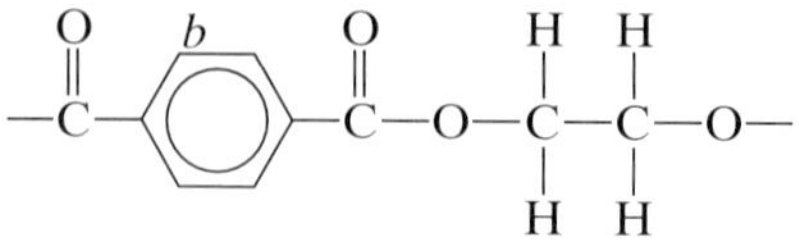

Polycarbonate

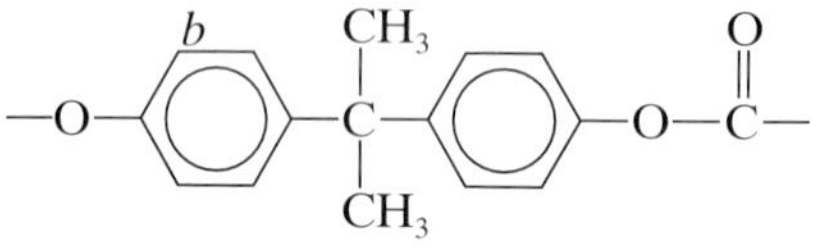

[b] The 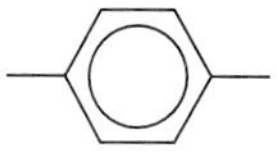 symbol in the backbone chain denotes an aromatic ring as

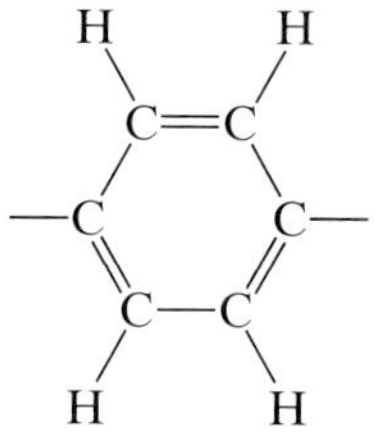

synthesized from smaller molecules, not all polymer chains will grow to the same length; this results in a distribution of chain lengths or molecular weights. Ordinarily, an average molecular weight is specified, which may be determined by the measurement of various physical properties such as viscosity and osmotic pressure.

There are several ways of defining average molecular weight. The number-average molecular weight $\overline{M}_n$ is obtained by dividing the chains into a series of size ranges and then determining the number fraction of chains within each size range (Figure 4.3*a*). This number-average molecular weight is expressed as

$$\overline{M}_n = \Sigma x_i M_i \tag{4.3a}$$

where M_i represents the mean (middle) molecular weight of size range i, and x_i is the fraction of the total number of chains within the corresponding size range.

A weight-average molecular weight $\overline{M}_w$ is based on the weight fraction of molecules within the various size ranges (Figure 4.3*b*). It is calculated according to

$$\overline{M}_w = \Sigma w_i M_i \tag{4.3b}$$

where, again, M_i is the mean molecular weight within a size range, whereas w_i denotes the weight fraction of molecules within the same size interval. Computations for both number-average and weight-average molecular weights are carried out in Example Problem 4.1. A typical molecular weight distribution along with these molecular weight averages are shown in Figure 4.4.

An alternate way of expressing average chain size of a polymer is as the **degree of polymerization** n, which represents the average number of mer units in a chain.

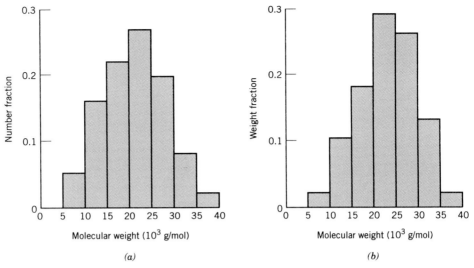

FIGURE 4.3 Hypothetical polymer molecule size distributions on the basis of (*a*) number and (*b*) weight fractions of molecules.

Both number-average (n_n) and weight-average (n_w) degrees of polymerization are possible, as follows:

$$n_n = \frac{\overline{M}_n}{\overline{m}} \tag{4.4a}$$

$$n_w = \frac{\overline{M}_w}{\overline{m}} \tag{4.4b}$$

where $\overline{M}_n$ and $\overline{M}_w$ are the number-average and weight-average molecular weights as defined above, while $\overline{m}$ is the mer molecular weight. For a copolymer (having

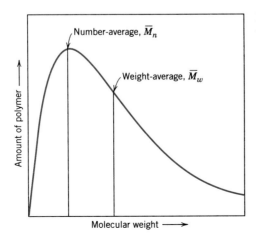

FIGURE 4.4 Distribution of molecular weights for a typical polymer.

two or more different mer units), $\overline{m}$ is determined from

$$\overline{m} = \Sigma f_j m_j \tag{4.5}$$

In this expression, f_j and m_j are, respectively, the chain fraction and molecular weight of mer j.

EXAMPLE PROBLEM 4.1

Assume that the molecular weight distributions shown in Figure 4.3 are for polyvinyl chloride. For this material, compute **(a)** the number-average molecular weight; **(b)** the number-average degree of polymerization; and **(c)** the weight-average molecular weight.

SOLUTION

(a) The data necessary for this computation, as taken from Figure 4.3a, are presented in Table 4.4a. According to Equation 4.3a, summation of all the $x_i M_i$ products (from the right-hand column) yields the number-average molecular weight, which in this case is 21,150 g/mol.

(b) To determine the number-average degree of polymerization (Equation 4.4a), it first becomes necessary to compute the mer molecular weight. For PVC, each mer consists of two carbon atoms, three hydrogen atoms, and a

Table 4.4a Data Used for Number-Average Molecular Weight Computations in Example Problem 4.1

Molecular Weight Range (g/mol)	Mean M_i (g/mol)	x_i	$x_i M_i$
5,000–10,000	7,500	0.05	375
10,000–15,000	12,500	0.16	2000
15,000–20,000	17,500	0.22	3850
20,000–25,000	22,500	0.27	6075
25,000–30,000	27,500	0.20	5500
30,000–35,000	32,500	0.08	2600
35,000–40,000	37,500	0.02	750
			$\overline{M}_n = 21{,}150$

Table 4.4b Data Used for Weight-Average Molecular Weight Computations in Example Problem 4.1

Molecular Weight Range (g/mol)	Mean M_i (g/mol)	w_i	$w_i M_i$
5,000–10,000	7,500	0.02	150
10,000–15,000	12,500	0.10	1250
15,000–20,000	17,500	0.18	3150
20,000–25,000	22,500	0.29	6525
25,000–30,000	27,500	0.26	7150
30,000–35,000	32,500	0.13	4225
35,000–40,000	37,500	0.02	750
			$\overline{M}_w = 23{,}200$

single chlorine atom (Table 4.3). Furthermore, the atomic weights of C, H, and Cl are, respectively, 12.01, 1.01, and 35.45 g/mol. Thus, for PVC

$$\overline{m} = 2(12.01 \text{ g/mol}) + 3(1.01 \text{ g/mol}) + 35.45 \text{ g/mol}$$

$$= 62.50 \text{ g/mol}$$

and

$$n_n = \frac{\overline{M}_n}{\overline{m}} = \frac{21{,}150 \text{ g/mol}}{62.50 \text{ g/mol}} = 338$$

(c) Table 4.4b shows the data for the weight-average molecular weight, as taken from Figure 4.3b. The $w_i M_i$ products for the several size intervals are tabulated in the right-hand column. The sum of these products (Equation 4.3b) yields a value of 23,200 g/mol for $\overline{M}_w$.

Various polymer characteristics are affected by the magnitude of the molecular weight. One of these is the melting or softening temperature; melting temperature is raised with increasing molecular weight (for $\overline{M}$ up to about 100,000 g/mol). At room temperature, polymers with very short chains (having molecular weights on the order of 100 g/mol) exist as liquids or gases. Those with molecular weights of approximately 1000 g/mol are waxy solids (such as paraffin wax) and soft resins. Solid polymers (sometimes termed *high polymers*), which are of prime interest here, commonly have molecular weights ranging between 10,000 and several million g/mol.

4.6 MOLECULAR SHAPE

There is no reason to suppose that polymer chain molecules are strictly straight, in the sense that the zigzag arrangement of the backbone atoms (Figure 4.1b) is disregarded. Single chain bonds are capable of rotation and bending in three dimensions. Consider the chain atoms in Figure 4.5a; a third carbon atom may lie at any point on the cone of revolution and still subtend about a 109° angle with the bond between the other two atoms. A straight chain segment results when

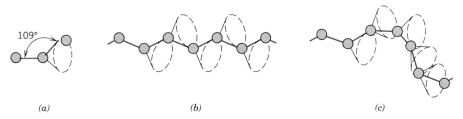

(a)   *(b)*   *(c)*

FIGURE 4.5 Schematic representations of how polymer chain shape is influenced by the positioning of backbone carbon atoms (solid circles). For (*a*), the rightmost atom may lie anywhere on the dashed circle and still subtend a 109° angle with the bond between the other two atoms. Straight and twisted chain segments are generated when the backbone atoms are situated as in (*b*) and (*c*), respectively. (From *Science and Engineering of Materials, 3rd edition,* by D. R. Askeland. © 1994. Reprinted with permission of Brooks/Cole, a division of Thomson Learning. Fax 800 730-2215.)

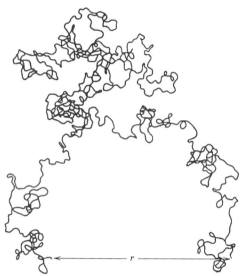

FIGURE 4.6 Schematic representation of a single polymer chain molecule that has numerous random kinks and coils produced by chain bond rotations. (From L. R. G. Treloar, *The Physics of Rubber Elasticity,* 2nd edition, Oxford University Press, Oxford, 1958, p. 47.)

successive chain atoms are positioned as in Figure 4.5*b*. On the other hand, chain bending and twisting are possible when there is a rotation of the chain atoms into other positions, as illustrated in Figure 4.5*c*.[2] Thus, a single chain molecule composed of many chain atoms might assume a shape similar to that represented schematically in Figure 4.6, having a multitude of bends, twists, and kinks.[3] Also indicated in this figure is the end-to-end distance of the polymer chain *r*; this distance is much smaller than the total chain length.

Polymers consist of large numbers of molecular chains, each of which may bend, coil, and kink in the manner of Figure 4.6. This leads to extensive intertwining and entanglement of neighboring chain molecules, a situation similar to that of a fishing line that has experienced backlash from a fishing reel. These random coils and molecular entanglements are responsible for a number of important characteristics of polymers, to include the large elastic extensions displayed by the rubber materials.

Some of the mechanical and thermal characteristics of polymers are a function of the ability of chain segments to experience rotation in response to applied stresses or thermal vibrations. Rotational flexibility is dependent on mer structure and chemistry. For example, the region of a chain segment that has a double bond (C=C) is rotationally rigid. Also, introduction of a bulky or large side group of atoms restricts rotational movement. For example, polystyrene molecules, which have a phenyl side group (Table 4.3), are more resistant to rotational motion than are polyethylene chains.

4.7 MOLECULAR STRUCTURE

The physical characteristics of a polymer depend not only on its molecular weight and shape, but also on differences in the structure of the molecular chains. Modern

---

[2] For some polymers, rotation of carbon backbone atoms within the cone may be hindered by bulky side group elements on neighboring chains.

[3] The term *conformation* is often used in reference to the physical outline of a molecule, or molecular shape, that can only be altered by rotation of chain atoms about single bonds.

polymer synthesis techniques permit considerable control over various structural possibilities. This section discusses several molecular structures including linear, branched, crosslinked, and network, in addition to various isomeric configurations.

LINEAR POLYMERS

Linear polymers are those in which the mer units are joined together end to end in single chains. These long chains are flexible and may be thought of as a mass of spaghetti, as represented schematically in Figure 4.7a, where each circle represents a mer unit. For linear polymers, there may be extensive van der Waals and hydrogen bonding between the chains. Some of the common polymers that form with linear structures are polyethylene, polyvinyl chloride, polystyrene, polymethyl methacrylate, nylon, and the fluorocarbons.

BRANCHED POLYMERS

Polymers may be synthesized in which side-branch chains are connected to the main ones, as indicated schematically in Figure 4.7b; these are fittingly called **branched polymers.** The branches, considered to be part of the main-chain molecule, result from side reactions that occur during the synthesis of the polymer. The chain packing efficiency is reduced with the formation of side branches, which results in a lowering of the polymer density. Those polymers that form linear structures may also be branched.

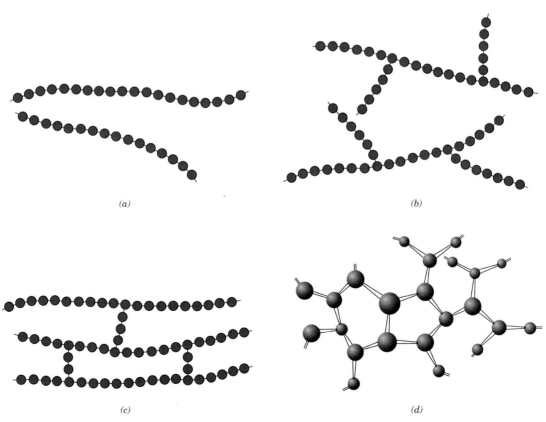

(a)

(b)

(c)

(d)

FIGURE 4.7 Schematic representations of (a) linear, (b) branched, (c) crosslinked, and (d) network (three-dimensional) molecular structures. Circles designate individual mer units.

CROSSLINKED POLYMERS

In **crosslinked polymers,** adjacent linear chains are joined one to another at various positions by covalent bonds, as represented in Figure 4.7c. The process of crosslinking is achieved either during synthesis or by a nonreversible chemical reaction that is usually carried out at an elevated temperature. Often, this crosslinking is accomplished by additive atoms or molecules that are covalently bonded to the chains. Many of the rubber elastic materials are crosslinked; in rubbers, this is called vulcanization, a process described in Section 8.19.

NETWORK POLYMERS

Trifunctional mer units, having three active covalent bonds, form three-dimensional networks (Figure 4.7d) and are termed **network polymers.** Actually, a polymer that is highly crosslinked may be classified as a network polymer. These materials have distinctive mechanical and thermal properties; the epoxies and phenol-formaldehyde belong to this group.

It should be pointed out that polymers are not usually of only one distinctive structural type. For example, a predominantly linear polymer might have some limited branching and crosslinking.

4.8 MOLECULAR CONFIGURATIONS (CD-ROM)

By way of summary of the preceding sections, polymer molecules may be characterized in terms of their size, shape, and structure. Molecular size is specified in terms of molecular weight (or degree of polymerization). Molecular shape relates to the degree of chain twisting, coiling, and bending. Molecular structure depends on the manner in which structural units are joined together. Linear, branched, crosslinked, and network structures are all possible, {in addition to several isomeric configurations (isotactic, syndiotactic, atactic, cis, and trans).} These molecular characteristics are presented in the taxonomic chart, Figure 4.8. It should be noted that some of the structural elements are not mutually exclusive of one another, and, in fact, it may be necessary to specify molecular structure in terms of more than one. For example, a linear polymer may also be isotactic.

4.9 THERMOPLASTIC AND THERMOSETTING POLYMERS

The response of a polymer to mechanical forces at elevated temperatures is related to its dominant molecular structure. And, in fact, one classification scheme for these materials is according to behavior with rising temperature. *Thermoplasts* (or **thermoplastic polymers**) and *thermosets* (or **thermosetting polymers**) are the two subdivisions. Thermoplasts soften when heated (and eventually liquefy) and harden when cooled—processes that are totally reversible and may be repeated. On a molecular level, as the temperature is raised, secondary bonding forces are diminished (by increased molecular motion) so that the relative movement of adjacent chains is facilitated when a stress is applied. Irreversible degradation results when the temperature of a molten thermoplastic polymer is raised to the point at which molecular vibrations become violent enough to break the primary covalent bonds. In addition, thermoplasts are relatively soft. Most linear polymers and those having

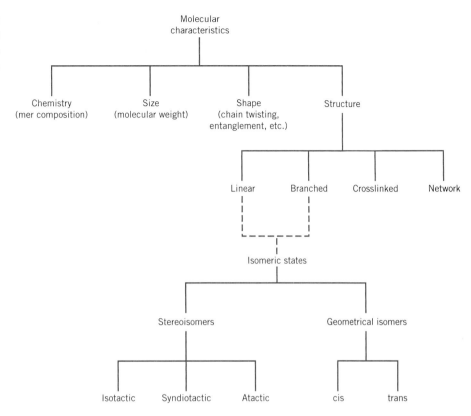

FIGURE 4.8
Classification scheme for the characteristics of polymer molecules.

some branched structures with flexible chains are thermoplastic. These materials are normally fabricated by the simultaneous application of heat and pressure.

Thermosetting polymers become permanently hard when heat is applied and do not soften upon subsequent heating. During the initial heat treatment, covalent crosslinks are formed between adjacent molecular chains; these bonds anchor the chains together to resist the vibrational and rotational chain motions at high temperatures. Crosslinking is usually extensive, in that 10 to 50% of the chain mer units are crosslinked. Only heating to excessive temperatures will cause severance of these crosslink bonds and polymer degradation. Thermoset polymers are generally harder and stronger than thermoplastics, and have better dimensional stability. Most of the crosslinked and network polymers, which include vulcanized rubbers, epoxies, and phenolic and some polyester resins, are thermosetting.

4.10 COPOLYMERS

Polymer chemists and scientists are continually searching for new materials that can be easily and economically synthesized and fabricated, with improved properties or better property combinations than are offered by the homopolymers heretofore discussed. One group of these materials are the copolymers.

Consider a copolymer that is composed of two mer units as represented by ● and ● in Figure 4.9. Depending on the polymerization process and the relative fractions of these mer types, different sequencing arrangements along the polymer chains are possible. For one, as depicted in Figure 4.9a, the two different units are randomly dispersed along the chain in what is termed a **random copolymer.** For an **alternating copolymer,** as the name suggests, the two mer units alternate chain

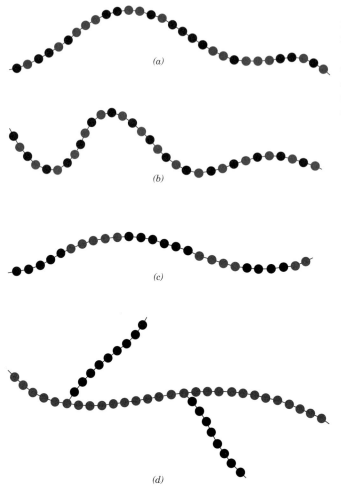

FIGURE 4.9 Schematic representations of (*a*) random, (*b*) alternating, (*c*) block, and (*d*) graft copolymers. The two different mer types are designated by black and colored circles.

positions, as illustrated in Figure 4.9*b*. A **block copolymer** is one in which identical mers are clustered in blocks along the chain (Figure 4.9*c*). And, finally, homopolymer side branches of one type may be grafted to homopolymer main chains that are composed of a different mer; such a material is termed a **graft copolymer** (Figure 4.9*d*).

Synthetic rubbers, discussed in Section 13.13, are often copolymers; chemical repeat units that are employed in some of these rubbers are contained in Table 4.5. Styrene–butadiene rubber (SBR) is a common random copolymer from which automobile tires are made. Nitrile rubber (NBR) is another random copolymer composed of acrylonitrile and butadiene. It is also highly elastic and, in addition, resistant to swelling in organic solvents; gasoline hoses are made of NBR.

4.11 POLYMER CRYSTALLINITY

The crystalline state may exist in polymeric materials. However, since it involves molecules instead of just atoms or ions, as with metals and ceramics, the atomic arrangements will be more complex for polymers. We think of **polymer crystallinity** as the packing of molecular chains so as to produce an ordered atomic array. Crystal structures may be specified in terms of unit cells, which are often quite complex. For example, Figure 4.10 shows the unit cell for polyethylene and its relationship

Table 4.5   Chemical Repeat Units That Are Employed in Copolymer Rubbers

Repeat Unit Name	*Repeat Unit Structure*	*Repeat Unit Name*	*Repeat Unit Structure*
Acrylonitrile	H H | | —C—C— | | H C≡N	*cis*-Isoprene	H CH₃ H H | | | | —C—C=C—C— | | H H
Styrene	H H | | —C—C— | | H ⬡	Isobutylene	H CH₃ | | —C—C— | | H CH₃
Butadiene	H H H H | | | | —C—C=C—C— | | H H	Dimethylsiloxane	CH₃ | —Si—O— | CH₃
Chloroprene	H Cl H H | | | | —C—C=C—C— | | H H		

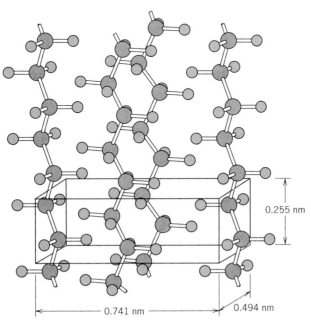

Figure 4.10
Arrangement of molecular chains in a unit cell for polyethylene. (Adapted from C. W. Bunn, *Chemical Crystallography,* Oxford University Press, Oxford, 1945, p. 233.)

0.255 nm

0.741 nm

0.494 nm

● C   ○ H

to the molecular chain structure; this unit cell has orthorhombic geometry (Table 3.6). Of course, the chain molecules also extend beyond the unit cell shown in the figure.

Molecular substances having small molecules (e.g., water and methane) are normally either totally crystalline (as solids) or totally amorphous (as liquids). As a consequence of their size and often complexity, polymer molecules are often only partially crystalline (or semicrystalline), having crystalline regions dispersed within the remaining amorphous material. Any chain disorder or misalignment will result in an amorphous region, a condition that is fairly common, since twisting, kinking, and coiling of the chains prevent the strict ordering of every segment of every chain. Other structural effects are also influential in determining the extent of crystallinity, as discussed below.

The degree of crystallinity may range from completely amorphous to almost entirely (up to about 95%) crystalline; by way of contrast, metal specimens are almost always entirely crystalline, whereas many ceramics are either totally crystalline or totally noncrystalline. Semicrystalline polymers are, in a sense, analogous to two-phase metal alloys, discussed in subsequent chapters.

The density of a crystalline polymer will be greater than an amorphous one of the same material and molecular weight, since the chains are more closely packed together for the crystalline structure. The degree of crystallinity by weight may be determined from accurate density measurements, according to

$$\% \text{ crystallinity} = \frac{\rho_c(\rho_s - \rho_a)}{\rho_s(\rho_c - \rho_a)} \times 100 \tag{4.10}$$

where ρ_s is the density of a specimen for which the percent crystallinity is to be determined, ρ_a is the density of the totally amorphous polymer, and ρ_c is the density of the perfectly crystalline polymer. The values of ρ_a and ρ_c must be measured by other experimental means.

The degree of crystallinity of a polymer depends on the rate of cooling during solidification as well as on the chain configuration. During crystallization upon cooling through the melting temperature, the chains, which are highly random and entangled in the viscous liquid, must assume an ordered configuration. For this to occur, sufficient time must be allowed for the chains to move and align themselves.

The molecular chemistry as well as chain configuration also influence the ability of a polymer to crystallize. Crystallization is not favored in polymers that are composed of chemically complex mer structures (e.g., polyisoprene). On the other hand, crystallization is not easily prevented in chemically simple polymers such as polyethylene and polytetrafluoroethylene, even for very rapid cooling rates.

For linear polymers, crystallization is easily accomplished because there are virtually no restrictions to prevent chain alignment. Any side branches interfere with crystallization, such that branched polymers never are highly crystalline; in fact, excessive branching may prevent any crystallization whatsoever. Most network and crosslinked polymers are almost totally amorphous; a few crosslinked polymers are partially crystalline. {With regard to stereoisomers, atactic polymers are difficult to crystallize; however, isotactic and syndiotactic polymers crystallize much more easily because the regularity of the geometry of the side groups facilitates the process of fitting together adjacent chains.} Also, the bulkier or larger the side-bonded groups of atoms, the less tendency there is for crystallization.

For copolymers, as a general rule, the more irregular and random the mer arrangements, the greater is the tendency for the development of noncrystallinity.

For alternating and block copolymers there is some likelihood of crystallization. On the other hand, random and graft copolymers are normally amorphous.

To some extent, the physical properties of polymeric materials are influenced by the degree of crystallinity. Crystalline polymers are usually stronger and more resistant to dissolution and softening by heat. Some of these properties are discussed in subsequent chapters.

4.12 POLYMER CRYSTALS

We shall now briefly discuss some of the models that have been proposed to describe the spatial arrangement of molecular chains in polymer crystals. One early model, accepted for many years, is the *fringed-micelle* model (Figure 4.11). It was proposed that a semicrystalline polymer consists of small crystalline regions (**crystallites,** or micelles), each having a precise alignment, which are embedded within the amorphous matrix composed of randomly oriented molecules. Thus a single chain molecule might pass through several crystallites as well as the intervening amorphous regions.

More recently, investigations centered on polymer single crystals grown from dilute solutions. These crystals are regularly shaped, thin platelets (or lamellae), approximately 10 to 20 nm thick, and on the order of 10 μm long. Frequently, these platelets will form a multilayered structure, like that shown in the electron micrograph of a single crystal of polyethylene, Figure 4.12. It is theorized that the molecular chains within each platelet fold back and forth on themselves, with folds occurring at the faces; this structure, aptly termed the **chain-folded model,** is illustrated schematically in Figure 4.13. Each platelet will consist of a number of molecules; however, the average chain length will be much greater than the thickness of the platelet.

Many bulk polymers that are crystallized from a melt form **spherulites.** As implied by the name, each spherulite may grow to be spherical in shape; one of them, as found in natural rubber, is shown in the transmission electron micrograph of the chapter-opening photograph for this chapter. The spherulite consists of an aggregate of ribbonlike chain-folded crystallites (lamellae) approximately 10 nm thick that radiate from the center outward. In this electron micrograph, these

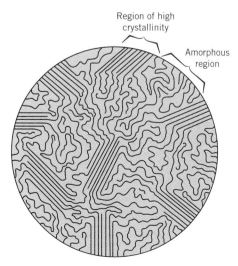

Region of high crystallinity

Amorphous region

FIGURE **4.11** Fringed-micelle model of a semicrystalline polymer, showing both crystalline and amorphous regions. (From H. W. Hayden, W. G. Moffatt, and J. Wulff, *The Structure and Properties of Materials,* Vol. III, *Mechanical Behavior.* Copyright © 1965 by John Wiley & Sons, New York. Reprinted by permission of John Wiley & Sons, Inc.)

FIGURE 4.12
Electron micrograph
of a polyethylene single
crystal. 20,000×. (From
A. Keller, R. H.
Doremus, B. W.
Roberts, and D.
Turnbull, Editors,
*Growth and Perfection
of Crystals.* General
Electric Company and
John Wiley & Sons,
Inc., 1958, p. 498.)

FIGURE 4.13
The chain-folded
structure for a plate-
shaped polymer
crystallite.

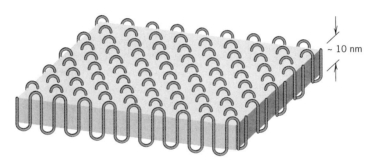

~ 10 nm

FIGURE 4.14
Schematic
representation of the
detailed structure of a
spherulite. (From John
C. Coburn, *Dielectric
Relaxation Processes in
Poly(ethylene
terephthalate)*,
Dissertation, University
of Utah, 1984.)

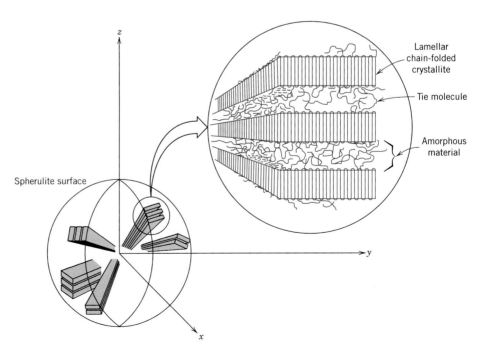

Lamellar
chain-folded
crystallite

Tie molecule

Amorphous
material

Spherulite surface

FIGURE 4.15
A transmission
photomicrograph (using
cross-polarized light)
showing the spherulite
structure of
polyethylene. Linear
boundaries form
between adjacent
spherulites, and within
each spherulite appears
a Maltese cross. 525×.
(Courtesy F. P. Price,
General Electric
Company.)

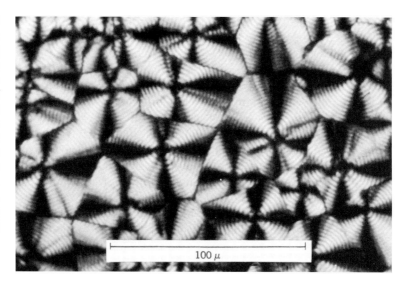

100 μ

lamellae appear as thin white lines. The detailed structure of a spherulite is illustrated schematically in Figure 4.14; shown here are the individual chain-folded lamellar crystals that are separated by amorphous material. Tie-chain molecules that act as connecting links between adjacent lamellae pass through these amorphous regions.

As the crystallization of a spherulitic structure nears completion, the extremities of adjacent spherulites begin to impinge on one another, forming more or less planar boundaries; prior to this time, they maintain their spherical shape. These boundaries are evident in Figure 4.15, which is a photomicrograph of polyethylene using cross-polarized light. A characteristic Maltese cross pattern appears within each spherulite.

Spherulites are considered to be the polymer analogue of grains in polycrystalline metals and ceramics. However, as discussed above, each spherulite is really composed of many different lamellar crystals and, in addition, some amorphous material. Polyethylene, polypropylene, polyvinyl chloride, polytetrafluoroethylene, and nylon form a spherulitic structure when they crystallize from a melt.

SUMMARY

Most polymeric materials are composed of very large molecules—chains of carbon atoms, to which are side-bonded various atoms or radicals. These macromolecules may be thought of as being composed of mers, smaller structural entities, which are repeated along the chain. Mer structures of some of the chemically simple polymers (e.g., polyethylene, polytetrafluoroethylene, polyvinyl chloride, and polypropylene) were presented.

Molecular weights for high polymers may be in excess of a million. Since all molecules are not of the same size, there is a distribution of molecular weights. Molecular weight is often expressed in terms of number and weight averages. Chain length may also be specified by degree of polymerization, the number of mer units per average molecule.

Several molecular characteristics that have an influence on the properties of polymers were discussed. Molecular entanglements occur when the chains assume twisted, coiled, and kinked shapes or contours. With regard to molecular structure, linear, branched, crosslinked, and network structures are possible, {in addition to isotactic, syndiotactic, and atactic stereoisomers, and the cis and trans geometrical isomers.} The copolymers include random, alternating, block, and graft types.

With regard to behavior at elevated temperatures, polymers are classified as either thermoplastic or thermosetting. The former have linear and branched structures; they soften when heated and harden when cooled. In contrast, thermosets, once having hardened, will not soften upon heating; their structures are crosslinked and network.

When the packing of molecular chains is such as to produce an ordered atomic arrangement, the condition of crystallinity is said to exist. In addition to being entirely amorphous, polymers may also exhibit varying degrees of crystallinity; for the latter case, crystalline regions are interdispersed within amorphous areas. Crystallinity is facilitated for polymers that are chemically simple and that have regular and symmetrical chain structures.

Polymer single crystals may be grown from dilute solutions as thin platelets and having chain-folded structures. Many semicrystalline polymers form spherulites; each spherulite consists of a collection of ribbonlike chain-folded lamellar crystallites that radiate outward from its center.

IMPORTANT TERMS AND CONCEPTS

Alternating copolymer
Atactic configuration
Bifunctional mer
Block copolymer
Branched polymer
Chain-folded model
Cis (structure)
Copolymer
Crosslinked polymer
Crystallite
Degree of polymerization
Graft copolymer

Homopolymer
Isomerism
Isotactic configuration
Linear polymer
Macromolecule
Mer
Molecular chemistry
Molecular structure
Molecular weight
Monomer
Network polymer
Polymer

Polymer crystallinity
Random copolymer
Saturated
Spherulite
Stereoisomerism
Syndiotactic configuration
Thermoplastic polymer
Thermosetting polymer
Trans (structure)
Trifunctional mer
Unsaturated

REFERENCES

Baer, E., "Advanced Polymers," *Scientific American,* Vol. 255, No. 4, October 1986, pp. 178–190.

Bovey, F. A. and F. H. Winslow (Editors), *Macromolecules: An Introduction to Polymer Science,* Academic Press, New York, 1979.

Cowie, J. M. G., *Polymers: Chemistry and Physics of Modern Materials,* 2nd edition, Chapman and Hall (USA), New York, 1991.

Engineered Materials Handbook, Vol. 2, *Engineering Plastics,* ASM International, Materials Park, OH, 1988.

McCrum, N. G., C. P. Buckley, and C. B. Bucknall,

Principles of Polymer Engineering, 2nd edition, Oxford University Press, Oxford, 1997. Chapters 0–6.

Rodriguez, F., *Principles of Polymer Systems,* 3rd edition, Hemisphere Publishing Company (Taylor & Francis), New York, 1989.

Rosen, S. L., *Fundamental Principles of Polymeric Materials,* 2nd edition, John Wiley & Sons, New York, 1993.

Rudin, A., *The Elements of Polymer Science and Engineering: An Introductory Text for Engineers and Chemists,* Academic Press, New York, 1982.

Schultz, J., *Polymer Materials Science,* Prentice-Hall, Englewood Cliffs, NJ, 1974.

Seymour, R. B. and C. E. Carraher, Jr., *Polymer Chemistry, An Introduction,* 3rd edition, Marcel Dekker, Inc., New York, 1992.

Sperling, L. H., *Introduction to Physical Polymer Science,* 2nd edition, John Wiley & Sons, New York, 1992.

Young, R. J. and P. Lovell, *Introduction to Polymers,* 2nd edition, Chapman and Hall, London, 1991.

QUESTIONS AND PROBLEMS

Note: To solve those problems having an asterisk (*) by their numbers, consultation of supplementary topics [appearing only on the CD-ROM (and not in print)] will probably be necessary.

4.1 Differentiate between polymorphism and isomerism.

4.2 On the basis of the structures presented in this chapter, sketch mer structures for the following polymers: **(a)** polyvinyl fluoride, **(b)** polychlorotrifluoroethylene, and **(c)** polyvinyl alcohol.

4.3 Compute mer molecular weights for the following: **(a)** polyvinyl chloride, **(b)** polyethylene terephthalate, **(c)** polycarbonate, and **(d)** polydimethylsiloxane.

4.4 The number-average molecular weight of a polypropylene is 1,000,000 g/mol. Compute the number-average degree of polymerization.

4.5 **(a)** Compute the mer molecular weight of polystyrene.

(b) Compute the weight-average molecular weight for a polystyrene for which the weight-average degree of polymerization is 25,000.

4.6 Below, molecular weight data for a polypropylene material are tabulated. Compute **(a)** the number-average molecular weight, **(b)** the weight-average molecular weight, **(c)** the number-average degree of polymerization, and **(d)** the weight-average degree of polymerization.

Molecular Weight Range (g/mol)	x_i	w_i
8,000–16,000	0.05	0.02
16,000–24,000	0.16	0.10
24,000–32,000	0.24	0.20
32,000–40,000	0.28	0.30
40,000–48,000	0.20	0.27
48,000–56,000	0.07	0.11

4.7 Below, molecular weight data for some polymer are tabulated. Compute **(a)** the number-average molecular weight, and **(b)** the weight-average molecular weight. **(c)** If it is known that this material's weight-average degree of polymerization is 780, which one of the polymers listed in Table 4.3 is this polymer? Why? **(d)** What is this material's number-average degree of polymerization?

Molecular Weight Range (g/mol)	x_i	w_i
15,000–30,000	0.04	0.01
30,000–45,000	0.07	0.04
45,000–60,000	0.16	0.11
60,000–75,000	0.26	0.24
75,000–90,000	0.24	0.27
90,000–105,000	0.12	0.16
105,000–120,000	0.08	0.12
120,000–135,000	0.03	0.05

4.8 Is it possible to have a polymethyl methacrylate homopolymer with the following molecular weight data and a weight-average degree of polymerization of 585? Why or why not?

Molecular Weight Range (g/mol)	x_i	w_i
8,000–20,000	0.04	0.01
20,000–32,000	0.10	0.05
32,000–44,000	0.16	0.12
44,000–56,000	0.26	0.25
56,000–68,000	0.23	0.27
68,000–80,000	0.15	0.21
80,000–92,000	0.06	0.09

4.9 High-density polyethylene may be chlorinated by inducing the random substitution of chlorine atoms for hydrogen.

(a) Determine the concentration of Cl (in wt%) that must be added if this substitution occurs for 5% of all the original hydrogen atoms.

(b) In what ways does this chlorinated poly-ethylene differ from polyvinyl chloride?

4.10 What is the difference between *configuration* and *conformation* in relation to polymer chains?

4.11 For a linear polymer molecule, the total chain length L depends on the bond length between chain atoms d, the total number of bonds in the molecule N, and the angle between adjacent backbone chain atoms θ, as follows:

$$L = Nd \sin\left(\frac{\theta}{2}\right) \qquad (4.11)$$

Furthermore, the average end-to-end distance for a series of polymer molecules r in Figure 4.6 is equal to

$$r = d\sqrt{N} \qquad (4.12)$$

A linear polytetrafluoroethylene has a num-ber-average molecular weight of 500,000 g/mol; compute average values of L and r for this material.

4.12 Using the definitions for total chain molecule length L (Equation 4.11) and average chain end-to-end distance r (Equation 4.12), for a linear polyethylene determine **(a)** the num-ber-average molecular weight for $L = 2500$ nm; and **(b)** the number-average molecular weight for $r = 20$ nm.

4.13 Make comparisons of thermoplastic and thermosetting polymers **(a)** on the basis of mechanical characteristics upon heating, and **(b)** according to possible molecular struc-tures.

4.14 Some of the polyesters may be either ther-moplastic or thermosetting. Suggest one rea-son for this.

4.15 **(a)** Is it possible to grind up and reuse phe-nol-formaldehyde? Why or why not?

(b) Is it possible to grind up and reuse poly-propylene? Why or why not?

4.16* Sketch portions of a linear polystyrene mole-cule that are **(a)** syndiotactic, **(b)** atactic, and **(c)** isotactic.

4.17* Sketch cis and trans mer structures for **(a)** butadiene, and **(b)** chloroprene.

4.18 Sketch the mer structure for each of the following alternating copolymers: **(a)** poly (butadiene-chloroprene), **(b)** poly(styrene-methyl methacrylate), and **(c)** poly(acryloni-trile-vinyl chloride).

4.19 The number-average molecular weight of a poly(styrene-butadiene) alternating copoly-mer is 1,350,000 g/mol; determine the aver-age number of styrene and butadiene mer units per molecule.

4.20 Calculate the number-average molecular weight of a random nitrile rubber [poly(acry-lonitrile-butadiene) copolymer] in which the fraction of butadiene mers is 0.30; assume that this concentration corresponds to a number-average degree of polymerization of 2000.

4.21 An alternating copolymer is known to have a number-average molecular weight of 250,000 g/mol and a number-average degree of polymerization of 3420. If one of the mers is styrene, which of ethylene, propylene, tet-rafluoroethylene, and vinyl chloride is the other mer? Why?

4.22 **(a)** Determine the ratio of butadiene to sty-rene mers in a copolymer having a weight-average molecular weight of 350,000 g/mol and weight-average degree of polymeriza-tion of 4425.

(b) Which type(s) of copolymer(s) will this copolymer be, considering the following pos-sibilities: random, alternating, graft, and block? Why?

4.23 Crosslinked copolymers consisting of 60 wt% ethylene and 40 wt% propylene may have elastic properties similar to those for natural rubber. For a copolymer of this composition, determine the fraction of both mer types.

4.24 A random poly(isobutylene-isoprene) co-polymer has a weight-average molecular weight of 200,000 g/mol and a weight-aver-age degree of polymerization of 3000. Com-pute the fraction of isobutylene and isoprene mers in this copolymer.

4.25 **(a)** Compare the crystalline state in metals and polymers.

(b) Compare the noncrystalline state as it applies to polymers and ceramic glasses.

4.26 Explain briefly why the tendency of a polymer to crystallize decreases with increasing molecular weight.

4.27* For each of the following pairs of polymers, do the following: (1) state whether or not it is possible to determine if one polymer is more likely to crystallize than the other; (2) if it is possible, note which is the more likely and then cite reason(s) for your choice; and (3) if it is not possible to decide, then state why.

(a) Linear and syndiotactic polyvinyl chloride; linear and isotactic polystyrene.

(b) Network phenol-formaldehyde; linear and heavily crosslinked *cis*-isoprene.

(c) Linear polyethylene; lightly branched isotactic polypropylene.

(d) Alternating poly(styrene-ethylene) copolymer; random poly(vinyl chloride-tetra-fluoroethylene) copolymer.

4.28 Compute the density of totally crystalline polyethylene. The orthorhombic unit cell for polyethylene is shown in Figure 4.10; also, the equivalent of two ethylene mer units is contained within each unit cell.

4.29 The density of totally crystalline polypropylene at room temperature is 0.946 g/cm^3. Also, at room temperature the unit cell for this material is monoclinic with lattice parameters

$a = 0.666$ nm $\quad \alpha = 90°$
$b = 2.078$ nm $\quad \beta = 99.62°$
$c = 0.650$ nm $\quad \gamma = 90°$

If the volume of a monoclinic unit cell, V_{mono}, is a function of these lattice parameters as

$$V_{mono} = abc \sin \beta$$

determine the number of mer units per unit cell.

4.30 The density and associated percent crystallinity for two polytetrafluoroethylene materials are as follows:

ρ (g/cm^3)	Crystallinity (%)
2.144	51.3
2.215	74.2

(a) Compute the densities of totally crystalline and totally amorphous polytetrafluoroethylene.

(b) Determine the percent crystallinity of a specimen having a density of 2.26 g/cm^3.

4.31 The density and associated percent crystallinity for two nylon 6,6 materials are as follows:

ρ (g/cm^3)	Crystallinity (%)
1.188	67.3
1.152	43.7

(a) Compute the densities of totally crystalline and totally amorphous nylon 6,6.

(b) Determine the density of a specimen having 55.4% crystallinity.

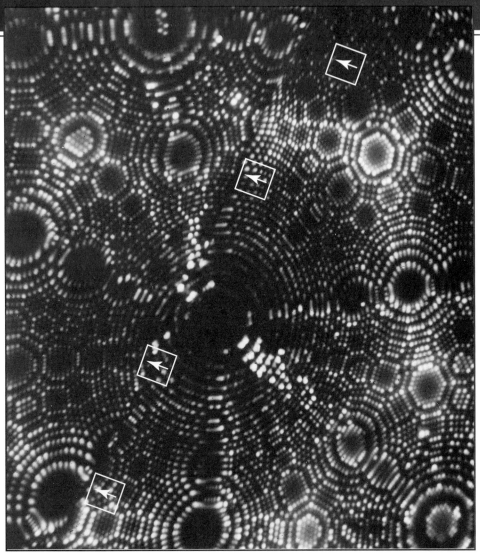

A field ion micrograph taken at the tip of a pointed tungsten specimen. Field ion microscopy is a sophisticated and fascinating technique that permits observation of individual atoms in a solid, which are represented by white spots. The symmetry and regularity of the atom arrangements are evident from the positions of the spots in this micrograph. A disruption of this symmetry occurs along a grain boundary, which is traced by the arrows. Approximately 3,460,000×. (Photomicrograph courtesy of J. J. Hren and R. W. Newman.)

Why Study *Imperfections in Solids?*

The properties of some materials are profoundly influenced by the presence of imperfections. Consequently, it is important to have a knowledge about the types of imperfections that exist, and the roles they play in affecting the behavior of materials. For example, the mechanical properties of pure metals experience significant alterations when alloyed (i.e., when impurity atoms are added)—e.g., sterling silver (92.5% silver-7.5% copper) is much harder and stronger than pure silver (Section 8.10).

Also, integrated circuit microelectronic devices found in our computers, calculators, and home appliances function because of highly controlled concentrations of specific impurities that are incorporated into small, localized regions of semiconducting materials (Sections 12.11 {and 12.14}).

5.1 INTRODUCTION

For a crystalline solid we have tacitly assumed that perfect order exists throughout the material on an atomic scale. However, such an idealized solid does not exist; all contain large numbers of various defects or imperfections. As a matter of fact, many of the properties of materials are profoundly sensitive to deviations from crystalline perfection; the influence is not always adverse, and often specific characteristics are deliberately fashioned by the introduction of controlled amounts or numbers of particular defects, as detailed in succeeding chapters.

By "crystalline defect" is meant a lattice irregularity having one or more of its dimensions on the order of an atomic diameter. Classification of crystalline imperfections is frequently made according to geometry or dimensionality of the defect. Several different imperfections are discussed in this chapter, including point defects (those associated with one or two atomic positions), linear (or one-dimensional) defects, as well as interfacial defects, or boundaries, which are two-dimensional. Impurities in solids are also discussed, since impurity atoms may exist as point defects. {Finally, techniques for the microscopic examination of defects and the structure of materials are briefly described.}

POINT DEFECTS

5.2 POINT DEFECTS IN METALS

The simplest of the point defects is a **vacancy,** or vacant lattice site, one normally occupied from which an atom is missing (Figure 5.1). All crystalline solids contain vacancies and, in fact, it is not possible to create such a material that is free of these defects. The necessity of the existence of vacancies is explained using principles of thermodynamics; in essence, the presence of vacancies increases the entropy (i.e., the randomness) of the crystal.

The equilibrium number of vacancies N_v for a given quantity of material depends on and increases with temperature according to

$$N_v = N \exp\left(-\frac{Q_v}{kT}\right)$$

(5.1)

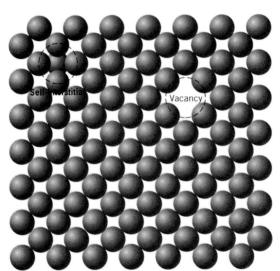

FIGURE 5.1 Two-dimensional representations of a vacancy and a self-interstitial. (Adapted from W. G. Moffatt, G. W. Pearsall, and J. Wulff, *The Structure and Properties of Materials*, Vol. I, *Structure*, p. 77. Copyright © 1964 by John Wiley & Sons, New York. Reprinted by permission of John Wiley & Sons, Inc.)

In this expression, N is the total number of atomic sites, Q_v is the energy required for the formation of a vacancy, T is the absolute temperature[1] in kelvins, and k is the gas or **Boltzmann's constant.** The value of k is 1.38×10^{-23} J/atom-K, or 8.62×10^{-5} eV/atom-K, depending on the units of Q_v.[2] Thus, the number of vacancies increases exponentially with temperature; that is, as T in Equation 5.1 increases, so does also the expression $\exp -(Q_v/kT)$. For most metals, the fraction of vacancies N_v/N just below the melting temperature is on the order of 10^{-4}; that is, one lattice site out of 10,000 will be empty. As ensuing discussions indicate, a number of other material parameters have an exponential dependence on temperature similar to that of Equation 5.1.

A **self-interstitial** is an atom from the crystal that is crowded into an interstitial site, a small void space that under ordinary circumstances is not occupied. This kind of defect is also represented in Figure 5.1. In metals, a self-interstitial introduces relatively large distortions in the surrounding lattice because the atom is substantially larger than the interstitial position in which it is situated. Consequently, the formation of this defect is not highly probable, and it exists in very small concentrations, which are significantly lower than for vacancies.

EXAMPLE PROBLEM 5.1

Calculate the equilibrium number of vacancies per cubic meter for copper at 1000°C. The energy for vacancy formation is 0.9 eV/atom; the atomic weight and density (at 1000°C) for copper are 63.5 g/mol and 8.40 g/cm³, respectively.

SOLUTION

This problem may be solved by using Equation 5.1; it is first necessary, however, to determine the value of N, the number of atomic sites per cubic meter for

---

[1] Absolute temperature in kelvins (K) is equal to °C + 273.

[2] Boltzmann's constant per mole of atoms becomes the gas constant R; in such a case $R = 8.31$ J/mol-K, or 1.987 cal/mol-K.

copper, from its atomic weight A_{Cu}, its density ρ, and Avogadro's number N_A, according to

$$N = \frac{N_A \rho}{A_{Cu}} \tag{5.2}$$

$$= \frac{(6.023 \times 10^{23} \text{ atoms/mol})(8.40 \text{ g/cm}^3)(10^6 \text{ cm}^3/\text{m}^3)}{63.5 \text{ g/mol}}$$

$$= 8.0 \times 10^{28} \text{ atoms/m}^3$$

Thus, the number of vacancies at 1000°C (1273 K) is equal to

$$N_v = N \exp\left(-\frac{Q_v}{kT}\right)$$

$$= (8.0 \times 10^{28} \text{ atoms/m}^3) \exp\left[-\frac{(0.9 \text{ eV})}{(8.62 \times 10^{-5} \text{ eV/K})(1273 \text{ K})}\right]$$

$$= 2.2 \times 10^{25} \text{ vacancies/m}^3$$

5.3 POINT DEFECTS IN CERAMICS

Point defects also may exist in ceramic compounds. As with metals, both vacancies and interstitials are possible; however, since ceramic materials contain ions of at least two kinds, defects for each ion type may occur. For example, in NaCl, Na interstitials and vacancies and Cl interstitials and vacancies may exist. It is highly improbable that there would be appreciable concentrations of anion (Cl^-) interstitials. The anion is relatively large, and to fit into a small interstitial position, substantial strains on the surrounding ions must be introduced. Anion and cation vacancies and a cation interstitial are represented in Figure 5.2.

The expression **defect structure** is often used to designate the types and concentrations of atomic defects in ceramics. Because the atoms exist as charged ions, when defect structures are considered, conditions of electroneutrality must be maintained.

FIGURE 5.2
Schematic representations of cation and anion vacancies and a cation interstitial. (From W. G. Moffatt, G. W. Pearsall, and J. Wulff, *The Structure and Properties of Materials,* Vol. 1, *Structure,* p. 78. Copyright © 1964 by John Wiley & Sons, New York. Reprinted by permission of John Wiley & Sons, Inc.)

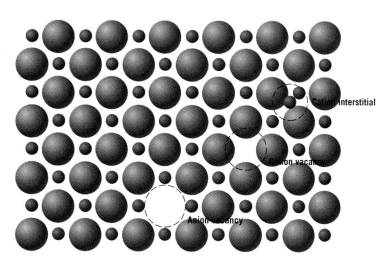

FIGURE 5.3
Schematic diagram
showing Frenkel and
Schottky defects in
ionic solids. (From W.
G. Moffatt, G. W.
Pearsall, and J. Wulff,
*The Structure and
Properties of Materials,*
Vol. 1, *Structure,* p. 78.
Copyright © 1964 by
John Wiley & Sons,
New York. Reprinted
by permission of John
Wiley & Sons, Inc.)

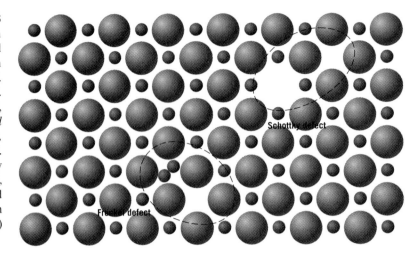

Electroneutrality is the state that exists when there are equal numbers of positive and negative charges from the ions. As a consequence, defects in ceramics do not occur alone. One such type of defect involves a cation–vacancy and a cation–interstitial pair. This is called a **Frenkel defect** (Figure 5.3). It might be thought of as being formed by a cation leaving its normal position and moving into an interstitial site. There is no change in charge because the cation maintains the same positive charge as an interstitial.

Another type of defect found in AX materials is a cation vacancy–anion vacancy pair known as a **Schottky defect,** also schematically diagrammed in Figure 5.3. This defect might be thought of as being created by removing one cation and one anion from the interior of the crystal and then placing them both at an external surface. Since both cations and anions have the same charge, and since for every anion vacancy there exists a cation vacancy, the charge neutrality of the crystal is maintained.

The ratio of cations to anions is not altered by the formation of either a Frenkel or a Schottky defect. If no other defects are present, the material is said to be stoichiometric. **Stoichiometry** may be defined as a state for ionic compounds wherein there is the exact ratio of cations to anions as predicted by the chemical formula. For example, NaCl is stoichiometric if the ratio of Na^+ ions to Cl^- ions is exactly $1:1$. A ceramic compound is *nonstoichiometric* if there is any deviation from this exact ratio.

Nonstoichiometry may occur for some ceramic materials in which two valence (or ionic) states exist for one of the ion types. Iron oxide (wüstite, FeO) is one such material, for the iron can be present in both Fe^{2+} and Fe^{3+} states; the number of each of these ion types depends on temperature and the ambient oxygen pressure. The formation of an Fe^{3+} ion disrupts the electroneutrality of the crystal by introducing an excess $+1$ charge, which must be offset by some type of defect. This may be accomplished by the formation of one Fe^{2+} vacancy (or the removal of two positive charges) for every two Fe^{3+} ions that are formed (Figure 5.4). The crystal is no longer stoichiometric because there is one more O ion than Fe ion; however, the crystal remains electrically neutral. This phenomenon is fairly common in iron oxide, and, in fact, its chemical formula is often written as $Fe_{1-x}O$ (where x is some

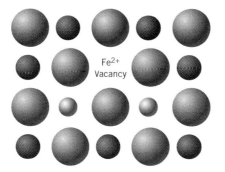 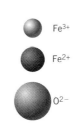

FIGURE 5.4 Schematic representation of an Fe^{2+} vacancy in FeO that results from the formation of two Fe^{3+} ions.

small and variable fraction substantially less than unity) to indicate a condition of nonstoichiometry with a deficiency of Fe.

5.4 IMPURITIES IN SOLIDS

IMPURITIES IN METALS

A pure metal consisting of only one type of atom just isn't possible; impurity or foreign atoms will always be present, and some will exist as crystalline point defects. In fact, even with relatively sophisticated techniques, it is difficult to refine metals to a purity in excess of 99.9999%. At this level, on the order of 10^{22} to 10^{23} impurity atoms will be present in one cubic meter of material. Most familiar metals are not highly pure; rather, they are **alloys,** in which impurity atoms have been added intentionally to impart specific characteristics to the material. Ordinarily alloying is used in metals to improve mechanical strength and corrosion resistance. For example, sterling silver is a 92.5% silver–7.5% copper alloy. In normal ambient environments, pure silver is highly corrosion resistant, but also very soft. Alloying with copper enhances the mechanical strength significantly, without depreciating the corrosion resistance appreciably.

The addition of impurity atoms to a metal will result in the formation of a **solid solution** and/or a new *second phase*, depending on the kinds of impurity, their concentrations, and the temperature of the alloy. The present discussion is concerned with the notion of a solid solution; treatment of the formation of a new phase is deferred to Chapter 10.

Several terms relating to impurities and solid solutions deserve mention. With regard to alloys, **solute** and **solvent** are terms that are commonly employed. "Solvent" represents the element or compound that is present in the greatest amount; on occasion, solvent atoms are also called *host atoms.* "Solute" is used to denote an element or compound present in a minor concentration.

SOLID SOLUTIONS

A solid solution forms when, as the solute atoms are added to the host material, the crystal structure is maintained, and no new structures are formed. Perhaps it is useful to draw an analogy with a liquid solution. If two liquids, soluble in each other (such as water and alcohol) are combined, a liquid solution is produced as the molecules intermix, and its composition is homogeneous throughout. A solid solution is also compositionally homogeneous; the impurity atoms are randomly and uniformly dispersed within the solid.

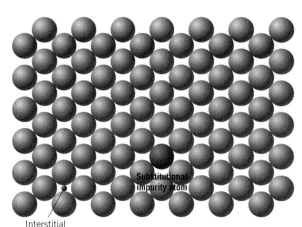

FIGURE 5.5 Two-dimensional schematic representations of substitutional and interstitial impurity atoms. (Adapted from W. G. Moffatt, G. W. Pearsall, and J. Wulff, *The Structure and Properties of Materials,* Vol. I, *Structure,* p. 77. Copyright © 1964 by John Wiley & Sons, New York. Reprinted by permission of John Wiley & Sons, Inc.)

Substitutional
impurity atom

Interstitial
impurity atom

Impurity point defects are found in solid solutions, of which there are two types: **substitutional** and **interstitial.** For substitutional, solute or impurity atoms replace or substitute for the host atoms (Figure 5.5). There are several features of the solute and solvent atoms that determine the degree to which the former dissolves in the latter; these are as follows:

1. *Atomic size factor.* Appreciable quantities of a solute may be accommodated in this type of solid solution only when the difference in atomic radii between the two atom types is less than about ±15%. Otherwise the solute atoms will create substantial lattice distortions and a new phase will form.

2. *Crystal structure.* For appreciable solid solubility the crystal structures for metals of both atom types must be the same.

3. *Electronegativity.* The more electropositive one element and the more electronegative the other, the greater is the likelihood that they will form an intermetallic compound instead of a substitutional solid solution.

4. *Valences.* Other factors being equal, a metal will have more of a tendency to dissolve another metal of higher valency than one of a lower valency.

An example of a substitutional solid solution is found for copper and nickel. These two elements are completely soluble in one another at all proportions. With regard to the aforementioned rules that govern degree of solubility, the atomic radii for copper and nickel are 0.128 and 0.125 nm, respectively, both have the FCC crystal structure, and their electronegativities are 1.9 and 1.8 (Figure 2.7); finally, the most common valences are +1 for copper (although it sometimes can be +2) and +2 for nickel.

For interstitial solid solutions, impurity atoms fill the voids or interstices among the host atoms (see Figure 5.5). For metallic materials that have relatively high atomic packing factors, these interstitial positions are relatively small. Consequently, the atomic diameter of an interstitial impurity must be substantially smaller than that of the host atoms. Normally, the maximum allowable concentration of interstitial impurity atoms is low (less than 10%). Even very small impurity atoms are ordinarily larger than the interstitial sites, and as a consequence they introduce some lattice strains on the adjacent host atoms. Problem 5.9 calls for determination of the radii of impurity atoms (in terms of R, the host atom radius) that will just fit into interstitial positions without introducing any lattice strains for both FCC and BCC crystal structures.

Carbon forms an interstitial solid solution when added to iron; the maximum concentration of carbon is about 2%. The atomic radius of the carbon atom is much less than that for iron: 0.071 nm versus 0.124 nm.

IMPURITIES IN CERAMICS

Impurity atoms can form solid solutions in ceramic materials much as they do in metals. Solid solutions of both substitutional and interstitial types are possible. For an interstitial, the ionic radius of the impurity must be relatively small in comparison to the anion. Since there are both anions and cations, a substitutional impurity will substitute for the host ion to which it is most similar in an electrical sense: if the impurity atom normally forms a cation in a ceramic material, it most probably will substitute for a host cation. For example, in sodium chloride, impurity Ca^{2+} and O^{2-} ions would most likely substitute for Na^+ and Cl^- ions, respectively. Schematic representations for cation and anion substitutional as well as interstitial impurities are shown in Figure 5.6. To achieve any appreciable solid solubility of substituting impurity atoms, the ionic size and charge must be very nearly the same as those of one of the host ions. For an impurity ion having a charge different from the host ion for which it substitutes, the crystal must compensate for this difference in charge so that electroneutrality is maintained with the solid. One way this is accomplished is by the formation of lattice defects—vacancies or interstitials of both ion types, as discussed above.

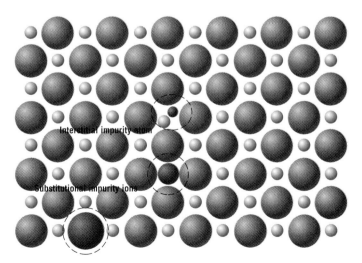

FIGURE 5.6 Schematic representations of interstitial, anion-substitutional, and cation-substitutional impurity atoms in an ionic compound. (Adapted from W. G. Moffatt, G. W. Pearsall, and J. Wulff, *The Structure and Properties of Materials,* Vol. 1, *Structure,* p. 78. Copyright © 1964 by John Wiley & Sons, New York. Reprinted by permission of John Wiley & Sons, Inc.)

EXAMPLE PROBLEM 5.2

If electroneutrality is to be preserved, what point defects are possible in NaCl when a Ca^{2+} substitutes for an Na^+ ion? How many of these defects exist for every Ca^{2+} ion?

SOLUTION

Replacement of an Na^+ by a Ca^{2+} ion introduces one extra positive charge. Electroneutrality is maintained when either a single positive charge is eliminated or another single negative charge is added. Removal of a positive charge is accomplished by the formation of one Na^+ vacancy. Alternatively, a Cl^- interstitial will supply an additional negative charge, negating the effect of each Ca^{2+} ion. However, as mentioned above, the formation of this defect is highly unlikely.

5.5 POINT DEFECTS IN POLYMERS

It should be noted that the defect concept is different in polymers (than in metals and ceramics) as a consequence of the chainlike macromolecules and the nature of the crystalline state for polymers. Point defects similar to those found in metals have been observed in crystalline regions of polymeric materials; these include vacancies and interstitial atoms and ions. Chain ends are considered to be defects inasmuch as they are chemically dissimilar to normal chain units; vacancies are also associated with the chain ends. Impurity atoms/ions or groups of atoms/ions may be incorporated in the molecular structure as interstitials; they may also be associated with main chains or as short side branches.

5.6 SPECIFICATION OF COMPOSITION

It is often necessary to express the **composition** (or *concentration*)[3] of an alloy in terms of its constituent elements. The two most common ways to specify composition are weight (or mass) percent and atom percent. The basis for **weight percent** (wt%) is the weight of a particular element relative to the total alloy weight. For an alloy that contains two hypothetical atoms denoted by 1 and 2, the concentration of 1 in wt%, C_1, is defined as

$$C_1 = \frac{m_1}{m_1 + m_2} \times 100 \tag{5.3}$$

where m_1 and m_2 represent the weight (or mass) of elements 1 and 2, respectively. The concentration of 2 would be computed in an analogous manner.

The basis for **atom percent** (at%) calculations is the number of moles of an element in relation to the total moles of the elements in the alloy. The number of moles in some specified mass of a hypothetical element 1, n_{m1}, may be computed

---

[3] The terms *composition* and *concentration* will be assumed to have the same meaning in this book (i.e., the relative content of a specific element or constituent in an alloy) and will be used interchangeably.

as follows:

$$n_{m1} = \frac{m_1'}{A_1} \tag{5.4}$$

Here, m_1' and A_1 denote the mass (in grams) and atomic weight, respectively, for element 1.

Concentration in terms of atom percent of element 1 in an alloy containing 1 and 2 atoms, C_1', is defined by[4]

$$C_1' = \frac{n_{m1}}{n_{m1} + n_{m2}} \times 100 \tag{5.5}$$

In like manner, the atom percent of 2 may be determined.

Atom percent computations also can be carried out on the basis of the number of atoms instead of moles, since one mole of all substances contains the same number of atoms.

COMPOSITION CONVERSIONS (CD-ROM)

MISCELLANEOUS IMPERFECTIONS

5.7 DISLOCATIONS—LINEAR DEFECTS

 A *dislocation* is a linear or one-dimensional defect around which some of the atoms are misaligned. One type of dislocation is represented in Figure 5.7: an extra portion

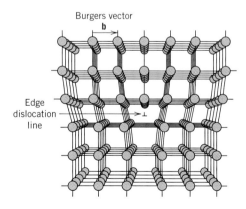

Burgers vector

Edge dislocation line

FIGURE 5.7 The atom positions around an edge dislocation; extra half-plane of atoms shown in perspective. (Adapted from A. G. Guy, *Essentials of Materials Science,* McGraw-Hill Book Company, New York, 1976, p. 153.)

---

[4] In order to avoid confusion in notations and symbols that are being used in this section, it should be pointed out that the prime (as in C_1' and m_1') is used to designate both composition, in atom percent, as well as mass of material in units of grams.

of a plane of atoms, or half-plane, the edge of which terminates within the crystal. This is termed an **edge dislocation;** it is a linear defect that centers around the line that is defined along the end of the extra half-plane of atoms. This is sometimes termed the **dislocation line,** which, for the edge dislocation in Figure 5.7, is perpendicular to the plane of the page. Within the region around the dislocation line there is some localized lattice distortion. The atoms above the dislocation line in Figure 5.7 are squeezed together, and those below are pulled apart; this is reflected in the slight curvature for the vertical planes of atoms as they bend around this extra half-plane. The magnitude of this distortion decreases with distance away from the dislocation line; at positions far removed, the crystal lattice is virtually perfect. Sometimes the edge dislocation in Figure 5.7 is represented by the symbol ⊥, which also indicates the position of the dislocation line. An edge dislocation may also be formed by an extra half-plane of atoms that is included in the bottom portion of the crystal; its designation is a ⊤.

Another type of dislocation, called a **screw dislocation,** exists, which may be thought of as being formed by a shear stress that is applied to produce the distortion shown in Figure 5.8a: the upper front region of the crystal is shifted one atomic distance to the right relative to the bottom portion. The atomic distortion associated with a screw dislocation is also linear and along a dislocation line, line AB in Figure 5.8b. The screw dislocation derives its name from the spiral or helical path or ramp that is traced around the dislocation line by the atomic planes of atoms. Sometimes the symbol ↻ is used to designate a screw dislocation.

Most dislocations found in crystalline materials are probably neither pure edge nor pure screw, but exhibit components of both types; these are termed **mixed dislocations.** All three dislocation types are represented schematically in Figure 5.9; the lattice distortion that is produced away from the two faces is mixed, having varying degrees of screw and edge character.

The magnitude and direction of the lattice distortion associated with a dislocation is expressed in terms of a **Burgers vector,** denoted by a **b**. Burgers vectors are indicated in Figures 5.7 and 5.8 for edge and screw dislocations, respectively. Furthermore, the nature of a dislocation (i.e., edge, screw, or mixed) is defined by the relative orientations of dislocation line and Burgers vector. For an edge, they are perpendicular (Figure 5.7), whereas for a screw, they are parallel (Figure 5.8); they are neither perpendicular nor parallel for a mixed dislocation. Also, even though a dislocation changes direction and nature within a crystal (e.g., from edge to mixed to screw), the Burgers vector will be the same at all points along its line. For example, all positions of the curved dislocation in Figure 5.9 will have the Burgers vector shown. For metallic materials, the Burgers vector for a dislocation will point in a close-packed crystallographic direction and will be of magnitude equal to the interatomic spacing.

Dislocations can be observed in crystalline materials using electron-microscopic techniques. In Figure 5.10, a high-magnification transmission electron micrograph, the dark lines are the dislocations.

Virtually all crystalline materials contain some dislocations that were introduced during solidification, during plastic deformation, and as a consequence of thermal stresses that result from rapid cooling. Dislocations are involved in the plastic deformation of these materials, as discussed in Chapter 8. Dislocations have been observed in polymeric materials; however, some controversy exists as to the nature of dislocation structures in polymers and the mechanism(s) by which polymers plastically deform.

FIGURE 5.8 (a) A screw dislocation within a crystal. (b) The screw dislocation in (a) as viewed from above. The dislocation line extends along line AB. Atom positions above the slip plane are designated by open circles, those below by solid circles. (Figure (b) from W. T. Read, Jr., *Dislocations in Crystals*, McGraw-Hill Book Company, New York, 1953.)

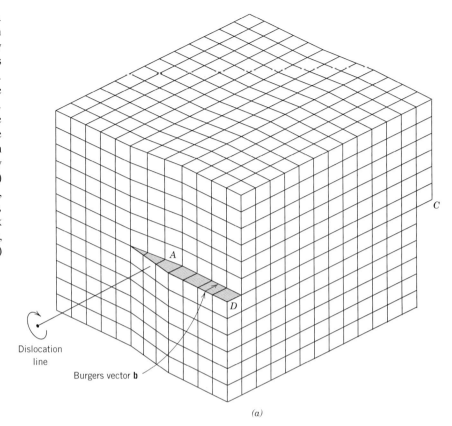

Dislocation line

Burgers vector **b**

(a)

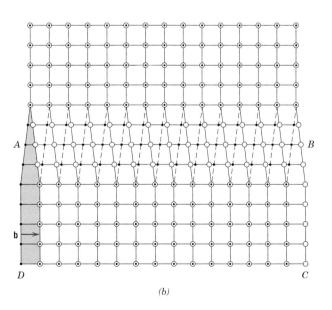

(b)

FIGURE 5.9 (a)
Schematic
representation of a
dislocation that has
edge, screw, and mixed
character. (b) Top view,
where open circles
denote atom positions
above the slip plane.
Solid circles, atom
positions below. At
point A, the dislocation
is pure screw, while at
point B, it is pure edge.
For regions in between
where there is
curvature in the
dislocation line, the
character is mixed edge
and screw. (Figure (b)
from W. T. Read, Jr.,
Dislocations in Crystals,
McGraw-Hill Book
Company, New York,
1953.)

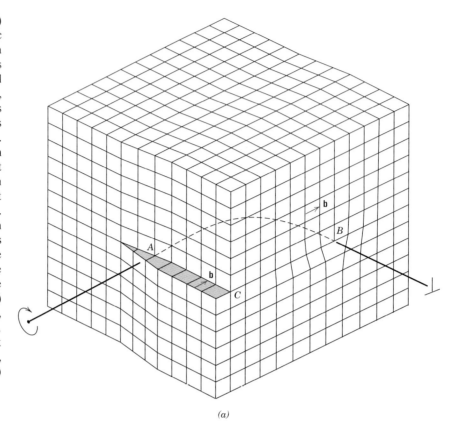

(a)

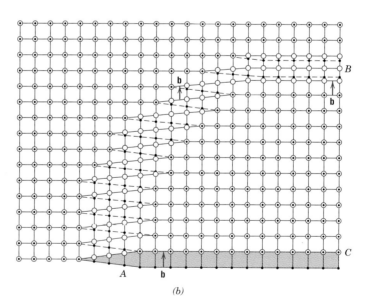

(b)

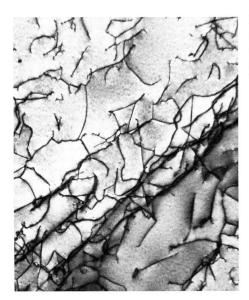

FIGURE 5.10   A transmission electron micrograph of a titanium alloy in which the dark lines are dislocations. 51,450×. (Courtesy of M. R. Plichta, Michigan Technological University.)

5.8 INTERFACIAL DEFECTS

Interfacial defects are boundaries that have two dimensions and normally separate regions of the materials that have different crystal structures and/or crystallographic orientations. These imperfections include external surfaces, grain boundaries, twin boundaries, stacking faults, and phase boundaries.

EXTERNAL SURFACES

One of the most obvious boundaries is the external surface, along which the crystal structure terminates. Surface atoms are not bonded to the maximum number of nearest neighbors, and are therefore in a higher energy state than the atoms at interior positions. The bonds of these surface atoms that are not satisfied give rise to a surface energy, expressed in units of energy per unit area (J/m^2 or erg/cm^2). To reduce this energy, materials tend to minimize, if at all possible, the total surface area. For example, liquids assume a shape having a minimum area—the droplets become spherical. Of course, this is not possible with solids, which are mechanically rigid.

GRAIN BOUNDARIES

Another interfacial defect, the grain boundary, was introduced in Section 3.17 as the boundary separating two small grains or crystals having different crystallographic orientations in polycrystalline materials. A grain boundary is represented schematically from an atomic perspective in Figure 5.11. Within the boundary region, which is probably just several atom distances wide, there is some atomic mismatch in a transition from the crystalline orientation of one grain to that of an adjacent one.

Various degrees of crystallographic misalignment between adjacent grains are possible (Figure 5.11). When this orientation mismatch is slight, on the order of a few degrees, then the term *small-* (or *low-*) *angle grain boundary* is used. These boundaries can be described in terms of dislocation arrays. One simple small-angle grain boundary is formed when edge dislocations are aligned in the manner of Figure 5.12. This type is called a *tilt boundary;* the angle of misorientation, θ, is

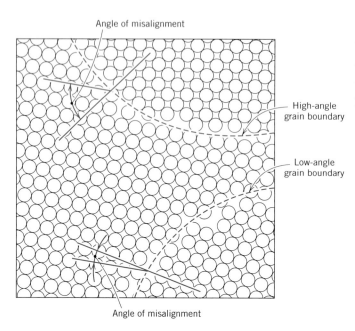

Angle of misalignment

High-angle
grain boundary

Low-angle
grain boundary

Angle of misalignment

FIGURE 5.11 Schematic diagram showing low- and high-angle grain boundaries and the adjacent atom positions.

also indicated in the figure. When the angle of misorientation is parallel to the boundary, a *twist boundary* results, which can be described by an array of screw dislocations.

The atoms are bonded less regularly along a grain boundary (e.g., bond angles are longer), and consequently, there is an interfacial or grain boundary energy

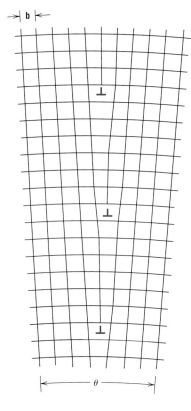

b

⊥

⊥

⊥

θ

FIGURE 5.12 Demonstration of how a tilt boundary having an angle of misorientation θ results from an alignment of edge dislocations.

similar to the surface energy described above. The magnitude of this energy is a function of the degree of misorientation, being larger for high-angle boundaries. Grain boundaries are more chemically reactive than the grains themselves as a consequence of this boundary energy. Furthermore, impurity atoms often preferentially segregate along these boundaries because of their higher energy state. The total interfacial energy is lower in large or coarse-grained materials than in fine-grained ones, since there is less total boundary area in the former. Grains grow at elevated temperatures to reduce the total boundary energy, a phenomenon explained in Section 8.14.

In spite of this disordered arrangement of atoms and lack of regular bonding along grain boundaries, a polycrystalline material is still very strong; cohesive forces within and across the boundary are present. Furthermore, the density of a polycrystalline specimen is virtually identical to that of a single crystal of the same material.

TWIN BOUNDARIES

A *twin boundary* is a special type of grain boundary across which there is a specific mirror lattice symmetry; that is, atoms on one side of the boundary are located in mirror image positions of the atoms on the other side (Figure 5.13). The region of material between these boundaries is appropriately termed a *twin*. Twins result from atomic displacements that are produced from applied mechanical shear forces (mechanical twins), and also during annealing heat treatments following deformation (annealing twins). Twinning occurs on a definite crystallographic plane and in a specific direction, both of which depend on the crystal structure. Annealing twins are typically found in metals that have the FCC crystal structure, while mechanical twins are observed in BCC and HCP metals. {The role of mechanical twins in the deformation process is discussed in Section 8.8. Annealing twins may be observed in the photomicrograph of the polycrystalline brass specimen shown in Figure 5.15c. The twins correspond to those regions having relatively straight and parallel sides and a different visual contrast than the untwinned regions of the grains within which they reside. An explanation for the variety of textural contrasts in this photomicrograph is provided in Section 5.12.}

MISCELLANEOUS INTERFACIAL DEFECTS

Other possible interfacial defects include stacking faults, phase boundaries, and ferromagnetic domain walls. Stacking faults are found in FCC metals when there is an interruption in the *ABCABCABC* . . . stacking sequence of close-packed planes (Section 3.15). Phase boundaries exist in multiphase materials (Section 10.3)

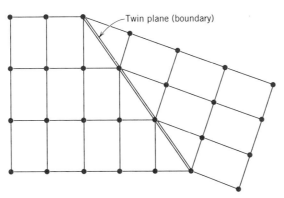

Twin plane (boundary)

FIGURE 5.13 Schematic diagram showing a twin plane or boundary and the adjacent atom positions (dark circles).

across which there is a sudden change in physical and/or chemical characteristics. {For ferromagnetic and ferrimagnetic materials, the boundary that separates regions having different directions of magnetization is termed a domain wall, which is discussed in Section 18.7.}

With regard to polymeric materials, the surfaces of chain-folded layers (Figure 4.14) are considered to be interfacial defects, as are boundaries between two adjacent crystalline regions.

Associated with each of the defects discussed in this section is an interfacial energy, the magnitude of which depends on boundary type, and which will vary from material to material. Normally, the interfacial energy will be greatest for external surfaces and least for domain walls.

5.9 BULK OR VOLUME DEFECTS

Other defects exist in all solid materials that are much larger than those heretofore discussed. These include pores, cracks, foreign inclusions, and other phases. They are normally introduced during processing and fabrication steps. Some of these defects and their effects on the properties of materials are discussed in subsequent chapters.

5.10 ATOMIC VIBRATIONS

Every atom in a solid material is vibrating very rapidly about its lattice position within the crystal. In a sense, these vibrations may be thought of as imperfections or defects. At any instant of time not all atoms vibrate at the same frequency and amplitude, nor with the same energy. At a given temperature there will exist a distribution of energies for the constituent atoms about an average energy. Over time the vibrational energy of any specific atom will also vary in a random manner. With rising temperature, this average energy increases, and, in fact, the temperature of a solid is really just a measure of the average vibrational activity of atoms and molecules. At room temperature, a typical vibrational frequency is on the order of 10^{13} vibrations per second, whereas the amplitude is a few thousandths of a nanometer.

Many properties and processes in solids are manifestations of this vibrational atomic motion. For example, melting occurs when the vibrations are vigorous enough to rupture large numbers of atomic bonds. {A more detailed discussion of atomic vibrations and their influence on the properties of materials is presented in Chapter 17.}

MICROSCOPIC EXAMINATION

5.11 GENERAL

On occasion it is necessary or desirable to examine the structural elements and defects that influence the properties of materials. Some structural elements are of *macroscopic* dimensions, that is, are large enough to be observed with the unaided eye. For example, the shape and average size or diameter of the grains for a polycrystalline specimen are important structural characteristics. Macroscopic grains are often evident on aluminum streetlight posts and also on garbage cans. Relatively large grains having different textures are clearly visible on the surface of the sectioned lead ingot shown in Figure 5.14. However, in most materials the constituent

FIGURE 5.14 High-purity polycrystalline lead ingot in which the individual grains may be discerned. 0.7×. (Reproduced with permission from *Metals Handbook*, Vol. 9, 9th edition, *Metallography and Microstructures*, American Society for Metals, Metals Park, OH, 1985.)

grains are of *microscopic* dimensions, having diameters that may be on the order of microns,[5] and their details must be investigated using some type of microscope. Grain size and shape are only two features of what is termed the **microstructure;** these and other microstructural characteristics are discussed in subsequent chapters.

Optical, electron, and scanning probe microscopes are commonly used in **microscopy.** These instruments aid in investigations of the microstructural features of all material types. Some of these techniques employ photographic equipment in conjunction with the microscope; the photograph on which the image is recorded is called a **photomicrograph.** In addition, some microstructural images are computer generated and/or enhanced.

Microscopic examination is an extremely useful tool in the study and characterization of materials. Several important applications of microstructural examinations are as follows: to ensure that the associations between the properties and structure (and defects) are properly understood; to predict the properties of materials once these relationships have been established; to design alloys with new property combinations; to determine whether or not a material has been correctly heat treated; and to ascertain the mode of mechanical fracture. {Several techniques that are commonly used in such investigations are discussed next.}

5.12 MICROSCOPIC TECHNIQUES (CD-ROM)

5.13 GRAIN SIZE DETERMINATION

The **grain size** is often determined when the properties of a polycrystalline material are under consideration. In this regard, there exist a number of techniques by which size is specified in terms of average grain volume, diameter, or area. Grain size may be estimated by using an intercept method, described as follows. Straight lines all the same length are drawn through several photomicrographs that show the

---

[5] A micron (μm), sometimes called a micrometer, is 10^{-6} m.

grain structure. The grains intersected by each line segment are counted; the line length is then divided by an average of the number of grains intersected, taken over all the line segments. The average grain diameter is found by dividing this result by the linear magnification of the photomicrographs.

Probably the most common method utilized, however, is that devised by the American Society for Testing and Materials (ASTM).[6] The ASTM has prepared several standard comparison charts, all having different average grain sizes. To each is assigned a number ranging from 1 to 10, which is termed the *grain size number;* the larger this number, the smaller the grains. A specimen must be properly prepared to reveal the grain structure, which is photographed at a magnification of $100\times$. Grain size is expressed as the grain size number of the chart that most nearly matches the grains in the micrograph. Thus, a relatively simple and convenient visual determination of grain size number is possible. Grain size number is used extensively in the specification of steels.

The rationale behind the assignment of the grain size number to these various charts is as follows. Let n represent the grain size number, and N the average number of grains per square inch at a magnification of $100\times$. These two parameters are related to each other through the expression

$$N = 2^{n-1} \tag{5.16}$$

SUMMARY

All solid materials contain large numbers of imperfections or deviations from crystalline perfection. The several types of imperfection are categorized on the basis of their geometry and size. Point defects are those associated with one or two atomic positions; in metals these include vacancies (or vacant lattice sites), self-interstitials (host atoms that occupy interstitial sites), and impurity atoms.

With regard to atomic point defects in ceramics, interstitials and vacancies for each anion and cation type are possible. These imperfections often occur in pairs as Frenkel and Schottky defects to ensure that crystal electroneutrality is maintained.

A solid solution may form when impurity atoms are added to a solid, in which case the original crystal structure is retained and no new phases are formed. For substitutional solid solutions, impurity atoms substitute for host atoms, and appreciable solubility is possible only when atomic diameters and electronegativities for both atom types are similar, when both elements have the same crystal structure, and when the impurity atoms have a valence that is the same as or less than the host material. Interstitial solid solutions form for relatively small impurity atoms that occupy interstitial sites among the host atoms.

For ceramic materials, the addition of impurity atoms may result in the formation of substitutional or interstitial solid solutions. Any charge imbalance created by the impurity ions may be compensated by the generation of host ion vacancies or interstitials.

Composition of an alloy may be specified in weight percent or atom percent. The basis for weight percent computations is the weight (or mass) of each alloy constituent relative to the total alloy weight. Atom percents are calculated in terms of the number of moles for each constituent relative to the total moles of all the

---

[6] ASTM Standard E 112, "Standard Methods for Estimating the Average Grain Size for Metals."

elements in the alloy. {Equations were provided for the conversion of one composition scheme to another.}

Dislocations are one-dimensional crystalline defects of which there are two pure types: edge and screw. An edge may be thought of in terms of the lattice distortion along the end of an extra half-plane of atoms; a screw, as a helical planar ramp. For mixed dislocations, components of both pure edge and screw are found. The magnitude and direction of lattice distortion associated with a dislocation is specified by its Burgers vector. The relative orientations of Burgers vector and dislocation line are (1) perpendicular for edge, (2) parallel for screw, and (3) neither perpendicular nor parallel for mixed.

Other imperfections include interfacial defects [external surfaces, grain boundaries (both small- and high-angle), twin boundaries, etc.], volume defects (cracks, pores, etc.), and atomic vibrations. Each type of imperfection has some influence on the properties of a material.

Many of the important defects and structural elements of materials are of microscopic dimensions, and observation is possible only with the aid of a microscope. {Both optical and electron microscopes are employed, usually in conjunction with photographic equipment. Transmissive and reflective modes are possible for each microscope type; preference is dictated by the nature of the specimen as well as the structural element or defect to be examined.}

{More recent scanning probe microscopic techniques have been developed that generate topographical maps representing the surface features and characteristics of the specimen. Examinations on the atomic and molecular levels are possible using these techniques.}

Grain size of polycrystalline materials is frequently determined using photomicrographic techniques. Two methods are commonly employed: intercept and standard comparison charts.

IMPORTANT TERMS AND CONCEPTS

Alloy	Imperfection	Screw dislocation
Atom percent	Interstitial solid solution	Self-interstitial
Atomic vibration	Microscopy	Solid solution
Boltzmann's constant	Microstructure	Solute
Burgers vector	Mixed dislocation	Solvent
Composition	Photomicrograph	Stoichiometry
Defect structure	Point defect	Substitutional solid solution
Dislocation line	Scanning electron microscope (SEM)	Transmission electron microscope (TEM)
Edge dislocation		
Electroneutrality	Scanning probe microscope (SPM)	Vacancy
Frenkel defect		Weight percent
Grain size	Schottky defect	

REFERENCES

ASM Handbook, Vol. 9, *Metallography and Microstructures,* ASM International, Materials Park, OH, 1985.

Barsoum, M. W., *Fundamentals of Ceramics,* The McGraw-Hill Companies, Inc., New York, 1997.

Chiang, Y. M., D. P. Birnie, III, and W. D. Kingery, *Physical Ceramics: Principles for Ceramic Science and Engineering*, John Wiley & Sons, Inc., New York, 1997.

Kingery, W. D., H. K. Bowen, and D. R. Uhlmann, *Introduction to Ceramics*, 2nd edition, John Wiley & Sons, New York, 1976. Chapters 4 and 5.

Moffatt, W. G., G. W. Pearsall, and J. Wulff, *The Structure and Properties of Materials*, Vol. 1, *Structure*, John Wiley & Sons, New York, 1964.

Phillips, V. A., *Modern Metallographic Techniques and Their Applications,* Wiley-Interscience, New York, 1971.

Van Bueren, H. G., *Imperfections in Crystals,* North-Holland Publishing Co., Amsterdam (Wiley-Interscience, New York), 1960.

Vander Voort, G. F., *Metallography, Principles and Practice,* McGraw-Hill Book Co., New York, 1984.

QUESTIONS AND PROBLEMS

Note: To solve those problems having an asterisk (*) by their numbers, consultation of supplementary topics [appearing only on the CD-ROM (and not in print)] will probably be necessary.

5.1 Calculate the fraction of atom sites that are vacant for lead at its melting temperature of 327°C (600°F). Assume an energy for vacancy formation of 0.55 eV/atom.

5.2 Calculate the number of vacancies per cubic meter in iron at 850°C. The energy for vacancy formation is 1.08 eV/atom. Furthermore, the density and atomic weight for Fe are 7.65 g/cm^3 and 55.85 g/mol, respectively.

5.3 Calculate the energy for vacancy formation in silver, given that the equilibrium number of vacancies at 800°C (1073 K) is 3.6×10^{23} m^{-3}. The atomic weight and density (at 800°C) for silver are, respectively, 107.9 g/mol and 9.5 g/cm^3.

5.4 Calculate the number of atoms per cubic meter in aluminum.

5.5 Would you expect Frenkel defects for anions to exist in ionic ceramics in relatively large concentrations? Why or why not?

5.6 In your own words, briefly define the term "stoichiometric."

5.7 If cupric oxide (CuO) is exposed to reducing atmospheres at elevated temperatures, some of the Cu^{2+} ions will become Cu$^+$.

(a) Under these conditions, name one crystalline defect that you would expect to form in order to maintain charge neutrality.

(b) How many Cu$^+$ ions are required for the creation of each defect?

(c) How would you express the chemical formula for this nonstoichiometric material?

5.8 Below, atomic radius, crystal structure, electronegativity, and the most common valence are tabulated, for several elements; for those that are nonmetals, only atomic radii are indicated.

Element	Atomic Radius (nm)	Crystal Structure	Electro-nega-tivity	Valence
Cu	0.1278	FCC	1.9	+2
C	0.071			
H	0.046			
O	0.060			
Ag	0.1445	FCC	1.9	+1
Al	0.1431	FCC	1.5	+3
Co	0.1253	HCP	1.8	+2
Cr	0.1249	BCC	1.6	+3
Fe	0.1241	BCC	1.8	+2
Ni	0.1246	FCC	1.8	+2
Pd	0.1376	FCC	2.2	+2
Pt	0.1387	FCC	2.2	+2
Zn	0.1332	HCP	1.6	+2

Which of these elements would you expect to form the following with copper:
(a) A substitutional solid solution having complete solubility?
(b) A substitutional solid solution of incomplete solubility?
(c) An interstitial solid solution?

5.9 For both FCC and BCC crystal structures, there are two different types of interstitial sites. In each case, one site is larger than the other, which site is normally occupied by impurity atoms. For FCC, this larger one is located at the center of each edge of the unit cell; it is termed an octahedral interstitial site. On the other hand, with BCC the larger site type is found at $0, \frac{1}{2}, \frac{1}{4}$ positions—that is, lying on {100} faces, and situated midway between two unit cell edges on this face and one-quarter of the distance between the other two unit cell edges; it is termed a tetrahedral interstitial site. For both FCC and BCC crystal structures, compute the radius r of an impurity atom that will just fit into one of these sites in terms of the atomic radius R of the host atom.

5.10 **(a)** Suppose that Li_2O is added as an impurity to CaO. If the Li^+ substitutes for Ca^{2+}, what kind of vacancies would you expect to form? How many of these vacancies are created for every Li^+ added?

(b) Suppose that $CaCl_2$ is added as an impurity to CaO. If the Cl^- substitutes for O^{2-}, what kind of vacancies would you expect to form? How many of the vacancies are created for every Cl^- added?

5.11 What point defects are possible for MgO as an impurity in Al_2O_3? How many Mg^{2+} ions must be added to form each of these defects?

5.12 What is the composition, in atom percent, of an alloy that consists of 30 wt% Zn and 70 wt% Cu?

5.13 What is the composition, in weight percent, of an alloy that consists of 6 at% Pb and 94 at% Sn?

5.14 Calculate the composition, in weight percent, of an alloy that contains 218.0 kg titanium, 14.6 kg of aluminum, and 9.7 kg of vanadium.

5.15 What is the composition, in atom percent, of an alloy that contains 98 g tin and 65 g of lead?

5.16 What is the composition, in atom percent, of an alloy that contains 99.7 lb_m copper, 102 lb_m zinc, and 2.1 lb_m lead?

5.17* Derive the following equations:
(a) Equation 5.7a.
(b) Equation 5.9a.
(c) Equation 5.10a.
(d) Equation 5.11b.

5.18* What is the composition, in atom percent, of an alloy that consists of 97 wt% Fe and 3 wt% Si?

5.19* Convert the atom percent composition in Problem 5.16 to weight percent.

5.20* The concentration of carbon in an iron-carbon alloy is 0.15 wt%. What is the concentration in kilograms of carbon per cubic meter of alloy?

5.21* Determine the approximate density of a high-leaded brass that has a composition of 64.5 wt% Cu, 33.5 wt% Zn, and 2 wt% Pb.

5.22* For a solid solution consisting of two elements (designated as 1 and 2), sometimes it is desirable to determine the number of atoms per cubic centimeter of one element in a solid solution, N_1, given the concentration of that element specified in weight percent, C_1. This computation is possible using the following expression:

$$N_1 = \frac{N_A C_1}{\frac{C_1 A_1}{\rho_1} + \frac{A_1}{\rho_2}(100 - C_1)} \qquad (5.17)$$

where

$$N_A = \text{Avogadro's number}$$
$$\rho_1 \text{ and } \rho_2 = \text{densities of the two elements}$$
$$A_1 = \text{the atomic weight of element 1}$$

Derive Equation 5.17 using Equation 5.2 and expressions contained in Section 5.3.

5.23 Gold forms a substitutional solid solution with silver. Compute the number of gold atoms per cubic centimeter for a silver-gold alloy that contains 10 wt% Au and 90 wt% Ag. The densities of pure gold and silver are 19.32 and 10.49 g/cm^3, respectively.

5.24 Germanium forms a substitutional solid solution with silicon. Compute the number of germanium atoms per cubic centimeter for a germanium-silicon alloy that contains 15 wt% Ge and 85 wt% Si. The densities of pure germanium and silicon are 5.32 and 2.33 g/cm^3, respectively.

5.25* Sometimes it is desirable to be able to determine the weight percent of one element, C_1, that will produce a specified concentration in terms of the number of atoms per cubic centimeter, N_1, for an alloy composed of two types of atoms. This computation is possible using the following expression:

$$C_1 = \frac{100}{1 + \frac{N_A \rho_2}{N_1 A_1} - \frac{\rho_2}{\rho_1}} \qquad (5.18)$$

where

$$N_A = \text{Avogadro's number}$$

$$\rho_1 \text{ and } \rho_2 = \text{densities of the two elements}$$

$$A_1 \text{ and } A_2 = \text{the atomic weights of the two elements}$$

Derive Equation 5.18 using Equation 5.2 and expressions contained in Section 5.3.

5.26 Molybdenum forms a substitutional solid solution with tungsten. Compute the weight percent of molybdenum that must be added to tungsten to yield an alloy that contains 1.0×10^{22} Mo atoms per cubic centimeter. The densities of pure Mo and W are 10.22 and 19.30 g/cm^3, respectively.

5.27 Niobium forms a substitutional solid solution with vanadium. Compute the weight percent of niobium that must be added to vanadium to yield an alloy that contains 1.55×10^{22} Nb atoms per cubic centimeter. The densities of pure Nb and V are 8.57 and 6.10 g/cm^3, respectively.

5.28 Copper and platinum both have the FCC crystal structure and Cu forms a substitutional solid solution for concentrations up to approximately 6 wt% Cu at room temperature. Compute the unit cell edge length for a 95 wt% Pt-5 wt% Cu alloy.

5.29 Cite the relative Burgers vector–dislocation line orientations for edge, screw, and mixed dislocations.

5.30 For both FCC and BCC crystal structures, the Burgers vector **b** may be expressed as

$$\mathbf{b} = \frac{a}{2}[hkl]$$

where a is the unit cell edge length and $[hkl]$ is the crystallographic direction having the greatest linear atomic density.

(a) What are the Burgers vector representations for FCC, BCC, and simple cubic crystal structures? See Problems 3.70 and 3.71 at the end of Chapter 3.

(b) If the magnitude of the Burgers vector $|\mathbf{b}|$ is

$$|\mathbf{b}| = \frac{a}{2}(h^2 + k^2 + l^2)^{1/2}$$

determine the values of $|\mathbf{b}|$ for aluminum and tungsten. You may want to consult Table 3.1.

5.31 **(a)** The surface energy of a single crystal depends on the crystallographic orientation with respect to the surface. Explain why this is so.

(b) For an FCC crystal, such as aluminum, would you expect the surface energy for a (100) plane to be greater or less than that for a (111) plane? Why?

5.32 **(a)** For a given material, would you expect the surface energy to be greater than, the same as, or less than the grain boundary energy? Why?

(b) The grain boundary energy of a low-angle grain boundary is less than for a high-angle one. Why is this so?

5.33 **(a)** Briefly describe a twin and a twin boundary.

(b) Cite the difference between mechanical and annealing twins.

5.34 For each of the following stacking sequences found in FCC metals, cite the type of planar defect that exists:

(a) . . . $A B C A B C B A C B A$ . . .

(b) . . . $A B C A B C B C A B C$ . . .

Now, copy the stacking sequences and indicate the position(s) of planar defect(s) with a vertical dashed line.

5.35* Using the intercept method, determine the average grain size, in millimeters, of the specimen whose microstructure is shown in Figure 5.16*b*; assume that the magnification is 100×, and use at least seven straight-line segments.

5.36 Employing the intercept technique, determine the average grain size for the steel specimen whose microstructure is shown in Figure 10.27*a*; use at least seven straight-line segments.

5.37* **(a)** For an ASTM grain size of 4, approximately how many grains would there be per square inch in a micrograph taken at a magnification of 100×?

(b) Estimate the grain size number for the photomicrograph in Figure 5.16*b*, assuming a magnification of 100×.

5.38 A photomicrograph was taken of some metal at a magnification of 100× and it was determined that the average number of grains per square inch is 10. Compute the ASTM grain size number for this alloy.

Design Problems

5.D1* Aluminum-lithium alloys have been developed by the aircraft industry in order to reduce the weight and improve the performance of its aircraft. A commercial aircraft skin material having a density of 2.55 g/cm³ is desired. Compute the concentration of Li (in wt%) that is required.

5.D2* Iron and vanadium both have the BCC crystal structure and V forms a substitutional solid solution in Fe for concentrations up to approximately 20 wt% V at room temperature. Determine the concentration in weight percent of V that must be added to iron to yield a unit cell edge length of 0.289 nm.

Chapter 6 / Diffusion

*P*hotograph of a steel gear that has been "case hardened." The outer surface layer was selectively hardened by a high-temperature heat treatment during which carbon from the surrounding atmosphere diffused into the surface. The "case" appears as the dark outer rim of that segment of the gear that has been sectioned. Actual size. (Photograph courtesy of Surface Division Midland-Ross.)

Why Study *Diffusion?*

Materials of all types are often heat treated to improve their properties. The phenomena that occur during a heat treatment almost always involve atomic diffusion. Often an enhancement of diffusion rate is desired; on occasion measures are taken to reduce it. Heat-treating temperatures and times, and/or cooling rates are often predictable using the mathematics of diffusion and appropriate diffusion constants. The steel gear shown on this page has been case hardened (Section 9.14); that is, its hardness and resistance to failure by fatigue have been enhanced by diffusing excess carbon or nitrogen into the outer surface layer.

6.1 INTRODUCTION

Many reactions and processes that are important in the treatment of materials rely on the transfer of mass either within a specific solid (ordinarily on a microscopic level) or from a liquid, a gas, or another solid phase. This is necessarily accomplished by **diffusion,** the phenomenon of material transport by atomic motion. This chapter discusses the atomic mechanisms by which diffusion occurs, the mathematics of diffusion, and the influence of temperature and diffusing species on the rate of diffusion.

The phenomenon of diffusion may be demonstrated with the use of a *diffusion couple,* which is formed by joining bars of two different metals together so that there is intimate contact between the two faces; this is illustrated for copper and nickel in Figure 6.1, which includes schematic representations of atom positions and composition across the interface. This couple is heated for an extended period at an elevated temperature (but below the melting temperature of both metals), and cooled to room temperature. Chemical analysis will reveal a condition similar to that represented in Figure 6.2, namely, pure copper and nickel at the two extremities of the couple, separated by an alloyed region. Concentrations of both metals vary with position as shown in Figure 6.2c. This result indicates that copper atoms have migrated or diffused into the nickel, and that nickel has diffused into copper. This process, whereby atoms of one metal diffuse into another, is termed **interdiffusion,** or **impurity diffusion.**

Interdiffusion may be discerned from a macroscopic perspective by changes in concentration which occur over time, as in the example for the Cu–Ni diffusion couple. There is a net drift or transport of atoms from high to low concentration regions. Diffusion also occurs for pure metals, but all atoms exchanging positions are of the same type; this is termed **self-diffusion.** Of course, self-diffusion is not normally subject to observation by noting compositional changes.

6.2 DIFFUSION MECHANISMS

From an atomic perspective, diffusion is just the stepwise migration of atoms from lattice site to lattice site. In fact, the atoms in solid materials are in constant motion, rapidly changing positions. For an atom to make such a move, two conditions must be met: (1) there must be an empty adjacent site, and (2) the atom must have sufficient energy to break bonds with its neighbor atoms and then cause some lattice distortion during the displacement. This energy is vibrational in nature (Section 5.10). At a specific temperature some small fraction of the total number of atoms

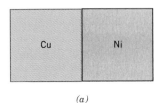

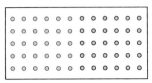

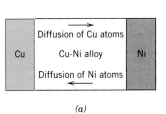

FIGURE 6.1 (*a*) A copper–nickel diffusion couple before a high-temperature heat treatment. (*b*) Schematic representations of Cu (colored circles) and Ni (gray circles) atom locations within the diffusion couple. (*c*) Concentrations of copper and nickel as a function of position across the couple.

(*a*)

(*b*)

(*c*)

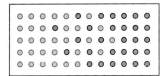

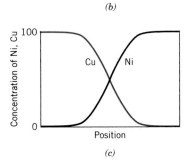

FIGURE 6.2 (*a*) A copper–nickel diffusion couple after a high-temperature heat treatment, showing the alloyed diffusion zone. (*b*) Schematic representations of Cu (colored circles) and Ni (gray circles) atom locations within the couple. (*c*) Concentrations of copper and nickel as a function of position across the couple.

(*a*)

(*b*)

(*c*)

is capable of diffusive motion, by virtue of the magnitudes of their vibrational energies. This fraction increases with rising temperature.

Several different models for this atomic motion have been proposed; of these possibilities, two dominate for metallic diffusion.

VACANCY DIFFUSION

One mechanism involves the interchange of an atom from a normal lattice position to an adjacent vacant lattice site or vacancy, as represented schematically in Figure 6.3a. This mechanism is aptly termed **vacancy diffusion.** Of course, this process necessitates the presence of vacancies, and the extent to which vacancy diffusion can occur is a function of the number of these defects that are present; significant concentrations of vacancies may exist in metals at elevated temperatures (Section 5.2). Since diffusing atoms and vacancies exchange positions, the diffusion of atoms in one direction corresponds to the motion of vacancies in the opposite direction. Both self-diffusion and interdiffusion occur by this mechanism; for the latter, the impurity atoms must substitute for host atoms.

INTERSTITIAL DIFFUSION

The second type of diffusion involves atoms that migrate from an interstitial position to a neighboring one that is empty. This mechanism is found for interdiffusion of impurities such as hydrogen, carbon, nitrogen, and oxygen, which have atoms that are small enough to fit into the interstitial positions. Host or substitutional impurity atoms rarely form interstitials and do not normally diffuse via this mechanism. This phenomenon is appropriately termed **interstitial diffusion** (Figure 6.3b).

In most metal alloys, interstitial diffusion occurs much more rapidly than diffusion by the vacancy mode, since the interstitial atoms are smaller, and thus more

FIGURE 6.3 Schematic representations of (a) vacancy diffusion and (b) interstitial diffusion.

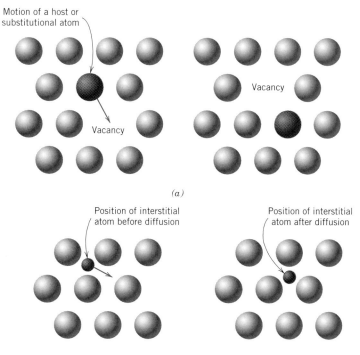

Motion of a host or substitutional atom

Vacancy

Vacancy

(a)

Position of interstitial atom before diffusion

Position of interstitial atom after diffusion

(b)

mobile. Furthermore, there are more empty interstitial positions than vacancies; hence, the probability of interstitial atomic movement is greater than for vacancy diffusion.

6.3 STEADY-STATE DIFFUSION

Diffusion is a time-dependent process—that is, in a macroscopic sense, the quantity of an element that is transported within another is a function of time. Often it is necessary to know how fast diffusion occurs, or the rate of mass transfer. This rate is frequently expressed as a **diffusion flux** (J), defined as the mass (or, equivalently, the number of atoms) M diffusing through and perpendicular to a unit cross-sectional area of solid per unit of time. In mathematical form, this may be represented as

$$J = \frac{M}{At} \tag{6.1a}$$

where A denotes the area across which diffusion is occurring and t is the elapsed diffusion time. In differential form, this expression becomes

$$J = \frac{1}{A}\frac{dM}{dt} \tag{6.1b}$$

The units for J are kilograms or atoms per meter squared per second (kg/m^2-s or atoms/m^2-s).

If the diffusion flux does not change with time, a steady-state condition exists. One common example of **steady-state diffusion** is the diffusion of atoms of a gas through a plate of metal for which the concentrations (or pressures) of the diffusing species on both surfaces of the plate are held constant. This is represented schematically in Figure 6.4a.

When concentration C is plotted versus position (or distance) within the solid x, the resulting curve is termed the **concentration profile;** the slope at a particular

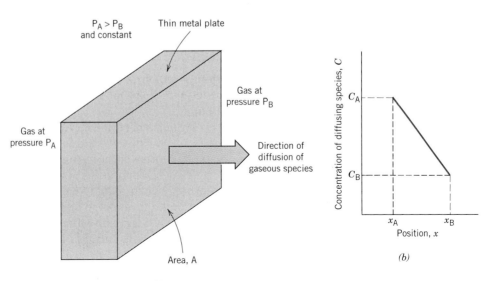

FIGURE 6.4
(a) Steady-state diffusion across a thin plate. (b) A linear concentration profile for the diffusion situation in (a).

point on this curve is the **concentration gradient:**

$$\text{concentration gradient} = \frac{dC}{dx} \tag{6.2a}$$

In the present treatment, the concentration profile is assumed to be linear, as depicted in Figure 6.4*b*, and

$$\text{concentration gradient} = \frac{\Delta C}{\Delta x} = \frac{C_A - C_B}{x_A - x_B} \tag{6.2b}$$

For diffusion problems, it is sometimes convenient to express concentration in terms of mass of diffusing species per unit volume of solid (kg/m^3 or g/cm^3).[1]

The mathematics of steady-state diffusion in a single (x) direction is relatively simple, in that the flux is proportional to the concentration gradient through the expression

$$J = -D\frac{dC}{dx} \tag{6.3}$$

The constant of proportionality D is called the **diffusion coefficient,** which is expressed in square meters per second. The negative sign in this expression indicates that the direction of diffusion is down the concentration gradient, from a high to a low concentration. Equation 6.3 is sometimes called **Fick's first law.**

Sometimes the term **driving force** is used in the context of what compels a reaction to occur. For diffusion reactions, several such forces are possible; but when diffusion is according to Equation 6.3, the concentration gradient is the driving force.

One practical example of steady-state diffusion is found in the purification of hydrogen gas. One side of a thin sheet of palladium metal is exposed to the impure gas composed of hydrogen and other gaseous species such as nitrogen, oxygen, and water vapor. The hydrogen selectively diffuses through the sheet to the opposite side, which is maintained at a constant and lower hydrogen pressure.

EXAMPLE PROBLEM 6.1

A plate of iron is exposed to a carburizing (carbon-rich) atmosphere on one side and a decarburizing (carbon-deficient) atmosphere on the other side at 700°C (1300°F). If a condition of steady state is achieved, calculate the diffusion flux of carbon through the plate if the concentrations of carbon at positions of 5 and 10 mm (5×10^{-3} and 10^{-2} m) beneath the carburizing surface are 1.2 and 0.8 kg/m^3, respectively. Assume a diffusion coefficient of 3×10^{-11} m^2/s at this temperature.

SOLUTION

Fick's first law, Equation 6.3, is utilized to determine the diffusion flux. Substitution of the values above into this expression yields

$$J = -D\frac{C_A - C_B}{x_A - x_B} = -(3 \times 10^{-11}\ m^2/s)\frac{(1.2 - 0.8)\ kg/m^3}{(5 \times 10^{-3} - 10^{-2})\ m}$$

$$= 2.4 \times 10^{-9}\ kg/m^2\text{-}s$$

---

[1] Conversion of concentration from weight percent to mass per unit volume (in kg/m^3) is possible using {Equation 5.9.}

6.4 NONSTEADY-STATE DIFFUSION

Most practical diffusion situations are nonsteady-state ones. That is, the diffusion flux and the concentration gradient at some particular point in a solid vary with time, with a net accumulation or depletion of the diffusing species resulting. This is illustrated in Figure 6.5, which shows concentration profiles at three different diffusion times. Under conditions of nonsteady state, use of Equation 6.3 is no longer convenient; instead, the partial differential equation

$$\frac{\partial C}{\partial t} = \frac{\partial}{\partial x}\left(D \frac{\partial C}{\partial x}\right) \tag{6.4a}$$

known as **Fick's second law,** is used. If the diffusion coefficient is independent of composition (which should be verified for each particular diffusion situation), Equation 6.4a simplifies to

$$\frac{\partial C}{\partial t} = D \frac{\partial^2 C}{\partial x^2} \tag{6.4b}$$

Solutions to this expression (concentration in terms of both position and time) are possible when physically meaningful boundary conditions are specified. Comprehensive collections of these are given by Crank, and Carslaw and Jaeger (see References).

One practically important solution is for a semi-infinite solid[2] in which the surface concentration is held constant. Frequently, the source of the diffusing species is a gas phase, the partial pressure of which is maintained at a constant value. Furthermore, the following assumptions are made:

1. Before diffusion, any of the diffusing solute atoms in the solid are uniformly distributed with concentration of C_0.

2. The value of x at the surface is zero and increases with distance into the solid.

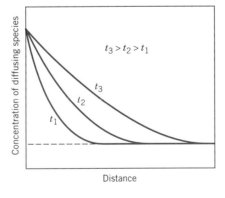

FIGURE 6.5 Concentration profiles for nonsteady-state diffusion taken at three different times, t_1, t_2, and t_3.

---

[2] A bar of solid is considered to be semi-infinite if none of the diffusing atoms reaches the bar end during the time over which diffusion takes place. A bar of length l is considered to be semi-infinite when $l > 10\sqrt{Dt}$.

3. The time is taken to be zero the instant before the diffusion process begins.

These boundary conditions are simply stated as

For $t = 0$, $C = C_0$ at $0 \leq x \leq \infty$

For $t > 0$, $C = C_s$ (the constant surface concentration) at $x = 0$

$C = C_0$ at $x = \infty$

Application of these boundary conditions to Equation 6.4b yields the solution

$$\frac{C_x - C_0}{C_s - C_0} = 1 - \mathrm{erf}\left(\frac{x}{2\sqrt{Dt}}\right) \qquad (6.5)$$

where C_x represents the concentration at depth x after time t. The expression $\mathrm{erf}(x/2\sqrt{Dt})$ is the Gaussian error function,[3] values of which are given in mathematical tables for various $x/2\sqrt{Dt}$ values; a partial listing is given in Table 6.1. The concentration parameters that appear in Equation 6.5 are noted in Figure 6.6, a concentration profile taken at a specific time. Equation 6.5 thus demonstrates the relationship between concentration, position, and time, namely, that C_x, being a function of the dimensionless parameter $x/\sqrt{Dt}$, may be determined at any time and position if the parameters C_0, C_s, and D are known.

Suppose that it is desired to achieve some specific concentration of solute, C_1, in an alloy; the left-hand side of Equation 6.5 now becomes

$$\frac{C_1 - C_0}{C_s - C_0} = \text{constant}$$

Table 6.1   Tabulation of Error Function Values

z	$\mathrm{erf}(z)$	z	$\mathrm{erf}(z)$	z	$\mathrm{erf}(z)$
0	0	0.55	0.5633	1.3	0.9340
0.025	0.0282	0.60	0.6039	1.4	0.9523
0.05	0.0564	0.65	0.6420	1.5	0.9661
0.10	0.1125	0.70	0.6778	1.6	0.9763
0.15	0.1680	0.75	0.7112	1.7	0.9838
0.20	0.2227	0.80	0.7421	1.8	0.9891
0.25	0.2763	0.85	0.7707	1.9	0.9928
0.30	0.3286	0.90	0.7970	2.0	0.9953
0.35	0.3794	0.95	0.8209	2.2	0.9981
0.40	0.4284	1.0	0.8427	2.4	0.9993
0.45	0.4755	1.1	0.8802	2.6	0.9998
0.50	0.5205	1.2	0.9103	2.8	0.9999

---

[3] This Gaussian error function is defined by

$$\mathrm{erf}(z) = \frac{2}{\sqrt{\pi}} \int_0^z e^{-y^2}\, dy$$

where $x/2\sqrt{Dt}$ has been replaced by the variable z.

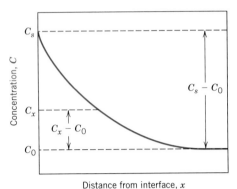

FIGURE 6.6 Concentration profile for nonsteady-state diffusion; concentration parameters relate to Equation 6.5.

This being the case, the right-hand side of this same expression is also a constant, and subsequently

$$\frac{x}{2\sqrt{Dt}} = \text{constant} \qquad (6.6a)$$

or

$$\frac{x^2}{Dt} = \text{constant} \qquad (6.6b)$$

Some diffusion computations are thus facilitated on the basis of this relationship, as demonstrated in Example Problem 6.3.

EXAMPLE PROBLEM 6.2

For some applications, it is necessary to harden the surface of a steel (or iron-carbon alloy) above that of its interior. One way this may be accomplished is by increasing the surface concentration of carbon in a process termed **carburizing;** the steel piece is exposed, at an elevated temperature, to an atmosphere rich in a hydrocarbon gas, such as methane (CH_4).

Consider one such alloy that initially has a uniform carbon concentration of 0.25 wt% and is to be treated at 950°C (1750°F). If the concentration of carbon at the surface is suddenly brought to and maintained at 1.20 wt%, how long will it take to achieve a carbon content of 0.80 wt% at a position 0.5 mm below the surface? The diffusion coefficient for carbon in iron at this temperature is 1.6×10^{-11} m²/s; assume that the steel piece is semi-infinite.

SOLUTION

Since this is a nonsteady-state diffusion problem in which the surface composition is held constant, Equation 6.5 is used. Values for all the parameters in this expression except time t are specified in the problem as follows:

$$C_0 = 0.25 \text{ wt\% C}$$

$$C_s = 1.20 \text{ wt\% C}$$

$$C_x = 0.80 \text{ wt\% C}$$

$$x = 0.50 \text{ mm} = 5 \times 10^{-4} \text{ m}$$

$$D = 1.6 \times 10^{-11} \text{ m}^2/\text{s}$$

Thus,

$$\frac{C_x - C_0}{C_s - C_0} = \frac{0.80 - 0.25}{1.20 - 0.25} = 1 - \mathrm{erf}\left[\frac{(5 \times 10^{-4}\,\mathrm{m})}{2\sqrt{(1.6 \times 10^{-11}\,\mathrm{m^2/s})(t)}}\right]$$

$$0.4210 = \mathrm{erf}\left(\frac{62.5\,\mathrm{s^{1/2}}}{\sqrt{t}}\right)$$

We must now determine from Table 6.1 the value of z for which the error function is 0.4210. An interpolation is necessary, as

z	erf(z)
0.35	0.3794
z	0.4210
0.40	0.4284

$$\frac{z - 0.35}{0.40 - 0.35} = \frac{0.4210 - 0.3794}{0.4284 - 0.3794}$$

or

$$z = 0.392$$

Therefore,

$$\frac{62.5\,\mathrm{s^{1/2}}}{\sqrt{t}} = 0.392$$

and solving for t,

$$t = \left(\frac{62.5\,\mathrm{s^{1/2}}}{0.392}\right)^2 = 25{,}400\,\mathrm{s} = 7.1\,\mathrm{h}$$

EXAMPLE PROBLEM 6.3

The diffusion coefficients for copper in aluminum at 500 and 600°C are 4.8×10^{-14} and 5.3×10^{-13} m^2/s, respectively. Determine the approximate time at 500°C that will produce the same diffusion result (in terms of concentration of Cu at some specific point in Al) as a 10-h heat treatment at 600°C.

SOLUTION

This is a diffusion problem in which Equation 6.6b may be employed. The composition in both diffusion situations will be equal at the same position (i.e., x is also a constant), thus

$$Dt = \mathrm{constant} \tag{6.7}$$

at both temperatures. That is,

$$D_{500}t_{500} = D_{600}t_{600}$$

or

$$t_{500} = \frac{D_{600}t_{600}}{D_{500}} = \frac{(5.3 \times 10^{-13}\,\mathrm{m^2/s})(10\,\mathrm{h})}{4.8 \times 10^{-14}\,\mathrm{m^2/s}} = 110.4\,\mathrm{h}$$

6.5 FACTORS THAT INFLUENCE DIFFUSION

DIFFUSING SPECIES

The magnitude of the diffusion coefficient D is indicative of the rate at which atoms diffuse. Coefficients, both self- and interdiffusion, for several metallic systems are listed in Table 6.2. The diffusing species as well as the host material influence the diffusion coefficient. For example, there is a significant difference in magnitude between self- and carbon interdiffusion in α iron at 500°C, the D value being greater for the carbon interdiffusion (3.0×10^{-21} vs. 2.4×10^{-12} m²/s). This comparison also provides a contrast between rates of diffusion via vacancy and interstitial modes as discussed above. Self-diffusion occurs by a vacancy mechanism, whereas carbon diffusion in iron is interstitial.

TEMPERATURE

Temperature has a most profound influence on the coefficients and diffusion rates. For example, for the self-diffusion of Fe in α-Fe, the diffusion coefficient increases approximately six orders of magnitude (from 3.0×10^{-21} to 1.8×10^{-15} m²/s) in rising temperature from 500 to 900°C (Table 6.2). The temperature dependence of diffusion coefficients is related to temperature according to

$$D = D_0 \exp\left(-\frac{Q_d}{RT}\right) \qquad (6.8)$$

where

D_0 = a temperature-independent preexponential (m²/s)

Q_d = the **activation energy** for diffusion (J/mol, cal/mol, or eV/atom)

Table 6.2   A Tabulation of Diffusion Data

Diffusing Species	Host Metal	$D_0(m^2/s)$	Activation Energy Q_d		Calculated Values	
			kJ/mol	eV/atom	T(°C)	$D(m^2/s)$
Fe	α-Fe (BCC)	2.8×10^{-4}	251	2.60	500	3.0×10^{-21}
					900	1.8×10^{-15}
Fe	γ-Fe (FCC)	5.0×10^{-5}	284	2.94	900	1.1×10^{-17}
					1100	7.8×10^{-16}
C	α-Fe	6.2×10^{-7}	80	0.83	500	2.4×10^{-12}
					900	1.7×10^{-10}
C	γ-Fe	2.3×10^{-5}	148	1.53	900	5.9×10^{-12}
					1100	5.3×10^{-11}
Cu	Cu	7.8×10^{-5}	211	2.19	500	4.2×10^{-19}
Zn	Cu	2.4×10^{-5}	189	1.96	500	4.0×10^{-18}
Al	Al	2.3×10^{-4}	144	1.49	500	4.2×10^{-14}
Cu	Al	6.5×10^{-5}	136	1.41	500	4.1×10^{-14}
Mg	Al	1.2×10^{-4}	131	1.35	500	1.9×10^{-13}
Cu	Ni	2.7×10^{-5}	256	2.65	500	1.3×10^{-22}

Source: E. A. Brandes and G. B. Brook (Editors), *Smithells Metals Reference Book,* 7th edition, Butterworth-Heinemann, Oxford, 1992.

R = the gas constant, 8.31 J/mol-K, 1.987 cal/mol-K, or 8.62 × 10⁻⁵ eV/atom-K

T = absolute temperature (K)

The activation energy may be thought of as that energy required to produce the diffusive motion of one mole of atoms. A large activation energy results in a relatively small diffusion coefficient. Table 6.2 also contains a listing of D_0 and Q_d values for several diffusion systems.

Taking natural logarithms of Equation 6.8 yields

$$\ln D = \ln D_0 - \frac{Q_d}{R}\left(\frac{1}{T}\right) \tag{6.9a}$$

Or in terms of logarithms to the base 10

$$\log D = \log D_0 - \frac{Q_d}{2.3R}\left(\frac{1}{T}\right) \tag{6.9b}$$

Since D_0, Q_d, and R are all constants, Equation 6.9b takes on the form of an equation of a straight line:

$$y = b + mx$$

where y and x are analogous, respectively, to the variables $\log D$ and $1/T$. Thus, if $\log D$ is plotted versus the reciprocal of the absolute temperature, a straight line should result, having slope and intercept of $-Q_d/2.3R$ and $\log D_0$, respectively. This is, in fact, the manner in which the values of Q_d and D_0 are determined experimentally. From such a plot for several alloy systems (Figure 6.7), it may be noted that linear relationships exist for all cases shown.

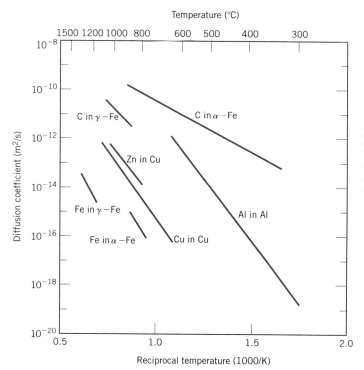

FIGURE 6.7 Plot of the logarithm of the diffusion coefficient versus the reciprocal of absolute temperature for several metals. [Data taken from E. A. Brandes and G. B. Brook (Editors), *Smithells Metals Reference Book,* 7th edition, Butterworth-Heinemann, Oxford, 1992.]

EXAMPLE PROBLEM 6.4

Using the data in Table 6.2, compute the diffusion coefficient for magnesium in aluminum at 550°C.

SOLUTION

This diffusion coefficient may be determined by applying Equation 6.8; the values of D_0 and Q_d from Table 6.2 are 1.2×10^{-4} m²/s and 131 kJ/mol, respectively. Thus,

$$D = (1.2 \times 10^{-4} \text{ m}^2/\text{s}) \exp \left[-\frac{(131{,}000 \text{ J/mol})}{(8.31 \text{ J/mol-K})(550 + 273 \text{ K})} \right]$$

$$= 5.8 \times 10^{-13} \text{ m}^2/\text{s}$$

EXAMPLE PROBLEM 6.5

In Figure 6.8 is shown a plot of the logarithm (to the base 10) of the diffusion coefficient versus reciprocal of absolute temperature, for the diffusion of copper in gold. Determine values for the activation energy and the preexponential.

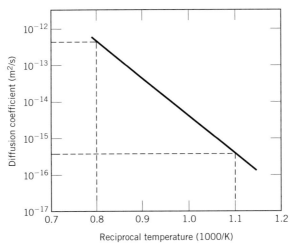

FIGURE 6.8   Plot of the logarithm of the diffusion coefficient versus the reciprocal of absolute temperature for the diffusion of copper in gold.

SOLUTION

From Equation 6.9b the slope of the line segment in Figure 6.8 is equal to $-Q_d/2.3R$, and the intercept at $1/T = 0$ gives the value of log D_0. Thus, the activation energy may be determined as

$$Q_d = -2.3R \text{ (slope)} = -2.3R \left[\frac{\Delta (\log D)}{\Delta \left(\dfrac{1}{T} \right)} \right]$$

$$= -2.3R \left[\frac{\log D_1 - \log D_2}{\dfrac{1}{T_1} - \dfrac{1}{T_2}} \right]$$

where D_1 and D_2 are the diffusion coefficient values at $1/T_1$ and $1/T_2$, respectively. Let us arbitrarily take $1/T_1 = 0.8 \times 10^{-3}$ (K)$^{-1}$ and $1/T_2 = 1.1 \times 10^{-3}$ (K)$^{-1}$. We may now read the corresponding log D_1 and log D_2 values from the line segment in Figure 6.8.

[Before this is done, however, a parenthetic note of caution is offered. The vertical axis in Figure 6.8 is scaled logarithmically (to the base 10); however, the actual diffusion coefficient values are noted on this axis. For example, for $D = 10^{-14}$ m^2/s, the logarithm of D is -14.0 *not* 10^{-14}. Furthermore, this logarithmic scaling affects the readings between decade values; for example, at a location midway between 10^{-14} and 10^{-15}, the value is not 5×10^{-15}, but rather, $10^{-14.5} = 3.2 \times 10^{-15}$].

Thus, from Figure 6.8, at $1/T_1 = 0.8 \times 10^{-3}$ (K)$^{-1}$, log $D_1 = -12.40$, while for $1/T_2 = 1.1 \times 10^{-3}$ (K)$^{-1}$, log $D_2 = -15.45$, and the activation energy, as determined from the slope of the line segment in Figure 6.8, is

$$Q_d = -2.3R \left[\frac{\log D_1 - \log D_2}{\dfrac{1}{T_1} - \dfrac{1}{T_2}} \right]$$

$$= -2.3 \, (8.31 \text{ J/mol-K}) \left[\frac{-12.40 - (-15.45)}{0.8 \times 10^{-3} \, (\text{K})^{-1} - 1.1 \times 10^{-3} \, (\text{K})^{-1}} \right]$$

$$= 194{,}000 \text{ J/mol} = 194 \text{ kJ/mol}$$

Now, rather than trying to make a graphical extrapolation to determine D_0, a more accurate value is obtained analytically using Equation 6.9b, and a specific value of D (or log D) and its corresponding T (or $1/T$) from Figure 6.8. Since we know that log $D = -15.45$ at $1/T = 1.1 \times 10^{-3}$ (K)$^{-1}$, then

$$\log D_0 = \log D + \frac{Q_d}{2.3R} \left(\frac{1}{T} \right)$$

$$= -15.45 + \frac{(194{,}000 \text{ J/mol})(1.1 \times 10^{-3} \, [\text{K}]^{-1})}{(2.3)(8.31 \text{ J/mol-K})}$$

$$= -4.28$$

Thus, $D_0 = 10^{-4.28}$ m^2/s $= 5.2 \times 10^{-5}$ m^2/s.

DESIGN EXAMPLE 6.1

The wear resistance of a steel gear is to be improved by hardening its surface. This is to be accomplished by increasing the carbon content within an outer surface layer as a result of carbon diffusion into the steel; the carbon is to be supplied from an external carbon-rich gaseous atmosphere at an elevated and constant temperature. The initial carbon content of the steel is 0.20 wt%, whereas the surface concentration is to be maintained at 1.00 wt%. In order for this treatment to be effective, a carbon content of 0.60 wt% must be established at a position 0.75 mm below the surface. Specify an appropriate heat treatment in terms of temperature and time for temperatures between 900°C and 1050°C. Use data in Table 6.2 for the diffusion of carbon in γ-iron.

SOLUTION

Since this is a nonsteady-state diffusion situation, let us first of all employ Equation 6.5, utilizing the following values for the concentration parameters:

$$C_0 = 0.20 \text{ wt\% C}$$
$$C_s = 1.00 \text{ wt\% C}$$
$$C_x = 0.60 \text{ wt\% C}$$

Therefore

$$\frac{C_x - C_0}{C_s - C_0} = \frac{0.60 - 0.20}{1.00 - 0.20} = 1 - \text{erf}\left(\frac{x}{2\sqrt{Dt}}\right)$$

And thus

$$0.5 = \text{erf}\left(\frac{x}{2\sqrt{Dt}}\right)$$

Using an interpolation technique as demonstrated in Example Problem 6.2 and the data presented in Table 6.1

$$\frac{x}{2\sqrt{Dt}} = 0.4747 \qquad (6.10)$$

The problem stipulates that $x = 0.75 \text{ mm} = 7.5 \times 10^{-4} \text{ m}$. Therefore

$$\frac{7.5 \times 10^{-4} \text{ m}}{2\sqrt{Dt}} = 0.4747$$

This leads to

$$Dt = 6.24 \times 10^{-7} \text{ m}^2$$

Furthermore, the diffusion coefficient depends on temperature according to Equation 6.8; and, from Table 6.2 for the diffusion of carbon in γ-iron, $D_0 = 2.3 \times 10^{-5}$ m²/s and $Q_d = 148,000$ J/mol. Hence

$$Dt = D_0 \exp\left(-\frac{Q_d}{RT}\right)(t) = 6.24 \times 10^{-7} \text{ m}^2$$

$$(2.3 \times 10^{-5} \text{ m}^2/\text{s}) \exp\left[-\frac{148,000 \text{ J/mol}}{(8.31 \text{ J/mol-K})(T)}\right](t) = 6.24 \times 10^{-7} \text{ m}^2$$

And solving for the time t

$$t \text{ (in s)} = \frac{0.0271}{\exp\left(-\dfrac{17,810}{T}\right)}$$

Thus, the required diffusion time may be computed for some specified temperature (in K). Below are tabulated t values for four different temperatures that lie within the range stipulated in the problem.

Temperature (°C)	Time	
	s	h
900	106,400	29.6
950	57,200	15.9
1000	32,300	9.0
1050	19,000	5.3

6.6 OTHER DIFFUSION PATHS

Atomic migration may also occur along dislocations, grain boundaries, and external surfaces. These are sometimes called *"short-circuit" diffusion paths* inasmuch as rates are much faster than for bulk diffusion. However, in most situations short-circuit contributions to the overall diffusion flux are insignificant because the cross-sectional areas of these paths are extremely small.

6.7 DIFFUSION IN IONIC AND POLYMERIC MATERIALS

We now extrapolate some of the diffusion principles discussed above to ionic and polymeric materials.

IONIC MATERIALS

For ionic compounds, the situation is more complicated than for metals inasmuch as it is necessary to consider the diffusive motion of two types of ions that have opposite charges. Diffusion in these materials occurs by a vacancy mechanism (Figure 6.3a). And, as we noted in Section 5.3, in order to maintain charge neutrality in an ionic material, the following may be said about vacancies: (1) ion vacancies occur in pairs [as with Schottky defects (Figure 5.3)], (2) they form in nonstoichiometric compounds (Figure 5.4), and (3) they are created by substitutional impurity ions having different charge states than the host ions (Example Problem 5.2). In any event, associated with the diffusive motion of a single ion is a transference of electrical charge. And in order to maintain localized charge neutrality in the vicinity of this moving ion, it is necessary that another species having an equal and opposite charge accompany the ion's diffusive motion. Possible charged species include another vacancy, an impurity atom, or an electronic carrier [i.e., a free electron or hole (Section 12.6)]. It follows that the rate of diffusion of these electrically charged couples is limited by the diffusion rate of the slowest moving species.

When an external electric field is applied across an ionic solid, the electrically charged ions migrate (i.e., diffuse) in response to forces that are brought to bear on them. And, as we discuss in Section 12.15, this ionic motion gives rise to an electric current. Furthermore, the electrical conductivity is a function of the diffusion coefficient (Equation 12.26). Consequently, much of the diffusion data for ionic solids comes from electrical conductivity measurements.

POLYMERIC MATERIALS

For polymeric materials, we are more interested in the diffusive motion of small foreign molecules (e.g., O_2, H_2O, CO_2, CH_4) between the molecular chains than in the diffusive motion of atoms within the chain structures. A polymer's permeability and absorption characteristics relate to the degree to which foreign substances diffuse into the material. Penetration of these foreign substances can lead to swelling and/or chemical reactions with the polymer molecules, and often to a depreciation of the material's mechanical and physical properties {(Section 16.11).}

Rates of diffusion are greater through amorphous regions than through crystalline regions; the structure of amorphous material is more "open." This diffusion mechanism may be considered to be analogous to interstitial diffusion in metals—that is, in polymers, diffusive movement from one open amorphous region to an adjacent open one.

Foreign molecule size also affects the diffusion rate: smaller molecules diffuse faster than larger ones. Furthermore, diffusion is more rapid for foreign molecules that are chemically inert than for those that react with the polymer.

For some applications low diffusion rates through polymeric materials are desirable, as with food and beverage packaging and with automobile tires and inner tubes. Polymer membranes are often used as filters to selectively separate one chemical species from another (or others) (e.g., the desalinization of water). In such instances it is normally the case that the diffusion rate of the substance to be filtered is significantly greater than that for the other substance(s).

SUMMARY

Solid-state diffusion is a means of mass transport within solid materials by stepwise atomic motion. The term "self-diffusion" refers to the migration of host atoms; for impurity atoms, the term "interdiffusion" is used. Two mechanisms are possible: vacancy and interstitial. For a given host metal, interstitial atomic species generally diffuse more rapidly.

For steady-state diffusion, the concentration profile of the diffusing species is time independent, and the flux or rate is proportional to the negative of the concentration gradient according to Fick's first law. The mathematics for nonsteady state are described by Fick's second law, a partial differential equation. The solution for a constant surface composition boundary condition involves the Gaussian error function.

The magnitude of the diffusion coefficient is indicative of the rate of atomic motion, being strongly dependent on and increasing exponentially with increasing temperature.

Diffusion in ionic materials occurs by a vacancy mechanism; localized charge neutrality is maintained by the coupled diffusive motion of a charged vacancy and some other charged entity. In polymers, small molecules of foreign substances diffuse between molecular chains by an interstitial-type mechanism from one amorphous region to an adjacent one.

IMPORTANT TERMS AND CONCEPTS

Activation energy	Diffusion flux	Interstitial diffusion
Carburizing	Driving force	Nonsteady-state diffusion
Concentration gradient	Fick's first and second laws	Self-diffusion
Concentration profile	Interdiffusion (impurity dif-	Steady-state diffusion
Diffusion	fusion)	Vacancy diffusion
Diffusion coefficient		

REFERENCES

Borg, R. J. and G. J. Dienes (Editors), *An Introduction to Solid State Diffusion,* Academic Press, San Diego, 1988.

Brandes, E. A. and G. B. Brook (Editors), *Smithells Metals Reference Book,* 7th edition, Butterworth-Heinemann Ltd., Oxford, 1992.

Carslaw, H. S. and J. C. Jaeger, *Conduction of Heat in Solids,* 2nd edition, Clarendon Press, Oxford, 1986.

Crank, J., *The Mathematics of Diffusion,* 2nd edition, Clarendon Press, Oxford, 1980.

Girifalco, L. A., *Atomic Migration in Crystals,* Blaisdell Publishing Company, New York, 1964.

Shewmon, P. G., *Diffusion in Solids,* McGraw-Hill Book Company, New York, 1963. Reprinted by The Minerals, Metals and Materials Society, Warrendale, PA, 1989.

QUESTIONS AND PROBLEMS

Note: To solve those problems having an asterisk (*) by their numbers, consultation of supplementary topics [appearing only on the CD-ROM (and not in print)] will probably be necessary.

6.1 Briefly explain the difference between self-diffusion and interdiffusion.

6.2 Self-diffusion involves the motion of atoms that are all of the same type; therefore it is not subject to observation by compositional changes, as with interdiffusion. Suggest one way in which self-diffusion may be monitored.

6.3 **(a)** Compare interstitial and vacancy atomic mechanisms for diffusion.
(b) Cite two reasons why interstitial diffusion is normally more rapid than vacancy diffusion.

6.4 Briefly explain the concept of steady state as it applies to diffusion.

6.5 **(a)** Briefly explain the concept of a driving force.
(b) What is the driving force for steady-state diffusion?

6.6 The purification of hydrogen gas by diffusion through a palladium sheet was discussed in Section 6.3. Compute the number of kilograms of hydrogen that pass per hour through a 5-mm thick sheet of palladium having an area of 0.20 m^2 at 500°C. Assume a diffusion coefficient of 1.0×10^{-8} m^2/s, that the concentrations at the high- and low-pressure sides of the plate are 2.4 and 0.6 kg of hydrogen per cubic meter of palladium, and that steady-state conditions have been attained.

6.7 A sheet of steel 1.5 mm thick has nitrogen atmospheres on both sides at 1200°C and is permitted to achieve a steady-state diffusion condition. The diffusion coefficient for nitrogen in steel at this temperature is 6×10^{-11} m^2/s, and the diffusion flux is found to be 1.2×10^{-7} kg/m^2-s. Also, it is known that the concentration of nitrogen in the steel at the high-pressure surface is 4 kg/m^3. How far into the sheet from this high-pressure side will the concentration be 2.0 kg/m^3? Assume a linear concentration profile.

6.8* A sheet of BCC iron 1 mm thick was exposed to a carburizing gas atmosphere on one side

and a decarburizing atmosphere on the other side at 725°C. After having reached steady state, the iron was quickly cooled to room temperature. The carbon concentrations at the two surfaces of the sheet were determined to be 0.012 and 0.0075 wt%. Compute the diffusion coefficient if the diffusion flux is 1.4×10^{-8} kg/m^2-s. *Hint:* Use Equation 5.9 to convert the concentrations from weight percent to kilograms of carbon per cubic meter of iron.

6.9 When α-iron is subjected to an atmosphere of hydrogen gas, the concentration of hydrogen in the iron, C_H (in weight percent), is a function of hydrogen pressure, p_{H_2} (in MPa), and absolute temperature (T) according to

$$C_H = 1.34 \times 10^{-2} \sqrt{p_{H_2}} \exp\left(-\frac{27.2 \text{ kJ/mol}}{RT}\right)$$
(6.11)

Furthermore, the values of D_0 and Q_d for this diffusion system are 1.4×10^{-7} m^2/s and 13,400 J/mol, respectively. Consider a thin iron membrane 1 mm thick that is at 250°C. Compute the diffusion flux through this membrane if the hydrogen pressure on one side of the membrane is 0.15 MPa (1.48 atm), and on the other side 7.5 MPa (74 atm).

6.10 Show that

$$C_x = \frac{B}{\sqrt{Dt}} \exp\left(-\frac{x^2}{4Dt}\right)$$

is also a solution to Equation 6.4b. The parameter B is a constant, being independent of both x and t.

6.11 Determine the carburizing time necessary to achieve a carbon concentration of 0.45 wt% at a position 2 mm into an iron–carbon alloy that initially contains 0.20 wt% C. The surface concentration is to be maintained at 1.30 wt% C, and the treatment is to be conducted at 1000°C. Use the diffusion data for γ-Fe in Table 6.2.

6.12 An FCC iron–carbon alloy initially containing 0.35 wt% C is exposed to an oxygen-rich and virtually carbon-free atmosphere at 1400 K (1127°C). Under these circumstances the carbon diffuses from the alloy and reacts at the surface with the oxygen in the atmosphere; that is, the carbon concentration at the surface position is maintained essentially at 0 wt% C. (This process of carbon depletion is termed *decarburization*.) At what position will the carbon concentration be 0.15 wt% after a 10-h treatment? The value of D at 1400 K is 6.9×10^{-11} m^2/s.

6.13 Nitrogen from a gaseous phase is to be diffused into pure iron at 700°C. If the surface concentration is maintained at 0.1 wt% N, what will be the concentration 1 mm from the surface after 10 h? The diffusion coefficient for nitrogen in iron at 700°C is 2.5×10^{-11} m^2/s.

6.14 **(a)** Consider a diffusion couple composed of two semi-infinite solids of the same metal. Each side of the diffusion couple has a different concentration of the same elemental impurity; furthermore, each impurity level is constant throughout its side of the diffusion couple. Solve Fick's second law for this diffusion situation assuming that the diffusion coefficient for the impurity is independent of concentration, and for the following boundary conditions:

$$C = C_1 \text{ for } x < 0, \text{ and } t = 0$$
$$C = C_2 \text{ for } x > 0, \text{ and } t = 0$$

Here we take the $x = 0$ position to be at the initial diffusion couple boundary.

(b) Using the result of part a, consider a diffusion couple composed of two silver-gold alloys; these alloys have compositions of 98 wt% Ag-2 wt% Au and 95 wt% Ag-5 wt% Au. Determine the time this diffusion couple must be heated at 750°C (1023 K) in order for the composition to be 2.5 wt% Au at the 50 μm position into the 2 wt% Au side of the diffusion couple. Preexponential and activation energy values for Au diffusion in Ag are 8.5×10^{-5} m^2/s and 202,100 J/mol, respectively.

6.15 For a steel alloy it has been determined that a carburizing heat treatment of 10 h duration will raise the carbon concentration to 0.45 wt% at a point 2.5 mm from the surface.

Estimate the time necessary to achieve the same concentration at a 5.0-mm position for an identical steel and at the same carburizing temperature.

6.16 Cite the values of the diffusion coefficients for the interdiffusion of carbon in both α-iron (BCC) and γ-iron (FCC) at 900°C. Which is larger? Explain why this is the case.

6.17 Using the data in Table 6.2, compute the value of D for the diffusion of zinc in copper at 650°C.

6.18 At what temperature will the diffusion coefficient for the diffusion of copper in nickel have a value of 6.5×10^{-17} m^2/s? Use the diffusion data in Table 6.2.

6.19 The preexponential and activation energy for the diffusion of iron in cobalt are 1.1×10^{-5} m^2/s and 253,300 J/mol, respectively. At what temperature will the diffusion coefficient have a value of 2.1×10^{-14} m^2/s?

6.20 The activation energy for the diffusion of carbon in chromium is 111,000 J/mol. Calculate the diffusion coefficient at 1100 K (827°C), given that D at 1400 K (1127°C) is 6.25×10^{-11} m^2/s.

6.21 The diffusion coefficients for iron in nickel are given at two temperatures:

T(K)	D(m^2/s)
1273	9.4×10^{-16}
1473	2.4×10^{-14}

(a) Determine the values of D_0 and the activation energy Q_d.
(b) What is the magnitude of D at 1100°C (1373 K)?

6.22 The diffusion coefficients for silver in copper are given at two temperatures:

T(°C)	D(m^2/s)
650	5.5×10^{-16}
900	1.3×10^{-13}

(a) Determine the values of D_0 and Q_d.
(b) What is the magnitude of D at 875°C?

6.23 Below is shown a plot of the logarithm (to the base 10) of the diffusion coefficient versus reciprocal of the absolute temperature, for the diffusion of iron in chromium. Deter-

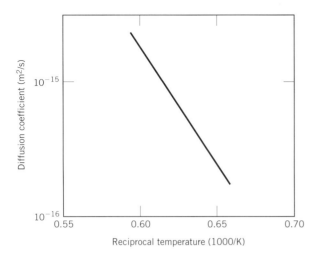

Diffusion coefficient (m²/s)

Reciprocal temperature (1000/K)

mine values for the activation energy and preexponential.

6.24 Carbon is allowed to diffuse through a steel plate 15 mm thick. The concentrations of carbon at the two faces are 0.65 and 0.30 kg C/m³ Fe, which are maintained constant. If the preexponential and activation energy are 6.2×10^{-7} m²/s and 80,000 J/mol, respectively, compute the temperature at which the diffusion flux is 1.43×10^{-9} kg/m²-s.

6.25 The steady-state diffusion flux through a metal plate is 5.4×10^{-10} kg/m²-s at a temperature of 727°C (1000 K) and when the concentration gradient is -350 kg/m⁴. Calculate the diffusion flux at 1027°C (1300 K) for the same concentration gradient and assuming an activation energy for diffusion of 125,000 J/mol.

6.26 At approximately what temperature would a specimen of γ-iron have to be carburized for 2 h to produce the same diffusion result as at 900°C for 15 h?

6.27 (a) Calculate the diffusion coefficient for copper in aluminum at 500°C.
(b) What time will be required at 600°C to produce the same diffusion result (in terms of concentration at a specific point) as for 10 h at 500°C?

6.28 A copper-nickel diffusion couple similar to that shown in Figure 6.1*a* is fashioned. After a 700-h heat treatment at 1100°C (1373 K) the concentration of Cu is 2.5 wt% at the 3.0-mm position within the nickel. At what

temperature must the diffusion couple need to be heated to produce this same concentration (i.e., 2.5 wt% Cu) at a 2.0-mm position after 700 h? The preexponential and activation energy for the diffusion of Cu in Ni are given in Table 6.2.

6.29 A diffusion couple similar to that shown in Figure 6.1*a* is prepared using two hypothetical metals A and B. After a 30-h heat treatment at 1000 K (and subsequently cooling to room temperature) the concentration of A in B is 3.2 wt% at the 15.5-mm position within metal B. If another heat treatment is conducted on an identical diffusion couple, only at 800 K for 30 h, at what position will the composition be 3.2 wt% A? Assume that the preexponential and activation energy for the diffusion coefficient are 1.8×10^{-5} m²/s and 152,000 J/mol, respectively.

6.30 The outer surface of a steel gear is to be hardened by increasing its carbon content. The carbon is to be supplied from an external carbon-rich atmosphere, which is maintained at an elevated temperature. A diffusion heat treatment at 850°C (1123 K) for 10 min increases the carbon concentration to 0.90 wt% at a position 1.0 mm below the surface. Estimate the diffusion time required at 650°C (923 K) to achieve this same concentration also at a 1.0-mm position. Assume that the surface carbon content is the same for both heat treatments, which is maintained constant. Use the diffusion data in Table 6.2 for C diffusion in α-Fe.

6.31 An FCC iron-carbon alloy initially containing 0.20 wt% C is carburized at an elevated temperature and in an atmosphere wherein the surface carbon concentration is maintained at 1.0 wt%. If after 49.5 h the concentration of carbon is 0.35 wt% at a position 4.0 mm below the surface, determine the temperature at which the treatment was carried out.

Design Problems

6.D1 It is desired to enrich the partial pressure of hydrogen in a hydrogen-nitrogen gas mixture for which the partial pressures of both gases are 0.1013 MPa (1 atm). It has been proposed to accomplish this by passing both gases

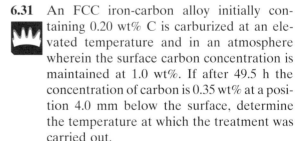

through a thin sheet of some metal at an elevated temperature; inasmuch as hydrogen diffuses through the plate at a higher rate than does nitrogen, the partial pressure of hydrogen will be higher on the exit side of the sheet. The design calls for partial pressures of 0.051 MPa (0.5 atm) and 0.01013 MPa (0.1 atm), respectively, for hydrogen and nitrogen. The concentrations of hydrogen and nitrogen (C_H and C_N, in mol/m^3) in this metal are functions of gas partial pressures (p_{H_2} and p_{N_2}, in MPa) and absolute temperature and are given by the following expressions:

$$C_H = 584 \sqrt{p_{H_2}} \exp\left(-\frac{27.8 \text{ kJ/mol}}{RT}\right)$$
(6.12a)

$$C_N = 2.75 \times 10^3 \sqrt{p_{N_2}} \exp\left(-\frac{37.6 \text{ kJ/mol}}{RT}\right)$$
(6.12b)

Furthermore, the diffusion coefficients for the diffusion of these gases in this metal are functions of the absolute temperature as follows:

$$D_H (\text{m}^2/\text{s}) = 1.4 \times 10^{-7} \exp\left(-\frac{13.4 \text{ kJ/mol}}{RT}\right)$$
(6.13a)

$$D_N (\text{m}^2/\text{s}) = 3.0 \times 10^{-7} \exp\left(-\frac{76.15 \text{ kJ/mol}}{RT}\right)$$
(6.13b)

Is it possible to purify hydrogen gas in this manner? If so, specify a temperature at which the process may be carried out, and also the thickness of metal sheet that would be required. If this procedure is not possible, then state the reason(s) why.

6.D2 A gas mixture is found to contain two diatomic A and B species for which the partial pressures of both are 0.1013 MPa (1 atm). This mixture is to be enriched in the partial pressure of the A species by passing both gases through a thin sheet of some metal at an elevated temperature. The resulting enriched mixture is to have a partial pressure of 0.051 MPa (0.5 atm) for gas A, and 0.0203 MPa (0.2 atm) for gas B. The concentrations of A and B (C_A and C_B, in mol/m^3) are functions

of gas partial pressures (p_{A_2} and p_{B_2}, in MPa) and absolute temperature according to the following expressions:

$$C_A = 500 \sqrt{p_{A_2}} \exp\left(-\frac{20.0 \text{ kJ/mol}}{RT}\right)$$
(6.14a)

$$C_B = 2.0 \times 10^3 \sqrt{p_{B_2}} \exp\left(-\frac{27.0 \text{ kJ/mol}}{RT}\right)$$
(6.14b)

Furthermore, the diffusion coefficients for the diffusion of these gases in the metal are functions of the absolute temperature as follows:

$$D_A (\text{m}^2/\text{s}) = 5.0 \times 10^{-7} \exp\left(-\frac{13.0 \text{ kJ/mol}}{RT}\right)$$
(6.15a)

$$D_B (\text{m}^2/\text{s}) = 3.0 \times 10^{-6} \exp\left(-\frac{21.0 \text{ kJ/mol}}{RT}\right)$$
(6.15b)

Is it possible to purify the A gas in this manner? If so, specify a temperature at which the process may be carried out, and also the thickness of metal sheet that would be required. If this procedure is not possible, then state the reason(s) why.

6.D3 The wear resistance of a steel shaft is to be improved by hardening its surface. This is to be accomplished by increasing the nitrogen content within an outer surface layer as a result of nitrogen diffusion into the steel. The nitrogen is to be supplied from an external nitrogen-rich gas at an elevated and constant temperature. The initial nitrogen content of the steel is 0.002 wt%, whereas the surface concentration is to be maintained at 0.50 wt%. In order for this treatment to be effective, a nitrogen content of 0.10 wt% must be established at a position 0.40 mm below the surface. Specify appropriate heat treatments in terms of temperature and time for temperatures between 475°C and 625°C. The preexponential and activation energy for the diffusion of nitrogen in iron are 3×10^{-7} m^2/s and 76,150 J/mol, respectively, over this temperature range.

Chapter 7 / Mechanical Properties

A modern Rockwell hardness tester. (Photograph courtesy of Wilson Instruments Division, Instron Corporation, originator of the Rockwell® Hardness Tester.)

Why Study *Mechanical Properties?*

It is incumbent on engineers to understand how the various mechanical properties are measured and what these properties represent; they may be called upon to design structures/components using prede-termined materials such that unacceptable levels of deformation and/or failure will not occur. We demonstrate this procedure with respect to the design of a tensile-testing apparatus in Design Example 7.1.

147

After studying this chapter you should be able to do the following:

1. Define engineering stress and engineering strain.
2. State Hooke's law, and note the conditions under which it is valid.
3. Define Poisson's ratio.
4. Given an engineering stress–strain diagram, determine (a) the modulus of elasticity, (b) the yield strength (0.002 strain offset), and (c) the tensile strength, and (d) estimate the percent elongation.
5. For the tensile deformation of a ductile cylindrical specimen, describe changes in specimen profile to the point of fracture.
6. Compute ductility in terms of both percent elongation and percent reduction of area for a material that is loaded in tension to fracture.

7. Compute the flexural strengths of ceramic rod specimens that have bent to fracture in three-point loading.
8. Make schematic plots of the three characteristic stress–strain behaviors observed for polymeric materials.
9. Name the two most common hardness-testing techniques; note two differences between them.
10. (a) Name and briefly describe the two different microhardness testing techniques, and (b) cite situations for which these techniques are generally used.
11. Compute the working stress for a ductile material.

7.1 INTRODUCTION

Many materials, when in service, are subjected to forces or loads; examples include the aluminum alloy from which an airplane wing is constructed and the steel in an automobile axle. In such situations it is necessary to know the characteristics of the material and to design the member from which it is made such that any resulting deformation will not be excessive and fracture will not occur. The mechanical behavior of a material reflects the relationship between its response or deformation to an applied load or force. Important mechanical properties are strength, hardness, ductility, and stiffness.

The mechanical properties of materials are ascertained by performing carefully designed laboratory experiments that replicate as nearly as possible the service conditions. Factors to be considered include the nature of the applied load and its duration, as well as the environmental conditions. It is possible for the load to be tensile, compressive, or shear, and its magnitude may be constant with time, or it may fluctuate continuously. Application time may be only a fraction of a second, or it may extend over a period of many years. Service temperature may be an important factor.

Mechanical properties are of concern to a variety of parties (e.g., producers and consumers of materials, research organizations, government agencies) that have differing interests. Consequently, it is imperative that there be some consistency in the manner in which tests are conducted, and in the interpretation of their results. This consistency is accomplished by using standardized testing techniques. Establishment and publication of these standards are often coordinated by professional societies. In the United States the most active organization is the American Society for Testing and Materials (ASTM). Its *Annual Book of ASTM Standards* comprises numerous volumes, which are issued and updated yearly; a large number of these standards relate to mechanical testing techniques. Several of these are referenced by footnote in this and subsequent chapters.

The role of structural engineers is to determine stresses and stress distributions within members that are subjected to well-defined loads. This may be accomplished by experimental testing techniques and/or by theoretical and mathematical stress analyses. These topics are treated in traditional stress analysis and strength of materials texts.

Materials and metallurgical engineers, on the other hand, are concerned with producing and fabricating materials to meet service requirements as predicted by these stress analyses. This necessarily involves an understanding of the relationships between the microstructure (i.e., internal features) of materials and their mechanical properties.

Materials are frequently chosen for structural applications because they have desirable combinations of mechanical characteristics. This chapter discusses the stress–strain behaviors of metals, ceramics, and polymers and the related mechanical properties; it also examines their other important mechanical characteristics. Discussions of the microscopic aspects of deformation mechanisms and methods to strengthen and regulate the mechanical behaviors are deferred to Chapter 8.

7.2 CONCEPTS OF STRESS AND STRAIN

If a load is static or changes relatively slowly with time and is applied uniformly over a cross section or surface of a member, the mechanical behavior may be ascertained by a simple stress–strain test; these are most commonly conducted for metals at room temperature. There are three principal ways in which a load may be applied: namely, tension, compression, and shear (Figures 7.1a, b, c). In engineering practice many loads are torsional rather than pure shear; this type of loading is illustrated in Figure 7.1d.

TENSION TESTS[1]

One of the most common mechanical stress–strain tests is performed in *tension.* As will be seen, the tension test can be used to ascertain several mechanical properties of materials that are important in design. A specimen is deformed, usually to fracture, with a gradually increasing tensile load that is applied uniaxially along the long axis of a specimen. A standard tensile specimen is shown in Figure 7.2. Normally, the cross section is circular, but rectangular specimens are also used. During testing, deformation is confined to the narrow center region, which has a uniform cross section along its length. The standard diameter is approximately 12.8 mm (0.5 in.), whereas the reduced section length should be at least four times this diameter; 60 mm ($2\frac{1}{4}$ in.) is common. Gauge length is used in ductility computations, as discussed in Section 7.6; the standard value is 50 mm (2.0 in.). The specimen is mounted by its ends into the holding grips of the testing apparatus (Figure 7.3). The tensile testing machine is designed to elongate the specimen at a constant rate, and to continuously and simultaneously measure the instantaneous applied load (with a load cell) and the resulting elongations (using an extensometer). A stress–strain test typically takes several minutes to perform and is destructive; that is, the test specimen is permanently deformed and usually fractured.

The output of such a tensile test is recorded on a strip chart (or by a computer) as load or force versus elongation. These load–deformation characteristics are dependent on the specimen size. For example, it will require twice the load to produce the same elongation if the cross-sectional area of the specimen is doubled. To minimize these geometrical factors, load and elongation are normalized to the respective parameters of **engineering stress** and **engineering strain.** Engineering

---

[1] ASTM Standards E 8 and E 8M, "Standard Test Methods for Tension Testing of Metallic Materials."

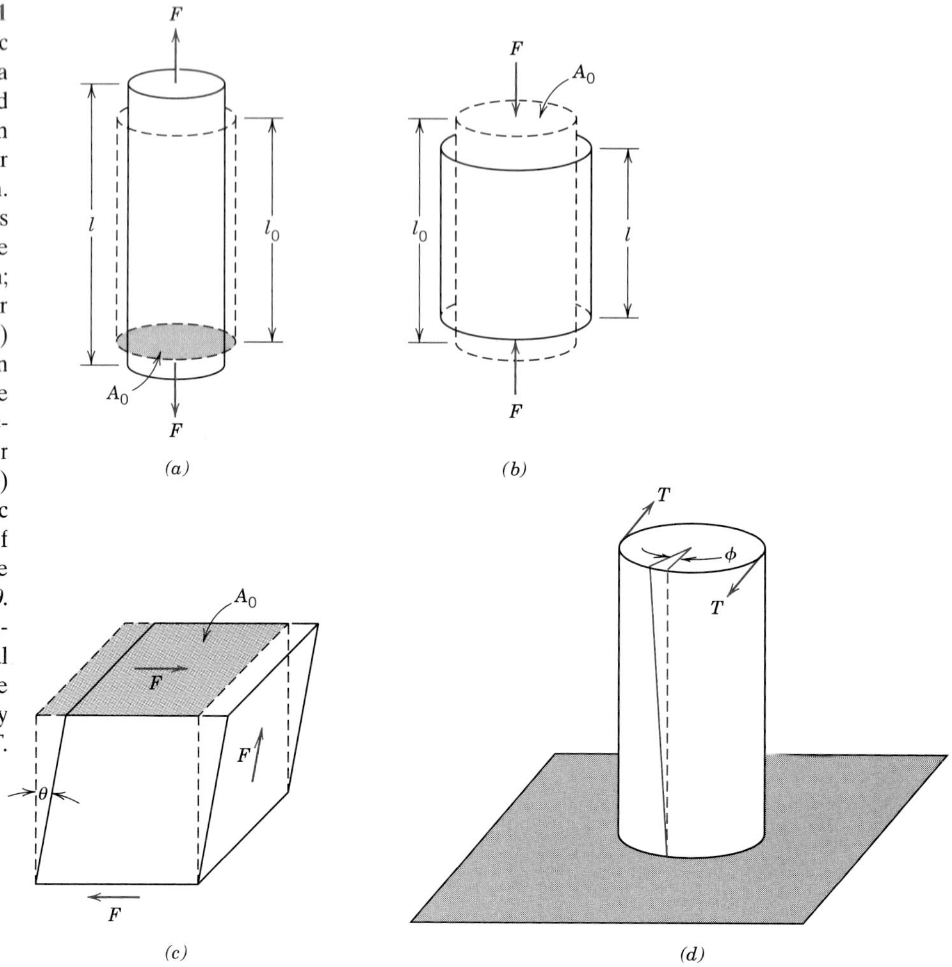

FIGURE 7.1 (a) Schematic illustration of how a tensile load produces an elongation and positive linear strain. Dashed lines represent the shape before deformation; solid lines, after deformation. (b) Schematic illustration of how a compressive load produces contraction and a negative linear strain. (c) Schematic representation of shear strain γ, where $\gamma = \tan \theta$. (d) Schematic representation of torsional deformation (i.e., angle of twist ϕ) produced by an applied torque T.

stress σ is defined by the relationship

$$\sigma = \frac{F}{A_0} \qquad (7.1)$$

in which F is the instantaneous load applied perpendicular to the specimen cross section, in units of newtons (N) or pounds force (lb_f), and A_0 is the original cross-sectional area before any load is applied (m^2 or $in.^2$). The units of engineering

FIGURE 7.2 A standard tensile specimen with circular cross section.

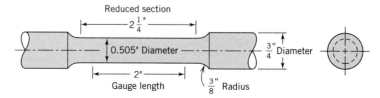

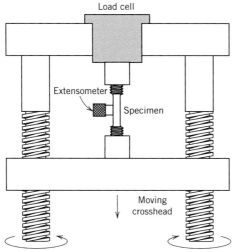

Load cell

Extensometer

Specimen

Moving
crosshead

FIGURE 7.3 Schematic representation of the apparatus used to conduct tensile stress–strain tests. The specimen is elongated by the moving crosshead; load cell and extensometer measure, respectively, the magnitude of the applied load and the elongation. (Adapted from H. W. Hayden, W. G. Moffatt, and J. Wulff, *The Structure and Properties of Materials,* Vol. III, *Mechanical Behavior,* p. 2. Copyright © 1965 by John Wiley & Sons, New York. Reprinted by permission of John Wiley & Sons, Inc.)

stress (referred to subsequently as just stress) are megapascals, MPa (SI) (where 1 MPa = 10^6 N/m^2), and pounds force per square inch, psi (Customary U.S.).[2]

Engineering strain ϵ is defined according to

$$\epsilon = \frac{l_i - l_0}{l_0} = \frac{\Delta l}{l_0} \tag{7.2}$$

in which l_0 is the original length before any load is applied, and l_i is the instantaneous length. Sometimes the quantity $l_i - l_0$ is denoted as Δl, and is the deformation elongation or change in length at some instant, as referenced to the original length. Engineering strain (subsequently called just strain) is unitless, but meters per meter or inches per inch are often used; the value of strain is obviously independent of the unit system. Sometimes strain is also expressed as a percentage, in which the strain value is multiplied by 100.

COMPRESSION TESTS[3]

Compression stress–strain tests may be conducted if in-service forces are of this type. A compression test is conducted in a manner similar to the tensile test, except that the force is compressive and the specimen contracts along the direction of the stress. Equations 7.1 and 7.2 are utilized to compute compressive stress and strain, respectively. By convention, a compressive force is taken to be negative, which yields a negative stress. Furthermore, since l_0 is greater than l_i, compressive strains computed from Equation 7.2 are necessarily also negative. Tensile tests are more common because they are easier to perform; also, for most materials used in structural applications, very little additional information is obtained from compressive tests. Compressive tests are used when a material's behavior under large and perma-

---

[2] Conversion from one system of stress units to the other is accomplished by the relationship 145 psi = 1 MPa.

[3] ASTM Standard E 9, "Standard Test Methods of Compression Testing of Metallic Materials at Room Temperature."

nent (i.e., plastic) strains is desired, as in manufacturing applications, or when the material is brittle in tension.

SHEAR AND TORSIONAL TESTS[4]

For tests performed using a pure shear force as shown in Figure 7.1c, the shear stress τ is computed according to

$$\tau = \frac{F}{A_0} \tag{7.3}$$

where F is the load or force imposed parallel to the upper and lower faces, each of which has an area of A_0. The shear strain γ is defined as the tangent of the strain angle θ, as indicated in the figure. The units for shear stress and strain are the same as for their tensile counterparts.

Torsion is a variation of pure shear, wherein a structural member is twisted in the manner of Figure 7.1d; torsional forces produce a rotational motion about the longitudinal axis of one end of the member relative to the other end. Examples of torsion are found for machine axles and drive shafts, and also for twist drills. Torsional tests are normally performed on cylindrical solid shafts or tubes. A shear stress τ is a function of the applied torque T, whereas shear strain γ is related to the angle of twist, ϕ in Figure 7.1d.

GEOMETRIC CONSIDERATIONS OF THE STRESS STATE

Stresses that are computed from the tensile, compressive, shear, and torsional force states represented in Figure 7.1 act either parallel or perpendicular to planar faces of the bodies represented in these illustrations. It should be noted that the stress state is a function of the orientations of the planes upon which the stresses are taken to act. For example, consider the cylindrical tensile specimen of Figure 7.4

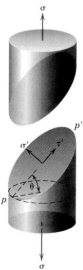

FIGURE 7.4 Schematic representation showing normal (σ') and shear (τ') stresses that act on a plane oriented at an angle θ relative to the plane taken perpendicular to the direction along which a pure tensile stress (σ) is applied.

---

[4] ASTM Standard E 143, "Standard Test for Shear Modulus."

that is subjected to a tensile stress σ applied parallel to its axis. Furthermore, consider also the plane p-p' that is oriented at some arbitrary angle θ relative to the plane of the specimen end-face. Upon this plane p-p', the applied stress is no longer a pure tensile one. Rather, a more complex stress state is present that consists of a tensile (or normal) stress σ' that acts normal to the p-p' plane, and, in addition, a shear stress τ' that acts parallel to this plane; both of these stresses are represented in the figure. Using mechanics of materials principles,[5] it is possible to develop equations for σ' and τ' in terms of σ and θ, as follows:

$$\sigma' = \sigma \cos^2 \theta = \sigma \left(\frac{1 + \cos 2\theta}{2} \right) \tag{7.4a}$$

$$\tau' = \sigma \sin \theta \cos \theta = \sigma \left(\frac{\sin 2\theta}{2} \right) \tag{7.4b}$$

These same mechanics principles allow the transformation of stress components from one coordinate system to another coordinate system that has a different orientation. Such treatments are beyond the scope of the present discussion.

ELASTIC DEFORMATION

7.3 STRESS–STRAIN BEHAVIOR

The degree to which a structure deforms or strains depends on the magnitude of an imposed stress. For most metals that are stressed in tension and at relatively low levels, stress and strain are proportional to each other through the relationship

$$\sigma = E\epsilon \tag{7.5}$$

This is known as Hooke's law, and the constant of proportionality E (GPa or psi)[6] is the **modulus of elasticity,** or *Young's modulus.* For most typical metals the magnitude of this modulus ranges between 45 GPa (6.5×10^6 psi), for magnesium, and 407 GPa (59×10^6 psi), for tungsten. The moduli of elasticity are slightly higher for ceramic materials, which range between about 70 and 500 GPa (10×10^6 and 70×10^6 psi). Polymers have modulus values that are smaller than both metals and ceramics, and which lie in the range 0.007 and 4 GPa (10^3 and 0.6×10^6 psi). Room

temperature modulus of elasticity values for a number of metals, ceramics, and polymers are presented in Table 7.1. A more comprehensive modulus list is provided in Table B.2, Appendix B.

Deformation in which stress and strain are proportional is called **elastic deformation;** a plot of stress (ordinate) versus strain (abscissa) results in a linear relationship, as shown in Figure 7.5. The slope of this linear segment corresponds to the modulus of elasticity E. This modulus may be thought of as stiffness, or a material's resistance to elastic deformation. The greater the modulus, the stiffer the material, or the smaller the elastic strain that results from the application of a given stress. The modulus is an important design parameter used for computing elastic deflections.

Elastic deformation is nonpermanent, which means that when the applied load is released, the piece returns to its original shape. As shown in the stress–strain

---

[5] See, for example, W. F. Riley, L. D. Sturges, and D. H. Morris, *Mechanics of Materials,* 5th edition, John Wiley & Sons, New York, 1999.
[6] The SI unit for the modulus of elasticity is gigapascal, GPa, where 1 GPa $= 10^9$ N/m$^2 =$ 10^3 MPa.

Table 7.1   Room-Temperature Elastic and Shear Moduli, and Poisson's Ratio for Various Materials

Material	Modulus of Elasticity		Shear Modulus		Poisson's Ratio
	GPa	10^6 psi	GPa	10^6 psi	
Metal Alloys					
Tungsten	407	59	160	23.2	0.28
Steel	207	30	83	12.0	0.30
Nickel	207	30	76	11.0	0.31
Titanium	107	15.5	45	6.5	0.34
Copper	110	16	46	6.7	0.34
Brass	97	14	37	5.4	0.34
Aluminum	69	10	25	3.6	0.33
Magnesium	45	6.5	17	2.5	0.35
Ceramic Materials					
Aluminum oxide (Al_2O_3)	393	57	—	—	0.22
Silicon carbide (SiC)	345	50	—	—	0.17
Silicon nitride (Si_3N_4)	304	44	—	—	0.30
Spinel ($MgAl_2O_4$)	260	38	—	—	—
Magnesium oxide (MgO)	225	33	—	—	0.18
Zirconia[a]	205	30	—	—	0.31
Mullite ($3Al_2O_3$-$2SiO_2$)	145	21	—	—	0.24
Glass–ceramic (Pyroceram)	120	17	—	—	0.25
Fused silica (SiO_2)	73	11	—	—	0.17
Soda–lime glass	69	10	—	—	0.23
Polymers[b]					
Phenol-formaldehyde	2.76–4.83	0.40–0.70	—	—	—
Polyvinyl chloride (PVC)	2.41–4.14	0.35–0.60	—	—	0.38
Polyester (PET)	2.76–4.14	0.40–0.60	—	—	—
Polystyrene (PS)	2.28–3.28	0.33–0.48	—	—	0.33
Polymethyl methacrylate (PMMA)	2.24–3.24	0.33–0.47	—	—	—
Polycarbonate (PC)	2.38	0.35	—	—	0.36
Nylon 6,6	1.58–3.80	0.23–0.55	—	—	0.39
Polypropylene (PP)	1.14–1.55	0.17–0.23	—	—	—
Polyethylene—high density (HDPE)	1.08	0.16	—	—	—
Polytetrafluoroethylene (PTFE)	0.40–0.55	0.058–0.080	—	—	0.46
Polyethylene—low density (LDPE)	0.17–0.28	0.025–0.041	—	—	—

[a] Partially stabilized with 3 mol% Y_2O_3.

[b] **Source:** *Modern Plastics Encyclopedia '96.* Copyright 1995, The McGraw-Hill Companies. Reprinted with permission.

plot (Figure 7.5), application of the load corresponds to moving from the origin up and along the straight line. Upon release of the load, the line is traversed in the opposite direction, back to the origin.

There are some materials (e.g., gray cast iron, concrete, and many polymers) for which this initial elastic portion of the stress–strain curve is not linear (Figure

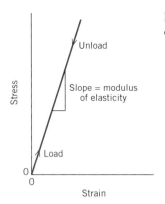

FIGURE 7.5 Schematic stress–strain diagram showing linear elastic deformation for loading and unloading cycles.

7.6); hence, it is not possible to determine a modulus of elasticity as described above. For this nonlinear behavior, either *tangent* or *secant modulus* is normally used. Tangent modulus is taken as the slope of the stress–strain curve at some specified level of stress, while secant modulus represents the slope of a secant drawn from the origin to some given point of the σ–ϵ curve. The determination of these moduli is illustrated in Figure 7.6.

On an atomic scale, macroscopic elastic strain is manifested as small changes in the interatomic spacing and the stretching of interatomic bonds. As a consequence, the magnitude of the modulus of elasticity is a measure of the resistance to separation of adjacent atoms/ions/molecules, that is, the interatomic bonding forces. Furthermore, this modulus is proportional to the slope of the interatomic force–separation curve (Figure 2.8a) at the equilibrium spacing:

$$E \propto \left(\frac{dF}{dr} \right)_{r_0} \tag{7.6}$$

Figure 7.7 shows the force–separation curves for materials having both strong and weak interatomic bonds; the slope at r_0 is indicated for each.

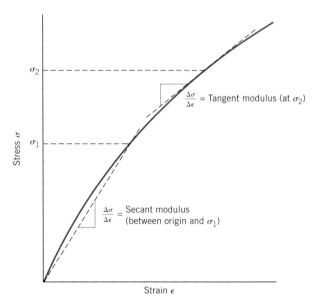

FIGURE 7.6 Schematic stress–strain diagram showing nonlinear elastic behavior, and how secant and tangent moduli are determined.

FIGURE 7.7 Force versus interatomic separation for weakly and strongly bonded atoms. The magnitude of the modulus of elasticity is proportional to the slope of each curve at the equilibrium interatomic separation r_0.

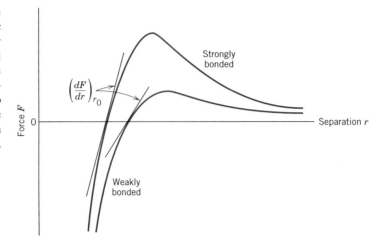

Differences in modulus values between metals, ceramics, and polymers are a direct consequence of the different types of atomic bonding that exist for the three materials types. Furthermore, with increasing temperature, the modulus of elasticity diminishes for all but some of the rubber materials; this effect is shown for several metals in Figure 7.8.

As would be expected, the imposition of compressive, shear, or torsional stresses also evokes elastic behavior. The stress–strain characteristics at low stress levels are virtually the same for both tensile and compressive situations, to include the magnitude of the modulus of elasticity. Shear stress and strain are proportional to each other through the expression

$$\tau = G\gamma \tag{7.7}$$

FIGURE 7.8 Plot of modulus of elasticity versus temperature for tungsten, steel, and aluminum. (Adapted from K. M. Ralls, T. H. Courtney, and J. Wulff, *Introduction to Materials Science and Engineering.* Copyright © 1976 by John Wiley & Sons, New York. Reprinted by permission of John Wiley & Sons, Inc.)

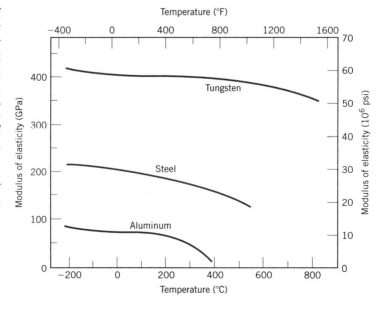

where G is the *shear modulus,* the slope of the linear elastic region of the shear stress–strain curve. Table 7.1 also gives the shear moduli for a number of the common metals.

7.4 ANELASTICITY

Up to this point, it has been assumed that elastic deformation is time independent, that is, that an applied stress produces an instantaneous elastic strain that remains constant over the period of time the stress is maintained. It has also been assumed that upon release of the load the strain is totally recovered, that is, that the strain immediately returns to zero. In most engineering materials, however, there will also exist a time-dependent elastic strain component. That is, elastic deformation will continue after the stress application, and upon load release some finite time is required for complete recovery. This time-dependent elastic behavior is known as **anelasticity,** and it is due to time-dependent microscopic and atomistic processes that are attendant to the deformation. For metals the anelastic component is normally small and is often neglected. However, for some polymeric materials its magnitude is significant; in this case it is termed *viscoelastic behavior,* {which is the discussion topic of Section 7.15.}

EXAMPLE PROBLEM 7.1

A piece of copper originally 305 mm (12 in.) long is pulled in tension with a stress of 276 MPa (40,000 psi). If the deformation is entirely elastic, what will be the resultant elongation?

SOLUTION

Since the deformation is elastic, strain is dependent on stress according to Equation 7.5. Furthermore, the elongation Δl is related to the original length l_0 through Equation 7.2. Combining these two expressions and solving for Δl yields

$$\sigma = \epsilon E = \left(\frac{\Delta l}{l_0}\right) E$$

$$\Delta l = \frac{\sigma l_0}{E}$$

The values of σ and l_0 are given as 276 MPa and 305 mm, respectively, and the magnitude of E for copper from Table 7.1 is 110 GPa (16×10^6 psi). Elongation is obtained by substitution into the expression above as

$$\Delta l = \frac{(276 \text{ MPa})(305 \text{ mm})}{110 \times 10^3 \text{ MPa}} = 0.77 \text{ mm } (0.03 \text{ in.})$$

7.5 ELASTIC PROPERTIES OF MATERIALS

When a tensile stress is imposed on virtually all materials, an elastic elongation and accompanying strain ϵ_z result in the direction of the applied stress (arbitrarily taken to be the z direction), as indicated in Figure 7.9. As a result of this elongation,

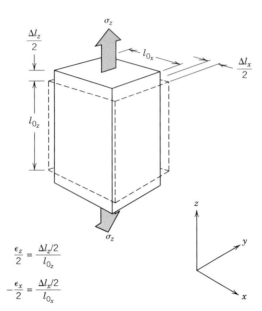

FIGURE 7.9 Axial (z) elongation (positive strain) and lateral (x and y) contractions (negative strains) in response to an imposed tensile stress. Solid lines represent dimensions after stress application; dashed lines, before.

there will be constrictions in the lateral (x and y) directions perpendicular to the applied stress; from these contractions, the compressive strains ϵ_x and ϵ_y may be determined. If the applied stress is uniaxial (only in the z direction), and the material is isotropic, then $\epsilon_x = \epsilon_y$. A parameter termed **Poisson's ratio** ν is defined as the ratio of the lateral and axial strains, or

$$\nu = -\frac{\epsilon_x}{\epsilon_z} = -\frac{\epsilon_y}{\epsilon_z} \tag{7.8}$$

The negative sign is included in the expression so that ν will always be positive, since ϵ_x and ϵ_z will always be of opposite sign. Theoretically, Poisson's ratio for isotropic materials should be $\frac{1}{4}$; furthermore, the maximum value for ν (or that value for which there is no net volume change) is 0.50. For many metals and other alloys, values of Poisson's ratio range between 0.25 and 0.35. Table 7.1 shows ν values for several common materials; a more comprehensive list is given in Table B.3, Appendix B.

For isotropic materials, shear and elastic moduli are related to each other and to Poisson's ratio according to

$$E = 2G(1 + \nu) \tag{7.9}$$

In most metals G is about $0.4E$; thus, if the value of one modulus is known, the other may be approximated.

Many materials are elastically anisotropic; that is, the elastic behavior (e.g., the magnitude of E) varies with crystallographic direction (see Table 3.7). For these materials the elastic properties are completely characterized only by the specification of several elastic constants, their number depending on characteristics of the crystal structure. Even for isotropic materials, for complete characterization of the elastic properties, at least two constants must be given. Since the grain orientation is random in most polycrystalline materials, these may be considered to be isotropic;

inorganic ceramic glasses are also isotropic. The remaining discussion of mechanical behavior assumes isotropy and polycrystallinity (for metals and crystalline ceramics) because such is the character of most engineering materials.

EXAMPLE PROBLEM 7.2

A tensile stress is to be applied along the long axis of a cylindrical brass rod that has a diameter of 10 mm (0.4 in.). Determine the magnitude of the load required to produce a 2.5×10^{-3} mm (10^{-4} in.) change in diameter if the deformation is entirely elastic.

SOLUTION

This deformation situation is represented in the accompanying drawing.

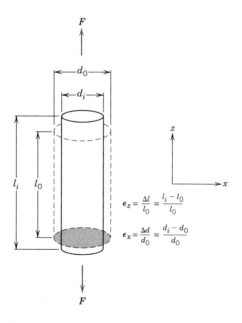

$$\epsilon_z = \frac{\Delta l}{l_0} = \frac{l_i - l_0}{l_0}$$

$$\epsilon_x = \frac{\Delta d}{d_0} = \frac{d_i - d_0}{d_0}$$

When the force F is applied, the specimen will elongate in the z direction and at the same time experience a reduction in diameter, Δd, of 2.5×10^{-3} mm in the x direction. For the strain in the x direction,

$$\epsilon_x = \frac{\Delta d}{d_0} = \frac{-2.5 \times 10^{-3} \text{ mm}}{10 \text{ mm}} = -2.5 \times 10^{-4}$$

which is negative, since the diameter is reduced.

It next becomes necessary to calculate the strain in the z direction using Equation 7.8. The value for Poisson's ratio for brass is 0.34 (Table 7.1), and thus

$$\epsilon_z = -\frac{\epsilon_x}{\nu} = -\frac{(-2.5 \times 10^{-4})}{0.34} = 7.35 \times 10^{-4}$$

The applied stress may now be computed using Equation 7.5 and the modulus of elasticity, given in Table 7.1 as 97 GPa (14×10^6 psi), as

$$\sigma = \epsilon_z E = (7.35 \times 10^{-4})(97 \times 10^3 \text{ MPa}) = 71.3 \text{ MPa}$$

Finally, from Equation 7.1, the applied force may be determined as

$$F = \sigma A_0 = \sigma \left(\frac{d_0}{2}\right)^2 \pi$$

$$= (71.3 \times 10^6 \text{ N/m}^2) \left(\frac{10 \times 10^{-3} \text{ m}}{2}\right)^2 \pi = 5600 \text{ N (1293 lb}_f)$$

MECHANICAL BEHAVIOR—METALS

For most metallic materials, elastic deformation persists only to strains of about 0.005. As the material is deformed beyond this point, the stress is no longer proportional to strain (Hooke's law, Equation 7.5, ceases to be valid), and permanent, nonrecoverable, or **plastic deformation** occurs. Figure 7.10a plots schematically the tensile stress–strain behavior into the plastic region for a typical metal. The transition from elastic to plastic is a gradual one for most metals; some curvature results at the onset of plastic deformation, which increases more rapidly with rising stress.

From an atomic perspective, plastic deformation corresponds to the breaking of bonds with original atom neighbors and then reforming bonds with new neighbors as large numbers of atoms or molecules move relative to one another; upon removal of the stress they do not return to their original positions. This permanent deformation for metals is accomplished by means of a process called slip, which involves the motion of dislocations as discussed in Section 8.2.

7.6 TENSILE PROPERTIES

YIELDING AND YIELD STRENGTH

Most structures are designed to ensure that only elastic deformation will result when a stress is applied. It is therefore desirable to know the stress level at which

FIGURE 7.10
(a) Typical stress–strain behavior for a metal showing elastic and plastic deformations, the proportional limit P, and the yield strength σ_y, as determined using the 0.002 strain offset method.
(b) Representative stress–strain behavior found for some steels demonstrating the yield point phenomenon.

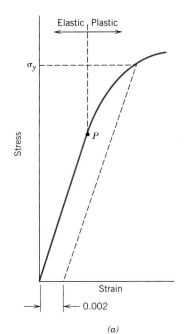

(a)

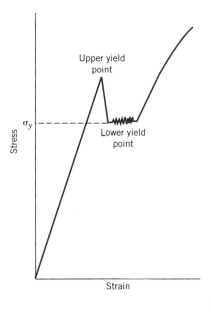

(b)

plastic deformation begins, or where the phenomenon of **yielding** occurs. For metals that experience this gradual elastic–plastic transition, the point of yielding may be determined as the initial departure from linearity of the stress–strain curve; this is sometimes called the **proportional limit,** as indicated by point P in Figure 7.10a. In such cases the position of this point may not be determined precisely. As a consequence, a convention has been established wherein a straight line is constructed parallel to the elastic portion of the stress–strain curve at some specified strain offset, usually 0.002. The stress corresponding to the intersection of this line and the stress–strain curve as it bends over in the plastic region is defined as the **yield strength** σ_y.[7] This is demonstrated in Figure 7.10a. Of course, the units of yield strength are MPa or psi.[8]

For those materials having a nonlinear elastic region (Figure 7.6), use of the strain offset method is not possible, and the usual practice is to define the yield strength as the stress required to produce some amount of strain (e.g., $\epsilon = 0.005$).

Some steels and other materials exhibit the tensile stress–strain behavior as shown in Figure 7.10b. The elastic–plastic transition is very well defined and occurs abruptly in what is termed a *yield point phenomenon.* At the upper yield point, plastic deformation is initiated with an actual decrease in stress. Continued deformation fluctuates slightly about some constant stress value, termed the lower yield point; stress subsequently rises with increasing strain. For metals that display this effect, the yield strength is taken as the average stress that is associated with the lower yield point, since it is well defined and relatively insensitive to the testing procedure.[9] Thus, it is not necessary to employ the strain offset method for these materials.

The magnitude of the yield strength for a metal is a measure of its resistance to plastic deformation. Yield strengths may range from 35 MPa (5000 psi) for a low-strength aluminum to over 1400 MPa (200,000 psi) for high-strength steels.

TENSILE STRENGTH

After yielding, the stress necessary to continue plastic deformation in metals increases to a maximum, point M in Figure 7.11, and then decreases to the eventual fracture, point F. The **tensile strength** TS (MPa or psi) is the stress at the maximum on the engineering stress–strain curve (Figure 7.11). This corresponds to the maximum stress that can be sustained by a structure in tension; if this stress is applied and maintained, fracture will result. All deformation up to this point is uniform throughout the narrow region of the tensile specimen. However, at this maximum stress, a small constriction or neck begins to form at some point, and all subsequent deformation is confined at this neck, as indicated by the schematic specimen insets in Figure 7.11. This phenomenon is termed "necking," and fracture ultimately occurs at the neck. The fracture strength corresponds to the stress at fracture.

Tensile strengths may vary anywhere from 50 MPa (7000 psi) for an aluminum to as high as 3000 MPa (450,000 psi) for the high-strength steels. Ordinarily, when the strength of a metal is cited for design purposes, the yield strength is used. This is because by the time a stress corresponding to the tensile strength has been

---

[7] "Strength" is used in lieu of "stress" because strength is a property of the metal, whereas stress is related to the magnitude of the applied load.

[8] For Customary U.S. units, the unit of kilopounds per square inch (ksi) is sometimes used for the sake of convenience, where

$$1 \text{ ksi} = 1000 \text{ psi}$$

[9] It should be pointed out that to observe the yield point phenomenon, a "stiff" tensile-testing apparatus must be used; by stiff is meant that there is very little elastic deformation of the machine during loading.

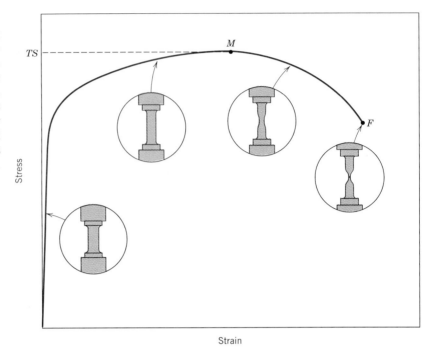

applied, often a structure has experienced so much plastic deformation that it is useless. Furthermore, fracture strengths are not normally specified for engineering design purposes.

EXAMPLE PROBLEM 7.3

From the tensile stress–strain behavior for the brass specimen shown in Figure 7.12, determine the following:

(a) The modulus of elasticity.

(b) The yield strength at a strain offset of 0.002.

(c) The maximum load that can be sustained by a cylindrical specimen having an original diameter of 12.8 mm (0.505 in.).

(d) The change in length of a specimen originally 250 mm (10 in.) long that is subjected to a tensile stress of 345 MPa (50,000 psi).

SOLUTION

(a) The modulus of elasticity is the slope of the elastic or initial linear portion of the stress–strain curve. The strain axis has been expanded in the inset, Figure 7.12, to facilitate this computation. The slope of this linear region is the rise over the run, or the change in stress divided by the corresponding change in strain; in mathematical terms,

$$E = \text{slope} = \frac{\Delta\sigma}{\Delta\epsilon} = \frac{\sigma_2 - \sigma_1}{\epsilon_2 - \epsilon_1} \tag{7.10}$$

Inasmuch as the line segment passes through the origin, it is convenient to take both σ_1 and ϵ_1 as zero. If σ_2 is arbitrarily taken as 150 MPa, then ϵ_2 will have

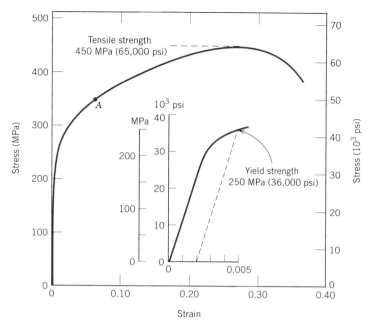

FIGURE 7.12 The stress–strain behavior for the brass specimen discussed in Example Problem 7.3.

a value of 0.0016. Therefore,

$$E = \frac{(150 - 0)\ \text{MPa}}{0.0016 - 0} = 93.8\ \text{GPa}\ (13.6 \times 10^6\ \text{psi})$$

which is very close to the value of 97 GPa (14×10^6 psi) given for brass in Table 7.1.

(b) The 0.002 strain offset line is constructed as shown in the inset; its intersection with the stress–strain curve is at approximately 250 MPa (36,000 psi), which is the yield strength of the brass.

(c) The maximum load that can be sustained by the specimen is calculated by using Equation 7.1, in which σ is taken to be the tensile strength, from Figure 7.12, 450 MPa (65,000 psi). Solving for F, the maximum load, yields

$$F = \sigma A_0 = \sigma \left(\frac{d_0}{2}\right)^2 \pi$$

$$= (450 \times 10^6\ \text{N/m}^2) \left(\frac{12.8 \times 10^{-3}\ \text{m}}{2}\right)^2 \pi = 57{,}900\ \text{N}\ (13{,}000\ \text{lb}_\text{f})$$

(d) To compute the change in length, Δl, in Equation 7.2, it is first necessary to determine the strain that is produced by a stress of 345 MPa. This is accomplished by locating the stress point on the stress–strain curve, point A, and reading the corresponding strain from the strain axis, which is approximately 0.06. Inasmuch as $l_0 = 250$ mm, we have

$$\Delta l = \epsilon l_0 = (0.06)(250\ \text{mm}) = 15\ \text{mm}\ (0.6\ \text{in.})$$

DUCTILITY

Ductility is another important mechanical property. It is a measure of the degree of plastic deformation that has been sustained at fracture. A material that experiences very little or no plastic deformation upon fracture is termed *brittle*. The tensile stress–strain behaviors for both ductile and brittle materials are schematically illustrated in Figure 7.13.

Ductility may be expressed quantitatively as either *percent elongation* or *percent reduction in area*. The percent elongation %EL is the percentage of plastic strain at fracture, or

$$\%\text{EL} = \left(\frac{l_f - l_0}{l_0} \right) \times 100 \qquad (7.11)$$

where l_f is the fracture length[10] and l_0 is the original gauge length as above. Inasmuch as a significant proportion of the plastic deformation at fracture is confined to the neck region, the magnitude of %EL will depend on specimen gauge length. The shorter l_0, the greater is the fraction of total elongation from the neck and, consequently, the higher the value of %EL. Therefore, l_0 should be specified when percent elongation values are cited; it is commonly 50 mm (2 in.).

Percent reduction in area %RA is defined as

$$\%\text{RA} = \left(\frac{A_0 - A_f}{A_0} \right) \times 100 \qquad (7.12)$$

where A_0 is the original cross-sectional area and A_f is the cross-sectional area at the point of fracture.[10] Percent reduction in area values are independent of both l_0 and A_0. Furthermore, for a given material the magnitudes of %EL and %RA will, in general, be different. Most metals possess at least a moderate degree of ductility at room temperature; however, some become brittle as the temperature is lowered (Section 9.8).

A knowledge of the ductility of materials is important for at least two reasons. First, it indicates to a designer the degree to which a structure will deform plastically

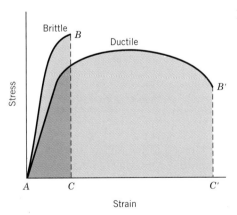

FIGURE 7.13 Schematic representations of tensile stress–strain behavior for brittle and ductile materials loaded to fracture.

---

[10] Both l_f and A_f are measured subsequent to fracture, and after the two broken ends have been repositioned back together.

before fracture. Second, it specifies the degree of allowable deformation during fabrication operations. We sometimes refer to relatively ductile materials as being "forgiving," in the sense that they may experience local deformation without fracture should there be an error in the magnitude of the design stress calculation.

Brittle materials are *approximately* considered to be those having a fracture strain of less than about 5%.

Thus, several important mechanical properties of metals may be determined from tensile stress–strain tests. Table 7.2 presents some typical room-temperature

Table 7.2 Room-Temperature Mechanical Properties (in Tension) for Various Materials

Material	Yield Strength		Tensile Strength		Ductility, %EL [in 50 mm (2 in.)][a]
	MPa	ksi	MPa	ksi	
Metal Alloys[b]					
Molybdenum	565	82	655	95	35
Titanium	450	65	520	75	25
Steel (1020)	180	26	380	55	25
Nickel	138	20	480	70	40
Iron	130	19	262	38	45
Brass (70 Cu–30 Zn)	75	11	300	44	68
Copper	69	10	200	29	45
Aluminum	35	5	90	13	40
Ceramic Materials[c]					
Zirconia (ZrO_2)[d]	—	—	800–1500	115–215	—
Silicon nitride (Si_3N_4)	—	—	250–1000	35–145	—
Aluminum oxide (Al_2O_3)	—	—	275–700	40–100	—
Silicon carbide (SiC)	—	—	100–820	15–120	—
Glass–ceramic (Pyroceram)	—	—	247	36	—
Mullite ($3Al_2O_3$-$2SiO_2$)	—	—	185	27	—
Spinel ($MgAl_2O_4$)	—	—	110–245	16–36	—
Fused silica (SiO_2)	—	—	110	16	—
Magnesium oxide (MgO)[e]	—	—	105	15	—
Soda–lime glass	—	—	69	10	—
Polymers					
Nylon 6,6	44.8–82.8	6.5–12	75.9–94.5	11.0–13.7	15–300
Polycarbonate (PC)	62.1	9.0	62.8–72.4	9.1–10.5	110–150
Polyester (PET)	59.3	8.6	48.3–72.4	7.0–10.5	30–300
Polymethyl methacrylate (PMMA)	53.8–73.1	7.8–10.6	48.3–72.4	7.0–10.5	2.0–5.5
Polyvinyl chloride (PVC)	40.7–44.8	5.9–6.5	40.7–51.7	5.9–7.5	40–80
Phenol-formaldehyde	—	—	34.5–62.1	5.0–9.0	1.5–2.0
Polystyrene (PS)	—	—	35.9–51.7	5.2–7.5	1.2–2.5
Polypropylene (PP)	31.0–37.2	4.5–5.4	31.0–41.4	4.5–6.0	100–600
Polyethylene—high density (HDPE)	26.2–33.1	3.8–4.8	22.1–31.0	3.2–4.5	10–1200
Polytetrafluoroethylene (PTFE)	—	—	20.7–34.5	3.0–5.0	200–400
Polyethylene—low density (LDPE)	9.0–14.5	1.3–2.1	8.3–31.4	1.2–4.55	100–650

[a] For polymers, percent elongation at break.
[b] Property values are for metal alloys in an annealed state.
[c] The tensile strength of ceramic materials is taken as flexural strength (Section 7.10).
[d] Partially stabilized with 3 mol% Y_2O_3.
[e] Sintered and containing approximately 5% porosity.

FIGURE 7.14
Engineering stress–
strain behavior for iron
at three temperatures.

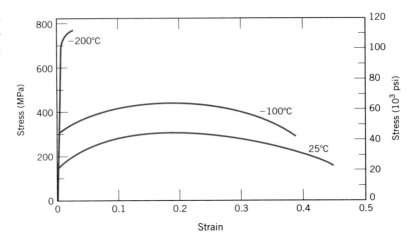

values of yield strength, tensile strength, and ductility for several common metals (and also for a number of polymers and ceramics). These properties are sensitive to any prior deformation, the presence of impurities, and/or any heat treatment to which the metal has been subjected. The modulus of elasticity is one mechanical parameter that is insensitive to these treatments. As with modulus of elasticity, the magnitudes of both yield and tensile strengths decline with increasing temperature; just the reverse holds for ductility—it usually increases with temperature. Figure 7.14 shows how the stress–strain behavior of iron varies with temperature.

RESILIENCE

Resilience is the capacity of a material to absorb energy when it is deformed elastically and then, upon unloading, to have this energy recovered. The associated property is the *modulus of resilience, U_r*, which is the strain energy per unit volume required to stress a material from an unloaded state up to the point of yielding.

Computationally, the modulus of resilience for a specimen subjected to a uniaxial tension test is just the area under the engineering stress–strain curve taken to yielding (Figure 7.15), or

$$U_r = \int_0^{\epsilon_y} \sigma \, d\epsilon \qquad (7.13a)$$

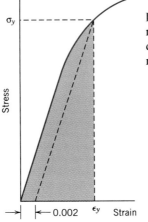

FIGURE 7.15 Schematic representation showing how modulus of resilience (corresponding to the shaded area) is determined from the tensile stress–strain behavior of a material.

Assuming a linear elastic region,

$$U_r = \tfrac{1}{2}\sigma_y\epsilon_y \qquad (7.13b)$$

in which ϵ_y is the strain at yielding.

The units of resilience are the product of the units from each of the two axes of the stress–strain plot. For SI units, this is joules per cubic meter (J/m^3, equivalent to Pa), whereas with Customary U.S. units it is inch-pounds force per cubic inch (in.-lb_f/in.^{3}, equivalent to psi). Both joules and inch-pounds force are units of energy, and thus this area under the stress–strain curve represents energy absorption per unit volume (in cubic meters or cubic inches) of material.

Incorporation of Equation 7.5 into Equation 7.13b yields

$$U_r = \tfrac{1}{2}\sigma_y\epsilon_y = \tfrac{1}{2}\sigma_y\left(\frac{\sigma_y}{E}\right) = \frac{\sigma_y^2}{2E} \qquad (7.14)$$

Thus, resilient materials are those having high yield strengths and low moduli of elasticity; such alloys would be used in spring applications.

TOUGHNESS

Toughness is a mechanical term that is used in several contexts; loosely speaking, it is a measure of the ability of a material to absorb energy up to fracture. Specimen geometry as well as the manner of load application are important in toughness determinations. For dynamic (high strain rate) loading conditions and when a notch (or point of stress concentration) is present, *notch toughness* is assessed by using an impact test, as discussed in Section 9.8. Furthermore, fracture toughness is a property indicative of a material's resistance to fracture when a crack is present (Section 9.5).

For the static (low strain rate) situation, toughness may be ascertained from the results of a tensile stress–strain test. It is the area under the σ–ϵ curve up to the point of fracture. The units for toughness are the same as for resilience (i.e., energy per unit volume of material). For a material to be tough, it must display both strength and ductility; and often, ductile materials are tougher than brittle ones. This is demonstrated in Figure 7.13, in which the stress–strain curves are plotted for both material types. Hence, even though the brittle material has higher yield and tensile strengths, it has a lower toughness than the ductile one, by virtue of lack of ductility; this is deduced by comparing the areas ABC and $AB'C'$ in Figure 7.13.

7.7 TRUE STRESS AND STRAIN

From Figure 7.11, the decline in the stress necessary to continue deformation past the maximum, point M, seems to indicate that the metal is becoming weaker. This is not at all the case; as a matter of fact, it is increasing in strength. However, the cross-sectional area is decreasing rapidly within the neck region, where deformation is occurring. This results in a reduction in the load-bearing capacity of the specimen. The stress, as computed from Equation 7.1, is on the basis of the original cross-sectional area before any deformation, and does not take into account this diminution in area at the neck.

Sometimes it is more meaningful to use a true stress–true strain scheme. **True stress** σ_T is defined as the load F divided by the instantaneous cross-sectional area

A_i over which deformation is occurring (i.e., the neck, past the tensile point), or

$$\sigma_T = \frac{F}{A_i} \tag{7.15}$$

Furthermore, it is occasionally more convenient to represent strain as **true strain** ϵ_T, defined by

$$\epsilon_T = \ln \frac{l_i}{l_0} \tag{7.16}$$

If no volume change occurs during deformation, that is, if

$$A_i l_i = A_0 l_0 \tag{7.17}$$

true and engineering stress and strain are related according to

$$\sigma_T = \sigma(1 + \epsilon) \tag{7.18a}$$

$$\epsilon_T = \ln(1 + \epsilon) \tag{7.18b}$$

Equations 7.18a and 7.18b are valid only to the onset of necking; beyond this point true stress and strain should be computed from actual load, cross-sectional area, and gauge length measurements.

A schematic comparison of engineering and true stress–strain behavior is made in Figure 7.16. It is worth noting that the true stress necessary to sustain increasing strain continues to rise past the tensile point M'.

Coincident with the formation of a neck is the introduction of a complex stress state within the neck region (i.e., the existence of other stress components in addition to the axial stress). As a consequence, the correct stress (*axial*) within the neck is slightly lower than the stress computed from the applied load and neck cross-sectional area. This leads to the "corrected" curve in Figure 7.16.

For some metals and alloys the region of the true stress-strain curve from the onset of plastic deformation to the point at which necking begins may be approximated by

$$\sigma_T = K\epsilon_T^n \tag{7.19}$$

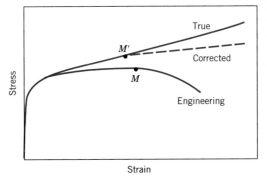

FIGURE 7.16 A comparison of typical tensile engineering stress–strain and true stress–strain behaviors. Necking begins at point M on the engineering curve, which corresponds to M' on the true curve. The "corrected" true stress–strain curve takes into account the complex stress state within the neck region.

Table 7.3   Tabulation of *n* and *K* Values (Equation 7.19) for Several Alloys

Material	n	K MPa	K psi
Low-carbon steel (annealed)	0.26	530	77,000
Alloy steel (Type 4340, annealed)	0.15	640	93,000
Stainless steel (Type 304, annealed)	0.45	1275	185,000
Aluminum (annealed)	0.20	180	26,000
Aluminum alloy (Type 2024, heat treated)	0.16	690	100,000
Copper (annealed)	0.54	315	46,000
Brass (70Cu–30Zn, annealed)	0.49	895	130,000

Source: From *Manufacturing Processes for Engineering Materials* by Seope Kalpakjian, © 1997. Reprinted by permission of Prentice-Hall, Inc., Upper Saddle River, NJ.

In this expression, K and n are constants, which values will vary from alloy to alloy, and will also depend on the condition of the material (i.e., whether it has been plastically deformed, heat treated, etc.). The parameter n is often termed the *strain-hardening exponent* and has a value less than unity. Values of n and K for several alloys are contained in Table 7.3.

EXAMPLE PROBLEM 7.4

A cylindrical specimen of steel having an original diameter of 12.8 mm (0.505 in.) is tensile tested to fracture and found to have an engineering fracture strength σ_f of 460 MPa (67,000 psi). If its cross-sectional diameter at fracture is 10.7 mm (0.422 in.), determine:

(a) The ductility in terms of percent reduction in area.

(b) The true stress at fracture.

SOLUTION

(a) Ductility is computed using Equation 7.12, as

$$\%RA = \frac{\left(\dfrac{12.8 \text{ mm}}{2}\right)^2 \pi - \left(\dfrac{10.7 \text{ mm}}{2}\right)^2 \pi}{\left(\dfrac{12.8 \text{ mm}}{2}\right)^2 \pi} \times 100$$

$$= \frac{128.7 \text{ mm}^2 - 89.9 \text{ mm}^2}{128.7 \text{ mm}^2} \times 100 = 30\%$$

(b) True stress is defined by Equation 7.15, where in this case the area is taken as the fracture area A_f. However, the load at fracture must first be computed from the fracture strength as

$$F = \sigma_f A_0 = (460 \times 10^6 \text{ N/m}^2)(128.7 \text{ mm}^2) \left(\frac{1 \text{ m}^2}{10^6 \text{ mm}^2} \right) = 59,200 \text{ N}$$

Thus, the true stress is calculated as

$$\sigma_T = \frac{F}{A_f} = \frac{59,200 \text{ N}}{(89.9 \text{ mm}^2) \left(\dfrac{1 \text{ m}^2}{10^6 \text{ mm}^2} \right)}$$

$$= 6.6 \times 10^8 \text{ N/m}^2 = 660 \text{ MPa } (95,700 \text{ psi})$$

EXAMPLE PROBLEM 7.5

Compute the strain-hardening exponent n in Equation 7.19 for an alloy in which a true stress of 415 MPa (60,000 psi) produces a true strain of 0.10; assume a value of 1035 MPa (150,000 psi) for K.

SOLUTION

This requires some algebraic manipulation of Equation 7.19 so that n becomes the dependent parameter. This is accomplished by taking logarithms and rearranging. Solving for n yields

$$n = \frac{\log \sigma_T - \log K}{\log \epsilon_T}$$

$$= \frac{\log(415 \text{ MPa}) - \log(1035 \text{ MPa})}{\log(0.1)} = 0.40$$

7.8 ELASTIC RECOVERY DURING PLASTIC DEFORMATION

Upon release of the load during the course of a stress–strain test, some fraction of the total deformation is recovered as elastic strain. This behavior is demonstrated in Figure 7.17, a schematic engineering stress–strain plot. During the unloading cycle, the curve traces a near straight-line path from the point of unloading (point D), and its slope is virtually identical to the modulus of elasticity, or parallel to the initial elastic portion of the curve. The magnitude of this elastic strain, which is regained during unloading, corresponds to the strain recovery, as shown in Figure 7.17. If the load is reapplied, the curve will traverse essentially the same linear portion in the direction opposite to unloading; yielding will again occur at the unloading stress level where the unloading began. There will also be an elastic strain recovery associated with fracture.

7.9 COMPRESSIVE, SHEAR, AND TORSIONAL DEFORMATION

Of course, metals may experience plastic deformation under the influence of applied compressive, shear, and torsional loads. The resulting stress–strain behavior into

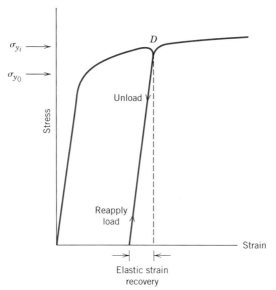

Figure 7.17 Schematic tensile stress–strain diagram showing the phenomena of elastic strain recovery and strain hardening. The initial yield strength is designated as σ_{y_0}; σ_{y_i} is the yield strength after releasing the load at point D, and then upon reloading.

the plastic region will be similar to the tensile counterpart (Figure 7.10a: yielding and the associated curvature). However, for compression, there will be no maximum, since necking does not occur; furthermore, the mode of fracture will be different from that for tension.

MECHANICAL BEHAVIOR—CERAMICS

Ceramic materials are somewhat limited in applicability by their mechanical properties, which in many respects are inferior to those of metals. The principal drawback is a disposition to catastrophic fracture in a brittle manner with very little energy absorption. In this section we explore the salient mechanical characteristics of these materials and how these properties are measured.

7.10 FLEXURAL STRENGTH

The stress–strain behavior of brittle ceramics is not usually ascertained by a tensile test as outlined in Section 7.2, for three reasons. First, it is difficult to prepare and test specimens having the required geometry. Second, it is difficult to grip brittle materials without fracturing them; and third, ceramics fail after only about 0.1% strain, which necessitates that tensile specimens be perfectly aligned in order to avoid the presence of bending stresses, which are not easily calculated. Therefore, a more suitable transverse bending test is most frequently employed, in which a rod specimen having either a circular or rectangular cross section is bent until fracture using a three- or four-point loading technique;[11] the three-point loading scheme is illustrated in Figure 7.18. At the point of loading, the top surface of the specimen is placed in a state of compression, whereas the bottom surface is in tension. Stress is computed from the specimen thickness, the bending moment, and

---

[11] ASTM Standard C 1161, "Standard Test Method for Flexural Strength of Advanced Ceramics at Ambient Temperature."

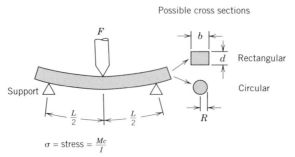

Possible cross sections

F

b

d Rectangular

R Circular

Support

$\dfrac{L}{2}$ $\dfrac{L}{2}$

$\sigma = \text{stress} = \dfrac{Mc}{I}$

where M = maximum bending moment
       c = distance from center of specimen
            to outer fibers
       I = moment of inertia of cross section
       F = applied load

	M	c	I	σ
Rectangular	$\dfrac{FL}{4}$	$\dfrac{d}{2}$	$\dfrac{bd^3}{12}$	$\dfrac{3FL}{2bd^2}$
Circular	$\dfrac{FL}{4}$	R	$\dfrac{\pi R^4}{4}$	$\dfrac{FL}{\pi R^3}$

FIGURE 7.18 A three-point loading scheme for measuring the stress–strain behavior and flexural strength of brittle ceramics, including expressions for computing stress for rectangular and circular cross sections.

the moment of inertia of the cross section; these parameters are noted in Figure 7.18 for rectangular and circular cross sections. The maximum tensile stress (as determined using these stress expressions) exists at the bottom specimen surface directly below the point of load application. Since the tensile strengths of ceramics are about one-tenth of their compressive strengths, and since fracture occurs on the tensile specimen face, the flexure test is a reasonable substitute for the tensile test.

The stress at fracture using this flexure test is known as the **flexural strength,** *modulus of rupture, fracture strength,* or the *bend strength,* an important mechanical parameter for brittle ceramics. For a rectangular cross section, the flexural strength σ_{fs} is equal to

$$\sigma_{fs} = \frac{3F_f L}{2bd^2} \tag{7.20a}$$

where F_f is the load at fracture, L is the distance between support points, and the other parameters are as indicated in Figure 7.18. When the cross section is circular, then

$$\sigma_{fs} = \frac{F_f L}{\pi R^3} \tag{7.20b}$$

R being the specimen radius.

Characteristic flexural strength values for several ceramic materials are given in Table 7.2. Since, during bending, a specimen is subjected to both compressive and tensile stresses, the magnitude of its flexural strength is greater than the tensile fracture strength. Furthermore, σ_{fs} will depend on specimen size; as explained in Section 9.6, with increasing specimen volume (under stress) there is an increase in the probability of the existence of a crack-producing flaw and, consequently, a decrease in flexural strength.

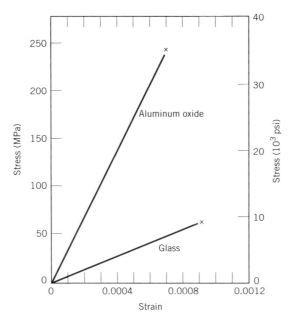

FIGURE 7.19 Typical stress–strain behavior to fracture for aluminum oxide and glass.

7.11 ELASTIC BEHAVIOR

The elastic stress–strain behavior for ceramic materials using these flexure tests is similar to the tensile test results for metals: a linear relationship exists between stress and strain. Figure 7.19 compares the stress–strain behavior to fracture for aluminum oxide (alumina) and glass. Again, the slope in the elastic region is the modulus of elasticity; also, the moduli of elasticity for ceramic materials are slightly higher than for metals (Table 7.2 and Table B.2, Appendix B). From Figure 7.19 it may be noted that neither of the materials experiences plastic deformation prior to fracture.

7.12 INFLUENCE OF POROSITY ON THE MECHANICAL PROPERTIES OF CERAMICS (CD-ROM)

MECHANICAL BEHAVIOR—POLYMERS

7.13 STRESS–STRAIN BEHAVIOR

The mechanical properties of polymers are specified with many of the same parameters that are used for metals, that is, modulus of elasticity, and yield and tensile strengths. For many polymeric materials, the simple stress–strain test is employed for the characterization of some of these mechanical parameters.[12] The mechanical characteristics of polymers, for the most part, are highly sensitive to the rate of deformation (strain rate), the temperature, and the chemical nature of the environment (the presence of water, oxygen, organic solvents, etc.). Some modifications of the testing techniques and specimen configurations used for metals are necessary with polymers, especially for the highly elastic materials, such as rubbers.

---

[12] ASTM Standard D 638, "Standard Test Method for Tensile Properties of Plastics."

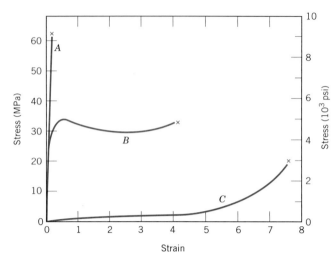

FIGURE 7.22 The stress–strain behavior for brittle (curve *A*), plastic (curve *B*), and highly elastic (elastomeric) (curve *C*) polymers.

Three typically different types of stress–strain behavior are found for polymeric materials, as represented in Figure 7.22. Curve *A* illustrates the stress–strain character for a brittle polymer, inasmuch as it fractures while deforming elastically. The behavior for the plastic material, curve *B*, is similar to that found for many metallic materials; the initial deformation is elastic, which is followed by yielding and a region of plastic deformation. Finally, the deformation displayed by curve *C* is totally elastic; this rubberlike elasticity (large recoverable strains produced at low stress levels) is displayed by a class of polymers termed the **elastomers.**

Modulus of elasticity (termed *tensile modulus* or sometimes just *modulus* for polymers) and ductility in percent elongation are determined for polymers in the same manner as for metals (Section 7.6). For plastic polymers (curve *B*, Figure 7.22), the yield point is taken as a maximum on the curve, which occurs just beyond the termination of the linear-elastic region (Figure 7.23); the stress at this maximum is the yield strength (σ_y). Furthermore, tensile strength (*TS*) corresponds to the stress at which fracture occurs (Figure 7.23); *TS* may be greater than or less than σ_y. Strength, for these plastic polymers, is normally taken as tensile strength. Table 7.2 and Tables B.2, B.3, and B.4 in Appendix B give these mechanical properties for a number of polymeric materials.

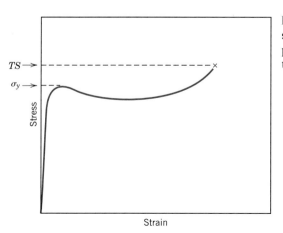

FIGURE 7.23 Schematic stress–strain curve for a plastic polymer showing how yield and tensile strengths are determined.

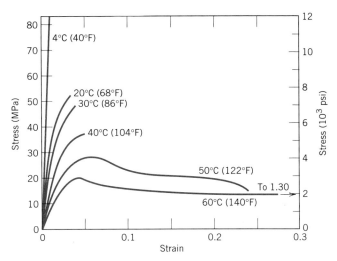

FIGURE 7.24   The influence of temperature on the stress–strain characteristics of polymethyl methacrylate. (From T. S. Carswell and H. K. Nason, "Effect of Environmental Conditions on the Mechanical Properties of Organic Plastics," *Symposium on Plastics,* American Society for Testing and Materials, Philadelphia, 1944. Copyright, ASTM. Reprinted with permission.)

Polymers are, in many respects, mechanically dissimilar to metals (and ceramic materials). For example, the modulus for highly elastic polymeric materials may be as low as 7 MPa (10^3 psi), but may run as high as 4 GPa (0.6×10^6 psi) for some of the very stiff polymers; modulus values for metals are much larger (Table 7.1). Maximum tensile strengths for polymers are on the order of 100 MPa (15,000 psi)—for some metal alloys 4100 MPa (600,000 psi). And, whereas metals rarely elongate plastically to more than 100%, some highly elastic polymers may experience elongations to as much as 1000%.

In addition, the mechanical characteristics of polymers are much more sensitive to temperature changes within the vicinity of room temperature. Consider the stress–strain behavior for polymethyl methacrylate (Plexiglas) at several temperatures between 4 and 60°C (40 and 140°F) (Figure 7.24). Several features of this figure are worth noting, as follows: increasing the temperature produces (1) a decrease in elastic modulus, (2) a reduction in tensile strength, and (3) an enhancement of ductility—at 4°C (40°F) the material is totally brittle, whereas considerable plastic deformation is realized at both 50 and 60°C (122 and 140°F).

The influence of strain rate on the mechanical behavior may also be important. In general, decreasing the rate of deformation has the same influence on the stress–strain characteristics as increasing the temperature; that is, the material becomes softer and more ductile.

7.14 MACROSCOPIC DEFORMATION

Some aspects of the macroscopic deformation of semicrystalline polymers deserve our attention. The tensile stress–strain curve for a semicrystalline material, which was initially unoriented, is shown in Figure 7.25; also included in the figure are

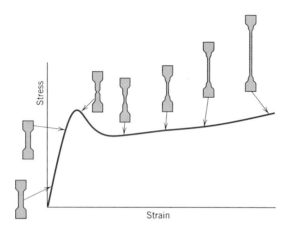

FIGURE 7.25 Schematic tensile stress–strain curve for a semicrystalline polymer. Specimen contours at several stages of deformation are included. (From Jerold M. Schultz, *Polymer Materials Science,* copyright © 1974, p. 488. Reprinted by permission of Prentice-Hall, Inc., Englewood Cliffs, NJ.)

schematic representations of specimen profile at various stages of deformation. Both upper and lower yield points are evident on the curve, which are followed by a near horizontal region. At the upper yield point, a small neck forms within the gauge section of the specimen. Within this neck, the chains become oriented (i.e., chain axes become aligned parallel to the elongation direction, a condition that is represented schematically in Figure 8.27e), which leads to localized strengthening. Consequently, there is a resistance to continued deformation at this point, and specimen elongation proceeds by the propagation of this neck region along the gauge length; the chain orientation phenomenon (Figure 8.27e) accompanies this neck extension. This tensile behavior may be contrasted to that found for ductile metals (Section 7.6), wherein once a neck has formed, all subsequent deformation is confined to within the neck region.

7.15 VISCOELASTICITY (CD-ROM)

HARDNESS AND OTHER MECHANICAL PROPERTY CONSIDERATIONS

7.16 HARDNESS

Another mechanical property that may be important to consider is **hardness,** which is a measure of a material's resistance to localized plastic deformation (e.g., a small dent or a scratch). Early hardness tests were based on natural minerals with a scale constructed solely on the ability of one material to scratch another that was softer. A qualitative and somewhat arbitrary hardness indexing scheme was devised, termed the Mohs scale, which ranged from 1 on the soft end for talc to 10 for diamond. Quantitative hardness techniques have been developed over the years in which a small indenter is forced into the surface of a material to be tested, under controlled conditions of load and rate of application. The depth or size of the resulting indentation is measured, which in turn is related to a hardness number; the softer the material, the larger and deeper the indentation, and the lower the hardness index number. Measured hardnesses are only relative (rather than absolute), and care should be exercised when comparing values determined by different techniques.

Hardness tests are performed more frequently than any other mechanical test for several reasons:

1. They are simple and inexpensive—ordinarily no special specimen need be prepared, and the testing apparatus is relatively inexpensive.

2. The test is nondestructive—the specimen is neither fractured nor excessively deformed; a small indentation is the only deformation.

3. Other mechanical properties often may be estimated from hardness data, such as tensile strength (see Figure 7.31).

ROCKWELL HARDNESS TESTS[13]

The Rockwell tests constitute the most common method used to measure hardness because they are so simple to perform and require no special skills. Several different scales may be utilized from possible combinations of various indenters and different loads, which permit the testing of virtually all metal alloys (as well as some polymers). Indenters include spherical and hardened steel balls having diameters of $\frac{1}{16}$, $\frac{1}{8}$, $\frac{1}{4}$, and $\frac{1}{2}$ in. (1.588, 3.175, 6.350, and 12.70 mm), and a conical diamond (Brale) indenter, which is used for the hardest materials.

With this system, a hardness number is determined by the difference in depth of penetration resulting from the application of an initial minor load followed by a larger major load; utilization of a minor load enhances test accuracy. On the basis of the magnitude of both major and minor loads, there are two types of tests: Rockwell and superficial Rockwell. For Rockwell, the minor load is 10 kg, whereas major loads are 60, 100, and 150 kg. Each scale is represented by a letter of the alphabet; several are listed with the corresponding indenter and load in Tables 7.4 and 7.5a. For superficial tests, 3 kg is the minor load; 15, 30, and 45 kg are the possible major load values. These scales are identified by a 15, 30, or 45 (according to load), followed by N, T, W, X, or Y, depending on indenter. Superficial tests are frequently performed on thin specimens. Table 7.5b presents several superficial scales.

When specifying Rockwell and superficial hardnesses, both hardness number and scale symbol must be indicated. The scale is designated by the symbol HR followed by the appropriate scale identification.[14] For example, 80 HRB represents a Rockwell hardness of 80 on the B scale, and 60 HR30W indicates a superficial hardness of 60 on the 30W scale.

For each scale, hardnesses may range up to 130; however, as hardness values rise above 100 or drop below 20 on any scale, they become inaccurate; and because the scales have some overlap, in such a situation it is best to utilize the next harder or softer scale.

Inaccuracies also result if the test specimen is too thin, if an indentation is made too near a specimen edge, or if two indentations are made too close to one another. Specimen thickness should be at least ten times the indentation depth, whereas allowance should be made for at least three indentation diameters between the center of one indentation and the specimen edge, or to the center of a second indentation. Furthermore, testing of specimens stacked one on top of another is not recommended. Also, accuracy is dependent on the indentation being made into a smooth flat surface.

---

[13] ASTM Standard E 18, "Standard Test Methods for Rockwell Hardness and Rockwell Superficial Hardness of Metallic Materials."

[14] Rockwell scales are also frequently designated by an R with the appropriate scale letter as a subscript, for example, R_C denotes the Rockwell C scale.

Table 7.4 Hardness Testing Techniques

Test	Indenter	Shape of Indentation		Load	Formula for Hardness Number[a]
		Side View	Top View		
Brinell	10-mm sphere of steel or tungsten carbide			P	$HB = \dfrac{2P}{\pi D[D - \sqrt{D^2 - d^2}]}$
Vickers microhardness	Diamond pyramid	136°		P	$HV = 1.854P/d_1^2$
Knoop microhardness	Diamond pyramid	$l/b = 7.11$ $b/t = 4.00$		P	$HK = 14.2P/l^2$
Rockwell and Superficial Rockwell	Diamond cone $\frac{1}{16}, \frac{1}{8}, \frac{1}{4}, \frac{1}{2}$ in. diameter steel spheres	120°		60 kg 100 kg 150 kg }Rockwell 15 kg 30 kg 45 kg }Superficial Rockwell	

[a] For the hardness formulas given, P (the applied load) is in kg, while D, d, d_1, and l are all in mm.
Source: Adapted from H. W. Hayden, W. G. Moffatt, and J. Wulff, *The Structure and Properties of Materials*, Vol. III, *Mechanical Behavior*. Copyright © 1965 by John Wiley & Sons, New York. Reprinted by permission of John Wiley & Sons, Inc.

Table 7.5a Rockwell Hardness Scales

Scale Symbol	Indenter	Major Load (kg)
A	Diamond	60
B	$\frac{1}{16}$ in. ball	100
C	Diamond	150
D	Diamond	100
E	$\frac{1}{8}$ in. ball	100
F	$\frac{1}{16}$ in. ball	60
G	$\frac{1}{16}$ in. ball	150
H	$\frac{1}{8}$ in. ball	60
K	$\frac{1}{8}$ in. ball	150

Table 7.5b Superficial Rockwell Hardness Scales

Scale Symbol	Indenter	Major Load (kg)
15N	Diamond	15
30N	Diamond	30
45N	Diamond	45
15T	$\frac{1}{16}$ in. ball	15
30T	$\frac{1}{16}$ in. ball	30
45T	$\frac{1}{16}$ in. ball	45
15W	$\frac{1}{8}$ in. ball	15
30W	$\frac{1}{8}$ in. ball	30
45W	$\frac{1}{8}$ in. ball	45

The modern apparatus for making Rockwell hardness measurements (see the chapter-opening photograph for this chapter) is automated and very simple to use; hardness is read directly, and each measurement requires only a few seconds.

The modern testing apparatus also permits a variation in the time of load application. This variable must also be considered in interpreting hardness data.

BRINELL HARDNESS TESTS[15]

In Brinell tests, as in Rockwell measurements, a hard, spherical indenter is forced into the surface of the metal to be tested. The diameter of the hardened steel (or tungsten carbide) indenter is 10.00 mm (0.394 in.). Standard loads range between 500 and 3000 kg in 500-kg increments; during a test, the load is maintained constant for a specified time (between 10 and 30 s). Harder materials require greater applied loads. The Brinell hardness number, HB, is a function of both the magnitude of the load and the diameter of the resulting indentation (see Table 7.4).[16] This diameter is measured with a special low-power microscope, utilizing a scale that is etched on the eyepiece. The measured diameter is then converted to the appropriate HB number using a chart; only one scale is employed with this technique.

Maximum specimen thickness as well as indentation position (relative to specimen edges) and minimum indentation spacing requirements are the same as for Rockwell tests. In addition, a well-defined indentation is required; this necessitates a smooth flat surface in which the indentation is made.

---

[15] ASTM Standard E 10, "Standard Test Method for Brinell Hardness of Metallic Materials."
[16] The Brinell hardness number is also represented by BHN.

KNOOP AND VICKERS MICROHARDNESS TESTS[17]

Two other hardness testing techniques are Knoop (pronounced $\overline{nup}$) and Vickers (sometimes also called diamond pyramid). For each test a very small diamond indenter having pyramidal geometry is forced into the surface of the specimen. Applied loads are much smaller than for Rockwell and Brinell, ranging between 1 and 1000 g. The resulting impression is observed under a microscope and measured; this measurement is then converted into a hardness number (Table 7.4). Careful specimen surface preparation (grinding and polishing) may be necessary to ensure a well-defined indentation that may be accurately measured. The Knoop and Vickers hardness numbers are designated by HK and HV, respectively,[18] and hardness scales for both techniques are approximately equivalent. Knoop and Vickers are referred to as microhardness testing methods on the basis of load and indenter size. Both are well suited for measuring the hardness of small, selected specimen regions; furthermore, Knoop is used for testing brittle materials such as ceramics.

There are other hardness-testing techniques that are frequently employed, but which will not be discussed here; these include ultrasonic microhardness, dynamic (Scleroscope), durometer (for plastic and elastomeric materials), and scratch hardness tests. These are described in references provided at the end of the chapter.

HARDNESS CONVERSION

The facility to convert the hardness measured on one scale to that of another is most desirable. However, since hardness is not a well-defined material property, and because of the experimental dissimilarities among the various techniques, a comprehensive conversion scheme has not been devised. Hardness conversion data have been determined experimentally and found to be dependent on material type and characteristics. The most reliable conversion data exist for steels, some of which are presented in Figure 7.30 for Knoop, Brinell, and two Rockwell scales; the Mohs scale is also included. Detailed conversion tables for various other metals and alloys are contained in ASTM Standard E 140, "Standard Hardness Conversion Tables for Metals." In light of the preceding discussion, care should be exercised in extrapolation of conversion data from one alloy system to another.

CORRELATION BETWEEN HARDNESS AND TENSILE STRENGTH

Both tensile strength and hardness are indicators of a metal's resistance to plastic deformation. Consequently, they are roughly proportional, as shown in Figure 7.31, on page 182, for tensile strength as a function of the HB for cast iron, steel, and brass. The same proportionality relationship does not hold for all metals, as Figure 7.31 indicates. As a rule of thumb for most steels, the HB and the tensile strength are related according to

$$TS(\text{MPa}) = 3.45 \times \text{HB} \qquad (7.25a)$$

$$TS(\text{psi}) = 500 \times \text{HB} \qquad (7.25b)$$

---

[17] ASTM Standard E 92, "Standard Test Method for Vickers Hardness of Metallic Materials," and ASTM Standard E 384, "Standard Test for Microhardness of Materials."
[18] Sometimes KHN and VHN are used to denote Knoop and Vickers hardness numbers, respectively.

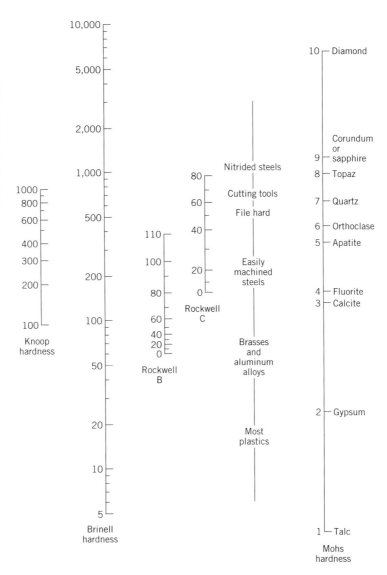

FIGURE 7.30
Comparison of several
hardness scales.
(Adapted from G. F.
Kinney, *Engineering
Properties and
Applications of Plastics,*
p. 202. Copyright
© 1957 by John
Wiley & Sons, New
York. Reprinted by
permission of John
Wiley & Sons, Inc.)

7.17 HARDNESS OF CERAMIC MATERIALS

One beneficial mechanical property of ceramics is their hardness, which is often utilized when an abrasive or grinding action is required; in fact, the hardest known materials are ceramics. A listing of a number of different ceramic materials according to Knoop hardness is contained in Table 7.6. Only ceramics having Knoop hardnesses of about 1000 or greater are utilized for their abrasive characteristics (Section 13.8).

7.18 TEAR STRENGTH AND HARDNESS OF POLYMERS

Mechanical properties that are sometimes influential in the suitability of a polymer for some particular application include tear resistance and hardness. The ability to resist tearing is an important property of some plastics, especially those used for thin films in packaging. *Tear strength,* the mechanical parameter that is measured,

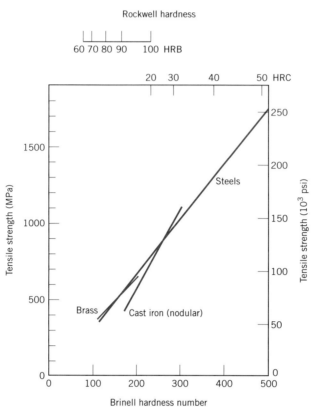

FIGURE 7.31 Relationships between hardness and tensile strength for steel, brass, and cast iron. (Data taken from *Metals Handbook: Properties and Selection: Irons and Steels,* Vol. 1, 9th edition, B. Bardes, Editor, American Society for Metals, 1978, pp. 36 and 461; and *Metals Handbook: Properties and Selection: Nonferrous Alloys and Pure Metals,* Vol. 2, 9th edition, H. Baker, Managing Editor, American Society for Metals, 1979, p. 327.)

is the energy required to tear apart a cut specimen that has a standard geometry. The magnitude of tensile and tear strengths are related.

Polymers are softer than metals and ceramics, and most hardness tests are conducted by penetration techniques similar to those described for metals in the previous section. Rockwell tests are frequently used for polymers.[19] Other indentation techniques employed are the Durometer and Barcol.[20]

Table 7.6   Approximate Knoop Hardness (100 g load) for Seven Ceramic Materials

Material	Approximate Knoop Hardness
Diamond (carbon)	7000
Boron carbide (B_4C)	2800
Silicon carbide (SiC)	2500
Tungsten carbide (WC)	2100
Aluminum oxide (Al_2O_3)	2100
Quartz (SiO_2)	800
Glass	550

[19] ASTM Standard D 785, "Rockwell Hardness of Plastics and Electrical Insulating Materials."

[20] ASTM Standard D 2240, "Standard Test Method for Rubber Property—Durometer Hardness;" and ASTM Standard D 2583, "Standard Test Method for Indentation of Rigid Plastics by Means of a Barcol Impressor."

PROPERTY VARIABILITY AND DESIGN/SAFETY FACTORS

7.19 VARIABILITY OF MATERIAL PROPERTIES

At this point it is worthwhile to discuss an issue that sometimes proves troublesome to many engineering students, namely, that measured material properties are not exact quantities. That is, even if we have a most precise measuring apparatus and a highly controlled test procedure, there will always be some scatter or variability in the data that are collected from specimens of the same material. For example, consider a number of identical tensile samples that are prepared from a single bar of some metal alloy, which samples are subsequently stress–strain tested in the same apparatus. We would most likely observe that each resulting stress–strain plot is slightly different from the others. This would lead to a variety of modulus of elasticity, yield strength, and tensile strength values. A number of factors lead to uncertainties in measured data. These include the test method, variations in specimen fabrication procedures, operator bias, and apparatus calibration. Furthermore, inhomogeneities may exist within the same lot of material, and/or slight compositional and other differences from lot to lot. Of course, appropriate measures should be taken to minimize the possibility of measurement error, and also to mitigate those factors that lead to data variability.

It should also be mentioned that scatter exists for other measured material properties such as density, electrical conductivity, and coefficient of thermal expansion.

It is important for the design engineer to realize that scatter and variability of materials properties are inevitable and must be dealt with appropriately. On occasion, data must be subjected to statistical treatments and probabilities determined. For example, instead of asking the question, "What is the fracture strength of this alloy?" the engineer should become accustomed to asking the question, "What is the probability of failure of this alloy under these given circumstances?"

It is often desirable to specify a typical value and degree of dispersion (or scatter) for some measured property; such is commonly accomplished by taking the average and the standard deviation, respectively.

COMPUTATION OF AVERAGE AND STANDARD DEVIATION VALUES (CD-ROM)

7.20 DESIGN/SAFETY FACTORS

There will always be uncertainties in characterizing the magnitude of applied loads and their associated stress levels for in-service applications; ordinarily load calculations are only approximate. Furthermore, as noted in the previous section, virtually all engineering materials exhibit a variability in their measured mechanical properties. Consequently, design allowances must be made to protect against unanticipated failure. One way this may be accomplished is by establishing, for the particular application, a **design stress,** denoted as σ_d. For static situations and when ductile materials are used, σ_d is taken as the calculated stress level σ_c (on the basis of the estimated maximum load) multiplied by a *design factor, N',* that is

$$\sigma_d = N'\sigma_c \tag{7.28}$$

where N' is greater than unity. Thus, the material to be used for the particular application is chosen so as to have a yield strength at least as high as this value of σ_d.

Alternatively, a **safe stress** or *working stress, σ_w*, is used instead of design stress. This safe stress is based on the yield strength of the material and is defined as the yield strength divided by a *factor of safety, N,* or

$$\sigma_w = \frac{\sigma_y}{N} \tag{7.29}$$

Utilization of design stress (Equation 7.28) is usually preferred since it is based on the anticipated maximum applied stress instead of the yield strength of the material; normally there is a greater uncertainty in estimating this stress level than in the specification of the yield strength. However, in the discussion of this text, we are concerned with factors that influence yield strengths, and not in the determination of applied stresses; therefore, the succeeding discussion will deal with working stresses and factors of safety.

The choice of an appropriate value of N is necessary. If N is too large, then component overdesign will result, that is, either too much material or a material having a higher-than-necessary strength will be used. Values normally range between 1.2 and 4.0. Selection of N will depend on a number of factors, including economics, previous experience, the accuracy with which mechanical forces and material properties may be determined, and, most important, the consequences of failure in terms of loss of life and/or property damage.

DESIGN EXAMPLE 7.1

A tensile-testing apparatus is to be constructed that must withstand a maximum load of 220,000 N (50,000 lb$_f$). The design calls for two cylindrical support posts, each of which is to support half of the maximum load. Furthermore, plain-carbon (1045) steel ground and polished shafting rounds are to be used; the minimum yield and tensile strengths of this alloy are 310 MPa (45,000 psi) and 565 MPa (82,000 psi), respectively. Specify a suitable diameter for these support posts.

SOLUTION

The first step in this design process is to decide on a factor safety, N, which then allows determination of a working stress according to Equation 7.29. In addition, to ensure that the apparatus will be safe to operate, we also want to minimize any elastic deflection of the rods during testing; therefore, a relatively conservative factor of safety is to be used, say $N = 5$. Thus, the working stress σ_w is just

$$\sigma_w = \frac{\sigma_y}{N}$$

$$= \frac{310 \text{ MPa}}{5} = 62 \text{ MPa (9000 psi)}$$

From the definition of stress, Equation 7.1,

$$A_0 = \left(\frac{d}{2}\right)^2 \pi = \frac{F}{\sigma_w}$$

where d is the rod diameter and F is the applied force; furthermore, each of the two rods must support half of the total force or 110,000 N (25,000 psi). Solving for

d leads to

$$d = 2\sqrt{\frac{F}{\pi\sigma_w}}$$

$$= 2\sqrt{\frac{110,000\ \text{N}}{\pi(62\times10^6\ \text{N/m}^2)}}$$

$$= 4.75\times10^{-2}\ \text{m} = 47.5\ \text{mm}\ (1.87\ \text{in.})$$

Therefore, the diameter of each of the two rods should be 47.5 mm or 1.87 in.

SUMMARY

A number of the important mechanical properties of materials have been discussed in this chapter. Concepts of stress and strain were first introduced. Stress is a measure of an applied mechanical load or force, normalized to take into account cross-sectional area. Two different stress parameters were defined—engineering stress and true stress. Strain represents the amount of deformation induced by a stress; both engineering and true strains are used.

Some of the mechanical characteristics of materials can be ascertained by simple stress–strain tests. There are four test types: tension, compression, torsion, and shear. Tensile are the most common. A material that is stressed first undergoes elastic, or nonpermanent, deformation, wherein stress and strain are proportional. The constant of proportionality is the modulus of elasticity for tension and compression, and is the shear modulus when the stress is shear. Poisson's ratio represents the negative ratio of transverse and longitudinal strains.

For metals, the phenomenon of yielding occurs at the onset of plastic or permanent deformation; yield strength is determined by a strain offset method from the stress–strain behavior, which is indicative of the stress at which plastic deformation begins. Tensile strength corresponds to the maximum tensile stress that may be sustained by a specimen, whereas percents elongation and reduction in area are measures of ductility—the amount of plastic deformation that has occurred at fracture. Resilience is the capacity of a material to absorb energy during elastic deformation; modulus of resilience is the area beneath the engineering stress–strain curve up to the yield point. Also, static toughness represents the energy absorbed during the fracture of a material, and is taken as the area under the entire engineering stress–strain curve. Ductile materials are normally tougher than brittle ones.

For the brittle ceramic materials, flexural strengths are determined by performing transverse bending tests to fracture. {Many ceramic bodies contain residual porosity, which is deleterious to both their moduli of elasticity and flexural strengths.}

On the basis of stress–strain behavior, polymers fall within three general classifications: brittle, plastic, and highly elastic. These materials are neither as strong nor as stiff as metals, and their mechanical properties are sensitive to changes in temperature and strain rate.

{Viscoelastic mechanical behavior, being intermediate between totally elastic and totally viscous, is displayed by a number of polymeric materials. It is characterized by the relaxation modulus, a time-dependent modulus of elasticity. The magnitude of the relaxation modulus is very sensitive to temperature; critical to the in-service temperature range for elastomers is this temperature dependence.}

Hardness is a measure of the resistance to localized plastic deformation. In several popular hardness-testing techniques (Rockwell, Brinell, Knoop, and Vickers) a small indenter is forced into the surface of the material, and an index number is determined on the basis of the size or depth of the resulting indentation. For many metals, hardness and tensile strength are approximately proportional to each other. In addition to their inherent brittleness, ceramic materials are distinctively hard. And polymers are relatively soft in comparison to the other material types.

Measured mechanical properties (as well as other material properties) are not exact and precise quantities, in that there will always be some scatter for the measured data. Typical material property values are commonly specified in terms of averages, whereas magnitudes of scatter may be expressed as standard deviations.

As a result of uncertainties in both measured mechanical properties and inservice applied stresses, safe or working stresses are normally utilized for design purposes. For ductile materials, safe stress is the ratio of the yield strength and a factor of safety.

IMPORTANT TERMS AND CONCEPTS

Anelasticity	Hardness	Tensile strength
Design stress	Modulus of elasticity	Toughness
Ductility	Plastic deformation	True strain
Elastic deformation	Poisson's ratio	True stress
Elastic recovery	Proportional limit	{Viscoelasticity}
Elastomer	{Relaxation modulus}	Yielding
Engineering strain	Resilience	Yield strength
Engineering stress	Safe stress	
Flexural strength	Shear	

REFERENCES

ASM Handbook, Vol. 8, *Mechanical Testing*, ASM International, Materials Park, OH, 1985.

Billmeyer, F. W., Jr., *Textbook of Polymer Science,* 3rd edition, Wiley-Interscience, New York, 1984. Chapter 11.

Boyer, H. E. (Editor), *Atlas of Stress–Strain Curves*, ASM International, Materials Park, OH, 1986.

Boyer, H. E. (Editor), *Hardness Testing,* ASM International, Materials Park, OH, 1987.

Davidge, R. W., *Mechanical Behaviour of Ceramics,* Cambridge University Press, Cambridge, 1979. Reprinted by TechBooks, Marietta, OH.

Dieter, G. E., *Mechanical Metallurgy,* 3rd edition, McGraw-Hill Book Co., New York, 1986.

Dowling, N. E., *Mechanical Behavior of Materials,* Prentice Hall, Inc., Englewood Cliffs, NJ, 1993.

Engineered Materials Handbook, Vol. 2, *Engineering Plastics,* ASM International, Materials Park, OH, 1988.

Engineered Materials Handbook, Vol. 4, *Ceramics and Glasses,* ASM International, Materials Park, OH, 1991.

Han, P. (Editor), *Tensile Testing,* ASM International, Materials Park, OH, 1992.

Harper, C. A. (Editor), *Handbook of Plastics, Elastomers and Composites,* 3rd edition, McGraw-Hill Book Company, New York, 1996.

Kingery, W. D., H. K. Bowen, and D. R. Uhlmann, *Introduction to Ceramics,* 2nd edition, John Wiley & Sons, New York, 1976. Chapters 14 and 15.

McClintock, F. A. and A. S. Argon, *Mechanical Behavior of Materials,* Addison-Wesley Publishing Co., Reading, MA, 1966. Reprinted by TechBooks, Marietta, OH.

Meyers, M. A. and K. K. Chawla, *Mechanical Metallurgy, Principles and Applications,* Prentice Hall, Inc., Englewood Cliffs, NJ, 1984.

Modern Plastics Encyclopedia, McGraw-Hill Book Company, New York. Revised and published annually.

Nielsen, L. E., *Mechanical Properties of Polymers and Composites,* 2nd edition, Marcel Dekker, New York, 1994.

Richerson, D. W., *Modern Ceramic Engineering,* 2nd edition, Marcel Dekker, New York, 1992.

Rosen, S. L., *Fundamental Principles of Polymeric Materials,* 2nd edition, John Wiley & Sons, New York, 1993.

Tobolsky, A. V., *Properties and Structures of Polymers,* John Wiley & Sons, New York, 1960. Advanced treatment.

Wachtman, J. B., *Mechanical Properties of Ceramics*, John Wiley & Sons, Inc., New York, 1996.

Ward, I. M. and D. W. Hadley, *An Introduction to the Mechanical Properties of Solid Polymers,* John Wiley & Sons, Chichester, UK, 1993.

Young, R. J. and P. Lovell, *Introduction to Polymers,* 2nd edition, Chapman and Hall, London, 1991.

QUESTIONS AND PROBLEMS

Note: To solve those problems having an asterisk (*) by their numbers, consultation of supplementary topics [appearing only on the CD-ROM (and not in print)] will probably be necessary.

7.1 Using mechanics of materials principles (i.e., equations of mechanical equilibrium applied to a free-body diagram), derive Equations 7.4a and 7.4b.

7.2 **(a)** Equations 7.4a and 7.4b are expressions for normal (σ') and shear (τ') stresses, respectively, as a function of the applied tensile stress (σ) and the inclination angle of the plane on which these stresses are taken (θ of Figure 7.4). Make a plot on which is presented the orientation parameters of these expressions (i.e., $\cos^2\theta$ and $\sin\theta\cos\theta$) versus θ.

(b) From this plot, at what angle of inclination is the normal stress a maximum?

(c) Also, at what inclination angle is the shear stress a maximum?

7.3 A specimen of aluminum having a rectangular cross section 10 mm $\times$ 12.7 mm (0.4 in. $\times$ 0.5 in.) is pulled in tension with 35,500 N (8000 lb$_f$) force, producing only elastic deformation. Calculate the resulting strain.

7.4 A cylindrical specimen of a titanium alloy having an elastic modulus of 107 GPa (15.5 $\times$ 10^6 psi) and an original diameter of 3.8 mm (0.15 in.) will experience only elastic deformation when a tensile load of 2000 N (450 lb$_f$) is applied. Compute the maximum length of the specimen before deformation if the maximum allowable elongation is 0.42 mm (0.0165 in.).

7.5 A steel bar 100 mm (4.0 in.) long and having a square cross section 20 mm (0.8 in.) on an edge is pulled in tension with a load of 89,000 N (20,000 lb$_f$), and experiences an elongation of 0.10 mm (4.0 $\times$ 10^{-3} in.). Assuming that the deformation is entirely elastic, calculate the elastic modulus of the steel.

7.6 Consider a cylindrical titanium wire 3.0 mm (0.12 in.) in diameter and 2.5 $\times$ 10^4 mm (1000 in.) long. Calculate its elongation when a load of 500 N (112 lb$_f$) is applied. Assume that the deformation is totally elastic.

7.7 For a bronze alloy, the stress at which plastic deformation begins is 275 MPa (40,000 psi), and the modulus of elasticity is 115 GPa (16.7 $\times$ 10^6 psi).

(a) What is the maximum load that may be applied to a specimen with a cross-sectional area of 325 mm^2 (0.5 in.^{2}) without plastic deformation?

(b) If the original specimen length is 115 mm (4.5 in.), what is the maximum length to which it may be stretched without causing plastic deformation?

7.8 A cylindrical rod of copper (E = 110 GPa,

FIGURE 7.33 Tensile stress–strain behavior for a plain carbon steel.

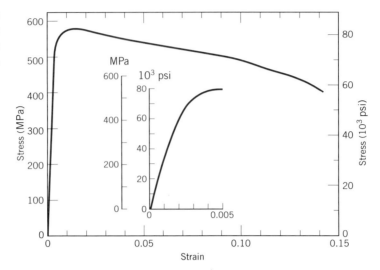

16 × 10⁶ psi) having a yield strength of 240 MPa (35,000 psi) is to be subjected to a load of 6660 N (1500 lb_f). If the length of the rod is 380 mm (15.0 in.), what must be the diameter to allow an elongation of 0.50 mm (0.020 in.)?

7.9 Consider a cylindrical specimen of a steel alloy (Figure 7.33) 10 mm (0.39 in.) in diameter and 75 mm (3.0 in.) long that is pulled in tension. Determine its elongation when a load of 23,500 N (5300 lb_f) is applied.

7.10 Figure 7.34 shows, for a gray cast iron, the tensile engineering stress–strain curve in the elastic region. Determine **(a)** the secant modulus taken to 35 MPa (5000 psi), and **(b)** the tangent modulus taken from the origin.

7.11 As was noted in Section 3.18, for single crystals of some substances, the physical properties are anisotropic, that is, they are dependent on crystallographic direction. One such property is the modulus of elasticity. For cubic single crystals, the modulus of elasticity in a general [uvw] direction, E_{uvw}, is described by the relationship

$$\frac{1}{E_{uvw}} = \frac{1}{E_{\langle100\rangle}} - 3\left(\frac{1}{E_{\langle100\rangle}} - \frac{1}{E_{\langle111\rangle}}\right)$$
$$(\alpha^2\beta^2 + \beta^2\gamma^2 + \gamma^2\alpha^2)$$

where $E_{\langle100\rangle}$ and $E_{\langle111\rangle}$ are the moduli of elasticity in [100] and [111] directions, respectively; α, β, and γ are the cosines of the

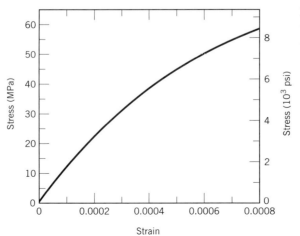

FIGURE 7.34 Tensile stress–strain behavior for a gray cast iron.

angles between [uvw] and the respective [100], [010], and [001] directions. Verify that the $E_{\langle 110 \rangle}$ values for aluminum, copper, and iron in Table 3.7 are correct.

7.12 In Section 2.6 it was noted that the net bonding energy E_N between two isolated positive and negative ions is a function of interionic distance r as follows:

$$E_N = -\frac{A}{r} + \frac{B}{r^n} \qquad (7.30)$$

where A, B, and n are constants for the particular ion pair. Equation 7.30 is also valid for the bonding energy between adjacent ions in solid materials. The modulus of elasticity E is proportional to the slope of the interionic force-separation curve at the equilibrium interionic separation; that is,

$$E \propto \left(\frac{dF}{dr} \right)_{r_0}$$

Derive an expression for the dependence of the modulus of elasticity on these A, B, and n parameters (for the two-ion system) using the following procedure:

1. Establish a relationship for the force F as a function of r, realizing that

$$F = \frac{dE_N}{dr}$$

2. Now take the derivative dF/dr.

3. Develop an expression for r_0, the equilibrium separation. Since r_0 corresponds to the value of r at the minimum of the E_N-versus-r-curve (Figure 2.8b), take the derivative dE_N/dr, set it equal to zero, and solve for r, which corresponds to r_0.

4. Finally, substitute this expression for r_0 into the relationship obtained by taking dF/dr.

7.13 Using the solution to Problem 7.12, rank the magnitudes of the moduli of elasticity for the following hypothetical X, Y, and Z materials from the greatest to the least. The appropriate A, B, and n parameters (Equation 7.30) for these three materials are tabulated below; they yield E_N in units of electron volts and r in nanometers:

Material	A	B	n
X	2.5	2×10^{-5}	8
Y	2.3	8×10^{-6}	10.5
Z	3.0	1.5×10^{-5}	9

7.14 A cylindrical specimen of aluminum having a diameter of 19 mm (0.75 in.) and length of 200 mm (8.0 in.) is deformed elastically in tension with a force of 48,800 N (11,000 lb$_f$). Using the data contained in Table 7.1, determine the following:

(a) The amount by which this specimen will elongate in the direction of the applied stress.

(b) The change in diameter of the specimen. Will the diameter increase or decrease?

7.15 A cylindrical bar of steel 10 mm (0.4 in.) in diameter is to be deformed elastically by application of a force along the bar axis. Using the data in Table 7.1, determine the force that will produce an elastic reduction of 3×10^{-3} mm (1.2×10^{-4} in.) in the diameter.

7.16 A cylindrical specimen of some alloy 8 mm (0.31 in.) in diameter is stressed elastically in tension. A force of 15,700 N (3530 lb$_f$) produces a reduction in specimen diameter of 5×10^{-3} mm (2×10^{-4} in.). Compute Poisson's ratio for this material if its modulus of elasticity is 140 GPa (20.3×10^6 psi).

7.17 A cylindrical specimen of a hypothetical metal alloy is stressed in compression. If its original and final diameters are 20.000 and 20.025 mm, respectively, and its final length is 74.96 mm, compute its original length if the deformation is totally elastic. The elastic and shear moduli for this alloy are 105 GPa and 39.7 GPa, respectively.

7.18 Consider a cylindrical specimen of some hypothetical metal alloy that has a diameter of 8.0 mm (0.31 in.). A tensile force of 1000 N (225 lb$_f$) produces an elastic reduction in diameter of 2.8×10^{-4} mm (1.10×10^{-5} in.). Compute the modulus of elasticity for this alloy, given that Poisson's ratio is 0.30.

7.19 A brass alloy is known to have a yield strength of 275 MPa (40,000 psi), a tensile strength of 380 MPa (55,000 psi), and an elastic modulus of 103 GPa (15.0×10^6 psi). A cylindrical specimen of this alloy 12.7 mm (0.50 in.) in diameter and 250 mm (10.0 in.) long is stressed in tension and found to elongate 7.6 mm (0.30 in.). On the basis of the information given, is it possible to compute the magnitude of the load that is necessary to produce this change in length? If so, calculate the load. If not, explain why.

7.20 A cylindrical metal specimen 15.0 mm (0.59 in.) in diameter and 150 mm (5.9 in.) long is to be subjected to a tensile stress of 50 MPa (7250 psi); at this stress level the resulting deformation will be totally elastic.

(a) If the elongation must be less than 0.072 mm (2.83×10^{-3} in.), which of the metals in Table 7.1 are suitable candidates? Why?

(b) If, in addition, the maximum permissible diameter decrease is 2.3×10^{-3} mm (9.1×10^{-5} in.), which of the metals in Table 7.1 may be used? Why?

7.21 Consider the brass alloy with stress–strain behavior shown in Figure 7.12. A cylindrical specimen of this material 6 mm (0.24 in.) in diameter and 50 mm (2 in.) long is pulled in tension with a force of 5000 N (1125 lb$_f$). If it is known that this alloy has a Poisson's ratio of 0.30, compute: **(a)** the specimen elongation, and **(b)** the reduction in specimen diameter.

7.22 Cite the primary differences between elastic, anelastic, and plastic deformation behaviors.

7.23 A cylindrical rod 100 mm long and having a diameter of 10.0 mm is to be deformed using a tensile load of 27,500 N. It must not experience either plastic deformation or a diameter reduction of more than 7.5×10^{-3} mm. Of the materials listed as follows, which are possible candidates? Justify your choice(s).

Material	Modulus of Elasticity (GPa)	Yield Strength (MPa)	Poisson's Ratio
Aluminum alloy	70	200	0.33
Brass alloy	101	300	0.35
Steel alloy	207	400	0.27
Titanium alloy	107	650	0.36

7.24 A cylindrical rod 380 mm (15.0 in.) long, having a diameter of 10.0 mm (0.40 in.), is to be subjected to a tensile load. If the rod is to experience neither plastic deformation nor an elongation of more than 0.9 mm (0.035 in.) when the applied load is 24,500 N (5500 lb$_f$), which of the four metals or alloys listed below are possible candidates? Justify your choice(s).

Material	Modulus of Elasticity (GPa)	Yield Strength (MPa)	Tensile Strength (MPa)
Aluminum alloy	70	255	420
Brass alloy	100	345	420
Copper	110	250	290
Steel alloy	207	450	550

7.25 Figure 7.33 shows the tensile engineering stress–strain behavior for a steel alloy.

(a) What is the modulus of elasticity?

(b) What is the proportional limit?

(c) What is the yield strength at a strain offset of 0.002?

(d) What is the tensile strength?

7.26 A cylindrical specimen of a brass alloy having a length of 60 mm (2.36 in.) must elongate only 10.8 mm (0.425 in.) when a tensile load of 50,000 N (11,240 lb$_f$) is applied. Under these circumstances, what must be the radius of the specimen? Consider this brass alloy to have the stress–strain behavior shown in Figure 7.12.

7.27 A load of 44,500 N (10,000 lb$_f$) is applied to a cylindrical specimen of steel (displaying the stress–strain behavior shown in Figure 7.33) that has a cross-sectional diameter of 10 mm (0.40 in.).

(a) Will the specimen experience elastic or plastic deformation? Why?

(b) If the original specimen length is 500 mm (20 in.), how much will it increase in length when this load is applied?

7.28 A bar of a steel alloy that exhibits the stress–strain behavior shown in Figure 7.33 is subjected to a tensile load; the specimen is 300 mm (12 in.) long, and of square cross section 4.5 mm (0.175 in.) on a side.

(a) Compute the magnitude of the load necessary to produce an elongation of 0.46 mm (0.018 in.).

(b) What will be the deformation after the load is released?

7.29 A cylindrical specimen of aluminum having a diameter of 0.505 in. (12.8 mm) and a gauge length of 2.000 in. (50.800 mm) is pulled in tension. Use the load–elongation characteristics tabulated below to complete problems a through f.

Load		Length	
lb_f	N	in.	mm
0	0	2.000	50.800
1,650	7,330	2.002	50.851
3,400	15,100	2.004	50.902
5,200	23,100	2.006	50.952
6,850	30,400	2.008	51.003
7,750	34,400	2.010	51.054
8,650	38,400	2.020	51.308
9,300	41,300	2.040	51.816
10,100	44,800	2.080	52.832
10,400	46,200	2.120	53.848
10,650	47,300	2.160	54.864
10,700	47,500	2.200	55.880
10,400	46,100	2.240	56.896
10,100	44,800	2.270	57.658
9,600	42,600	2.300	58.420
8,200	36,400	2.330	59.182
	Fracture		

(a) Plot the data as engineering stress versus engineering strain.

(b) Compute the modulus of elasticity.

(c) Determine the yield strength at a strain offset of 0.002.

(d) Determine the tensile strength of this alloy.

(e) What is the approximate ductility, in percent elongation?

(f) Compute the modulus of resilience.

7.30 A specimen of ductile cast iron having a rectangular cross section of dimensions 4.8 mm × 15.9 mm ($\frac{3}{16}$ in. × $\frac{5}{8}$ in.) is deformed in tension. Using the load-elongation data tabulated below, complete problems a through f.

Load		Length	
N	lb_f	mm	in.
0	0	75.000	2.953
4,740	1065	75.025	2.954
9,140	2055	75.050	2.955
12,920	2900	75.075	2.956
16,540	3720	75.113	2.957
18,300	4110	75.150	2.959
20,170	4530	75.225	2.962
22,900	5145	75.375	2.968
25,070	5635	75.525	2.973
26,800	6025	75.750	2.982
28,640	6440	76.500	3.012
30,240	6800	78.000	3.071
31,100	7000	79.500	3.130
31,280	7030	81.000	3.189
30,820	6930	82.500	3.248
29,180	6560	84.000	3.307
27,190	6110	85.500	3.366
24,140	5430	87.000	3.425
18,970	4265	88.725	3.493
	Fracture		

(a) Plot the data as engineering stress versus engineering strain.

(b) Compute the modulus of elasticity.

(c) Determine the yield strength at a strain offset of 0.002.

(d) Determine the tensile strength of this alloy.

(e) Compute the modulus of resilience.

(f) What is the ductility, in percent elongation?

7.31 A cylindrical metal specimen having an original diameter of 12.8 mm (0.505 in.) and gauge length of 50.80 mm (2.000 in.) is pulled in tension until fracture occurs. The diameter at the point of fracture is 6.60 mm (0.260 in.), and the fractured gauge length is 72.14 mm (2.840 in.). Calculate the ductility in terms of percent reduction in area and percent elongation.

7.32 Calculate the moduli of resilience for the materials having the stress–strain behaviors shown in Figures 7.12 and 7.33.

7.33 Determine the modulus of resilience for each of the following alloys:

	Yield Strength			Load		Length		Diameter	
Material	MPa	psi	lb_f	N	in.	mm	in.	mm
Steel alloy	550	80,000	10,400	46,100	2.240	56.896	0.461	11.71
Brass alloy	350	50,750	10,100	44,800	2.270	57.658	0.431	10.95
Aluminum alloy	250	36,250	9,600	42,600	2.300	58.420	0.418	10.62
Titanium alloy	800	116,000	8,200	36,400	2.330	59.182	0.370	9.40

Use modulus of elasticity values in Table 7.1.

7.34 A brass alloy to be used for a spring application must have a modulus of resilience of at least 0.75 MPa (110 psi). What must be its minimum yield strength?

7.35 **(a)** Make a schematic plot showing the tensile true stress–strain behavior for a typical metal alloy.

(b) Superimpose on this plot a schematic curve for the compressive true stress–strain behavior for the same alloy. Explain any difference between this curve and the one in part a.

(c) Now superimpose a schematic curve for the compressive engineering stress–strain behavior for this same alloy, and explain any difference between this curve and the one in part b.

7.36 Show that Equations 7.18a and 7.18b are valid when there is no volume change during deformation.

7.37 Demonstrate that Equation 7.16, the expression defining true strain, may also be represented by

$$\epsilon_T = \ln\left(\frac{A_0}{A_i}\right)$$

when specimen volume remains constant during deformation. Which of these two expressions is more valid during necking? Why?

7.38 Using the data in Problem 7.29 and Equations 7.15, 7.16, and 7.18a, generate a true stress–true strain plot for aluminum. Equation 7.18a becomes invalid past the point at which necking begins; therefore, measured diameters are given below for the last four data points, which should be used in true stress computations.

7.39 A tensile test is performed on a metal specimen, and it is found that a true plastic strain of 0.20 is produced when a true stress of 575 MPa (83,500 psi) is applied; for the same metal, the value of K in Equation 7.19 is 860 MPa (125,000 psi). Calculate the true strain that results from the application of a true stress of 600 MPa (87,000 psi).

7.40 For some metal alloy, a true stress of 415 MPa (60,175 psi) produces a plastic true strain of 0.475. How much will a specimen of this material elongate when a true stress of 325 MPa (46,125 psi) is applied if the original length is 300 mm (11.8 in.)? Assume a value of 0.25 for the strain-hardening exponent n.

7.41 The following true stresses produce the corresponding true plastic strains for a brass alloy:

True Stress (psi)	True Strain
50,000	0.10
60,000	0.20

What true stress is necessary to produce a true plastic strain of 0.25?

7.42 For a brass alloy, the following engineering stresses produce the corresponding plastic engineering strains, prior to necking:

Engineering Stress (MPa)	Engineering Strain
235	0.194
250	0.296

On the basis of this information, compute the *engineering* stress necessary to produce an *engineering* strain of 0.25.

7.43 Find the toughness (or energy to cause fracture) for a metal that experiences both elastic and plastic deformation. Assume Equation 7.5 for elastic deformation, that the modulus of elasticity is 172 GPa (25×10^6 psi), and that elastic deformation terminates at a strain of 0.01. For plastic deformation, assume that the relationship between stress and strain is described by Equation 7.19, in which the values for K and n are 6900 MPa (1×10^6 psi) and 0.30, respectively. Furthermore, plastic deformation occurs between strain values of 0.01 and 0.75, at which point fracture occurs.

7.44 For a tensile test, it can be demonstrated that necking begins when

$$\frac{d\sigma_T}{d\epsilon_T} = \sigma_T \qquad (7.31)$$

Using Equation 7.19, determine the value of the true strain at this onset of necking.

7.45 Taking the logarithm of both sides of Equation 7.19 yields

$$\log \sigma_T = \log K + n \log \epsilon_T \qquad (7.32)$$

Thus, a plot of $\log \sigma_T$ versus $\log \epsilon_T$ in the plastic region to the point of necking should yield a straight line having a slope of n and an intercept (at $\log \sigma_T = 0$) of $\log K$.

Using the appropriate data tabulated in Problem 7.29, make a plot of $\log \sigma_T$ versus $\log \epsilon_T$ and determine the values of n and K. It will be necessary to convert engineering stresses and strains to true stresses and strains using Equations 7.18a and 7.18b.

7.46 A cylindrical specimen of a brass alloy 7.5 mm (0.30 in.) in diameter and 90.0 mm (3.54 in.) long is pulled in tension with a force of 6000 N (1350 lb$_f$); the force is subsequently released.

(a) Compute the final length of the specimen at this time. The tensile stress–strain behavior for this alloy is shown in Figure 7.12.

(b) Compute the final specimen length when the load is increased to 16,500 N (3700 lb$_f$) and then released.

7.47 A steel specimen having a rectangular cross section of dimensions 19 mm $\times$ 3.2 mm ($\frac{3}{4}$ in. $\times \frac{1}{8}$ in.) has the stress–strain behavior

shown in Figure 7.33. If this specimen is subjected to a tensile force of 33,400 N (7,500 lb$_f$), then

(a) Determine the elastic and plastic strain values.

(b) If its original length is 460 mm (18 in.), what will be its final length after the load in part a is applied and then released?

7.48 A three-point bending test is performed on a glass specimen having a rectangular cross section of height d 5 mm (0.2 in.) and width b 10 mm (0.4 in.); the distance between support points is 45 mm (1.75 in.).

(a) Compute the flexural strength if the load at fracture is 290 N (65 lb$_f$).

(b) The point of maximum deflection Δy occurs at the center of the specimen and is described by

$$\Delta y = \frac{FL^3}{48EI}$$

where E is the modulus of elasticity and I the cross-sectional moment of inertia. Compute Δy at a load of 266 N (60 lb$_f$).

7.49 A circular specimen of MgO is loaded using a three-point bending mode. Compute the minimum possible radius of the specimen without fracture, given that the applied load is 425 N (95.5 lb$_f$), the flexural strength is 105 MPa (15,000 psi), and the separation between load points is 50 mm (2.0 in.).

7.50 A three-point bending test was performed on an aluminum oxide specimen having a circular cross section of radius 3.5 mm (0.14 in.); the specimen fractured at a load of 950 N (215 lb$_f$) when the distance between the support points was 50 mm (2.0 in.). Another test is to be performed on a specimen of this same material, but one that has a square cross section of 12 mm (0.47 in.) length on each edge. At what load would you expect this specimen to fracture if the support point separation is 40 mm (1.6 in.)?

7.51 **(a)** A three-point transverse bending test is conducted on a cylindrical specimen of aluminum oxide having a reported flexural strength of 390 MPa (56,600 psi). If the specimen radius is 2.5 mm (0.10 in.) and the support point separation distance is 30 mm (1.2 in.), predict whether or not you would expect the specimen to fracture when a load of 620 N (140 lb$_f$) is applied. Justify your prediction.

(b) Would you be 100% certain of the prediction in part a? Why or why not?

7.52* The modulus of elasticity for beryllium oxide (BeO) having 5 vol% porosity is 310 GPa (45×10^6 psi).

(a) Compute the modulus of elasticity for the nonporous material.

(b) Compute the modulus of elasticity for 10 vol% porosity.

7.53* The modulus of elasticity for boron carbide (B$_4$C) having 5 vol% porosity is 290 GPa (42×10^6 psi).

(a) Compute the modulus of elasticity for the nonporous material.

(b) At what volume percent porosity will the modulus of elasticity be 235 GPa (34×10^6 psi)?

7.54* Using the data in Table 7.2, do the following:

(a) Determine the flexural strength for nonporous MgO assuming a value of 3.75 for n in Equation 7.22.

(b) Compute the volume fraction porosity at which the flexural strength for MgO is 62 MPa (9000 psi).

7.55* The flexural strength and associated volume fraction porosity for two specimens of the same ceramic material are as follows:

σ_{fs} (MPa)	P
100	0.05
50	0.20

(a) Compute the flexural strength for a completely nonporous specimen of this material.

(b) Compute the flexural strength for a 0.1 volume fraction porosity.

7.56 From the stress–strain data for polymethyl methacrylate shown in Figure 7.24, determine the modulus of elasticity and tensile strength at room temperature [20°C (68°F)], and compare these values with those given in Tables 7.1 and 7.2.

7.57 When citing the ductility as percent elongation for semicrystalline polymers, it is not necessary to specify the specimen gauge length, as is the case with metals. Why is this so?

7.58* In your own words, briefly describe the phenomenon of viscoelasticity.

7.59* For some viscoelastic polymers that are subjected to stress relaxation tests, the stress decays with time according to

$$\sigma(t) = \sigma(0) \exp\left(-\frac{t}{\tau}\right) \qquad (7.33)$$

where $\sigma(t)$ and $\sigma(0)$ represent the time-dependent and initial (i.e., time $= 0$) stresses, respectively, and t and τ denote elapsed time and the relaxation time; τ is a time-independent constant characteristic of the material. A specimen of some viscoelastic polymer the stress relaxation of which obeys Equation 7.33 was suddenly pulled in tension to a measured strain of 0.6; the stress necessary to maintain this constant strain was measured as a function of time. Determine $E_r(10)$ for this material if the initial stress level was 2.76 MPa (400 psi), which dropped to 1.72 MPa (250 psi) after 60 s.

7.60* In Figure 7.35, the logarithm of $E_r(t)$ versus the logarithm of time is plotted for polyisobutylene at a variety of temperatures. Make a plot of log $E_r(10)$ versus temperature and then estimate the T_g.

7.61* On the basis of the curves in Figure 7.26, sketch schematic strain-time plots for the following polystyrene materials at the specified temperatures:

(a) Amorphous at 120°C.

(b) Crosslinked at 150°C.

(c) Crystalline at 230°C.

(d) Crosslinked at 50°C.

7.62* **(a)** Contrast the manner in which stress relaxation and viscoelastic creep tests are conducted.

(b) For each of these tests, cite the experimental parameter of interest and how it is determined.

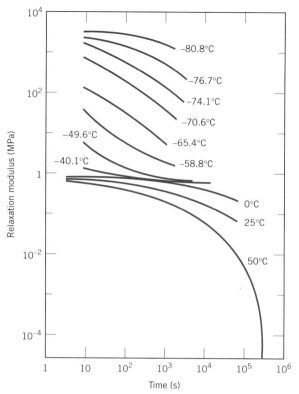

FIGURE 7.35 Logarithm of relaxation modulus versus logarithm of time for polyisobutylene between −80 and 50°C. (Adapted from E. Catsiff and A. V. Tobolsky, "Stress-Relaxation of Polyisobutylene in the Transition Region [1,2]," *J. Colloid Sci.,* **10,** 377 [1955]. Reprinted by permission of Academic Press, Inc.)

7.63* Make two schematic plots of the logarithm of relaxation modulus versus temperature for an amorphous polymer (curve *C* in Figure 7.29).

(a) On one of these plots demonstrate how the behavior changes with increasing molecular weight.

(b) On the other plot, indicate the change in behavior with increasing crosslinking.

7.64 (a) A 10-mm-diameter Brinell hardness indenter produced an indentation 1.62 mm in diameter in a steel alloy when a load of 500 kg was used. Compute the HB of this material.

(b) What will be the diameter of an indentation to yield a hardness of 450 HB when a 500 kg load is used?

7.65 Estimate the Brinell and Rockwell hardnesses for the following:

(a) The naval brass for which the stress–strain behavior is shown in Figure 7.12.

(b) The steel for which the stress–strain behavior is shown in Figure 7.33.

7.66 Using the data represented in Figure 7.31, specify equations relating tensile strength and Brinell hardness for brass and nodular cast iron, similar to Equations 7.25a and 7.25b for steels.

7.67 Cite five factors that lead to scatter in measured material properties.

7.68* Below are tabulated a number of Rockwell B hardness values that were measured on a single steel specimen. Compute average and standard deviation hardness values.

83.3	80.7	86.4
88.3	84.7	85.2
82.8	87.8	86.9
86.2	83.5	84.4
87.2	85.5	86.3

7.69 Upon what three criteria are factors of safety based?

7.70 Determine working stresses for the two alloys the stress–strain behaviors of which are shown in Figures 7.12 and 7.33.

Design Problems

7.D1 A large tower is to be supported by a series of steel wires. It is estimated that the load on each wire will be 11,100 N (2500 lb$_f$). Determine the minimum required wire diameter assuming a factor of safety of 2 and a yield strength of 1030 MPa (150,000 psi).

7.D2 **(a)** Gaseous hydrogen at a constant pressure of 1.013 MPa (10 atm) is to flow within the inside of a thin-walled cylindrical tube of nickel that has a radius of 0.1 m. The temperature of the tube is to be 300°C and the pressure of hydrogen outside of the tube will be maintained at 0.01013 MPa (0.1 atm). Calculate the minimum wall thickness if the diffusion flux is to be no greater than 1×10^{-7} mol/m^2-s. The concentration of hydrogen in the nickel, C_H (in moles hydrogen per m^3 of Ni) is a function of hydrogen pressure, p_{H_2} (in MPa) and absolute temperature (T) according to

$$C_H = 30.8 \sqrt{p_{H_2}} \exp\left(-\frac{12.3 \text{ kJ/mol}}{RT}\right)$$

$$(7.34)$$

Furthermore, the diffusion coefficient for the diffusion of H in Ni depends on temperature as

$$D_H(\text{m}^2/\text{s}) = 4.76 \times 10^{-7} \exp\left(-\frac{39.56 \text{ kJ/mol}}{RT}\right)$$

$$(7.35)$$

(b) For thin-walled cylindrical tubes that are pressurized, the circumferential stress is a function of the pressure difference across the wall (Δp), cylinder radius (r), and tube thickness (Δx) as

$$\sigma = \frac{r \Delta p}{4 \Delta x}$$

$$(7.36)$$

Compute the circumferential stress to which the walls of this pressurized cylinder are exposed.

(c) The room-temperature yield strength of Ni is 100 MPa (15,000 psi) and, furthermore, σ_y diminishes about 5 MPa for every 50°C rise in temperature. Would you expect the wall thickness computed in part (b) to be suitable for this Ni cylinder at 300°C? Why or why not?

(d) If this thickness is found to be suitable, compute the minimum thickness that could be used without any deformation of the tube walls. How much would the diffusion flux increase with this reduction in thickness? On the other hand, if the thickness determined in part (c) is found to be unsuitable, then specify a minimum thickness that you would use. In this case, how much of a diminishment in diffusion flux would result?

7.D3 Consider the steady-state diffusion of hydrogen through the walls of a cylindrical nickel tube as described in Problem 7.D2. One design calls for a diffusion flux of 5×10^{-8} mol/m^2-s, a tube radius of 0.125 m, and inside and outside pressures of 2.026 MPa (20 atm) and 0.0203 MPa (0.2 atm), respectively; the maximum allowable temperature is 450°C. Specify a suitable temperature and wall thickness to give this diffusion flux and yet ensure that the tube walls will not experience any permanent deformation.

7.D4 It is necessary to select a ceramic material to be stressed using a three-point loading scheme (Figure 7.18). The specimen must have a circular cross section and a radius of 2.5 mm (0.10 in.), and must not experience fracture or a deflection of more than 6.2×10^{-2} mm (2.4×10^{-3} in.) at its center when a load of 275 N (62 lb$_f$) is applied. If the distance between support points is 45 mm (1.77 in.), which of the ceramic materials in Tables 7.1 and 7.2 are candidates? The magnitude of the centerpoint deflection may be computed using the equation supplied in Problem 7.48.

Chapter 8 / Deformation and Strengthening Mechanisms

*I*n this photomicrograph of a lithium fluoride (LiF) single crystal, the small pyramidal pits represent those positions at which dislocations intersect the surface. The surface was polished and then chemically treated; these "etch pits" result from localized chemical attack around the dislocations and indicate the distribution of the dislocations. 750×. (Photomicrograph courtesy of W. G. Johnston, General Electric Co.)

Why Study *Deformation and Strengthening Mechanisms?*

With a knowledge of the nature of dislocations and the role they play in the plastic deformation process, we are able to understand the underlying mechanisms of the techniques that are used to strengthen and harden metals and their alloys; thus, it becomes possible to design and tailor the mechanical properties of materials—for example, the strength or toughness of a metal-matrix composite.

Also, understanding the mechanisms by which polymers elastically and plastically deform allows one to alter and control their moduli of elasticity and strengths (Sections 8.17 and 8.18).

After studying this chapter you should be able to do the following:

1. Describe edge and screw dislocation motion from an atomic perspective.
2. Describe how plastic deformation occurs by the motion of edge and screw dislocations in response to applied shear stresses.
3. Define slip system and cite one example.
4. Describe how the grain structure of a polycrystalline metal is altered when it is plastically deformed.
5. Explain how grain boundaries impede dislocation motion and why a metal having small grains is stronger than one having large grains.
6. Describe and explain solid-solution strengthening for substitutional impurity atoms in terms of lattice strain interactions with dislocations.
7. Describe and explain the phenomenon of strain hardening (or cold working) in terms of dislocations and strain field interactions.
8. Describe recrystallization in terms of both the alteration of microstructure and mechanical characteristics of the material.
9. Describe the phenomenon of grain growth from both macroscopic and atomic perspectives.
10. On the basis of slip considerations, explain why crystalline ceramic materials are normally brittle.
11. Describe/sketch the various stages in the plastic deformation of a semicrystalline (spherulitic) polymer.
12. Discuss the influence of the following factors on polymer tensile modulus and/or strength: (a) molecular weight, (b) degree of crystallinity, (c) predeformation, and (d) heat treating of undeformed materials.
13. Describe the molecular mechanism by which elastomeric polymers deform elastically.

8.1 INTRODUCTION

In this chapter we explore various deformation mechanisms that have been proposed to explain the deformation behaviors of metals, ceramics, and polymeric materials. Techniques that may be used to strengthen the various material types are described and explained in terms of these deformation mechanisms.

DEFORMATION MECHANISMS FOR METALS

Chapter 7 explained that metallic materials may experience two kinds of deformation: elastic and plastic. Plastic deformation is permanent, and strength and hardness are measures of a material's resistance to this deformation. On a microscopic scale, plastic deformation corresponds to the net movement of large numbers of atoms in response to an applied stress. During this process, interatomic bonds must be ruptured and then reformed. Furthermore, plastic deformation most often involves the motion of dislocations, linear crystalline defects that were introduced in Section 5.7. This section discusses the characteristics of dislocations and their involvement in plastic deformation. Sections 8.9, 8.10, and 8.11 present several techniques for strengthening single-phase metals, the mechanisms of which are described in terms of dislocations.

8.2 HISTORICAL

Early materials studies led to the computation of the theoretical strengths of perfect crystals, which were many times greater than those actually measured. During the 1930s it was theorized that this discrepancy in mechanical strengths could be explained by a type of linear crystalline defect that has since come to be known as a dislocation. It was not until the 1950s, however, that the existence of such dislocation defects was established by direct observation with the electron microscope. Since then, a theory of dislocations has evolved that explains many of the physical and mechanical phenomena in metals [as well as crystalline ceramics (Section 8.15)].

8.3 BASIC CONCEPTS OF DISLOCATIONS

Edge and screw are the two fundamental dislocation types. In an edge dislocation, localized lattice distortion exists along the end of an extra half-plane of atoms, which also defines the dislocation line (Figure 5.7). A screw dislocation may be thought of as resulting from shear distortion; its dislocation line passes through the center of a spiral, atomic plane ramp (Figure 5.8). Many dislocations in crystalline materials have both edge and screw components; these are mixed dislocations (Figure 5.9).

 Plastic deformation corresponds to the motion of large numbers of dislocations. An edge dislocation moves in response to a shear stress applied in a direction perpendicular to its line; the mechanics of dislocation motion are represented in Figure 8.1. Let the initial extra half-plane of atoms be plane *A*. When the shear stress is applied as indicated (Figure 8.1*a*), plane *A* is forced to the right; this in turn pushes the top halves of planes *B*, *C*, *D*, and so on, in the same direction. If the applied shear stress is of sufficient magnitude, the interatomic bonds of plane *B* are severed along the shear plane, and the upper half of plane *B* becomes the extra half-plane as plane *A* links up with the bottom half of plane *B* (Figure 8.1*b*). This process is subsequently repeated for the other planes, such that the extra half-plane, by discrete steps, moves from left to right by successive and repeated breaking of bonds and shifting by interatomic distances of upper half-planes. Before and after the movement of a dislocation through some particular region of the crystal, the atomic arrangement is ordered and perfect; it is only during the passage of the extra half-plane that the lattice structure is disrupted. Ultimately this extra half-plane may emerge

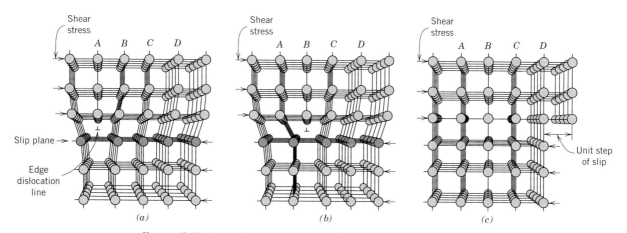

(a) (b) (c)

FIGURE 8.1 Atomic rearrangements that accompany the motion of an edge dislocation as it moves in response to an applied shear stress. (*a*) The extra half-plane of atoms is labeled *A*. (*b*) The dislocation moves one atomic distance to the right as *A* links up to the lower portion of plane *B*; in the process, the upper portion of *B* becomes the extra half-plane. (*c*) A step forms on the surface of the crystal as the extra half-plane exits. (Adapted from A. G. Guy, *Essentials of Materials Science*, McGraw-Hill Book Company, New York, 1976, p. 153.)

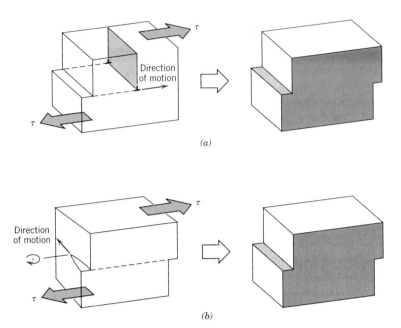

(a)

(b)

FIGURE 8.2 The formation of a step on the surface of a crystal by the motion of (a) an edge dislocation and (b) a screw dislocation. Note that for an edge, the dislocation line moves in the direction of the applied shear stress τ; for a screw, the dislocation line motion is perpendicular to the stress direction. (Adapted from H. W. Hayden, W. G. Moffatt, and J. Wulff, *The Structure and Properties of Materials,* Vol. III, *Mechanical Behavior,* p. 70. Copyright © 1965 by John Wiley & Sons, New York. Reprinted by permission of John Wiley & Sons, Inc.)

from the right surface of the crystal, forming an edge that is one atomic distance wide; this is shown in Figure 8.1c.

The process by which plastic deformation is produced by dislocation motion is termed **slip;** the crystallographic plane along which the dislocation line traverses is the *slip plane,* as indicated in Figure 8.1. Macroscopic plastic deformation simply corresponds to permanent deformation that results from the movement of dislocations, or slip, in response to an applied shear stress, as represented in Figure 8.2a.

Dislocation motion is analogous to the mode of locomotion employed by a caterpillar (Figure 8.3). The caterpillar forms a hump near its posterior end by pulling in its last pair of legs a unit leg distance. The hump is propelled forward by repeated lifting and shifting of leg pairs. When the hump reaches the anterior

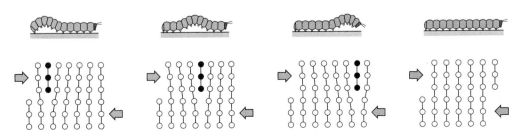

FIGURE 8.3 Representation of the analogy between caterpillar and dislocation motion.

end, the entire caterpillar has moved forward by the leg separation distance. The caterpillar hump and its motion correspond to the extra half-plane of atoms in the dislocation model of plastic deformation.

The motion of a screw dislocation in response to the applied shear stress is shown in Figure 8.2b; the direction of movement is perpendicular to the stress direction. For an edge, motion is parallel to the shear stress. However, the net plastic deformation for the motion of both dislocation types is the same (see Figure 8.2). The direction of motion of the mixed dislocation line is neither perpendicular nor parallel to the applied stress, but lies somewhere in between.

All metals and alloys contain some dislocations that were introduced during solidification, during plastic deformation, and as a consequence of thermal stresses that result from rapid cooling. The number of dislocations, or **dislocation density** in a material, is expressed as the total dislocation length per unit volume, or, equivalently, the number of dislocations that intersect a unit area of a random section. The units of dislocation density are millimeters of dislocation per cubic millimeter or just per square millimeter. Dislocation densities as low as 10^3 mm^{-2} are typically found in carefully solidified metal crystals. For heavily deformed metals, the density may run as high as 10^9 to 10^{10} mm^{-2}. Heat treating a deformed metal specimen can diminish the density to on the order of 10^5 to 10^6 mm^{-2}. By way of contrast, a typical dislocation density for ceramic materials is between 10^2 and 10^4 mm^{-2}; also, for silicon single crystals used in integrated circuits the value normally lies between 0.1 and 1 mm^{-2}.

8.4 CHARACTERISTICS OF DISLOCATIONS

Several characteristics of dislocations are important with regard to the mechanical properties of metals. These include strain fields that exist around dislocations, which are influential in determining the mobility of the dislocations, as well as their ability to multiply.

When metals are plastically deformed, some fraction of the deformation energy (approximately 5%) is retained internally; the remainder is dissipated as heat. The major portion of this stored energy is as strain energy associated with dislocations. Consider the edge dislocation represented in Figure 8.4. As already mentioned, some atomic lattice distortion exists around the dislocation line because of the presence of the extra half-plane of atoms. As a consequence, there are regions in which compressive, tensile, and shear **lattice strains** are imposed on the neighboring atoms. For example, atoms immediately above and adjacent to the dislocation line

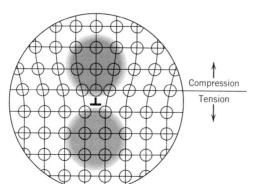

FIGURE 8.4 Regions of compression (dark) and tension (colored) located around an edge dislocation. (Adapted from W. G. Moffatt, G. W. Pearsall, and J. Wulff, *The Structure and Properties of Materials*, Vol. I, *Structure*, p. 85. Copyright © 1964 by John Wiley & Sons, New York. Reprinted by permission of John Wiley & Sons, Inc.)

Compression

Tension

are squeezed together. As a result, these atoms may be thought of as experiencing a compressive strain relative to atoms positioned in the perfect crystal and far removed from the dislocation; this is illustrated in Figure 8.4. Directly below the half-plane, the effect is just the opposite; lattice atoms sustain an imposed tensile strain, which is as shown. Shear strains also exist in the vicinity of the edge dislocation. For a screw dislocation, lattice strains are pure shear only. These lattice distortions may be considered to be strain fields that radiate from the dislocation line. The strains extend into the surrounding atoms, and their magnitudes decrease with radial distance from the dislocation.

The strain fields surrounding dislocations in close proximity to one another may interact such that forces are imposed on each dislocation by the combined interactions of all its neighboring dislocations. For example, consider two edge dislocations that have the same sign and the identical slip plane, as represented in Figure 8.5a. The compressive and tensile strain fields for both lie on the same side of the slip plane; the strain field interaction is such that there exists between these two isolated dislocations a mutual repulsive force that tends to move them apart. On the other hand, two dislocations of opposite sign and having the same slip plane will be attracted to one another, as indicated in Figure 8.5b, and dislocation annihilation will occur when they meet. That is, the two extra half-planes of atoms will align and become a complete plane. Dislocation interactions are possible between edge, screw, and/or mixed dislocations, and for a variety of orientations. These strain fields and associated forces are important in the strengthening mechanisms for metals.

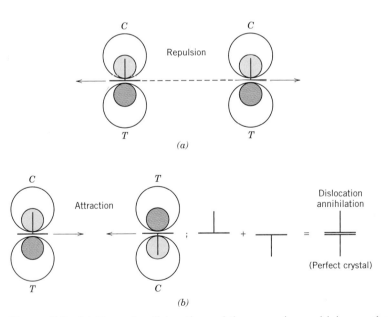

FIGURE 8.5   (a) Two edge dislocations of the same sign and lying on the same slip plane exert a repulsive force on each other; C and T denote compression and tensile regions, respectively. (b) Edge dislocations of opposite sign and lying on the same slip plane exert an attractive force on each other. Upon meeting, they annihilate each other and leave a region of perfect crystal. (Adapted from H. W. Hayden, W. G. Moffatt, and J. Wulff, *The Structure and Properties of Materials*, Vol. III, *Mechanical Behavior*, p. 75. Copyright © 1965 by John Wiley & Sons, New York. Reprinted by permission of John Wiley & Sons.)

During plastic deformation, the number of dislocations increases dramatically. We know that the dislocation density in a metal that has been highly deformed may be as high as 10^{10} mm^{-2}. One important source of these new dislocations is existing dislocations, which multiply; furthermore, grain boundaries, as well as internal defects and surface irregularities such as scratches and nicks, which act as stress concentrations, may serve as dislocation formation sites during deformation.

8.5 SLIP SYSTEMS

Dislocations do not move with the same degree of ease on all crystallographic planes of atoms and in all crystallographic directions. Ordinarily there is a preferred plane, and in that plane there are specific directions along which dislocation motion occurs. This plane is called the *slip plane;* it follows that the direction of movement is called the *slip direction.* This combination of the slip plane and the slip direction is termed the **slip system.** The slip system depends on the crystal structure of the metal and is such that the atomic distortion that accompanies the motion of a dislocation is a minimum. For a particular crystal structure, the slip plane is that plane having the most dense atomic packing, that is, has the greatest planar density. The slip direction corresponds to the direction, in this plane, that is most closely packed with atoms, that is, has the highest linear density. {Planar and linear atomic densities were discussed in Section 3.14.}

Consider, for example, the FCC crystal structure, a unit cell of which is shown in Figure 8.6a. There is a set of planes, the {111} family, all of which are closely packed. A (111)-type plane is indicated in the unit cell; in Figure 8.6b, this plane is positioned within the plane of the page, in which atoms are now represented as touching nearest neighbors.

Slip occurs along ⟨110⟩-type directions within the {111} planes, as indicated by arrows in Figure 8.6. Hence, {111}⟨110⟩ represents the slip plane and direction combination, or the slip system for FCC. Figure 8.6b demonstrates that a given slip plane may contain more than a single slip direction. Thus, several slip systems may exist for a particular crystal structure; the number of independent slip systems represents the different possible combinations of slip planes and directions. For example, for face-centered cubic, there are 12 slip systems: four unique {111} planes and, within each plane, three independent ⟨110⟩ directions.

The possible slip systems for BCC and HCP crystal structures are listed in Table 8.1. For each of these structures, slip is possible on more than one family of planes (e.g., {110}, {211}, and {321} for BCC). For metals having these two crystal structures, some slip systems are often operable only at elevated temperatures.

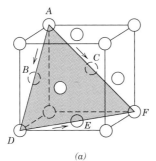

(a)

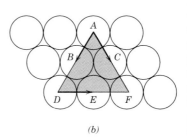

(b)

FIGURE 8.6 (a) A {111}⟨110⟩ slip system shown within an FCC unit cell. (b) The (111) plane from (a) and three ⟨110⟩ slip directions (as indicated by arrows) within that plane comprise possible slip systems.

Table 8.1   Slip Systems for Face-Centered Cubic, Body-Centered Cubic, and Hexagonal Close-Packed Metals

Metals	Slip Plane	Slip Direction	Number of Slip Systems
Face-Centered Cubic			
Cu, Al, Ni, Ag, Au	{111}	$\langle 1\bar{1}0 \rangle$	12
Body-Centered Cubic			
α-Fe, W, Mo	{110}	$\langle \bar{1}11 \rangle$	12
α-Fe, W	{211}	$\langle \bar{1}11 \rangle$	12
α-Fe, K	{321}	$\langle \bar{1}11 \rangle$	24
Hexagonal Close-Packed			
Cd, Zn, Mg, Ti, Be	{0001}	$\langle 11\bar{2}0 \rangle$	3
Ti, Mg, Zr	{10$\bar{1}$0}	$\langle 11\bar{2}0 \rangle$	3
Ti, Mg	{10$\bar{1}$1}	$\langle 11\bar{2}0 \rangle$	6

Metals with FCC or BCC crystal structures have a relatively large number of slip systems (at least 12). These metals are quite ductile because extensive plastic deformation is normally possible along the various systems. Conversely, HCP metals, having few active slip systems, are normally quite brittle.

8.6 SLIP IN SINGLE CRYSTALS (CD-ROM)

8.7 PLASTIC DEFORMATION OF POLYCRYSTALLINE METALS

For polycrystalline metals, because of the random crystallographic orientations of the numerous grains, the direction of slip varies from one grain to another. For each, dislocation motion occurs along the slip system that has the most favorable orientation (i.e., the highest shear stress). This is exemplified by a photomicrograph of a polycrystalline copper specimen that has been plastically deformed (Figure 8.10); before deformation the surface was polished. Slip lines[1] are visible, and it appears that two slip systems operated for most of the grains, as evidenced by two sets of parallel yet intersecting sets of lines. Furthermore, variation in grain orientation is indicated by the difference in alignment of the slip lines for the several grains.

Gross plastic deformation of a polycrystalline specimen corresponds to the comparable distortion of the individual grains by means of slip. During deformation, mechanical integrity and coherency are maintained along the grain boundaries; that is, the grain boundaries usually do not come apart or open up. As a consequence, each individual grain is constrained, to some degree, in the shape it may assume by its neighboring grains. The manner in which grains distort as a result of gross plastic deformation is indicated in Figure 8.11. Before deformation the grains are equiaxed, or have approximately the same dimension in all directions. For this

---

[1] Surface steps or ledges produced by dislocations (Figure 8.1c) that have exited from a grain and that appear as lines when viewed with a microscope are called *slip lines.*

FIGURE 8.10 Slip lines on the surface of a polycrystalline specimen of copper that was polished and subsequently deformed. 173×. (Photomicrograph courtesy of C. Brady, National Bureau of Standards.)

FIGURE 8.11 Alteration of the grain structure of a polycrystalline metal as a result of plastic deformation. (*a*) Before deformation the grains are equiaxed. (*b*) The deformation has produced elongated grains. 170×. (From W. G. Moffatt, G. W. Pearsall, and J. Wulff, *The Structure and Properties of Materials,* Vol. I, *Structure,* p. 140. Copyright © 1964 by John Wiley & Sons, New York. Reprinted by permission of John Wiley & Sons, Inc.)

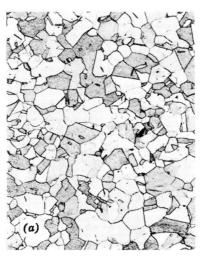

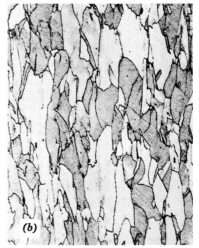

particular deformation, the grains become elongated along the direction in which the specimen was extended.

Polycrystalline metals are stronger than their single-crystal equivalents, which means that greater stresses are required to initiate slip and the attendant yielding. This is, to a large degree, also a result of geometrical constraints that are imposed on the grains during deformation. Even though a single grain may be favorably oriented with the applied stress for slip, it cannot deform until the adjacent and less favorably oriented grains are capable of slip also; this requires a higher applied stress level.

8.8 DEFORMATION BY TWINNING (CD-ROM)

MECHANISMS OF STRENGTHENING IN METALS

Metallurgical and materials engineers are often called on to design alloys having high strengths yet some ductility and toughness; ordinarily, ductility is sacrificed when an alloy is strengthened. Several hardening techniques are at the disposal of an engineer, and frequently alloy selection depends on the capacity of a material to be tailored with the mechanical characteristics required for a particular application.

Important to the understanding of strengthening mechanisms is the relation between dislocation motion and mechanical behavior of metals. Because macroscopic plastic deformation corresponds to the motion of large numbers of dislocations, *the ability of a metal to plastically deform depends on the ability of dislocations to move.* Since hardness and strength (both yield and tensile) are related to the ease with which plastic deformation can be made to occur, by reducing the mobility of dislocations, the mechanical strength may be enhanced; that is, greater mechanical forces will be required to initiate plastic deformation. In contrast, the more unconstrained the dislocation motion, the greater the facility with which a metal may deform, and the softer and weaker it becomes. Virtually all strengthening techniques rely on this simple principle: *restricting or hindering dislocation motion renders a material harder and stronger.*

The present discussion is confined to strengthening mechanisms for single-phase metals, by grain size reduction, solid-solution alloying, and strain hardening. Deformation and strengthening of multiphase alloys are more complicated, involving concepts yet to be discussed.

8.9 STRENGTHENING BY GRAIN SIZE REDUCTION

The size of the grains, or average grain diameter, in a polycrystalline metal influences the mechanical properties. Adjacent grains normally have different crystallographic orientations and, of course, a common grain boundary, as indicated in Figure 8.14. During plastic deformation, slip or dislocation motion must take place across this common boundary, say, from grain A to grain B in Figure 8.14. The grain boundary acts as a barrier to dislocation motion for two reasons:

1. Since the two grains are of different orientations, a dislocation passing into grain B will have to change its direction of motion; this becomes more difficult as the crystallographic misorientation increases.

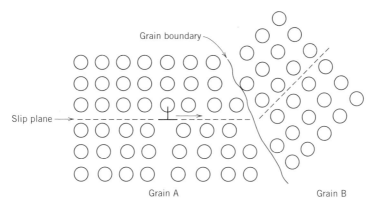

FIGURE 8.14 The motion of a dislocation as it encounters a grain boundary, illustrating how the boundary acts as a barrier to continued slip. Slip planes are discontinuous and change directions across the boundary. (From *A Textbook of Materials Technology* by Van Vlack, © 1973. Reprinted by permission of Prentice-Hall, Inc., Upper Saddle River, NJ.)

2. The atomic disorder within a grain boundary region will result in a discontinuity of slip planes from one grain into the other.

It should be mentioned that, for high-angle grain boundaries, it may not be the case that dislocations traverse grain boundaries during deformation; rather, a stress concentration ahead of a slip plane in one grain may activate sources of new dislocations in an adjacent grain.

A fine-grained material (one that has small grains) is harder and stronger than one that is coarse grained, since the former has a greater total grain boundary area to impede dislocation motion. For many materials, the yield strength σ_y varies with grain size according to

$$\sigma_y = \sigma_0 + k_y d^{-1/2} \tag{8.5}$$

In this expression, termed the *Hall-Petch equation, d* is the average grain diameter, and σ_0 and k_y are constants for a particular material. It should be noted that Equation 8.5 is not valid for both very large (i.e., coarse) grain and extremely fine grain polycrystalline materials. Figure 8.15 demonstrates the yield strength dependence on grain size for a brass alloy. Grain size may be regulated by the rate of solidification from the liquid phase, and also by plastic deformation followed by an appropriate heat treatment, as discussed in Section 8.14.

It should also be mentioned that grain size reduction improves not only strength, but also the toughness of many alloys.

Small-angle grain boundaries (Section 5.8) are not effective in interfering with the slip process because of the slight crystallographic misalignment across the boundary. On the other hand, twin boundaries (Section 5.8) will effectively block slip and increase the strength of the material. Boundaries between two different phases are also impediments to movements of dislocations; this is important in the strengthening of more complex alloys. The sizes and shapes of the constituent phases significantly affect the mechanical properties of multiphase alloys; these are the topics of discussion in Sections 11.7, 11.8, {and 15.1.}

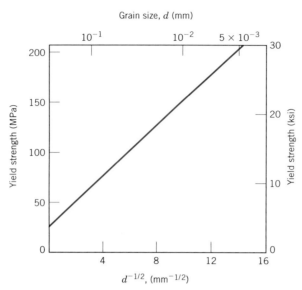

FIGURE 8.15 The influence of grain size on the yield strength of a 70 Cu–30 Zn brass alloy. Note that the grain diameter increases from right to left and is not linear. (Adapted from H. Suzuki, "The Relation Between the Structure and Mechanical Properties of Metals," Vol. II, *National Physical Laboratory, Symposium No. 15*, 1963, p. 524.)

8.10 SOLID-SOLUTION STRENGTHENING

Another technique to strengthen and harden metals is alloying with impurity atoms that go into either substitutional or interstitial solid solution. Accordingly, this is called **solid-solution strengthening.** High-purity metals are almost always softer and weaker than alloys composed of the same base metal. Increasing the concentration of the impurity results in an attendant increase in tensile and yield strengths, as indicated in Figures 8.16a and 8.16b for nickel in copper; the dependence of ductility on nickel concentration is presented in Figure 8.16c.

Alloys are stronger than pure metals because impurity atoms that go into solid solution ordinarily impose lattice strains on the surrounding host atoms. Lattice strain field interactions between dislocations and these impurity atoms result, and, consequently, dislocation movement is restricted. For example, an impurity atom that is smaller than a host atom for which it substitutes exerts tensile strains on the surrounding crystal lattice, as illustrated in Figure 8.17a. Conversely, a larger substitutional atom imposes compressive strains in its vicinity (Figure 8.18a). These solute atoms tend to diffuse to and segregate around dislocations in a way so as to reduce the overall strain energy, that is, to cancel some of the strain in the lattice surrounding a dislocation. To accomplish this, a smaller impurity atom is located where its tensile strain will partially nullify some of the dislocation's compressive strain. For the edge dislocation in Figure 8.17b, this would be adjacent to the dislocation line and above the slip plane. A larger impurity atom would be situated as in Figure 8.18b.

The resistance to slip is greater when impurity atoms are present because the overall lattice strain must increase if a dislocation is torn away from them. Furthermore, the same lattice strain interactions (Figures 8.17b and 8.18b) will exist between impurity atoms and dislocations that are in motion during plastic deformation. Thus, a greater applied stress is necessary to first initiate and then continue plastic deformation for solid-solution alloys, as opposed to pure metals; this is evidenced by the enhancement of strength and hardness.

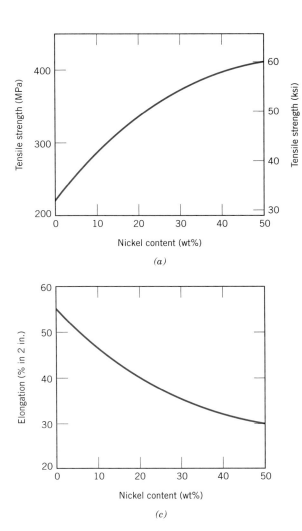

(a)

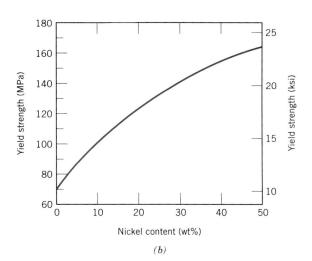

(b)

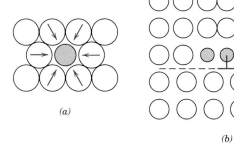

(c)

FIGURE 8.16 Variation with nickel content of (a) tensile strength, (b) yield strength, and (c) ductility (%EL) for copper-nickel alloys, showing strengthening.

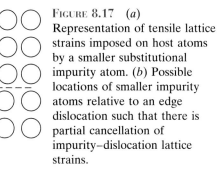

(a)

(b)

FIGURE 8.17 (a) Representation of tensile lattice strains imposed on host atoms by a smaller substitutional impurity atom. (b) Possible locations of smaller impurity atoms relative to an edge dislocation such that there is partial cancellation of impurity–dislocation lattice strains.

FIGURE 8.18 (*a*) Representation of compressive strains imposed on host atoms by a larger substitutional impurity atom. (*b*) Possible locations of larger impurity atoms relative to an edge dislocation such that there is partial cancellation of impurity–dislocation lattice strains.

(a)

(b)

8.11 STRAIN HARDENING

Strain hardening is the phenomenon whereby a ductile metal becomes harder and stronger as it is plastically deformed. Sometimes it is also called *work hardening,* or, because the temperature at which deformation takes place is "cold" relative to the absolute melting temperature of the metal, **cold working.** Most metals strain harden at room temperature.

It is sometimes convenient to express the degree of plastic deformation as *percent cold work* rather than as strain. Percent cold work (%CW) is defined as

$$\%CW = \left(\frac{A_0 - A_d}{A_0}\right) \times 100 \tag{8.6}$$

where A_0 is the original area of the cross section that experiences deformation, and A_d is the area after deformation.

Figures 8.19*a* and 8.19*b* demonstrate how steel, brass, and copper increase in yield and tensile strength with increasing cold work. The price for this enhancement of hardness and strength is in the ductility of the metal. This is shown in Figure 8.19*c*, in which the ductility, in percent elongation, experiences a reduction with increasing percent cold work for the same three alloys. The influence of cold work on the stress–strain behavior of a steel is vividly portrayed in Figure 8.20.

Strain hardening is demonstrated in a stress–strain diagram presented earlier (Figure 7.17). Initially, the metal with yield strength σ_{y_0} is plastically deformed to point D. The stress is released, then reapplied with a resultant new yield strength, σ_{y_i}. The metal has thus become stronger during the process because σ_{y_i} is greater than σ_{y_0}.

The strain hardening phenomenon is explained on the basis of dislocation–dislocation strain field interactions similar to those discussed in Section 8.4. The dislocation density in a metal increases with deformation or cold work, due to dislocation multiplication or the formation of new dislocations, as noted previously. Consequently, the average distance of separation between dislocations decreases—the dislocations are positioned closer together. On the average, dislocation–dislocation strain interactions are repulsive. The net result is that the motion of a dislocation is hindered by the presence of other dislocations. As the dislocation density increases, this resistance to dislocation motion by other dislocations becomes more pronounced. Thus, the imposed stress necessary to deform a metal increases with increasing cold work.

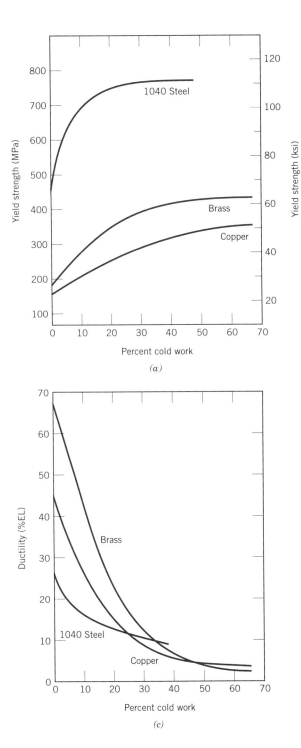

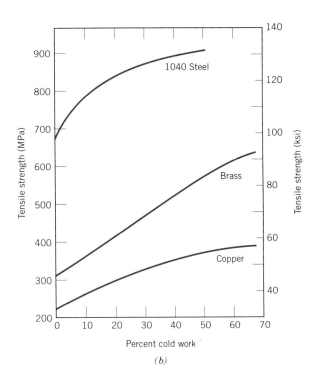

FIGURE 8.19   For 1040 steel, brass, and copper, (a) the increase in yield strength, (b) the increase in tensile strength, and (c) the decrease in ductility (%EL) with percent cold work. (Adapted from *Metals Handbook: Properties and Selection: Irons and Steels,* Vol. 1, 9th edition, B. Bardes, Editor, American Society for Metals, 1978, p. 226; and *Metals Handbook: Properties and Selection: Nonferrous Alloys and Pure Metals,* Vol. 2, 9th edition, H. Baker, Managing Editor, American Society for Metals, 1979, pp. 276 and 327.)

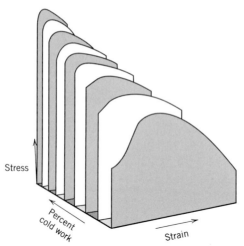

FIGURE 8.20 The influence of cold work on the stress–strain behavior for a low-carbon steel. (From *Metals Handbook: Properties and Selection: Irons and Steels*, Vol. 1, 9th edition, B. Bardes, Editor, American Society for Metals, 1978, p. 221.)

Strain hardening is often utilized commercially to enhance the mechanical properties of metals during fabrication procedures. The effects of strain hardening may be removed by an annealing heat treatment, {as discussed in Section 14.5.}

In passing, for the mathematical expression relating true stress and strain, Equation 7.19, the parameter *n* is called the *strain hardening exponent,* which is a measure of the ability of a metal to strain harden; the larger its magnitude, the greater the strain hardening for a given amount of plastic strain.

EXAMPLE PROBLEM 8.2

Compute the tensile strength and ductility (%EL) of a cylindrical copper rod if it is cold worked such that the diameter is reduced from 15.2 mm to 12.2 mm (0.60 in. to 0.48 in.).

SOLUTION

It is first necessary to determine the percent cold work resulting from the deformation. This is possible using Equation 8.6:

$$\%CW = \frac{\left(\dfrac{15.2 \text{ mm}}{2}\right)^2 \pi - \left(\dfrac{12.2 \text{ mm}}{2}\right)^2 \pi}{\left(\dfrac{15.2 \text{ mm}}{2}\right)^2 \pi} \times 100 = 35.6\%$$

The tensile strength is read directly from the curve for copper (Figure 8.19*b*) as 340 MPa (50,000 psi). From Figure 8.19*c*, the ductility at 35.6%CW is about 7%EL.

In summary, we have just discussed the three mechanisms that may be used to strengthen and harden single-phase metal alloys—strengthening by grain size reduction, solid solution strengthening, and strain hardening. Of course they may be used in conjunction with one another; for example, a solid-solution strengthened alloy may also be strain hardened.

RECOVERY, RECRYSTALLIZATION, AND GRAIN GROWTH

As outlined in the preceding paragraphs of this chapter, plastically deforming a polycrystalline metal specimen at temperatures that are low relative to its absolute melting temperature produces microstructural and property changes that include (1) a change in grain shape (Section 8.7), (2) strain hardening (Section 8.11), and (3) an increase in dislocation density (Section 8.4). Some fraction of the energy expended in deformation is stored in the metal as strain energy, which is associated with tensile, compressive, and shear zones around the newly created dislocations (Section 8.4). Furthermore, other properties such as electrical conductivity (Section 12.8) and corrosion resistance may be modified as a consequence of plastic deformation.

These properties and structures may revert back to the precold-worked states by appropriate heat treatment (sometimes termed an annealing treatment). Such restoration results from two different processes that occur at elevated temperatures: **recovery** and **recrystallization,** which may be followed by **grain growth.**

8.12 RECOVERY

During recovery, some of the stored internal strain energy is relieved by virtue of dislocation motion (in the absence of an externally applied stress), as a result of enhanced atomic diffusion at the elevated temperature. There is some reduction in the number of dislocations, and dislocation configurations (similar to that shown in Figure 5.12) are produced having low strain energies. In addition, physical properties such as electrical and thermal conductivities and the like are recovered to their precold-worked states.

8.13 RECRYSTALLIZATION

Even after recovery is complete, the grains are still in a relatively high strain energy state. Recrystallization is the formation of a new set of strain-free and equiaxed grains (i.e., having approximately equal dimensions in all directions) that have low dislocation densities and are characteristic of the precold-worked condition. The driving force to produce this new grain structure is the difference in internal energy between the strained and unstrained material. The new grains form as very small nuclei and grow until they completely replace the parent material, processes that involve short-range diffusion. Several stages in the recrystallization process are represented in Figures 8.21a to 8.21d; in these photomicrographs, the small speckled grains are those that have recrystallized. Thus, recrystallization of cold-worked metals may be used to refine the grain structure.

Also, during recrystallization, the mechanical properties that were changed as a result of cold working are restored to their precold-worked values; that is, the metal becomes softer, weaker, yet more ductile. Some heat treatments are designed to allow recrystallization to occur with these modifications in the mechanical characteristics {(Section 14.5).}

Recrystallization is a process the extent of which depends on both time and temperature. The degree (or fraction) of recrystallization increases with time, as may be noted in the photomicrographs shown in Figures 8.21a–d. The explicit time dependence of recrystallization is addressed in more detail in Section 11.3.

FIGURE 8.21 Photomicrographs showing several stages of the recrystallization and grain growth of brass. (*a*) Cold-worked (33%CW) grain structure. (*b*) Initial stage of recrystallization after heating 3 s at 580°C (1075°F); the very small grains are those that have recrystallized. (*c*) Partial replacement of cold-worked grains by recrystallized ones (4 s at 580°C). (*d*) Complete recrystallization (8 s at 580°C). (*e*) Grain growth after 15 min at 580°C. (*f*) Grain growth after 10 min at 700°C (1290°F). All photomicrographs 75×. (Photomicrographs courtesy of J. E. Burke, General Electric Company.)

(e)   (f)

FIGURE 8.21   (continued)

The influence of temperature is demonstrated in Figure 8.22, which plots tensile strength and ductility (at room temperature) of a brass alloy as a function of the temperature and for a constant heat treatment time of 1 h. The grain structures found at the various stages of the process are also presented schematically.

The recrystallization behavior of a particular metal alloy is sometimes specified in terms of a **recrystallization temperature,** the temperature at which recrystallization just reaches completion in 1 h. Thus, the recrystallization temperature for the brass alloy of Figure 8.22 is about 450°C (850°F). Typically, it is between one third and one half of the absolute melting temperature of a metal or alloy and depends on several factors, including the amount of prior cold work and the purity of the alloy. Increasing the percentage of cold work enhances the rate of recrystallization, with the result that the recrystallization temperature is lowered, and approaches a constant or limiting value at high deformations; this effect is shown in Figure 8.23. Furthermore, it is this limiting or minimum recrystallization temperature that is normally specified in the literature. There exists some critical degree of cold work below which recrystallization cannot be made to occur, as shown in the figure; normally, this is between 2 and 20% cold work.

Recrystallization proceeds more rapidly in pure metals than in alloys. Thus, alloying raises the recrystallization temperature, sometimes quite substantially. For pure metals, the recrystallization temperature is normally $0.3T_m$, where T_m is the absolute melting temperature; for some commercial alloys it may run as high as $0.7T_m$. Recrystallization and melting temperatures for a number of metals and alloys are listed in Table 8.2.

Plastic deformation operations are often carried out at temperatures above the recrystallization temperature in a process termed *hot working,* {described in Section 14.2.} The material remains relatively soft and ductile during deformation because it does not strain harden, and thus large deformations are possible.

FIGURE 8.22 The influence of annealing temperature on the tensile strength and ductility of a brass alloy. Grain size as a function of annealing temperature is indicated. Grain structures during recovery, recrystallization, and grain growth stages are shown schematically. (Adapted from G. Sachs and K. R. Van Horn, *Practical Metallurgy, Applied Metallurgy and the Industrial Processing of Ferrous and Nonferrous Metals and Alloys,* American Society for Metals, 1940, p. 139.)

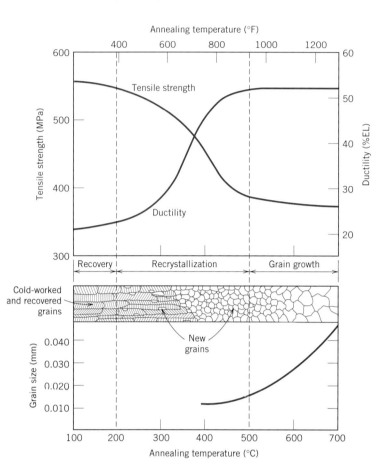

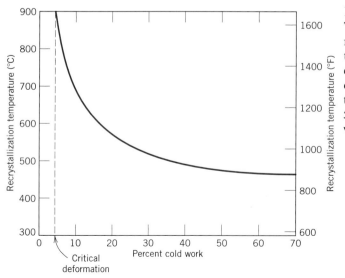

FIGURE 8.23 The variation of recrystallization temperature with percent cold work for iron. For deformations less than the critical (about 5%CW), recrystallization will not occur.

Table 8.2 Recrystallization and Melting Temperatures for Various Metals and Alloys

Metal	Recrystallization Temperature		Melting Temperature	
	°C	°F	°C	°F
Lead	−4	25	327	620
Tin	−4	25	232	450
Zinc	10	50	420	788
Aluminum (99.999 wt%)	80	176	660	1220
Copper (99.999 wt%)	120	250	1085	1985
Brass (60 Cu–40 Zn)	475	887	900	1652
Nickel (99.99 wt%)	370	700	1455	2651
Iron	450	840	1538	2800
Tungsten	1200	2200	3410	6170

DESIGN EXAMPLE 8.1

A cylindrical rod of noncold-worked brass having an initial diameter of 6.4 mm (0.25 in.) is to be cold worked by drawing such that the cross-sectional area is reduced. It is required to have a cold-worked yield strength of at least 345 MPa (50,000 psi) and a ductility in excess of 20%EL; in addition, a final diameter of 5.1 mm (0.20 in.) is necessary. Describe the manner in which this procedure may be carried out.

SOLUTION

Let us first consider the consequences (in terms of yield strength and ductility) of cold working in which the brass specimen diameter is reduced from 6.4 mm (designated by d_0) to 5.1 mm (d_i). The %CW may be computed from Equation 8.6 as

$$\%\text{CW} = \frac{\left(\dfrac{d_0}{2}\right)^2 \pi - \left(\dfrac{d_i}{2}\right)^2 \pi}{\left(\dfrac{d_0}{2}\right)^2 \pi} \times 100$$

$$= \frac{\left(\dfrac{6.4 \text{ mm}}{2}\right)^2 \pi - \left(\dfrac{5.1 \text{ mm}}{2}\right)^2 \pi}{\left(\dfrac{6.4 \text{ mm}}{2}\right)^2 \pi} \times 100 = 36.5\%\text{CW}$$

From Figures 8.19a and 8.19c, a yield strength of 410 MPa (60,000 psi) and a ductility of 8%EL are attained from this deformation. According to the stipulated criteria, the yield strength is satisfactory; however, the ductility is too low.

Another processing alternative is a partial diameter reduction, followed by a recrystallization heat treatment in which the effects of the cold work are nullified. The required yield strength, ductility, and diameter are achieved through a second drawing step.

Again, reference to Figure 8.19a indicates that 20%CW is required to give a yield strength of 345 MPa. On the other hand, from Figure 8.19c, ductilities greater than 20%EL are possible only for deformations of 23%CW or less. Thus during the final drawing operation, deformation must be between 20%CW and 23%CW. Let's take the average of these extremes, 21.5%CW, and then calculate the final diameter for the first drawing d_0', which becomes the original diameter for the second drawing. Again, using Equation 8.6,

$$21.5\%\text{CW} = \frac{\left(\dfrac{d_0'}{2}\right)^2 \pi - \left(\dfrac{5.1 \text{ mm}}{2}\right)^2 \pi}{\left(\dfrac{d_0'}{2}\right)^2 \pi} \times 100$$

Now, solving from d_0' from the expression above gives

$$d_0' = 5.8 \text{ mm (0.226 in.)}$$

8.14 GRAIN GROWTH

After recrystallization is complete, the strain-free grains will continue to grow if the metal specimen is left at the elevated temperature (Figures 8.21d–f); this phenomenon is called **grain growth.** Grain growth does not need to be preceded by recovery and recrystallization; it may occur in all polycrystalline materials, metals and ceramics alike.

An energy is associated with grain boundaries, as explained in Section 5.8. As grains increase in size, the total boundary area decreases, yielding an attendant reduction in the total energy; this is the driving force for grain growth.

Grain growth occurs by the migration of grain boundaries. Obviously, not all grains can enlarge, but large ones grow at the expense of small ones that shrink. Thus, the average grain size increases with time, and at any particular instant there will exist a range of grain sizes. Boundary motion is just the short-range diffusion of atoms from one side of the boundary to the other. The directions of boundary movement and atomic motion are opposite to each other, as shown in Figure 8.24.

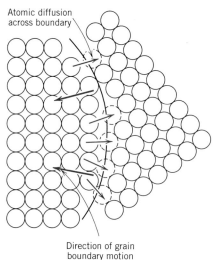

Atomic diffusion across boundary

Direction of grain boundary motion

FIGURE 8.24 Schematic representation of grain growth via atomic diffusion. (From *Elements of Materials Science and Engineering* by Van Vlack, © 1989. Reprinted by permission of Prentice-Hall, Inc., Upper Saddle River, NJ.)

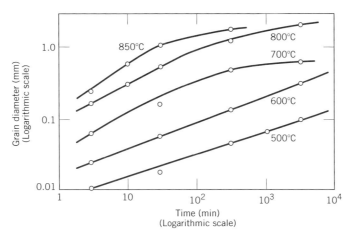

FIGURE 8.25 The logarithm of grain diameter versus the logarithm of time for grain growth in brass at several temperatures. (From J. E. Burke, "Some Factors Affecting the Rate of Grain Growth in Metals." Reprinted with permission from *Metallurgical Transactions*, Vol. 180, 1949, a publication of The Metallurgical Society of AIME, Warrendale, Pennsylvania.)

For many polycrystalline materials, the grain diameter d varies with time t according to the relationship

$$d^n - d_0^n = Kt \qquad (8.7)$$

where d_0 is the initial grain diameter at $t = 0$, and K and n are time-independent constants; the value of n is generally equal to or greater than 2.

The dependence of grain size on time and temperature is demonstrated in Figure 8.25, a plot of the logarithm of grain size as a function of the logarithm of time for a brass alloy at several temperatures. At lower temperatures the curves are linear. Furthermore, grain growth proceeds more rapidly as temperature increases; that is, the curves are displaced upward to larger grain sizes. This is explained by the enhancement of diffusion rate with rising temperature.

The mechanical properties at room temperature of a fine-grained metal are usually superior (i.e., higher strength and toughness) to those of coarse-grained ones. If the grain structure of a single-phase alloy is coarser than that desired, refinement may be accomplished by plastically deforming the material, then subjecting it to a recrystallization heat treatment, as described above.

DEFORMATION MECHANISMS FOR CERAMIC MATERIALS

Although at room temperature most ceramic materials suffer fracture before the onset of plastic deformation, a brief exploration into the possible mechanisms is worthwhile. Plastic deformation is different for crystalline and noncrystalline ceramics; however, each is discussed.

8.15 CRYSTALLINE CERAMICS

For crystalline ceramics, plastic deformation occurs, as with metals, by the motion of dislocations. One reason for the hardness and brittleness of these materials is the difficulty of slip (or dislocation motion). For crystalline ceramic materials for which the bonding is predominantly ionic, there are very few slip systems (crystallographic planes and directions within those planes) along which dislocations may move. This is a consequence of the electrically charged nature of the ions. For slip in some directions, ions of like charge are brought into close proximity to one another; because of electrostatic repulsion, this mode of slip is very restricted. This is not a problem in metals, since all atoms are electrically neutral.

On the other hand, for ceramics in which the bonding is highly covalent, slip is also difficult and they are brittle for the following reasons: (1) the covalent bonds are relatively strong; (2) there are also limited numbers of slip systems; and (3) dislocation structures are complex.

8.16 NONCRYSTALLINE CERAMICS

Plastic deformation does not occur by dislocation motion for noncrystalline ceramics because there is no regular atomic structure. Rather, these materials deform by *viscous flow*, the same manner in which liquids deform; the rate of deformation is proportional to the applied stress. In response to an applied shear stress, atoms or ions slide past one another by the breaking and reforming of interatomic bonds. However, there is no prescribed manner or direction in which this occurs, as with dislocations. Viscous flow on a macroscopic scale is demonstrated in Figure 8.26.

The characteristic property for viscous flow, **viscosity,** is a measure of a non-crystalline material's resistance to deformation. For viscous flow in a liquid that originates from shear stresses imposed by two flat and parallel plates, the viscosity η is the ratio of the applied shear stress τ and the change in velocity dv with distance dy in a direction perpendicular to and away from the plates, or

$$\eta = \frac{\tau}{dv/dy} = \frac{F/A}{dv/dy} \tag{8.8}$$

This scheme is represented in Figure 8.26.

The units for viscosity are poises (P) and pascal-seconds (Pa-s); 1 P = 1 dyne-s/cm^2, and 1 Pa-s = 1 N-s/m^2. Conversion from one system of units to the other is

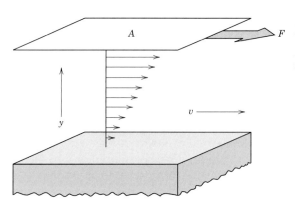

FIGURE 8.26 Representation of the viscous flow of a liquid or fluid glass in response to an applied shear force.

according to

$$10 \text{ P} = 1 \text{ Pa-s}$$

Liquids have relatively low viscosities; for example, the viscosity of water at room temperature is about 10^{-3} Pa-s. On the other hand, glasses have extremely large viscosities at ambient temperatures, which is accounted for by strong interatomic bonding. As the temperature is raised, the magnitude of the bonding is diminished, the sliding motion or flow of the atoms or ions is facilitated, and subsequently there is an attendant decrease in viscosity. {A discussion of the temperature dependence of viscosity for glasses is deferred to Section 14.7.}

MECHANISMS OF DEFORMATION AND FOR STRENGTHENING OF POLYMERS

An understanding of deformation mechanisms of polymers is important in order for us to be able to manage the mechanical characteristics of these materials. In this regard, deformation models for two different types of polymers—semicrystalline and elastomeric—deserve our attention. The stiffness and strength of semicrystalline materials are often important considerations; elastic and plastic deformation mechanisms are treated in the succeeding section, whereas methods used to stiffen and strengthen these materials are discussed in Section 8.18. On the other hand, elastomers are utilized on the basis of their unusual elastic properties; the deformation mechanism of elastomers is also treated.

8.17 DEFORMATION OF SEMICRYSTALLINE POLYMERS

Many semicrystalline polymers in bulk form will have the spherulitic structure described in Section 4.12. By way of review, let us repeat here that each spherulite consists of numerous chain-folded ribbons, or lamellae, that radiate outward from the center. Separating these lamellae are areas of amorphous material (Figure 4.14); adjacent lamellae are connected by tie chains that pass through these amorphous regions.

MECHANISM OF ELASTIC DEFORMATION

The mechanism of elastic deformation in semicrystalline polymers in response to tensile stresses is the elongation of the chain molecules from their stable conformations, in the direction of the applied stress, by the bending and stretching of the strong chain covalent bonds. In addition, there may be some slight displacement of adjacent molecules, which is resisted by relatively weak secondary or van der Waals bonds. Furthermore, inasmuch as semicrystalline polymers are composed of both crystalline and amorphous regions, they may, in a sense, be considered composite materials. As such, the elastic modulus may be taken as some combination of the moduli of crystalline and amorphous phases.

MECHANISM OF PLASTIC DEFORMATION

The mechanism of plastic deformation is best described by the interactions between lamellar and intervening amorphous regions in response to an applied tensile load. This process occurs in several stages, which are schematically diagrammed in Figure 8.27. Two adjacent chain-folded lamellae and the interlamel-

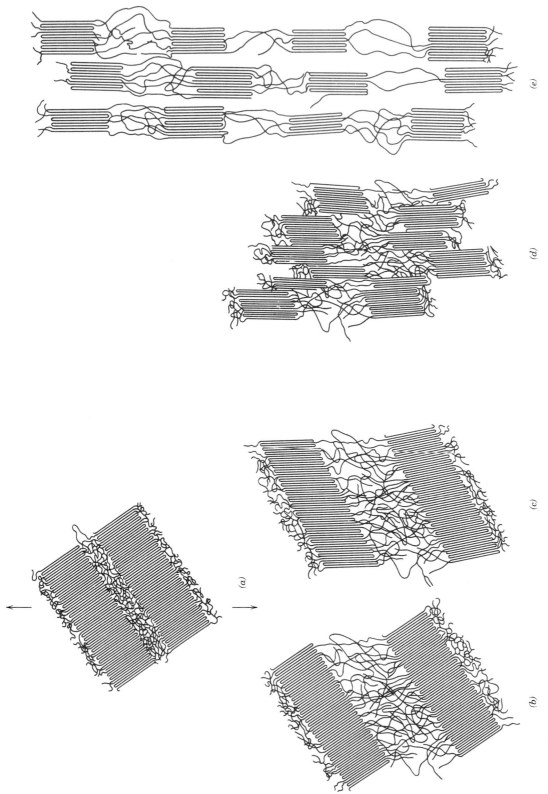

FIGURE 8.27 Stages in the deformation of a semicrystalline polymer. (*a*) Two adjacent chain-folded lamellae and interlamellar amorphous material before deformation. (*b*) Elongation of amorphous tie chains during the first stage of deformation. (*c*) Tilting of lamellar chain folds during the second stage. (*d*) Separation of crystalline block segments during the third stage. (*e*) Orientation of block segments and tie chains with the tensile axis in the final deformation stage. (From Jerold M. Schultz, *Polymer Materials Science,* copyright © 1974, pp. 500–501. Reprinted by permoission of Prentice-Hall, Inc., Englewood Cliffs, NJ.)

lar amorphous material, prior to deformation, are shown in Figure 8.27a. During the initial stage of deformation (Figure 8.27b) the chains in the amorphous regions slip past each other and align in the loading direction. This causes the lamellar ribbons simply to slide past one another as the tie chains within the amorphous regions become extended. Continued deformation in the second stage occurs by the tilting of the lamellae so that the chain folds become aligned with the tensile axis (Figure 8.27c). Next, crystalline block segments separate from the lamellae, which segments remain attached to one another by tie chains (Figure 8.27d). In the final stage (Figure 8.27e), the blocks and tie chains become oriented in the direction of the tensile axis. Thus appreciable tensile deformation of semicrystalline polymers produces a highly oriented structure. During deformation the spherulites experience shape changes for moderate levels of elongation. However, for large deformations, the spherulitic structure is virtually destroyed. Also, it is interesting to note that, to a large degree, the processes represented in Figure 8.27 are reversible. That is, if deformation is terminated at some arbitrary stage, and the specimen is heated to an elevated temperature near its melting point (i.e., annealed), the material will revert back to having the spherulitic structure that was characteristic of its undeformed state. Furthermore, the specimen will tend to shrink back to the shape it had prior to deformation; the extent of this shape and structural recovery will depend on the annealing temperature and also the degree of elongation.

8.18a FACTORS THAT INFLUENCE THE MECHANICAL PROPERTIES OF SEMICRYSTALLINE POLYMERS [DETAILED VERSION (CD-ROM)]

8.18b FACTORS THAT INFLUENCE THE MECHANICAL PROPERTIES OF SEMICRYSTALLINE POLYMERS (CONCISE VERSION)

A number of factors influence the mechanical characteristics of polymeric materials. For example, we have already discussed the effect of temperature and strain rate on stress–strain behavior (Section 7.13, Figure 7.24). Again, increasing the temperature or diminishing the strain rate leads to a decrease in the tensile modulus, a reduction in tensile strength, and an enhancement of ductility.

In addition, several structural/processing factors have decided influences on the mechanical behavior (i.e., strength and modulus) of polymeric materials. An increase in strength results whenever any restraint is imposed on the process illustrated in Figure 8.27; for example, a significant degree of intermolecular bonding or extensive chain entanglements inhibit relative chain motions. It should be noted that even though secondary intermolecular (e.g., van der Waals) bonds are much weaker than the primary covalent ones, significant intermolecular forces result from the formation of large numbers of van der Waals interchain bonds. Furthermore, the modulus rises as both the secondary bond strength and chain alignment increase. The mechanical behavior of polymers is affected by several structural/processing

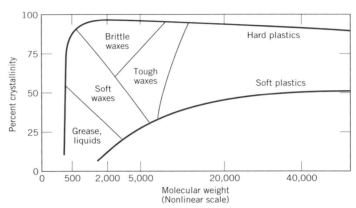

FIGURE 8.28 The influence of degree of crystallinity and molecular weight on the physical characteristics of polyethylene. (From R. B. Richards, "Polyethylene—Structure, Crystallinity and Properties," *J. Appl. Chem.*, **1**, 370, 1951.)

factors that include molecular weight, degree of crystallinity, and predeformation (drawing).

The magnitude of the tensile modulus does not seem to be influenced by molecular weight alterations. On the other hand, for many polymers, it has been observed that tensile strength increases with increasing molecular weight.

For a specific polymer, the degree of crystallinity can have a rather significant influence on the mechanical properties, since it affects the extent of intermolecular secondary bonding. It has been observed that, for semicrystalline polymers, tensile modulus increases significantly with degree of crystallinity; in most cases, strength is also enhanced, and the material becomes more brittle. The influences of chain chemistry and structure (branching, stereoisomerism, etc.) on degree of crystallinity were discussed in Chapter 4.

The effects of both percent crystallinity and molecular weight on the physical state of polyethylene are represented in Figure 8.28.

On a commercial basis, one important technique that is used to improve mechanical strength and tensile modulus is permanently deforming the polymer in tension. This procedure, sometimes called *drawing,* produces the neck extension illustrated schematically in Figure 7.25. It is an important stiffening and strengthening technique that is employed in the production of fibers and films {(Section 14.15).} During drawing, the molecular chains slip past one another and become highly oriented; for semicrystalline materials the chains assume conformations similar to those represented schematically in Figure 8.27e. It should also be noted that the mechanical properties of drawn polymers are normally anisotropic.

8.19 DEFORMATION OF ELASTOMERS

One of the fascinating properties of the elastomeric materials is their rubberlike elasticity. That is, they have the ability to be deformed to quite large deformations, and then elastically spring back to their original form. This behavior was probably first observed in natural rubber; however, the past few years have brought about the synthesis of a large number of elastomers with a wide variety of properties. Typical stress–strain characteristics of elastomeric materials are displayed in Figure 7.22, curve *C.* Their moduli of elasticity are quite small and, furthermore, vary with strain since the stress–strain curve is nonlinear.

In an unstressed state, an elastomer will be amorphous and composed of molecular chains that are highly twisted, kinked, and coiled. Elastic deformation, upon application of a tensile load, is simply the partial uncoiling, untwisting, and straightening, and the resultant elongation of the chains in the stress direction, a phenomenon represented in Figure 8.29. Upon release of the stress, the chains spring back to their prestressed conformations, and the macroscopic piece returns to its original shape.

The driving force for elastic deformation is a thermodynamic parameter called *entropy*, which is a measure of the degree of disorder within a system; entropy increases with increasing disorder. As an elastomer is stretched and the chains straighten and become more aligned, the system becomes more ordered. From this state, the entropy increases if the chains return to their original kinked and coiled contours. Two intriguing phenomena result from this entropic effect. First, when stretched, an elastomer experiences a rise in temperature; second, the modulus of elasticity increases with increasing temperature, which is opposite to the behavior found in other materials (see Figure 7.8).

Several criteria must be met in order for a polymer to be elastomeric: (1) It must not easily crystallize; elastomeric materials are amorphous, having molecular chains that are naturally coiled and kinked in the unstressed state. (2) Chain bond rotations must be relatively free in order for the coiled chains to readily respond to an applied force. (3) For elastomers to experience relatively large elastic deformations, the onset of plastic deformation must be delayed. Restricting the motions of chains past one another by crosslinking accomplishes this objective. The crosslinks act as anchor points between the chains and prevent chain slippage from occurring; the role of crosslinks in the deformation process is illustrated in Figure 8.29. Crosslinking in many elastomers is carried out in a process called vulcanization, as discussed below. (4) Finally, the elastomer must be above its glass transition temperature (Section 11.16). The lowest temperature at which rubberlike behavior persists for many of the common elastomers is between -50 and $-90°C$ (-60 and $-130°F$). Below its glass transition temperature, an elastomer becomes brittle such that its stress–strain behavior resembles curve *A* in Figure 7.22.

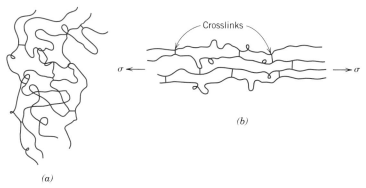

FIGURE 8.29 Schematic representation of crosslinked polymer chain molecules (*a*) in an unstressed state and (*b*) during elastic deformation in response to an applied tensile stress. (Adapted from Z. D. Jastrzebski, *The Nature and Properties of Engineering Materials*, 3rd edition. Copyright © 1987 by John Wiley & Sons, New York. Reprinted by permission of John Wiley & Sons, Inc.)

VULCANIZATION

The crosslinking process in elastomers is called **vulcanization,** which is achieved by a nonreversible chemical reaction, ordinarily carried out at an elevated temperature. In most vulcanizing reactions, sulfur compounds are added to the heated elastomer; chains of sulfur atoms bond with adjacent polymer backbone chains and crosslink them, which is accomplished according to the following reaction:

$$
\begin{array}{c}
\mathrm{H\ CH_3\ H\ H} \\
\mathrm{|\quad|\quad|\quad|} \\
\mathrm{-C-C=C-C-} \\
\mathrm{|\qquad\qquad|} \\
\mathrm{H\qquad\quad H}
\end{array}
\qquad
\begin{array}{c}
\mathrm{H\ CH_3\ H\ H} \\
\mathrm{|\quad|\quad|\quad|} \\
\mathrm{-C-C-C-C-} \\
\mathrm{|\quad|\quad|\quad|} \\
\mathrm{H\quad(S)_m\ (S)_n\ H}
\end{array}
$$

$$+ (m + n)\, \mathrm{S} \rightarrow \hspace{6cm} (8.10)$$

$$
\begin{array}{c}
\mathrm{H\qquad\quad H} \\
\mathrm{|\qquad\qquad|} \\
\mathrm{-C-C=C-C-} \\
\mathrm{|\quad|\quad|\quad|} \\
\mathrm{H\ CH_3\ H\ H}
\end{array}
\qquad
\begin{array}{c}
\mathrm{H\qquad\quad H} \\
\mathrm{|\qquad\qquad|} \\
\mathrm{-C-C-C-C-} \\
\mathrm{|\quad|\quad|\quad|} \\
\mathrm{H\ CH_3\ H\ H}
\end{array}
$$

in which the two crosslinks shown consist of m and n sulfur atoms. Crosslink main chain sites are carbon atoms that were doubly bonded before vulcanization, but, after vulcanization, have become singly bonded.

Unvulcanized rubber is soft and tacky, and has poor resistance to abrasion. Modulus of elasticity, tensile strength, and resistance to degradation by oxidation are all enhanced by vulcanization. The magnitude of the modulus of elasticity is directly proportional to the density of the crosslinks. Stress–strain curves for vulcanized and unvulcanized natural rubber are presented in Figure 8.30. To produce a rubber that is capable of large extensions without rupture of the primary chain bonds, there must be relatively few crosslinks, and these must be widely separated. Useful rubbers result when about 1 to 5 parts (by weight) of sulfur is added to 100 parts of rubber. Increasing the sulfur content further hardens the rubber and also reduces its extensibility. Also, since they are crosslinked, elastomeric materials are thermosetting in nature.

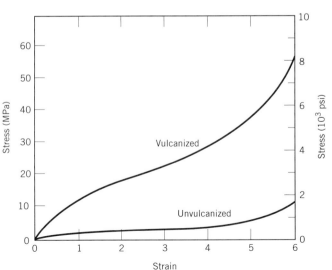

FIGURE 8.30
Stress–strain curves to 600% elongation for unvulcanized and vulcanized natural rubber.

SUMMARY

On a microscopic level, plastic deformation of metals corresponds to the motion of dislocations in response to an externally applied shear stress, a process termed "slip." Slip occurs on specific crystallographic planes and within these planes only in certain directions. A slip system represents a slip plane–slip direction combination, and operable slip systems depend on the crystal structure of the material.

{The critical resolved shear stress is the minimum shear stress required to initiate dislocation motion; the yield strength of a metal single crystal depends on both the magnitude of the critical resolved shear stress and the orientation of slip components relative to the direction of the applied stress.}

For polycrystalline metals, slip occurs within each grain along the slip systems that are most favorably oriented with the applied stress; furthermore, during deformation, grains change shape in such a manner that coherency at the grain boundaries is maintained.

{Under some circumstances limited plastic deformation may occur in BCC and HCP metals by mechanical twinning. Normally, twinning is important to the degree that accompanying crystallographic reorientations make the slip process more favorable.}

Since the ease with which a metal is capable of plastic deformation is a function of dislocation mobility, restricting dislocation motion increases hardness and strength. On the basis of this principle, three different strengthening mechanisms were discussed. Grain boundaries serve as barriers to dislocation motion; thus refining the grain size of a polycrystalline metal renders it harder and stronger. Solid solution strengthening results from lattice strain interactions between impurity atoms and dislocations. And, finally, as a metal is plastically deformed, the dislocation density increases, as does also the extent of repulsive dislocation–dislocation strain field interactions; strain hardening is just the enhancement of strength with increased plastic deformation.

The microstructural and mechanical characteristics of a plastically deformed metal specimen may be restored to their predeformed states by an appropriate heat treatment, during which recovery, recrystallization, and grain growth processes are allowed to occur. During recovery there is a reduction in dislocation density and alterations in dislocation configurations. Recrystallization is the formation of a new set of grains that are strain free; in addition, the material becomes softer and more ductile. Grain growth is the increase in average grain size of polycrystalline materials, which proceeds by grain boundary motion.

Any plastic deformation of crystalline ceramics is a result of dislocation motion; the brittleness of these materials is, in part, explained by the limited number of operable slip systems. The mode of plastic deformation for noncrystalline materials is by viscous flow; a material's resistance to deformation is expressed as viscosity. At room temperature, the viscosity of many noncrystalline ceramics is extremely high.

During the elastic deformation of a semicrystalline polymer that is stressed in tension, the constituent molecules elongate in the stress direction by the bending and stretching of covalent chain bonds. Slight molecular displacements are resisted by weak secondary bonds.

The mechanism of plastic deformation for semicrystalline polymers having the spherulitic structure was presented. Tensile deformation is thought to occur in several stages as both amorphous tie chains and chain-folded block segments (which separate from the ribbonlike lamellae) become oriented with the tensile axis. Also, during deformation the shapes of spherulites are altered (for moderate deformations); relatively large degrees of deformation lead to a complete destruction of the spherulites. Furthermore, the predeformed spherulitic structure and macroscopic shape may be virtually restored by annealing at an elevated temperature below the polymer's melting temperature.

The mechanical behavior of a polymer will be influenced by both inservice and structural/processing factors. With regard to the former, increasing the temperature and/or diminishing the strain rate leads to reductions in tensile modulus and tensile strength, and an enhancement of ductility. In addition, other factors that affect the mechanical properties include molecular weight, degree of crystallinity, predeformation drawing, and heat treating. The influence of each of these factors was discussed.

Large elastic extensions are possible for the elastomeric materials that are amorphous and lightly crosslinked. Deformation corresponds to the unkinking and uncoiling of chains in response to an applied tensile stress. Crosslinking is often achieved during a vulcanization process.

IMPORTANT TERMS AND CONCEPTS

Cold working	Recovery	Slip system
Critical resolved shear stress	Recrystallization	Solid-solution strengthening
Dislocation density	Recrystallization temperature	Strain hardening
Grain growth	Resolved shear stress	Viscosity
Lattice strain	Slip	Vulcanization

REFERENCES

Hirth, J. P. and J. Lothe, *Theory of Dislocations,* 2nd edition, Wiley-Interscience, New York, 1982. Reprinted by Krieger Publishing Company, Melbourne, FL, 1992.

Hull, D., *Introduction to Dislocations,* 3rd edition, Pergamon Press, Inc., Elmsford, NY, 1984.

Kingery, W. D., H. K. Bowen, and D. R. Uhlmann, *Introduction to Ceramics,* 2nd edition, John Wiley & Sons, New York, 1976. Chapter 14.

Read, W. T., Jr., *Dislocations in Crystals,* McGraw-Hill Book Company, New York, 1953.

Richerson, D. W., *Modern Ceramic Engineering,* 2nd edition, Marcel Dekker, New York, 1992. Chapter 5.

Schultz, J., *Polymer Materials Science,* Prentice-Hall, Englewood Cliffs, NJ, 1974.

Weertman, J. and J. R. Weertman, *Elementary Dislocation Theory,* The Macmillan Co., New York, 1964. Reprinted by Oxford University Press, Oxford, 1992.

QUESTIONS AND PROBLEMS

Note: To solve those problems having an asterisk (*) by their numbers, consultation of supplementary topics [appearing only on the CD-ROM (and not in print)] will probably be necessary.

8.1 To provide some perspective on the dimensions of atomic defects, consider a metal specimen that has a dislocation density of 10^4 mm^{-2}. Suppose that all the dislocations in 1000 mm^3 (1 cm^3) were somehow removed and linked end to end. How far (in miles) would this chain extend? Now suppose that the density is increased to 10^{10} mm^{-2} by cold working. What would be the chain length of dislocations in 1000 mm^3 of material?

8.2 Consider two edge dislocations of opposite sign and having slip planes that are separated by several atomic distances as indicated in the diagram. Briefly describe the defect that results when these two dislocations become aligned with each other.

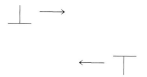

8.3 Is it possible for two screw dislocations of opposite sign to annihilate each other? Explain your answer.

8.4 For each of edge, screw, and mixed dislocations, cite the relationship between the direction of the applied shear stress and the direction of dislocation line motion.

8.5 (a) Define a slip system.

(b) Do all metals have the same slip system? Why or why not?

8.6* (a) Compare planar densities (Section 3.14) for the (100), (110), and (111) planes for FCC.

(b) Compare planar densities for the (100), (110), and (111) planes for BCC.

8.7 One slip system for the BCC crystal structure is {110}⟨111⟩. In a manner similar to Figure 8.6b sketch a {110}-type plane for the BCC structure, representing atom positions with circles. Now, using arrows, indicate two different ⟨111⟩ slip directions within this plane.

8.8 One slip system for the HCP crystal structure is {0001}⟨11$\overline{2}$0⟩. In a manner similar to Figure 8.6b, sketch a {0001}-type plane for the HCP structure, and using arrows, indicate three different ⟨11$\overline{2}$0⟩ slip directions within this plane. You might find Figure 3.22 helpful.

8.9* Explain the difference between resolved shear stress and critical resolved shear stress.

8.10* Sometimes $\cos\phi \cos\lambda$ in Equation 8.1 is termed the *Schmid factor*. Determine the magnitude of the Schmid factor for an FCC single crystal oriented with its [100] direction parallel to the loading axis.

8.11* Consider a metal single crystal oriented such that the normal to the slip plane and the slip direction are at angles of 43.1° and 47.9°, respectively, with the tensile axis. If the criti-

cal resolved shear stress is 20.7 MPa (3000 psi), will an applied stress of 45 MPa (6500 psi) cause the single crystal to yield? If not, what stress will be necessary?

8.12* A single crystal of aluminum is oriented for a tensile test such that its slip plane normal makes an angle of 28.1° with the tensile axis. Three possible slip directions make angles of 62.4°, 72.0°, and 81.1° with the same tensile axis.

(a) Which of these three slip directions is most favored?

(b) If plastic deformation begins at a tensile stress of 1.95 MPa (280 psi), determine the critical resolved shear stress for aluminum.

8.13* Consider a single crystal of silver oriented such that a tensile stress is applied along a [001] direction. If slip occurs on a (111) plane and in a [$\overline{1}$01] direction, and is initiated at an applied tensile stress of 1.1 MPa (160 psi), compute the critical resolved shear stress.

8.14* The critical resolved shear stress for iron is 27 MPa (4000 psi). Determine the maximum possible yield strength for a single crystal of Fe pulled in tension.

8.15* List four major differences between deformation by twinning and deformation by slip relative to mechanism, conditions of occurrence, and final result.

8.16 Briefly explain why small-angle grain boundaries are not as effective in interfering with the slip process as are high-angle grain boundaries.

8.17 Briefly explain why HCP metals are typically more brittle than FCC and BCC metals.

8.18 Describe in your own words the three strengthening mechanisms discussed in this chapter (i.e., grain size reduction, solid solution strengthening, and strain hardening). Be sure to explain how dislocations are involved in each of the strengthening techniques.

8.19 (a) From the plot of yield strength versus (grain diameter)$^{-1/2}$ for a 70 Cu–30 Zn cartridge brass, Figure 8.15, determine values for the constants σ_0 and k_y in Equation 8.5.

(b) Now predict the yield strength of this alloy when the average grain diameter is 1.0×10^{-3} mm.

8.20 The lower yield point for an iron that has an average grain diameter of 5×10^{-2} mm is 135 MPa (19,500 psi). At a grain diameter of 8×10^{-3} mm, the yield point increases to 260 MPa (37,500 psi). At what grain diameter will the lower yield point be 205 MPa (30,000 psi)?

8.21 If it is assumed that the plot in Figure 8.15 is for noncold-worked brass, determine the grain size of the alloy in Figure 8.19; assume its composition is the same as the alloy in Figure 8.15.

8.22 In the manner of Figures 8.17*b* and 8.18*b* indicate the location in the vicinity of an edge dislocation at which an interstitial impurity atom would be expected to be situated. Now briefly explain in terms of lattice strains why it would be situated at this position.

8.23 When making hardness measurements, what will be the effect of making an indentation very close to a preexisting indentation? Why?

8.24 **(a)** Show, for a tensile test, that

$$\%\mathrm{CW} = \left(\frac{\epsilon}{\epsilon + 1} \right) \times 100$$

if there is no change in specimen volume during the deformation process (i.e., $A_0 l_0 = A_d l_d$).

(b) Using the result of part a, compute the percent cold work experienced by naval brass (the stress–strain behavior of which is shown in Figure 7.12) when a stress of 400 MPa (58,000 psi) is applied.

8.25 Two previously undeformed cylindrical specimens of an alloy are to be strain hardened by reducing their cross-sectional areas (while maintaining their circular cross sections). For one specimen, the initial and deformed radii are 16 mm and 11 mm, respectively. The second specimen, with an initial radius of 12 mm, must have the same deformed hardness as the first specimen; compute the second specimen's radius after deformation.

8.26 Two previously undeformed specimens of the same metal are to be plastically deformed by reducing their cross-sectional areas. One has a circular cross section, and the other is rectangular; during deformation the circular cross section is to remain circular, and the rectangular is to remain as such. Their original and deformed dimensions are as follows:

	Circular (diameter, mm)	Rectangular (mm)
Original dimensions	15.2	125 × 175
Deformed dimensions	11.4	75 × 200

Which of these specimens will be the hardest after plastic deformation, and why?

8.27 A cylindrical specimen of cold-worked copper has a ductility (%EL) of 25%. If its cold-worked radius is 10 mm (0.40 in.), what was its radius before deformation?

8.28 **(a)** What is the approximate ductility (%EL) of a brass that has a yield strength of 275 MPa (40,000 psi)?

(b) What is the approximate Brinell hardness of a 1040 steel having a yield strength of 690 MPa (100,000 psi)?

8.29 Experimentally, it has been observed for single crystals of a number of metals that the critical resolved shear stress τ_{crss} is a function of the dislocation density ρ_D as

$$\tau_{\mathrm{crss}} = \tau_0 + A \sqrt{\rho_D}$$

where τ_0 and A are constants. For copper, the critical resolved shear stress is 2.10 MPa (305 psi) at a dislocation density of 10^5 mm^{-2}. If it is known that the value of A for copper is 6.35×10^{-3} MPa-mm (0.92 psi-mm), compute the τ_{crss} at a dislocation density of 10^7 mm^{-2}.

8.30 Briefly cite the differences between recovery and recrystallization processes.

8.31 Estimate the fraction of recrystallization from the photomicrograph in Figure 8.21*c*.

8.32 Explain the differences in grain structure for a metal that has been cold worked and one that has been cold worked and then recrystallized.

8.33 Briefly explain why some metals (e.g., lead and tin) do not strain harden when deformed at room temperature.

8.34 **(a)** What is the driving force for recrystallization?

(b) For grain growth?

8.35 **(a)** From Figure 8.25, compute the length of time required for the average grain diameter to increase from 0.01 to 0.1 mm at 500°C for this brass material.

(b) Repeat the calculation at 600°C.

8.36 The average grain diameter for a brass material was measured as a function of time at 650°C, which is tabulated below at two different times:

Time (min)	Grain Diameter (mm)
30	3.9×10^{-2}
90	6.6×10^{-2}

(a) What was the original grain diameter?

(b) What grain diameter would you predict after 150 min at 650°C?

8.37 An undeformed specimen of some alloy has an average grain diameter of 0.040 mm. You are asked to reduce its average grain diameter to 0.010 mm. Is this possible? If so, explain the procedures you would use and name the processes involved. If it is not possible, explain why.

8.38 Grain growth is strongly dependent on temperature (i.e., rate of grain growth increases with increasing temperature), yet temperature is not explicitly given as a part of Equation 8.7.

(a) Into which of the parameters in this expression would you expect temperature to be included?

(b) On the basis of your intuition, cite an explicit expression for this temperature dependence.

8.39 An uncold-worked brass specimen of average grain size 0.008 mm has a yield strength of 160 MPa (23,500 psi). Estimate the yield strength of this alloy after it has been heated to 600°C for 1000 s, if it is known that the value of k_y is 12.0 MPa-mm$^{1/2}$ (1740 psi-mm$^{1/2}$).

8.40 Cite one reason why ceramic materials are, in general, harder yet more brittle than metals.

8.41 In your own words, describe the mechanisms by which semicrystalline polymers **(a)** elas-

tically deform and **(b)** plastically deform, and **(c)** by which elastomers elastically deform.

8.42 Briefly explain how each of the following influences the tensile modulus of a semicrystalline polymer and why:

(a) molecular weight;

(b) degree of crystallinity;

(c) deformation by drawing;

(d) annealing of an undeformed material;

(e) annealing of a drawn material.

8.43* Briefly explain how each of the following influences the tensile or yield strength of a semicrystalline polymer and why:

(a) molecular weight;

(b) degree of crystallinity;

(c) deformation by drawing;

(d) annealing of an undeformed material.

8.44 Normal butane and isobutane have boiling temperatures of -0.5 and $-12.3°C$ (31.1 and 9.9°F), respectively. Briefly explain this behavior on the basis of their molecular structures, as presented in Section 4.2.

8.45* The tensile strength and number-average molecular weight for two polymethyl methacrylate materials are as follows:

Tensile Strength (MPa)	Number Average Molecular Weight (g/mol)
107	40,000
170	60,000

Estimate the tensile strength at a number-average molecular weight of 30,000 g/mol.

8.46* The tensile strength and number-average molecular weight for two polyethylene materials are as follows:

Tensile Strength (MPa)	Number Average Molecular Weight (g/mol)
85	12,700
150	28,500

Estimate the number-average molecular weight that is required to give a tensile strength of 195 MPa.

8.47* For each of the following pairs of polymers, do the following: (1) state whether or not it is possible to decide if one polymer has a higher tensile modulus than the other; (2) if this is possible, note which has the higher tensile modulus and then cite the reason(s) for your choice; and (3) if it is not possible to decide, then state why.

(a) Syndiotactic polystyrene having a number-average molecular weight of 400,000 g/mol; isotactic polystyrene having a number-average molecular weight of 650,000 g/mol.

(b) Branched and atactic polyvinyl chloride with a weight-average molecular weight of 100,000 g/mol; linear and isotactic polyvinyl chloride having a weight-average molecular weight of 75,000 g/mol.

(c) Random styrene-butadiene copolymer with 5% of possible sites crosslinked; block styrene-butadiene copolymer with 10% of possible sites crosslinked.

(d) Branched polyethylene with a number-average molecular weight of 100,000 g/mol; atactic polypropylene with a number-average molecular weight of 150,000 g/mol.

8.48* For each of the following pairs of polymers, do the following: (1) state whether or not it is possible to decide if one polymer has a higher tensile strength than the other; (2) if this is possible, note which has the higher tensile strengh and then cite the reason(s) for your choice; and (3) if it is not possible to decide, then state why.

(a) Syndiotactic polystyrene having a number-average molecular weight of 600,000 g/mol; isotactic polystyrene having a number-average molecular weight of 500,000 g/mol.

(b) Linear and isotactic polyvinyl chloride with a weight-average molecular weight of 100,000 g/mol; branched and atactic polyvinyl chloride having a weight-average molecular weight of 75,000 g/mol.

(c) Graft acrylonitrile-butadiene copolymer with 10% of possible sites crosslinked; alternating acrylonitrile-butadiene copolymer with 5% of possible sites crosslinked.

(d) Network polyester; lightly branched polytetrafluoroethylene.

8.49 Would you expect the tensile strength of polychlorotrifluoroethylene to be greater than,

the same as, or less than that of a polytetrafluoroethylene specimen having the same molecular weight and degree of crystallinity? Why?

8.50* For each of the following pairs of polymers, plot and label schematic stress–strain curves on the same graph (i.e., make separate plots for parts a, b, c, and d).

(a) Isotactic and linear polypropylene having a weight-average molecular weight of 120,000 g/mol; atactic and linear polypropylene having a weight-average molecular weight of 100,000 g/mol.

(b) Branched polyvinyl chloride having a number-average degree of polymerization of 2000; heavily crosslinked polyvinyl chloride having a number-average degree of polymerization of 2000.

(c) Poly(styrene-butadiene) random copolymer having a number-average molecular weight of 100,000 g/mol and 10% of the available sites crosslinked and tested at 20°C; poly(styrene-butadiene) random copolymer having a number-average molecular weight of 120,000 g/mol and 15% of the available sites crosslinked and tested at −85°C. *Hint:* poly(styrene-butadiene) copolymers may exhibit elastomeric behavior.

(d) Polyisoprene, molecular weight of 100,000 g/mol having 10% of available sites crosslinked; polyisoprene, molecular weight of 100,000 g/mol having 20% of available sites crosslinked. *Hint:* polyisoprene is a natural rubber that may display elastomeric behavior.

8.51 List the two molecular characteristics that are essential for elastomers.

8.52 Which of the following would you expect to be elastomers and which thermosetting polymers at room temperature? Justify each choice.

(a) Epoxy having a network structure.

(b) Lightly crosslinked poly(styrene-butadiene) random copolymer that has a glass-transition temperature of −50°C.

(c) Lightly branched and semicrystalline polytetrafluoroethylene that has a glass-transition temperature of −100°C.

(d) Heavily crosslinked poly(ethylene-propylene) random copolymer that has a glass-transition temperature of 0°C.

(e) Thermoplastic elastomer that has a glass-transition temperature of 75°C.

8.53 In terms of molecular structure, explain why phenol-formaldehyde (Bakelite) will not be an elastomer.

8.54 Ten kilograms of polybutadiene is vulcanized with 4.8 kg sulfur. What fraction of the possible crosslink sites is bonded to sulfur crosslinks, assuming that, on the average, 4.5 sulfur atoms participate in each crosslink?

8.55 Compute the weight percent sulfur that must be added to completely crosslink an alternating chloroprene-acrylonitrile copolymer, assuming that five sulfur atoms participate in each crosslink.

8.56 The vulcanization of polyisoprene is accomplished with sulfur atoms according to Equation 8.10. If 57 wt% sulfur is combined with polyisoprene, how many crosslinks will be associated with each isoprene mer if it is assumed that, on the average, six sulfur atoms participate in each crosslink?

8.57 For the vulcanization of polyisoprene, compute the weight percent of sulfur that must be added to ensure that 8% of possible sites will be crosslinked; assume that, on the average, three sulfur atoms are associated with each crosslink.

8.58 Demonstrate, in a manner similar to Equation 8.10, how vulcanization may occur in a chloroprene rubber.

Design Problems

8.D1 Determine whether or not it is possible to cold work steel so as to give a minimum Brinell hardness of 225 and at the same time have a ductility of at least 12%EL. Justify your decision.

8.D2 Determine whether or not it is possible to cold work brass so as to give a minimum Brinell hardness of 120 and at the same time have a ductility of at least 20%EL. Justify your decision.

8.D3 A cylindrical specimen of cold-worked steel has a Brinell hardness of 250.

(a) Estimate its ductility in percent elongation.

(b) If the specimen remained cylindrical during deformation and its uncold-worked radius was 5 mm (0.20 in.), determine its radius after deformation.

8.D4 It is necessary to select a metal alloy for an application that requires a yield strength of at least 345 MPa (50,000 psi) while maintaining a minimum ductility (%EL) of 20%. If the metal may be cold worked, decide which of the following are candidates: copper, brass, and a 1040 steel. Why?

8.D5 A cylindrical rod of 1040 steel originally 15.2 mm (0.60 in.) in diameter is to be cold worked by drawing; the circular cross section will be maintained during deformation. A cold-worked tensile strength in excess of 840 MPa (122,000 psi) and a ductility of at least 12%EL are desired. Furthermore, the final diameter must be 10 mm (0.40 in.). Explain how this may be accomplished.

8.D6 A cylindrical rod of copper originally 16.0 mm (0.625 in.) in diameter is to be cold worked by drawing; the circular cross section will be maintained during deformation. A cold-worked yield strength in excess of 250 MPa (36,250 psi) and a ductility of at least 12%EL are desired. Furthermore, the final diameter must be 11.3 mm (0.445 in.). Explain how this may be accomplished.

8.D7 A cylindrical 1040 steel rod having a minimum tensile strength of 865 MPa (125,000 psi), a ductility of at least 10%EL, and a final diameter of 6.0 mm (0.25 in.) is desired. Some 7.94 mm (0.313 in.) diameter 1040 steel stock, which has been cold worked 20%, is available. Describe the procedure you would follow to obtain this material. Assume that 1040 steel experiences cracking at 40%CW.

Chapter 9 / Failure

An oil tanker that fractured in a brittle manner by crack propagation around its girth. (Photography by Neal Boenzi. Reprinted with permission from *The New York Times*.)

Why Study *Failure?*

The design of a component or structure often calls upon the engineer to minimize the possibility of failure. Thus, it is important to understand the mechanics of the various failure modes—i.e., fracture, fatigue, and creep—and, in addition, be familiar with appropriate design principles that may be employed to prevent in-service failures. {For example, we discuss in Section 20.5 material selection and processing issues relating to the fatigue of an automobile valve spring.}

Learning Objectives

After studying this chapter you should be able to do the following:

1. Describe the mechanism of crack propagation for both ductile and brittle modes of fracture.
2. Explain why the strengths of brittle materials are much lower than predicted by theoretical calculations.
3. Define fracture toughness in terms of (a) a brief statement, and (b) an equation; define all parameters in this equation.
{4. Make distinctions between *stress intensity factor*, *fracture toughness*, and *plane strain fracture toughness*.}
5. Briefly explain why there is normally significant scatter in the fracture strength for identical specimens of the same ceramic material.
6. Briefly describe the phenomenon of *crazing.*
7. Name and describe the two impact fracture testing techniques.
8. Define fatigue and specify the conditions under which it occurs.
9. From a fatigue plot for some material, determine (a) the fatigue lifetime (at a specified stress level), and (b) the fatigue strength (at a specified number of cycles).
10. Define creep and specify the conditions under which it occurs.
11. Given a creep plot for some material, determine (a) the steady-state creep rate, and (b) the rupture lifetime.

9.1 INTRODUCTION

The failure of engineering materials is almost always an undesirable event for several reasons; these include human lives that are put in jeopardy, economic losses, and the interference with the availability of products and services. Even though the causes of failure and the behavior of materials may be known, prevention of failures is difficult to guarantee. The usual causes are improper materials selection and processing and inadequate design of the component or its misuse. It is the responsibility of the engineer to anticipate and plan for possible failure and, in the event that failure does occur, to assess its cause and then take appropriate preventive measures against future incidents.

Topics to be addressed in this chapter are the following: simple fracture (both ductile and brittle modes), fundamentals of fracture mechanics, impact fracture testing, the ductile-to-brittle transition, fatigue, and creep. These discussions include failure mechanisms, testing techniques, and methods by which failure may be prevented or controlled.

FRACTURE

9.2 FUNDAMENTALS OF FRACTURE

Simple fracture is the separation of a body into two or more pieces in response to an imposed stress that is static (i.e., constant or slowly changing with time) and at temperatures that are low relative to the melting temperature of the material. The applied stress may be tensile, compressive, shear, or torsional; the present discussion will be confined to fractures that result from uniaxial tensile loads. For engineering materials, two fracture modes are possible: **ductile** and **brittle.** Classification is based on the ability of a material to experience plastic deformation. Ductile materials typically exhibit substantial plastic deformation with high energy absorption before fracture. On the other hand, there is normally little or no plastic deformation with low energy absorption accompanying a brittle fracture. The tensile stress–strain behaviors of both fracture types may be reviewed in Figure 7.13.

"Ductile" and "brittle" are relative terms; whether a particular fracture is one mode or the other depends on the situation. Ductility may be quantified in terms

235

of percent elongation (Equation 7.11) and percent reduction in area (Equation 7.12). Furthermore, ductility is a function of temperature of the material, the strain rate, and the stress state. The disposition of normally ductile materials to fail in a brittle manner is discussed in Section 9.8.

Any fracture process involves two steps—crack formation and propagation—in response to an imposed stress. The mode of fracture is highly dependent on the mechanism of crack propagation. Ductile fracture is characterized by extensive plastic deformation in the vicinity of an advancing crack. Furthermore, the process proceeds relatively slowly as the crack length is extended. Such a crack is often said to be *stable*. That is, it resists any further extension unless there is an increase in the applied stress. In addition, there will ordinarily be evidence of appreciable gross deformation at the fracture surfaces (e.g., twisting and tearing). On the other hand, for brittle fracture, cracks may spread extremely rapidly, with very little accompanying plastic deformation. Such cracks may be said to be *unstable,* and crack propagation, once started, will continue spontaneously without an increase in magnitude of the applied stress.

Ductile fracture is almost always preferred for two reasons. First, brittle fracture occurs suddenly and catastrophically without any warning; this is a consequence of the spontaneous and rapid crack propagation. On the other hand, for ductile fracture, the presence of plastic deformation gives warning that fracture is imminent, allowing preventive measures to be taken. Second, more strain energy is required to induce ductile fracture inasmuch as ductile materials are generally tougher. Under the action of an applied tensile stress, most metal alloys are ductile, whereas ceramics are notably brittle, and polymers may exhibit both types of fracture.

9.3 DUCTILE FRACTURE

Ductile fracture surfaces will have their own distinctive features on both macroscopic and microscopic levels. Figure 9.1 shows schematic representations for two characteristic macroscopic fracture profiles. The configuration shown in Figure 9.1*a* is found for extremely soft metals, such as pure gold and lead at room temperature, and other metals, polymers, and inorganic glasses at elevated temperatures. These highly ductile materials neck down to a point fracture, showing virtually 100% reduction in area.

The most common type of tensile fracture profile for ductile metals is that represented in Figure 9.1*b*, which fracture is preceded by only a moderate amount

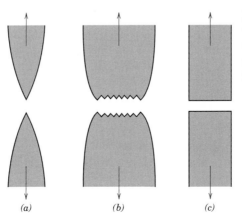

(a)   (b)   (c)

FIGURE 9.1 (*a*) Highly ductile fracture in which the specimen necks down to a point. (*b*) Moderately ductile fracture after some necking. (*c*) Brittle fracture without any plastic deformation.

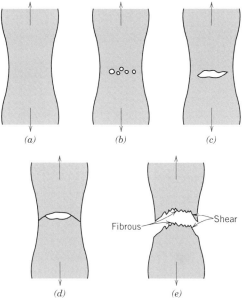

FIGURE 9.2 Stages in the cup-and-cone fracture. (*a*) Initial necking. (*b*) Small cavity formation. (*c*) Coalescence of cavities to form a crack. (*d*) Crack propagation. (*e*) Final shear fracture at a 45° angle relative to the tensile direction. (From K. M. Ralls, T. H. Courtney, and J. Wulff, *Introduction to Materials Science and Engineering*, p. 468. Copyright © 1976 by John Wiley & Sons, New York. Reprinted by permission of John Wiley & Sons, Inc.)

of necking. The fracture process normally occurs in several stages (Figure 9.2). First, after necking begins, small cavities, or microvoids, form in the interior of the cross section, as indicated in Figure 9.2*b*. Next, as deformation continues, these microvoids enlarge, come together, and coalesce to form an elliptical crack, which has its long axis perpendicular to the stress direction. The crack continues to grow in a direction parallel to its major axis by this microvoid coalescence process (Figure 9.2*c*). Finally, fracture ensues by the rapid propagation of a crack around the outer perimeter of the neck (Figure 9.2*d*), by shear deformation at an angle of about 45° with the tensile axis—this is the angle at which the shear stress is a maximum. Sometimes a fracture having this characteristic surface contour is termed a *cup-and-cone fracture* because one of the mating surfaces is in the form of a cup, the other like a cone. In this type of fractured specimen (Figure 9.3*a*), the central

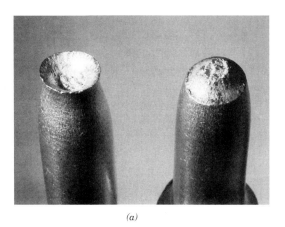

FIGURE 9.3 (*a*) Cup-and-cone fracture in aluminum. (*b*) Brittle fracture in a mild steel.

interior region of the surface has an irregular and fibrous appearance, which is indicative of plastic deformation.

FRACTOGRAPHIC STUDIES (CD-ROM)

9.4 BRITTLE FRACTURE

Brittle fracture takes place without any appreciable deformation, and by rapid crack propagation. The direction of crack motion is very nearly perpendicular to the direction of the applied tensile stress and yields a relatively flat fracture surface, as indicated in Figure 9.1c.

Fracture surfaces of materials that failed in a brittle manner will have their own distinctive patterns; any signs of gross plastic deformation will be absent. For example, in some steel pieces, a series of V-shaped "chevron" markings may form near the center of the fracture cross section that point back toward the crack initiation site (Figure 9.5a). Other brittle fracture surfaces contain lines or ridges that radiate from the origin of the crack in a fanlike pattern (Figure 9.5b). Often, both of these marking patterns will be sufficiently coarse to be discerned with the naked eye. For very hard and fine-grained metals, there will be no discernible fracture pattern. Brittle fracture in amorphous materials, such as ceramic glasses, yields a relatively shiny and smooth surface.

For most brittle crystalline materials, crack propagation corresponds to the successive and repeated breaking of atomic bonds along specific crystallographic planes; such a process is termed *cleavage*. This type of fracture is said to be **transgranular** (or *transcrystalline*), because the fracture cracks pass through the grains. Macroscopically, the fracture surface may have a grainy or faceted texture (Figure 9.3b), as a result of changes in orientation of the cleavage planes from grain to grain. This feature is more evident in the scanning electron micrograph shown in Figure 9.6a.

In some alloys, crack propagation is along grain boundaries; this fracture is termed **intergranular.** Figure 9.6b is a scanning electron micrograph showing a typical intergranular fracture, in which the three-dimensional nature of the grains may be seen. This type of fracture normally results subsequent to the occurrence of processes that weaken or embrittle grain boundary regions.

9.5a PRINCIPLES OF FRACTURE MECHANICS [DETAILED VERSION (CD-ROM)]

9.5b PRINCIPLES OF FRACTURE MECHANICS (CONCISE VERSION)

Brittle fracture of normally ductile materials, such as that shown in the chapter-opening photograph of this chapter, has demonstrated the need for a better understanding of the mechanisms of fracture. Extensive research endeavors over the past several decades have led to the evolution of the field of **fracture mechanics.** This subject allows quantification of the relationships between material properties, stress level, the presence of crack-producing flaws, and crack propagation mechanisms. Design engineers are now better equipped to anticipate, and thus prevent, structural failures. The present discussion centers on some of the fundamental principles of the mechanics of fracture.

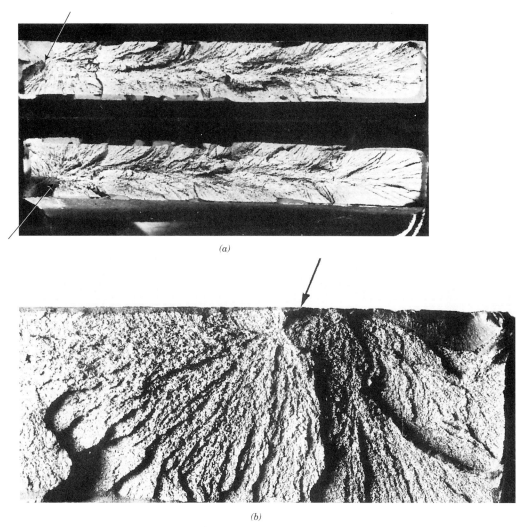

(a)

(b)

FIGURE 9.5   (a) Photograph showing V-shaped "chevron" markings characteristic of brittle fracture. Arrows indicate origin of crack. Approximately actual size. (From R. W. Hertzberg, *Deformation and Fracture Mechanics of Engineering Materials,* 3rd edition. Copyright © 1989 by John Wiley & Sons, New York. Reprinted by permission of John Wiley & Sons, Inc. Photograph courtesy of Roger Slutter, Lehigh University.) (b) Photograph of a brittle fracture surface showing radial fan-shaped ridges. Arrow indicates origin of crack. Approximately 2×. (Reproduced with permission from D. J. Wulpi, *Understanding How Components Fail,* American Society for Metals, Materials Park, OH, 1985.)

STRESS CONCENTRATION

The measured fracture strengths for most brittle materials are significantly lower than those predicted by theoretical calculations based on atomic bonding energies. This discrepancy is explained by the presence of very small, microscopic flaws or cracks that always exist under normal conditions at the surface and within the interior of a body of material. These flaws are a detriment to the fracture strength because an applied stress may be amplified or concentrated at the tip, the magnitude

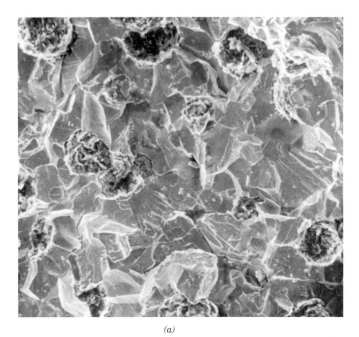

(a)

(b)

FIGURE 9.6 (a) Scanning electron fractograph of ductile cast iron showing a transgranular fracture surface. Magnification unknown. (From V. J. Colangelo and F. A. Heiser, *Analysis of Metallurgical Failures,* 2nd edition. Copyright © 1987 by John Wiley & Sons, New York. Reprinted by permission of John Wiley & Sons, Inc.) (b) Scanning electron fractograph showing an intergranular fracture surface. 50×. (Reproduced with permission from *ASM Handbook,* Vol. 12, *Fractography,* ASM International, Materials Park, OH, 1987.)

of this amplification depending on crack orientation and geometry. This phenomenon is demonstrated in Figure 9.7, a stress profile across a cross section containing an internal crack. As indicated by this profile, the magnitude of this localized stress diminishes with distance away from the crack tip. At positions far removed, the stress is just the nominal stress σ_0, or the load divided by the specimen cross-

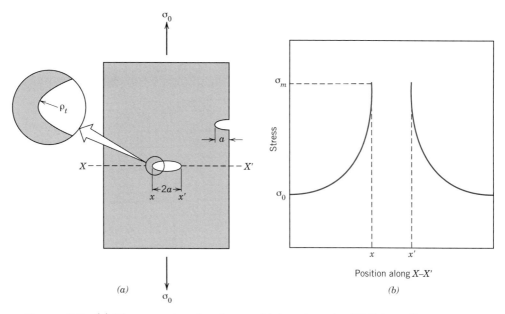

FIGURE 9.7 (a) The geometry of surface and internal cracks. (b) Schematic stress profile along the line X–X' in (a), demonstrating stress amplification at crack tip positions.

sectional area (perpendicular to this load). Due to their ability to amplify an applied stress in their locale, these flaws are sometimes called **stress raisers.**

If it is assumed that a crack has an elliptical shape and is oriented perpendicular to the applied stress, the maximum stress at the crack tip, σ_m, may be approximated by

$$\sigma_m = 2\sigma_0 \left(\frac{a}{\rho_t}\right)^{1/2} \tag{9.1b}$$

where σ_0 is the magnitude of the nominal applied tensile stress, ρ_t is the radius of curvature of the crack tip (Figure 9.7a), and a represents the length of a surface crack, or half of the length of an internal crack. For a relatively long microcrack that has a small tip radius of curvature, the factor $(a/\rho_t)^{1/2}$ may be very large. This will yield a value of σ_m that is many times the value of σ_0.

Sometimes the ratio σ_m/σ_0 is denoted as the *stress concentration factor K_t*:

$$K_t = \frac{\sigma_m}{\sigma_0} = 2\left(\frac{a}{\rho_t}\right)^{1/2} \tag{9.2}$$

which is simply a measure of the degree to which an external stress is amplified at the tip of a crack.

By way of comment, it should be said that stress amplification is not restricted to these microscopic defects; it may occur at macroscopic internal discontinuities (e.g., voids), at sharp corners, and at notches in large structures.

Furthermore, the effect of a stress raiser is more significant in brittle than in ductile materials. For a ductile material, plastic deformation ensues when the maximum stress exceeds the yield strength. This leads to a more uniform distribution of stress in the vicinity of the stress raiser and to the development of a maximum stress concentration factor less than the theoretical value. Such yielding and stress redistribution do not occur to any appreciable extent around flaws and discontinuities in brittle materials; therefore, essentially the theoretical stress concentration will result.

Using principles of fracture mechanics, it is possible to show that the critical stress σ_c required for crack propagation in a brittle material is described by the expression

$$\sigma_c = \left(\frac{2E\gamma_s}{\pi a}\right)^{1/2} \tag{9.3}$$

where

$$E = \text{modulus of elasticity}$$

$$\gamma_s = \text{specific surface energy}$$

$$a = \text{one half the length of an internal crack}$$

All brittle materials contain a population of small cracks and flaws that have a variety of sizes, geometries, and orientations. When the magnitude of a tensile stress at the tip of one of these flaws exceeds the value of this critical stress, a crack forms and then propagates, which results in fracture. Very small and virtually defect-free metallic and ceramic whiskers have been grown with fracture strengths that approach their theoretical values.

EXAMPLE PROBLEM 9.1

A relatively large plate of a glass is subjected to a tensile stress of 40 MPa. If the specific surface energy and modulus of elasticity for this glass are 0.3 J/m^2 and 69 GPa, respectively, determine the maximum length of a surface flaw that is possible without fracture.

SOLUTION

To solve this problem it is necessary to employ Equation 9.3. Rearrangement of this expression such that a is the dependent variable, and realizing that $\sigma = 40$ MPa, $\gamma_s = 0.3$ J/m^2, and $E = 69$ GPa leads to

$$a = \frac{2E\gamma_s}{\pi \sigma^2}$$

$$= \frac{(2)(69 \times 10^9 \text{ N/m}^2)(0.3 \text{ N/m})}{\pi(40 \times 10^6 \text{ N/m}^2)^2}$$

$$= 8.2 \times 10^{-6} \text{ m} = 0.0082 \text{ mm} = 8.2 \ \mu\text{m}$$

FRACTURE TOUGHNESS

Furthermore, using fracture mechanical principles, an expression has been developed that relates this critical stress for crack propagation (σ_c) to crack length (a) as

$$K_c = Y\sigma_c\sqrt{\pi a} \tag{9.9a}$$

In this expression K_c is the **fracture toughness,** a property that is a measure of a material's resistance to brittle fracture when a crack is present. Worth noting is that K_c has the unusual units of MPa$\sqrt{m}$ or psi$\sqrt{in.}$ (alternatively ksi$\sqrt{in.}$). Furthermore, Y is a dimensionless parameter or function that depends on both crack and specimen sizes and geometries, as well as the manner of load application.

Relative to this Y parameter, for planar specimens containing cracks that are much shorter than the specimen width, Y has a value of approximately unity. For example, for a plate of infinite width having a through-thickness crack (Figure 9.11a), $Y = 1.0$; whereas for a plate of semi-infinite width containing an edge crack of length a (Figure 9.11b), $Y \cong 1.1$. Mathematical expressions for Y have been determined for a variety of crack-specimen geometries; these expressions are often relatively complex.

For relatively thin specimens, the value of K_c will depend on specimen thickness. However, when specimen thickness is much greater than the crack dimensions, K_c becomes independent of thickness; under these conditions a condition of **plane strain** exists. By plane strain we mean that when a load operates on a crack in the manner represented in Figure 9.11a, there is no strain component perpendicular to the front and back faces. The K_c value for this thick-specimen situation is known as the **plane strain fracture toughness** K_{Ic}; furthermore, it is also defined by

$$K_{Ic} = Y\sigma\sqrt{\pi a} \tag{9.11}$$

K_{Ic} is the fracture toughness cited for most situations. The I (i.e., Roman numeral "one") subscript for K_{Ic} denotes that the plane strain fracture toughness is for mode I crack displacement, as illustrated in Figure 9.9a[5] (see page 244).

Brittle materials, for which appreciable plastic deformation is not possible in front of an advancing crack, have low K_{Ic} values and are vulnerable to catastrophic failure. On the other hand, K_{Ic} values are relatively large for ductile materials. Fracture mechanics is especially useful in predicting catastrophic failure in materials

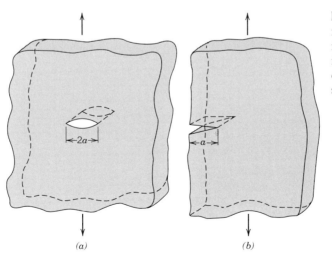

FIGURE 9.11 Schematic representations of (a) an interior crack in a plate of infinite width, and (b) an edge crack in a plate of semi-infinite width.

(a) (b)

[5] Two other crack displacement modes denoted by II and III and as illustrated in Figures 9.9b and 9.9c are also possible; however, mode I is most commonly encountered.

FIGURE 9.9 The three modes of crack surface displacement. (*a*) Mode I, opening or tensile mode; (*b*) mode II, sliding mode; and (*c*) mode III, tearing mode.

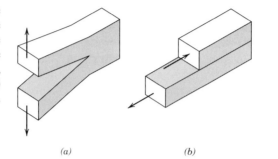
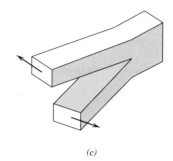

(a)   (b)   (c)

having intermediate ductilities. Plane strain fracture toughness values for a number of different materials are presented in Table 9.1; a more extensive list of K_{Ic} values is contained in Table B.5, Appendix B.

The plane strain fracture toughness K_{Ic} is a fundamental material property that depends on many factors, the most influential of which are temperature, strain rate, and microstructure. The magnitude of K_{Ic} diminishes with increasing strain rate and

Table 9.1   Room-Temperature Yield Strength and Plane Strain Fracture Toughness Data for Selected Engineering Materials

Material	Yield Strength		K_{Ic}	
	MPa	*ksi*	*MPa* $\sqrt{m}$	*ksi* $\sqrt{in.}$
Metals				
Aluminum Alloy[a] (7075-T651)	495	72	24	22
Aluminum Alloy[a] (2024-T3)	345	50	44	40
Titanium Alloy[a] (Ti-6Al-4V)	910	132	55	50
Alloy Steel[a] (4340 tempered @ 260°C)	1640	238	50.0	45.8
Alloy Steel[a] (4340 tempered @ 425°C)	1420	206	87.4	80.0
Ceramics				
Concrete	—	—	0.2–1.4	0.18–1.27
Soda-Lime Glass	—	—	0.7–0.8	0.64–0.73
Aluminum Oxide	—	—	2.7–5.0	2.5–4.6
Polymers				
Polystyrene (PS)	—	—	0.7–1.1	0.64–1.0
Polymethyl Methacrylate (PMMA)	53.8–73.1	7.8–10.6	0.7–1.6	0.64–1.5
Polycarbonate (PC)	62.1	9.0	2.2	2.0

[a] **Source:** Reprinted with permission, *Advanced Materials and Processes,* ASM International, © 1990.

decreasing temperature. Furthermore, an enhancement in yield strength wrought by solid solution or dispersion additions or by strain hardening generally produces a corresponding decrease in K_{Ic}. Furthermore, K_{Ic} normally increases with reduction in grain size as composition and other microstructural variables are maintained constant. Yield strengths are included for some of the materials listed in Table 9.1. Furthermore, K_{Ic} normally increases with reduction in grain size as composition and other microstructural variables are maintained constant.

Several different testing techniques are used to measure K_{Ic}.[6] Virtually any specimen size and shape consistent with mode I crack displacement may be utilized, and accurate values will be realized provided that the Y scale parameter in Equation 9.11 has been properly determined.

DESIGN USING FRACTURE MECHANICS

According to Equations 9.9a and 9.11, three variables must be considered relative to the possibility for fracture of some structural component—viz. the fracture toughness (K_c) or plane strain fracture toughness (K_{Ic}), the imposed stress (σ), and the flaw size (a), assuming, of course, that Y has been determined. When designing a component, it is first important to decide which of these variables are constrained by the application and which are subject to design control. For example, material selection (and hence K_c or K_{Ic}) is often dictated by factors such as density (for lightweight applications) or the corrosion characteristics of the environment. Or, the allowable flaw size is either measured or specified by the limitations of available flaw detection techniques. It is important to realize, however, that once any combination of two of the above parameters is prescribed, the third becomes fixed (Equations 9.9a and 9.11). For example, assume that K_{Ic} and the magnitude of a are specified by application constraints; therefore, the design (or critical) stress σ_c must be

$$\sigma_c \leq \frac{K_{Ic}}{Y\sqrt{\pi a}} \tag{9.13}$$

On the other hand, if stress level and plane strain fracture toughness are fixed by the design situation, then the maximum allowable flaw size a_c is

$$a_c = \frac{1}{\pi}\left(\frac{K_{Ic}}{\sigma Y}\right)^2 \tag{9.14}$$

A number of nondestructive test (NDT) techniques have been developed that permit detection and measurement of both internal and surface flaws. Such NDT methods are used to avoid the occurrence of catastrophic failure by examining structural components for defects and flaws that have dimensions approaching the critical size.

DESIGN EXAMPLE 9.1

Consider the thin-walled spherical tank of radius r and thickness t (Figure 9.15) that may be used as a pressure vessel.

---

[6] See for example ASTM Standard E 399, "Standard Test Method for Plane Strain Fracture Toughness of Metallic Materials."

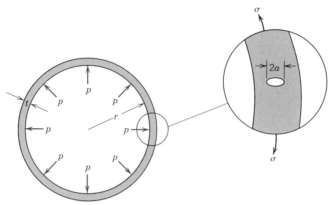

FIGURE 9.15 Schematic diagram showing the cross section of a spherical tank that is subjected to an internal pressure p, and that has a radial crack of length $2a$ in its wall.

(a) One design of such a tank calls for yielding of the wall material prior to failure as a result of the formation of a crack of critical size and its subsequent rapid propagation. Thus, plastic distortion of the wall may be observed and the pressure within the tank released before the occurrence of catastrophic failure. Consequently, materials having large critical crack lengths are desired. On the basis of this criterion, rank the metal alloys listed in Table B.5, Appendix B, as to critical crack size, from longest to shortest.

(b) An alternative design that is also often utilized with pressure vessels is termed *leak-before-break*. Using principles of fracture mechanics, allowance is made for the growth of a crack through the thickness of the vessel wall prior to the occurrence of rapid crack propagation (Figure 9.15). Thus, the crack will completely penetrate the wall without catastrophic failure, allowing for its detection by the leaking of pressurized fluid. With this criterion the critical crack length a_c (i.e., one-half of the total internal crack length) is taken to be equal to the pressure vessel thickness t. Allowance for $a_c = t$ instead of $a_c = t/2$ assures that fluid leakage will occur prior to the buildup of dangerously high pressures. Using this criterion, rank the metal alloys in Table B.5, Appendix B as to the maximum allowable pressure.

For this spherical pressure vessel, the circumferential wall stress σ is a function of the pressure p in the vessel and the radius r and wall thickness t according to

$$\sigma = \frac{pr}{2t} \tag{9.15}$$

For both parts (a) and (b) assume a condition of plane strain.

SOLUTION

(a) For the first design criterion, it is desired that the circumferential wall stress be less than the yield strength of the material. Substitution of σ_y for σ in Equation 9.11, and incorporation of a factor of safety N leads to

$$K_{Ic} = Y\left(\frac{\sigma_y}{N}\right)\sqrt{\pi a_c} \tag{9.16}$$

where a_c is the critical crack length. Solving for a_c yields the following expression:

$$a_c = \frac{N^2}{Y^2\pi}\left(\frac{K_{Ic}}{\sigma_y}\right)^2 \tag{9.17}$$

Table 9.2 Ranking of Several Metal Alloys Relative to Critical Crack Length (Yielding Criterion) for a Thin-Walled Spherical Pressure Vessel

Material	$\left(\dfrac{K_{Ic}}{\sigma_y}\right)^2$ (mm)
Medium carbon (1040) steel	43.1
AZ31B magnesium	19.6
2024 aluminum (T3)	16.3
Ti-5Al-2.5Sn titanium	6.6
4140 steel (tempered @ 482°C)	5.3
4340 steel (tempered @ 425°C)	3.8
Ti-6Al-4V titanium	3.7
17-7PH steel	3.4
7075 aluminum (T651)	2.4
4140 steel (tempered @ 370°C)	1.6
4340 Steel (tempered @ 260°C)	0.93

Therefore, the critical crack length is proportional to the square of the K_{Ic}-σ_y ratio, which is the basis for the ranking of the metal alloys in Table B.5. The ranking is provided in Table 9.2, where it may be seen that the medium carbon (1040) steel with the largest ratio has the longest critical crack length, and, therefore, is the most desirable material on the basis of this criterion.

(b) As stated previously, the leak-before-break criterion is just met when one-half of the internal crack length is equal to the thickness of the pressure vessel—i.e., when $a = t$. Substitution of $a = t$ into Equation 9.11 gives

$$K_{Ic} = Y\sigma\sqrt{\pi t} \tag{9.18}$$

And, from Equation 9.15

$$t = \frac{pr}{2\sigma} \tag{9.19}$$

The stress is replaced by the yield strength, inasmuch as the tank should be designed to contain the pressure without yielding; furthermore, substitution of Equation 9.19 into Equation 9.18, after some rearrangement, yields the following expression:

$$p = \frac{2}{Y^2\pi r}\left(\frac{K_{Ic}^2}{\sigma_y}\right) \tag{9.20}$$

Hence, for some given spherical vessel of radius r, the maximum allowable pressure consistent with this leak-before-break criterion is proportional to K_{Ic}^2/σ_y. The same several materials are ranked according to this ratio in Table 9.3; as may be noted, the medium carbon steel will contain the greatest pressures.

Of the eleven metal alloys that are listed in Table B.5, the medium carbon steel ranks first according to both yielding and leak-before-break criteria. For these

Table 9.3 Ranking of Several Metal Alloys Relative to Maximum Allowable Pressure (Leak-Before-Break Criterion) for a Thin-Walled Spherical Pressure Vessel

Material	$\dfrac{K_{Ic}^2}{\sigma_y}$ (MPa-m)
Medium carbon (1040) steel	11.2
4140 steel (tempered @ 482°C)	6.1
Ti-5Al-2.5Sn titanium	5.8
2024 aluminum (T3)	5.6
4340 steel (tempered @ 425°C)	5.4
17-7PH steel	4.4
AZ31B magnesium	3.9
Ti-6Al-4V titanium	3.3
4140 steel (tempered @ 370°C)	2.4
4340 steel (tempered @ 260°C)	1.5
7075 aluminum (T651)	1.2

reasons, many pressure vessels are constructed of medium carbon steels, when temperature extremes and corrosion need not be considered.

9.6 BRITTLE FRACTURE OF CERAMICS

At room temperature, both crystalline and noncrystalline ceramics almost always fracture before any plastic deformation can occur in response to an applied tensile load. Furthermore, the mechanics of brittle fracture and principles of fracture mechanics developed earlier in this chapter also apply to the fracture of this group of materials.

It should be noted that stress raisers in brittle ceramics may be minute surface or interior cracks (microcracks), internal pores, and grain corners, which are virtually impossible to eliminate or control. For example, even moisture and contaminants in the atmosphere can introduce surface cracks in freshly drawn glass fibers; these cracks deleteriously affect the strength. In addition, plane strain fracture toughness values for ceramic materials are smaller than for metals; typically they are below $10\ \text{MPa}\sqrt{\text{m}}$ ($9\ \text{ksi}\sqrt{\text{in.}}$). Values of K_{Ic} for several ceramic materials are included in Table 9.1 and Table B.5, Appendix B.

There is usually considerable variation and scatter in the fracture strength for many specimens of a specific brittle ceramic material. A distribution of fracture strengths for portland cement is shown in Figure 9.16. This phenomenon may be explained by the dependence of fracture strength on the probability of the existence of a flaw that is capable of initiating a crack. This probability varies from specimen to specimen of the same material and depends on fabrication technique and any subsequent treatment. Specimen size or volume also influences fracture strength; the larger the specimen, the greater this flaw existence probability, and the lower the fracture strength.

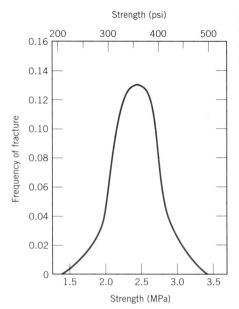

Strength (psi)

FIGURE 9.16 The frequency distribution of observed fracture strengths for a portland cement. (From W. Weibull, *Ing. Vetensk. Akad.,* Proc. 151, No. 153, 1939.)

For compressive stresses, there is no stress amplification associated with any existent flaws. For this reason, brittle ceramics display much higher strengths in compression than in tension (on the order of a factor of 10), and they are generally utilized when load conditions are compressive. Also, the fracture strength of a brittle ceramic may be enhanced dramatically by imposing residual compressive stresses at its surface. {One way this may be accomplished is by thermal tempering (see Section 14.7).}

Statistical theories have been developed that in conjunction with experimental data are used to determine the risk of fracture for a given material; a discussion of these is beyond the scope of the present treatment. However, due to the dispersion in the measured fracture strengths of brittle ceramic materials, average values and factors of safety as discussed in Sections 7.19 and 7.20 are not normally employed for design purposes.

STATIC FATIGUE (CD-ROM)

9.7 FRACTURE OF POLYMERS

The fracture strengths of polymeric materials are low relative to those of metals and ceramics. As a general rule, the mode of fracture in thermosetting polymers is brittle. In simple terms, associated with the fracture process is the formation of cracks at regions where there is a localized stress concentration (i.e., scratches, notches, and sharp flaws). Covalent bonds in the network or crosslinked structure are severed during fracture.

For thermoplastic polymers, both ductile and brittle modes are possible, and many of these materials are capable of experiencing a ductile-to-brittle transition. Factors that favor brittle fracture are a reduction in temperature, an increase in strain rate, the presence of a sharp notch, increased specimen thickness, and, in addition, a modification of the polymer structure (chemical, molecular, and/or

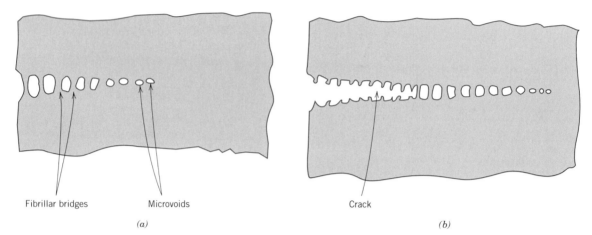

Fibrillar bridges   Microvoids   Crack

(a)   (b)

FIGURE 9.17 Schematic drawings of (a) a craze showing microvoids and fibrillar bridges, and (b) a craze followed by a crack. (From J. W. S. Hearle, *Polymers and Their Properties,* Vol. 1, *Fundamentals of Structure and Mechanics,* Ellis Horwood, Ltd., Chichester, West Sussex, England, 1982.)

microstructural). Glassy thermoplastics are brittle at relatively low temperatures; as the temperature is raised, they become ductile in the vicinity of their glass transition temperatures and experience plastic yielding prior to fracture. This behavior is demonstrated by the stress–strain characteristics of polymethyl methacrylate in Figure 7.24. At 4°C, PMMA is totally brittle, whereas at 60°C it becomes extremely ductile.

One phenomenon that frequently precedes fracture in some glassy thermoplastic polymers is *crazing*. Associated with crazes are regions of very localized yielding, which lead to the formation of small and interconnected microvoids (Figure 9.17a). Fibrillar bridges form between these microvoids wherein molecular chains become oriented. If the applied tensile load is sufficient, these bridges elongate and break, causing the microvoids to grow and coalesce; as the microvoids coalesce, cracks begin to form, as demonstrated in Figure 9.17b. A craze is different from a crack in that it can support a load across its face. Furthermore, this process of craze growth prior to cracking absorbs fracture energy and effectively increases the fracture toughness of the polymer. Crazes form at highly stressed regions associated with scratches, flaws, and molecular inhomogeneities; in addition, they propagate perpendicular to the applied tensile stress, and typically are 5 μm or less thick. Figure 9.18 is a photomicrograph in which a craze is shown.

 Principles of fracture mechanics developed in Section 9.5 also apply to brittle and quasi-brittle polymers. The magnitude of K_{Ic} will depend on characteristics of the polymer (i.e., molecular weight, percent crystallinity, etc.) as well as temperature, strain rate, and the external environment. Representative values of K_{Ic} for several polymers are included in Table 9.1 and Table B.5, Appendix B.

9.8 IMPACT FRACTURE TESTING

Prior to the advent of fracture mechanics as a scientific discipline, impact testing techniques were established so as to ascertain the fracture characteristics of materials. It was realized that the results of laboratory tensile tests could not be extrapolated to predict fracture behavior; for example, under some circumstances normally

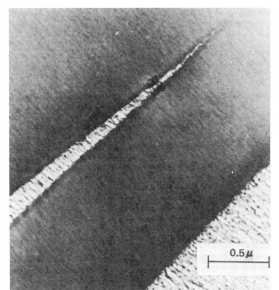

FIGURE 9.18 Photomicrograph of a craze in polyphenylene oxide. (From R. P. Kambour and R. E. Robertson, "The Mechanical Properties of Plastics," in *Polymer Science, A Materials Science Handbook,* A. D. Jenkins, Editor. Reprinted with permission of Elsevier Science Publishers.)

ductile metals fracture abruptly and with very little plastic deformation. Impact test conditions were chosen to represent those most severe relative to the potential for fracture, namely, (1) deformation at a relatively low temperature, (2) a high strain rate (i.e., rate of deformation), and (3) a triaxial stress state (which may be introduced by the presence of a notch).

IMPACT TESTING TECHNIQUES

Two standardized tests,[7] the **Charpy** and **Izod,** were designed and are still used to measure the **impact energy,** sometimes also termed *notch toughness.* The Charpy V-notch (CVN) technique is most commonly used in the United States. For both Charpy and Izod, the specimen is in the shape of a bar of square cross section, into which a V-notch is machined (Figure 9.19*a*). The apparatus for making V-notch impact tests is illustrated schematically in Figure 9.19*b*. The load is applied as an impact blow from a weighted pendulum hammer that is released from a cocked position at a fixed height *h*. The specimen is positioned at the base as shown. Upon release, a knife edge mounted on the pendulum strikes and fractures the specimen at the notch, which acts as a point of stress concentration for this high velocity impact blow. The pendulum continues its swing, rising to a maximum height *h'*, which is lower than *h*. The energy absorption, computed from the difference between *h* and *h'*, is a measure of the impact energy. The primary difference between the Charpy and Izod techniques lies in the manner of specimen support, as illustrated in Figure 9.19*b*. Furthermore, these are termed impact tests in light of the manner of load application. Variables including specimen size and shape as well as notch configuration and depth influence the test results.

Both plane strain fracture toughness and these impact tests determine the fracture properties of materials. The former are quantitative in nature, in that a specific property of the material is determined (i.e., K_{Ic}). The results of the impact

---

[7] ASTM Standard E 23, "Standard Test Methods for Notched Bar Impact Testing of Metallic Materials."

FIGURE 9.19 (*a*) Specimen used for Charpy and Izod impact tests. (*b*) A schematic drawing of an impact testing apparatus. The hammer is released from fixed height *h* and strikes the specimen; the energy expended in fracture is reflected in the difference between *h* and the swing height *h'*. Specimen placements for both Charpy and Izod tests are also shown. (Figure (*b*) adapted from H. W. Hayden, W. G. Moffatt, and J. Wulff, *The Structure and Properties of Materials,* Vol. III, *Mechanical Behavior,* p. 13. Copyright © 1965 by John Wiley & Sons, New York. Reprinted by permission of John Wiley & Sons, Inc.)

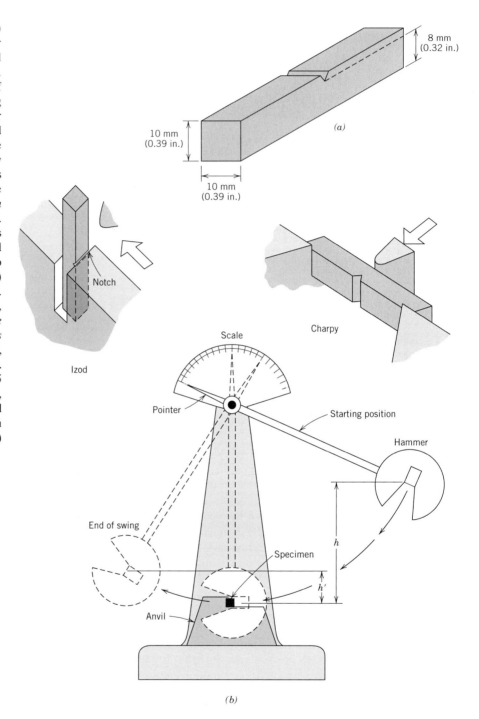

tests, on the other hand, are more qualitative and are of little use for design purposes. Impact energies are of interest mainly in a relative sense and for making comparisons—absolute values are of little significance. Attempts have been made to correlate plane strain fracture toughnesses and CVN energies, with only limited success. Plane strain fracture toughness tests are not as simple to perform as impact tests; furthermore, equipment and specimens are more expensive.

DUCTILE-TO-BRITTLE TRANSITION

One of the primary functions of Charpy and Izod tests is to determine whether or not a material experiences a **ductile-to-brittle transition** with decreasing temperature and, if so, the range of temperatures over which it occurs. The ductile-to-brittle transition is related to the temperature dependence of the measured impact energy absorption. This transition is represented for a steel by curve *A* in Figure 9.20. At higher temperatures the CVN energy is relatively large, in correlation with a ductile mode of fracture. As the temperature is lowered, the impact energy drops suddenly over a relatively narrow temperature range, below which the energy has a constant but small value; that is, the mode of fracture is brittle.

Alternatively, appearance of the failure surface is indicative of the nature of fracture, and may be used in transition temperature determinations. For ductile fracture this surface appears fibrous or dull (or of shear character); conversely, totally brittle surfaces have a granular (shiny) texture (or cleavage character). Over the ductile-to-brittle transition, features of both types will exist (Figure 9.21). Frequently, the percent shear fracture is plotted as a function of temperature—curve *B* in Figure 9.20.

For many alloys there is a range of temperatures over which the ductile-to-brittle transition occurs (Figure 9.20); this presents some difficulty in specifying a single ductile-to-brittle transition temperature. No explicit criterion has been established, and so this temperature is often defined as that temperature at which the CVN energy assumes some value (e.g., 20 J or 15 ft-lb$_f$), or corresponding to some given fracture appearance (e.g., 50% fibrous fracture). Matters are further complicated inasmuch as a different transition temperature may be realized for each of these criteria. Perhaps the most conservative transition temperature is that at which the fracture surface becomes 100% fibrous; on this basis, the transition temperature is approximately 110°C (230°F) for the steel alloy that is the subject of Figure 9.20.

FIGURE 9.20 Temperature dependence of the Charpy V-notch impact energy (curve *A*) and percent shear fracture (curve *B*) for an A283 steel. (Reprinted from *Welding Journal.* Used by permission of the American Welding Society.)

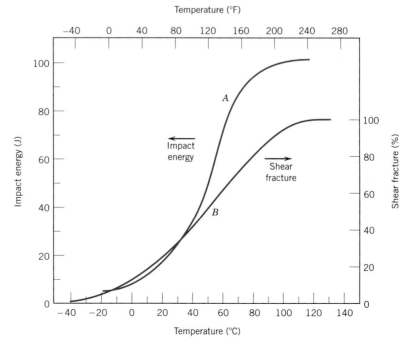

-59   -12   4   16   24   79

FIGURE 9.21   Photograph of fracture surfaces of A36 steel Charpy V-notch specimens tested at indicated temperatures (in °C). (From R. W. Hertzberg, *Deformation and Fracture Mechanics of Engineering Materials,* 3rd edition, Fig. 9.6, p. 329. Copyright © 1989 by John Wiley & Sons, Inc., New York. Reprinted by permission of John Wiley & Sons, Inc.)

Structures constructed from alloys that exhibit this ductile-to-brittle behavior should be used only at temperatures above the transition temperature, to avoid brittle and catastrophic failure. Classic examples of this type of failure occurred, with disastrous consequences, during World War II when a number of welded transport ships, away from combat, suddenly and precipitously split in half. The vessels were constructed of a steel alloy that possessed adequate ductility according to room-temperature tensile tests. The brittle fractures occurred at relatively low ambient temperatures, at about 4°C (40°F), in the vicinity of the transition temperature of the alloy. Each fracture crack originated at some point of stress concentration, probably a sharp corner or fabrication defect, and then propagated around the entire girth of the ship.

Not all metal alloys display a ductile-to-brittle transition. Those having FCC crystal structures (including aluminum- and copper-based alloys) remain ductile even at extremely low temperatures. However, BCC and HCP alloys experience this transition. For these materials the transition temperature is sensitive to both alloy composition and microstructure. For example, decreasing the average grain size of steels results in a lowering of the transition temperature. Hence, refining the grain size both strengthens (Section 8.9) and toughens steels. In contrast, increasing the carbon content, while increasing the strength of steels, also raises the CVN transition of steels, as indicated in Figure 9.22.

Izod or Charpy tests are also conducted to assess impact strength of polymeric materials. As with metals, polymers may exhibit ductile or brittle fracture under impact loading conditions, depending on the temperature, specimen size, strain rate, and mode of loading, as discussed in the preceding section. Both semicrystalline and amorphous polymers are brittle at low temperatures, and both have relatively low impact strengths. However, they experience a ductile-to-brittle transition over a relatively narrow temperature range, similar to that shown for a steel in Figure 9.20. Of course, impact strength undergoes a gradual decrease at still higher temperatures as the polymer begins to soften. Ordinarily, the two impact characteristics most sought after are a high impact strength at the ambient temperature and a ductile-to-brittle transition temperature that lies below room temperature.

Most ceramics also experience a ductile-to-brittle transition, which occurs only at elevated temperatures—ordinarily in excess of 1000°C (1850°F).

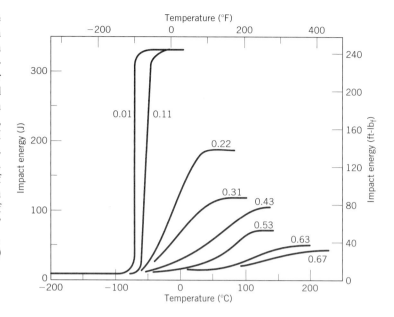

FIGURE 9.22 Influence of carbon content on the Charpy V-notch energy-versus-temperature behavior for steel. (Reprinted with permission from ASM International, Metals Park, OH 44073-9989, USA; J. A. Reinbolt and W. J. Harris, Jr., "Effect of Alloying Elements on Notch Toughness of Pearlitic Steels," *Transactions of ASM,* Vol. 43, 1951.)

FATIGUE

Fatigue is a form of failure that occurs in structures subjected to dynamic and fluctuating stresses (e.g., bridges, aircraft, and machine components). Under these circumstances it is possible for failure to occur at a stress level considerably lower than the tensile or yield strength for a static load. The term "fatigue" is used because this type of failure normally occurs after a lengthy period of repeated stress or strain cycling. Fatigue is important inasmuch as it is the single largest cause of failure in metals, estimated to comprise approximately 90% of all metallic failures; polymers and ceramics (except for glasses) are also susceptible to this type of failure. Furthermore, it is catastrophic and insidious, occurring very suddenly and without warning.

Fatigue failure is brittlelike in nature even in normally ductile metals, in that there is very little, if any, gross plastic deformation associated with failure. The process occurs by the initiation and propagation of cracks, and ordinarily the fracture surface is perpendicular to the direction of an applied tensile stress.

9.9 CYCLIC STRESSES

The applied stress may be axial (tension-compression), flexural (bending), or torsional (twisting) in nature. In general, three different fluctuating stress–time modes are possible. One is represented schematically by a regular and sinusoidal time dependence in Figure 9.23a, wherein the amplitude is symmetrical about a mean zero stress level, for example, alternating from a maximum tensile stress (σ_{max}) to a minimum compressive stress (σ_{min}) of equal magnitude; this is referred to as a *reversed stress cycle*. Another type, termed *repeated stress cycle,* is illustrated in Figure 9.23b; the maxima and minima are asymmetrical relative to the zero stress level. Finally, the stress level may vary randomly in amplitude and frequency, as exemplified in Figure 9.23c.

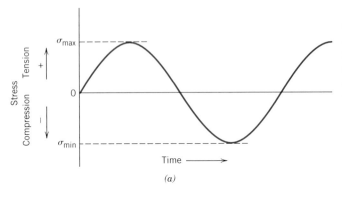

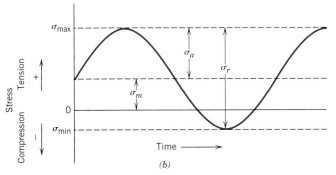

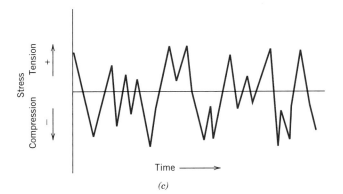

FIGURE 9.23 Variation of stress with time that accounts for fatigue failures. (*a*) Reversed stress cycle, in which the stress alternates from a maximum tensile stress (+) to a maximum compressive stress (−) of equal magnitude. (*b*) Repeated stress cycle, in which maximum and minimum stresses are asymmetrical relative to the zero stress level; mean stress σ_m, range of stress σ_r, and stress amplitude σ_a are indicated. (*c*) Random stress cycle.

Also indicated in Figure 9.23*b* are several parameters used to characterize the fluctuating stress cycle. The stress amplitude alternates about a *mean stress* σ_m, defined as the average of the maximum and minimum stresses in the cycle, or

$$\sigma_m = \frac{\sigma_{\max} + \sigma_{\min}}{2} \tag{9.21}$$

Furthermore, the *range of stress* σ_r is just the difference between $\sigma_{\max}$ and $\sigma_{\min}$, namely,

$$\sigma_r = \sigma_{\max} - \sigma_{\min} \tag{9.22}$$

Stress amplitude σ_a is just one half of this range of stress, or

$$\sigma_a = \frac{\sigma_r}{2} = \frac{\sigma_{max} - \sigma_{min}}{2} \tag{9.23}$$

Finally, the *stress ratio R* is just the ratio of minimum and maximum stress amplitudes:

$$R = \frac{\sigma_{min}}{\sigma_{max}} \tag{9.24}$$

By convention, tensile stresses are positive and compressive stresses are negative. For example, for the reversed stress cycle, the value of R is -1.

9.10 THE *S–N* CURVE

As with other mechanical characteristics, the fatigue properties of materials can be determined from laboratory simulation tests.[8] A test apparatus should be designed to duplicate as nearly as possible the service stress conditions (stress level, time frequency, stress pattern, etc.). A schematic diagram of a rotating-bending test apparatus, commonly used for fatigue testing, is shown in Figure 9.24; the compression and tensile stresses are imposed on the specimen as it is simultaneously bent and rotated. Tests are also frequently conducted using an alternating uniaxial tension-compression stress cycle.

A series of tests are commenced by subjecting a specimen to the stress cycling at a relatively large maximum stress amplitude (σ_{max}), usually on the order of two thirds of the static tensile strength; the number of cycles to failure is counted. This procedure is repeated on other specimens at progressively decreasing maximum stress amplitudes. Data are plotted as stress S versus the logarithm of the number N of cycles to failure for each of the specimens. The values of S are normally taken as stress amplitudes (σ_a, Equation 9.23); on occasion, σ_{max} or σ_{min} values may be used.

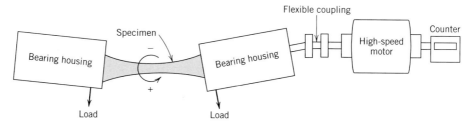

FIGURE 9.24 Schematic diagram of fatigue testing apparatus for making rotating-bending tests. (From *Materials Science in Engineering,* fourth edition, by Keyser, Carl A., © 1986. Reprinted by permission of Prentice-Hall, Inc., Upper Saddle River, NJ.)

---

[8] See ASTM Standard E 466, "Standard Practice for Conducting Constant Amplitude Axial Fatigue Tests of Metallic Materials," and ASTM Standard E 468, "Standard Practice for Presentation of Constant Amplitude Fatigue Test Results for Metallic Materials."

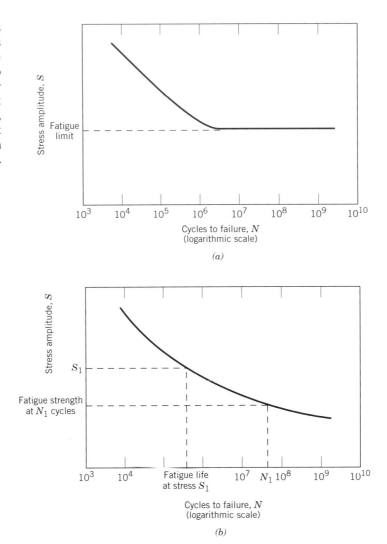

FIGURE 9.25 Stress amplitude (S) versus logarithm of the number of cycles to fatigue failure (N) for (a) a material that displays a fatigue limit, and (b) a material that does not display a fatigue limit.

Two distinct types of S–N behavior are observed, which are represented schematically in Figure 9.25. As these plots indicate, the higher the magnitude of the stress, the smaller the number of cycles the material is capable of sustaining before failure. For some ferrous (iron base) and titanium alloys, the S–N curve (Figure 9.25a) becomes horizontal at higher N values; or, there is a limiting stress level, called the **fatigue limit** (also sometimes the *endurance limit*), below which fatigue failure will not occur. This fatigue limit represents the largest value of fluctuating stress that will *not* cause failure for essentially an infinite number of cycles. For many steels, fatigue limits range between 35 and 60% of the tensile strength.

Most nonferrous alloys (e.g., aluminum, copper, magnesium) do not have a fatigue limit, in that the S–N curve continues its downward trend at increasingly greater N values (Figure 9.25b). Thus, fatigue will ultimately occur regardless of the magnitude of the stress. For these materials, the fatigue response is specified as **fatigue strength,** which is defined as the stress level at which failure will occur for some specified number of cycles (e.g., 10^7 cycles). The determination of fatigue strength is also demonstrated in Figure 9.25*b*.

Another important parameter that characterizes a material's fatigue behavior is **fatigue life** N_f. It is the number of cycles to cause failure at a specified stress level, as taken from the *S–N* plot (Figure 9.25*b*).

Unfortunately, there always exists considerable scatter in fatigue data, that is, a variation in the measured N value for a number of specimens tested at the same stress level. This may lead to significant design uncertainties when fatigue life and/or fatigue limit (or strength) are being considered. The scatter in results is a consequence of the fatigue sensitivity to a number of test and material parameters that are impossible to control precisely. These parameters include specimen fabrication and surface preparation, metallurgical variables, specimen alignment in the apparatus, mean stress, and test frequency.

Fatigue *S–N* curves similar to those shown in Figure 9.25 represent "best fit" curves which have been drawn through average-value data points. It is a little unsettling to realize that approximately one half of the specimens tested actually failed at stress levels lying nearly 25% below the curve (as determined on the basis of statistical treatments).

Several statistical techniques have been developed to specify fatigue life and fatigue limit in terms of probabilities. One convenient way of representing data treated in this manner is with a series of constant probability curves, several of which are plotted in Figure 9.26. The *P* value associated with each curve represents the probability of failure. For example, at a stress of 200 MPa (30,000 psi), we would expect 1% of the specimens to fail at about 10^6 cycles and 50% to fail at about 2×10^7 cycles, and so on. It should be remembered that *S–N* curves represented in the literature are normally average values, unless noted otherwise.

The fatigue behaviors represented in Figures 9.25*a* and 9.25*b* may be classified into two domains. One is associated with relatively high loads that produce not only elastic strain but also some plastic strain during each cycle. Consequently, fatigue lives are relatively short; this domain is termed *low-cycle fatigue* and occurs at less than about 10^4 to 10^5 cycles. For lower stress levels wherein deformations are totally elastic, longer lives result. This is called *high-cycle fatigue* inasmuch as relatively large numbers of cycles are required to produce

FIGURE 9.26   Fatigue *S–N* probability of failure curves for a 7075-T6 aluminum alloy; *P* denotes the probability of failure. (From G. M. Sinclair and T. J. Dolan, *Trans., ASME,* **75,** 1953, p. 867. Reprinted with permission of the American Society of Mechanical Engineers.)

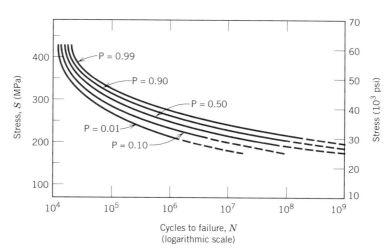

fatigue failure. High-cycle fatigue is associated with fatigue lives greater than about 10^4 to 10^5 cycles.

9.11 FATIGUE IN POLYMERIC MATERIALS

Polymers may experience fatigue failure under conditions of cyclic loading. As with metals, fatigue occurs at stress levels that are low relative to the yield strength. Fatigue data are plotted in the same manner for both types of material, and the resulting curves have the same general shape. Fatigue curves for several common polymers are shown in Figure 9.27, as stress versus the number of cycles to failure (on a logarithmic scale). Some polymers have a fatigue limit. As would be expected, fatigue strengths and fatigue limits for polymeric materials are much lower than for metals.

The fatigue behavior of polymers is much more sensitive to loading frequency than for metals. Cycling polymers at high frequencies and/or relatively large stresses can cause localized heating; consequently, failure may be due to a softening of the material rather than as a result of typical fatigue processes.

9.12a CRACK INITIATION AND PROPAGATION [DETAILED VERSION (CD-ROM)]

9.12b CRACK INITIATION AND PROPAGATION (CONCISE VERSION)

The process of fatigue failure is characterized by three distinct steps: (1) crack initiation, wherein a small crack forms at some point of high stress concentration; (2) crack propagation, during which this crack advances incrementally with each stress cycle; and (3) final failure, which occurs very rapidly once the advancing

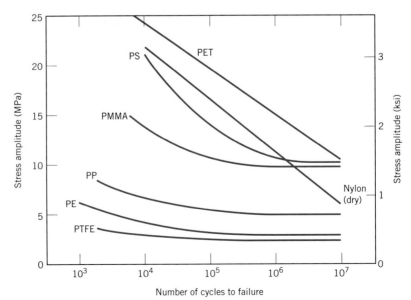

FIGURE 9.27 Fatigue curves (stress amplitude versus the number of cycles to failure) for polyethylene terephthalate (PET), nylon, polystyrene (PS), polymethyl methacrylate (PMMA), polypropylene (PP), polyethylene (PE), and polytetrafluoroethylene (PTFE). The testing frequency was 30 Hz. (From M. N. Riddell, "A Guide to Better Testing of Plastics," *Plast. Eng.,* Vol. 30, No. 4, p. 78, 1974.)

crack has reached a critical size. Cracks associated with fatigue failure almost always initiate (or nucleate) on the surface of a component at some point of stress concentration. Crack nucleation sites include surface scratches, sharp fillets, keyways, threads, dents, and the like. In addition, cyclic loading can produce microscopic surface discontinuities resulting from dislocation slip steps which may also act as stress raisers, and therefore as crack initiation sites.

The region of a fracture surface that formed during the crack propagation step may be characterized by two types of markings termed *beachmarks* and *striations*. Both of these features indicate the position of the crack tip at some point in time and appear as concentric ridges that expand away from the crack initiation site(s), frequently in a circular or semicircular pattern. Beachmarks (sometimes also called "clamshell marks") are of macroscopic dimensions (Figure 9.30), and may be observed with the unaided eye. These markings are found for components that experienced interruptions during the crack propagation stage—for example, a machine that operated only during normal work-shift hours. Each beachmark band represents a period of time over which crack growth occurred.

On the other hand, fatigue striations are microscopic in size and subject to observation with the electron microscope (either TEM or SEM). Figure 9.31 is an electron fractograph which shows this feature. Each striation is thought to represent the advance distance of a crack front during a single load cycle. Striation width depends on, and increases with, increasing stress range.

At this point it should be emphasized that although both beachmarks and striations are fatigue fracture surface features having similar appearances, they are nevertheless different, both in origin and size. There may be literally thousands of striations within a single beachmark.

Often the cause of failure may be deduced after examination of the failure surfaces. The presence of beachmarks and/or striations on a fracture surface con-

FIGURE 9.30 Fracture surface of a rotating steel shaft that experienced fatigue failure. Beachmark ridges are visible in the photograph. (Reproduced with permission from D. J. Wulpi, *Understanding How Components Fail,* American Society for Metals, Materials Park, OH, 1985.)

FIGURE 9.31 Transmission electron fractograph showing fatigue striations in aluminum. Magnification unknown. (From V. J. Colangelo and F. A. Heiser, *Analysis of Metallurgical Failures,* 2nd edition. Copyright © 1987 by John Wiley & Sons, New York. Reprinted by permission of John Wiley & Sons, Inc.)

firms that the cause of failure was fatigue. Nevertheless, the absence of either or both does not exclude fatigue as the cause of failure.

One final comment regarding fatigue failure surfaces: Beachmarks and striations will not appear on that region over which the rapid failure occurs. Rather, the rapid failure may be either ductile or brittle; evidence of plastic deformation will be present for ductile, and absent for brittle, failure. This region of failure may be noted in Figure 9.32.

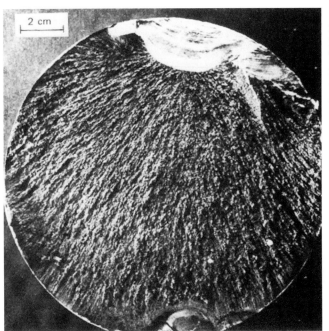

FIGURE 9.32 Fatigue failure surface. A crack formed at the top edge. The smooth region also near the top corresponds to the area over which the crack propagated slowly. Rapid failure occurred over the area having a dull and fibrous texture (the largest area). Approximately 0.5 ×. (Reproduced by permission from *Metals Handbook: Fractography and Atlas of Fractographs,* Vol. 9, 8th edition, H. E. Boyer, Editor, American Society for Metals, 1974.)

9.13 CRACK PROPAGATION RATE (CD-ROM)

9.14 FACTORS THAT AFFECT FATIGUE LIFE

As was mentioned in Section 9.10, the fatigue behavior of engineering materials is highly sensitive to a number of variables. Some of these factors include mean stress level, geometrical design, surface effects, and metallurgical variables, as well as the environment. This section is devoted to a discussion of these factors and, in addition, to measures that may be taken to improve the fatigue resistance of structural components.

MEAN STRESS

The dependence of fatigue life on stress amplitude is represented on the S–N plot. Such data are taken for a constant mean stress σ_m, often for the reversed cycle situation ($\sigma_m = 0$). Mean stress, however, will also affect fatigue life, which influence may be represented by a series of S–N curves, each measured at a different σ_m; this is depicted schematically in Figure 9.36. As may be noted, increasing the mean stress level leads to a decrease in fatigue life.

SURFACE EFFECTS

For many common loading situations, the maximum stress within a component or structure occurs at its surface. Consequently, most cracks leading to fatigue failure originate at surface positions, specifically at stress amplification sites. Therefore, it has been observed that fatigue life is especially sensitive to the condition and configuration of the component surface. Numerous factors influence fatigue resistance, the proper management of which will lead to an improvement in fatigue life. These include design criteria as well as various surface treatments.

Design Factors

The design of a component can have a significant influence on its fatigue characteristics. Any notch or geometrical discontinuity can act as a stress raiser and fatigue crack initiation site; these design features include grooves, holes, keyways, threads, and so on. The sharper the discontinuity (i.e., the smaller the radius of curvature),

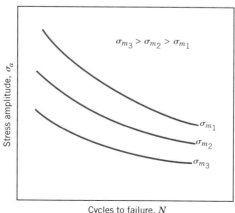

FIGURE 9.36 Demonstration of influence of mean stress σ_m on S–N fatigue behavior.

$\sigma_{m_3} > \sigma_{m_2} > \sigma_{m_1}$

Stress amplitude, σ_a

σ_{m_1}

σ_{m_2}

σ_{m_3}

Cycles to failure, N
(logarithmic scale)

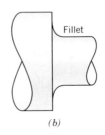

FIGURE 9.37 Demonstration of how design can reduce stress amplification. (*a*) Poor design: sharp corner. (*b*) Good design: fatigue lifetime improved by incorporating rounded fillet into a rotating shaft at the point where there is a change in diameter.

(a) *(b)*

the more severe the stress concentration. The probability of fatigue failure may be reduced by avoiding (when possible) these structural irregularities, or by making design modifications whereby sudden contour changes leading to sharp corners are eliminated—for example, calling for rounded fillets with large radii of curvature at the point where there is a change in diameter for a rotating shaft (Figure 9.37).

Surface Treatments

During machining operations, small scratches and grooves are invariably introduced into the workpiece surface by cutting tool action. These surface markings can limit the fatigue life. It has been observed that improving the surface finish by polishing will enhance fatigue life significantly.

One of the most effective methods of increasing fatigue performance is by imposing residual compressive stresses within a thin outer surface layer. Thus, a surface tensile stress of external origin will be partially nullified and reduced in magnitude by the residual compressive stress. The net effect is that the likelihood of crack formation and therefore of fatigue failure is reduced.

Residual compressive stresses are commonly introduced into ductile metals mechanically by localized plastic deformation within the outer surface region. Commercially, this is often accomplished by a process termed *shot peening*. Small, hard particles (shot) having diameters within the range of 0.1 to 1.0 mm are projected at high velocities onto the surface to be treated. The resulting deformation induces compressive stresses to a depth of between one quarter and one half of the shot diameter. The influence of shot peening on the fatigue behavior of steel is demonstrated schematically in Figure 9.38.

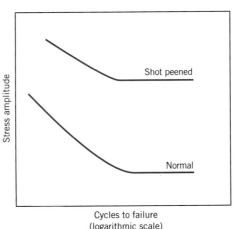

FIGURE 9.38 Schematic *S–N* fatigue curves for normal and shot-peened steel.

FIGURE 9.39 Photomicrograph showing both core (bottom) and carburized outer case (top) regions of a case-hardened steel. The case is harder as attested by the smaller microhardness indentation. 100×. (From R. W. Hertzberg, *Deformation and Fracture Mechanics of Engineering Materials,* 3rd edition. Copyright © 1989 by John Wiley & Sons, New York. Reprinted by permission of John Wiley & Sons, Inc.)

Case hardening is a technique whereby both surface hardness and fatigue life are enhanced for steel alloys. This is accomplished by a carburizing or nitriding process whereby a component is exposed to a carbonaceous or nitrogenous atmosphere at an elevated temperature. A carbon- or nitrogen-rich outer surface layer (or "case") is introduced by atomic diffusion from the gaseous phase. The case is normally on the order of 1 mm deep and is harder than the inner core of material. (The influence of carbon content on hardness for Fe–C alloys is demonstrated in Figure 11.21a.) The improvement of fatigue properties results from increased hardness within the case, as well as the desired residual compressive stresses the formation of which attends the carburizing or nitriding process. A carbon-rich outer case may be observed for the gear shown in the chapter-opening photograph for Chapter 6; it appears as a dark outer rim within the sectioned segment. The increase in case hardness is demonstrated in the photomicrograph appearing in Figure 9.39. The dark and elongated diamond shapes are Knoop microhardness indentations. The upper indentation, lying within the carburized layer, is smaller than the core indentation.

9.15 ENVIRONMENTAL EFFECTS (CD-ROM)

CREEP

Materials are often placed in service at elevated temperatures and exposed to static mechanical stresses (e.g., turbine rotors in jet engines and steam generators that experience centrifugal stresses, and high-pressure steam lines). Deformation under such circumstances is termed **creep.** Defined as the time-dependent and permanent deformation of materials when subjected to a constant load or stress, creep is normally an undesirable phenomenon and is often the limiting factor in the lifetime of a part. It is observed in all materials types; for metals it becomes important only for temperatures greater than about $0.4 T_m$ (T_m = absolute melting temperature).

9.16 GENERALIZED CREEP BEHAVIOR

A typical creep test[10] consists of subjecting a specimen to a constant load or stress while maintaining the temperature constant; deformation or strain is measured and plotted as a function of elapsed time. Most tests are the constant load type, which yield information of an engineering nature; constant stress tests are employed to provide a better understanding of the mechanisms of creep.

Figure 9.40 is a schematic representation of the typical constant load creep behavior of metals. Upon application of the load there is an instantaneous deformation, as indicated in the figure, which is mostly elastic. The resulting creep curve consists of three regions, each of which has its own distinctive strain–time feature. *Primary* or *transient creep* occurs first, typified by a continuously decreasing creep rate; that is, the slope of the curve diminishes with time. This suggests that the material is experiencing an increase in creep resistance or strain hardening (Section 8.11)—deformation becomes more difficult as the material is strained. For *secondary creep*, sometimes termed *steady-state creep*, the rate is constant; that is, the plot becomes linear. This is often the stage of creep that is of the longest duration. The constancy of creep rate is explained on the basis of a balance between the competing processes of strain hardening and recovery, recovery (Section 8.12) being the process whereby a material becomes softer and retains its ability to experience deformation. Finally, for *tertiary creep,* there is an acceleration of the rate and ultimate failure. This failure is frequently termed *rupture* and results from microstructural and/or metallurgical changes; for example, grain boundary separation, and the formation of internal cracks, cavities, and voids. Also, for tensile loads, a neck may form at some point within the deformation region. These all lead to a decrease in the effective cross-sectional area and an increase in strain rate.

For metallic materials most creep tests are conducted in uniaxial tension using a specimen having the same geometry as for tensile tests (Figure 7.2). On the other hand, uniaxial compression tests are more appropriate for brittle materials; these provide a better measure of the intrinsic creep properties inasmuch as there is no stress amplification and crack propagation, as with tensile loads. Compressive test specimens are usually right cylinders or parallelepipeds having length-to-diameter

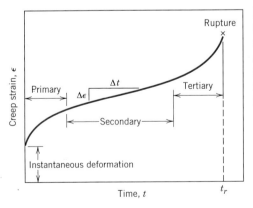

FIGURE 9.40 Typical creep curve of strain versus time at constant stress and constant elevated temperature. The minimum creep rate $\Delta\epsilon/\Delta t$ is the slope of the linear segment in the secondary region. Rupture lifetime t_r is the total time to rupture.

[10] ASTM Standard E 139, "Standard Practice for Conducting Creep, Creep-Rupture, and Stress-Rupture Tests of Metallic Materials."

ratios ranging from about 2 to 4. For most materials creep properties are virtually independent of loading direction.

Possibly the most important parameter from a creep test is the slope of the secondary portion of the creep curve ($\Delta\epsilon/\Delta t$ in Figure 9.40); this is often called the minimum or *steady-state creep rate* $\dot{\epsilon}_s$. It is the engineering design parameter that is considered for long-life applications, such as a nuclear power plant component that is scheduled to operate for several decades, and when failure or too much strain is not an option. On the other hand, for many relatively short-life creep situations (e.g., turbine blades in military aircraft and rocket motor nozzles), *time to rupture*, or the *rupture lifetime* t_r, is the dominant design consideration; it is also indicated in Figure 9.40. Of course, for its determination, creep tests must be conducted to the point of failure; these are termed *creep rupture* tests. Thus, a knowledge of these creep characteristics of a material allows the design engineer to ascertain its suitability for a specific application.

9.17a STRESS AND TEMPERATURE EFFECTS [DETAILED VERSION (CD-ROM)]

9.17b STRESS AND TEMPERATURE EFFECTS (CONCISE VERSION)

Both temperature and the level of the applied stress influence the creep characteristics (Figure 9.41). At a temperature substantially below $0.4\,T_m$, and after the initial deformation, the strain is virtually independent of time. With either increasing stress or temperature, the following will be noted: (1) the instantaneous strain at the time of stress application increases; (2) the steady-state creep rate is increased; and (3) the rupture lifetime is diminished.

The results of creep rupture tests are most commonly presented as the logarithm of stress versus the logarithm of rupture lifetime. Figure 9.42 is one such plot for a nickel alloy in which a linear relationship can be seen to exist at each temperature. For some alloys and over relatively large stress ranges, nonlinearity in these curves is observed.

Both temperature and stress effects on the steady-state creep rate are represented graphically as logarithm of stress versus logarithm of $\dot{\epsilon}_s$ for tests conducted

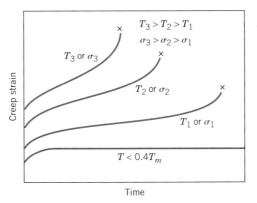

FIGURE 9.41   Influence of stress σ and temperature T on creep behavior.

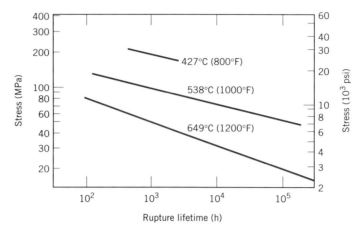

FIGURE 9.42 Stress (logarithmic scale) versus rupture lifetime (logarithmic scale) for a low carbon–nickel alloy at three temperatures. (From *Metals Handbook: Properties and Selection: Stainless Steels, Tool Materials and Special-Purpose Metals*, Vol. 3, 9th edition, D. Benjamin, Senior Editor, American Society for Metals, 1980, p. 130.)

at a variety of temperatures. Figure 9.43 shows data that were collected at three temperatures for the same nickel alloy. Clearly, a straight line segment is drawn at each temperature.

9.18 DATA EXTRAPOLATION METHODS (CD-ROM)

9.19 ALLOYS FOR HIGH-TEMPERATURE USE

There are several factors that affect the creep characteristics of metals. These include melting temperature, elastic modulus, and grain size. In general, the higher the melting temperature, the greater the elastic modulus, and the larger the grain

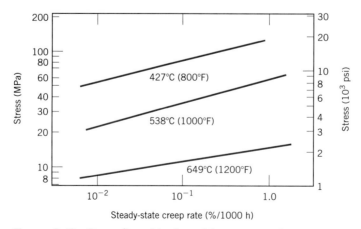

FIGURE 9.43 Stress (logarithmic scale) versus steady-state creep rate (logarithmic scale) for a low carbon–nickel alloy at three temperatures. (From *Metals Handbook: Properties and Selection: Stainless Steels, Tool Materials and Special-Purpose Metals*, Vol. 3, 9th edition, D. Benjamin, Senior Editor, American Society for Metals, 1980, p. 131.)

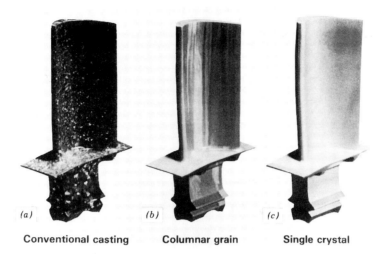

Conventional casting **Columnar grain** **Single crystal**

FIGURE 9.45 (*a*) Polycrystalline turbine blade that was produced by a conventional casting technique. High-temperature creep resistance is improved as a result of an oriented columnar grain structure (*b*) produced by a sophisticated directional solidification technique. Creep resistance is further enhanced when single-crystal blades (*c*) are used. (Courtesy of Pratt & Whitney.)

size, the better is a material's resistance to creep. Relative to grain size, smaller grains permit more grain-boundary sliding, which results in higher creep rates. This effect may be contrasted to the influence of grain size on the mechanical behavior at low temperatures [i.e., increase in both strength (Section 8.9) and toughness (Section 9.8)].

Stainless steels (Section 13.2), the refractory metals and the superalloys (Section 13.3) are especially resilient to creep and are commonly employed in high-temperature service applications. The creep resistance of the cobalt and nickel superalloys is enhanced by solid-solution alloying, and also by the addition of a dispersed phase which is virtually insoluble in the matrix. In addition, advanced processing techniques have been utilized; one such technique is directional solidification, which produces either highly elongated grains or single-crystal components (Figure 9.45). Another is the controlled unidirectional solidification of alloys having specially designed compositions wherein two-phase composites result.

9.20 CREEP IN CERAMIC AND POLYMERIC MATERIALS

Ceramic materials often experience creep deformation as a result of exposure to stresses (usually compressive) at elevated temperatures. In general, the time-deformation creep behavior of ceramics is similar to that of metals (Figure 9.40); however, creep occurs at higher temperatures in ceramics.

Viscoelastic creep is the term used to denote the creep phenomenon in polymeric materials. {It is one of the topics of discussion in Section 7.15.}

SUMMARY

Fracture, in response to tensile loading and at relatively low temperatures, may occur by ductile and brittle modes, both of which involve the formation and propagation of cracks. For ductile fracture, evidence will exist of gross plastic deformation at the fracture surface. In tension, highly ductile metals will neck down to essentially a point fracture; cup-and-cone mating fracture surfaces result for moderate ductility. {Microscopically, dimples (spherical and parabolic) are produced.} Cracks in ductile materials are said to be stable (i.e., resist extension without an increase in applied stress); and inasmuch as fracture is noncatastrophic, this fracture mode is almost always preferred.

For brittle fracture, cracks are unstable, and the fracture surface is relatively flat and perpendicular to the direction of the applied tensile load. Chevron and ridgelike patterns are possible, which indicate the direction of crack propagation. Transgranular (through-grain) and intergranular (between-grain) fractures are found in brittle polycrystalline materials.

The discipline of fracture mechanics allows for a better understanding of the fracture process and provides for structural design wherein the probability of failure is minimized. The significant discrepancy between actual and theoretical fracture strengths of brittle materials is explained by the existence of small flaws that are capable of amplifying an applied tensile stress in their vicinity, leading ultimately to crack formation. Stress amplification is greatest for long flaws that have small tip radii of curvature. Fracture ensues when the theoretical cohesive strength is exceeded at the tip of one of these flaws. Consideration of elastic strain and crack surface energies gives rise to an expression for a crack propagation critical stress in brittle materials; this parameter is a function of elastic modulus, specific surface energy, and crack length.

{The stress distributions in front of an advancing crack may be expressed in terms of position (as radial and angular coordinates) as well as stress intensity factor.} The critical value of the stress intensity factor (i.e., that at which fracture occurs) is termed the fracture toughness, which is related to stress level, crack length, and a geometrical factor. The fracture toughness of a material is indicative of its resistance to brittle fracture when a crack is present. It depends on specimen thickness, and, for relatively thick specimens (i.e., conditions of plane strain), is termed the plane strain fracture toughness. This parameter is the one normally cited for design purposes; its value is relatively large for ductile materials (and small for brittle ones), and is a function of microstructure, strain rate, and temperature. With regard to designing against the possibility of fracture, consideration must be given to material (its fracture toughness), the stress level, and the flaw size detection limit.

At room temperature, virtually all ceramics are brittle. Microcracks, the presence of which is very difficult to control, result in amplification of applied tensile stresses and account for relatively low fracture strengths (flexural strengths). This amplification does not occur with compressive loads, and, consequently, ceramics are stronger in compression.

Fracture strengths of polymeric materials are also low relative to metals. Both brittle and ductile fracture modes are possible, and some thermoplastic materials experience a ductile-to-brittle transition with a lowering of temperature, an increase in strain rate, and/or an alteration of specimen thickness or geometry. In some glassy thermoplastics, the crack formation process may be preceded by crazing; crazing can lead to an increase in ductility and toughness of the material.

Qualitatively, the fracture behavior of materials may be determined using Charpy and Izod impact testing techniques; impact energy (or notch toughness) is measured for specimens into which a V-shaped notch has been machined. On the basis of the temperature dependence of this impact energy (or appearance of the fracture surface), it is possible to ascertain whether or not a material experiences a ductile-to-brittle transition and the temperature range over which such a transition occurs. Metal alloys having BCC and HCP crystal structures experience this transition, and, for structural applications, should be used at temperatures in excess of this transition range.

Fatigue is a common type of catastrophic failure wherein the applied stress level fluctuates with time. Test data are plotted as stress versus the logarithm of the number of cycles to failure. For many materials, the number of cycles to failure increases continuously with diminishing stress. Fatigue strength represents the failure stress for a specified number of cycles. For some steels and titanium alloys, stress ceases to decrease with, and becomes independent of, the number of cycles; fatigue limit is the magnitude of this constant stress level, below which fatigue will not occur even for virtually an infinite number of cycles. Another fatigue property is fatigue life, which, for a specific stress, is the number of cycles to failure.

As a result of significant scatter in measured fatigue data, statistical analyses are performed that lead to specification of fatigue life and limit in terms of probabilities.

The processes of fatigue crack initiation and propagation were discussed. Cracks normally nucleate on the surface of a component at some point of stress concentration. Propagation proceeds in two stages, which are characterized by propagation direction and rate. {The mechanism for the more rapid stage II corresponds to a repetitive plastic blunting and sharpening process at the advancing crack tip.}

Two characteristic fatigue surface features are beachmarks and striations. Beachmarks form on components that experience applied stress interruptions; they normally may be observed with the naked eye. Fatigue striations are of microscopic dimensions, and each is thought to represent the crack tip advance distance over a single load cycle.

{An analytical expression was proposed for fatigue crack propagation rate in terms of the stress intensity range at the crack tip. Integration of the expression yields an equation whereby fatigue life may be estimated.}

Measures that may be taken to extend fatigue life include (1) reducing the mean stress level, (2) eliminating sharp surface discontinuities, (3) improving the surface finish by polishing, (4) imposing surface residual compressive stresses by shot peening, and (5) case hardening by using a carburizing or nitriding process.

{The fatigue behavior of materials may also be affected by the environment. Thermal stresses may be induced in components that are exposed to elevated temperature fluctuations and when thermal expansion and/or contraction is restrained; fatigue for these conditions is termed thermal fatigue. The presence of a chemically active environment may lead to a reduction in fatigue life for corrosion fatigue; small pit crack nucleation sites form on the component surface as a result of chemical reactions.}

The time-dependent plastic deformation of materials subjected to a constant load (or stress) and temperatures greater than about $0.4T_m$ is termed creep. A typical creep curve (strain versus time) will normally exhibit three distinct regions. For transient (or primary) creep, the rate (or slope) diminishes with time. The plot becomes linear (i.e., creep rate is constant) in the steady-state (or secondary) region. And finally, deformation accelerates for tertiary creep, just prior to failure (or rupture). Important design parameters available from such a plot include the steady-state creep rate (slope of the linear region) and rupture lifetime.

Both temperature and applied stress level influence creep behavior. Increasing either of these parameters produces the following effects: (1) an increase in the instantaneous initial deformation, (2) an increase in the steady-state creep rate, and (3) a diminishment of the rupture lifetime. {Analytical expressions were presented which relate $\dot{\epsilon}_s$ to both temperature and stress. Creep mechanisms may be discerned on the basis of steady-state rate stress exponent and creep activation energy values.}

{Extrapolation of creep test data to lower temperature–longer time regimes is possible using the Larson–Miller parameter.}

Metal alloys that are especially resistant to creep have high elastic moduli and melting temperatures; these include the superalloys, the stainless steels, and the refractory metals. Various processing techniques are employed to improve the creep properties of these materials.

The creep phenomenon is also observed in ceramic and polymeric materials.

IMPORTANT TERMS AND CONCEPTS

Brittle fracture	Fatigue life	Izod test
Case hardening	Fatigue limit	Plane strain
Charpy test	Fatigue strength	Plane strain fracture toughness
Corrosion fatigue	Fracture mechanics	Stress intensity factor
Creep	Fracture toughness	Stress raiser
Ductile fracture	Impact energy	Thermal fatigue
Ductile-to-brittle transition	Intergranular fracture	Transgranular fracture
Fatigue		

REFERENCES

ASM Handbook, Vol. 11, *Failure Analysis and Prevention,* ASM International, Materials Park, OH, 1986.

ASM Handbook, Vol. 12, *Fractography,* ASM International, Materials Park, OH, 1987.

Boyer, H. E. (Editor), *Atlas of Creep and Stress–Rupture Curves,* ASM International, Materials Park, OH, 1988.

Boyer, H. E. (Editor), *Atlas of Fatigue Curves,* ASM International, Materials Park, OH, 1986.

Colangelo, V. J. and F. A. Heiser, *Analysis of Metallurgical Failures*, 2nd edition, John Wiley & Sons, New York, 1987.

Collins, J. A., *Failure of Materials in Mechanical Design,* 2nd edition, John Wiley & Sons, New York, 1993.

Courtney, T. H., *Mechanical Behavior of Materials,* McGraw-Hill Book Co., New York, 1990.

Davidge, R. W., *Mechanical Behaviour of Ceramics,* Cambridge University Press, Cambridge, 1979. Reprinted by TechBooks, Marietta, OH.

Dieter, G. E., *Mechanical Metallurgy,* 3rd edition, McGraw-Hill Book Co., New York, 1986.

Esaklul, K. A., *Handbook of Case Histories in Failure Analysis,* ASM International, Materials Park, OH, 1992 and 1993. In two volumes.

Fatigue Data Book: Light Structural Alloys, ASM International, Materials Park, OH, 1995.

Hertzberg, R. W., *Deformation and Fracture Mechanics of Engineering Materials,* 4th edition, John Wiley & Sons, New York, 1996.

Murakami, Y. (Editor), *Stress Intensity Factors Handbook,* Pergamon Press, Oxford, 1987. In two volumes.

Tetelman, A. S. and A. J. McEvily, *Fracture of Structural Materials,* John Wiley & Sons, New York, 1967. Reprinted by Books on Demand, Ann Arbor, MI.

Wachtman, J. B., *Mechanical Properties of Ceramics,* John Wiley & Sons, Inc., New York, 1996.

Ward, I. M. and D. W. Hadley, *An Introduction to the Mechanical Properties of Solid Polymers,* John Wiley & Sons, Chichester, UK, 1993.

Wulpi, D. J., *Understanding How Components Fail,* American Society for Metals, Materials Park, OH, 1985.

Young, R. J. and P. Lovell, *Introduction to Polymers,* 2nd edition, Chapman and Hall, London, 1991.

QUESTIONS AND PROBLEMS

Note: To solve those problems having an asterisk (*) by their numbers, consultation of supplementary topics [appearing only on the CD-ROM (and not in print)] will probably be necessary.

9.1 Cite at least two situations in which the possibility of failure is part of the design of a component or product.

9.2* Estimate the theoretical cohesive strengths of the ceramic materials listed in Table 7.1.

9.3 What is the magnitude of the maximum stress that exists at the tip of an internal crack having a radius of curvature of 2.5×10^{-4} mm (10^{-5} in.) and a crack length of 2.5×10^{-2} mm (10^{-3} in.) when a tensile stress of 170 MPa (25,000 psi) is applied?

9.4 Estimate the theoretical fracture strength of a brittle material if it is known that fracture occurs by the propagation of an elliptically shaped surface crack of length 0.25 mm (0.01 in.) and having a tip radius of curvature of 1.2×10^{-3} mm (4.7×10^{-5} in.) when a stress of 1200 MPa (174,000 psi) is applied.

9.5* A specimen of a ceramic material having a modulus of elasticity of 300 GPa (43.5×10^6 psi) is pulled in tension with a stress of 900 MPa (130,000 psi). Will the specimen fail if its "most severe flaw" is an internal crack that has a length of 0.30 mm (0.012 in.) and a tip radius of curvature of 5×10^{-4} mm (2×10^{-5} in.)? Why or why not?

9.6 Briefly explain **(a)** why there may be significant scatter in the fracture strength for some given ceramic material, and **(b)** why fracture strength increases with decreasing specimen size.

9.7 The tensile strength of brittle materials may be determined using a variation of Equation 9.1b. Compute the critical crack tip radius for an Al_2O_3 specimen that experiences tensile fracture at an applied stress of 275 MPa (40,000 psi). Assume a critical surface crack length of 2×10^{-3} mm and a theoretical fracture strength of $E/10$, where E is the modulus of elasticity.

9.8 If the specific surface energy for soda-lime glass is 0.30 J/m^2, using data contained in Table 7.1, compute the critical stress required for the propagation of a surface crack of length 0.05 mm.

9.9 A polystyrene component must not fail when a tensile stress of 1.25 MPa (180 psi) is applied. Determine the maximum allowable surface crack length if the surface energy of polystyrene is 0.50 J/m^2 (2.86×10^{-3} in.-lb_f/$in.^2$). Assume a modulus of elasticity of 3.0 GPa (0.435×10^6 psi).

9.10* The parameter K in Equations 9.7a, 9.7b, and 9.7c is a function of the applied nominal stress σ and crack length a as

$$K = \sigma \sqrt{\pi a}$$

Compute the magnitudes of the normal stresses σ_x and σ_y in front of a surface crack of length 2.5 mm (0.10 in.) (as depicted in Figure 9.10) in response to a nominal tensile stress of 75 MPa (10,875 psi) at the following positions:

(a) $r = 0.15$ mm (6.0×10^{-3} in.), $\theta = 30°$
(b) $r = 0.15$ mm (6.0×10^{-3} in.), $\theta = 60°$
(c) $r = 0.75$ mm (3.0×10^{-2} in.), $\theta = 30°$
(d) $r = 0.75$ mm (3.0×10^{-2} in.), $\theta = 60°$

9.11* The parameter K in Equations 9.7a, 9.7b, and 9.7c is defined in the previous problem.

(a) For a surface crack of length 3.0 mm (0.118 in.), determine the radial position at an angle θ of 45° at which the normal stress σ_x is 110 MPa (16,000 psi) when the magnitude of the nominal applied stress is 100 MPa (14,500 psi).

(b) Compute the normal stress σ_y at this same position.

9.12* A portion of a tensile specimen is shown as follows:

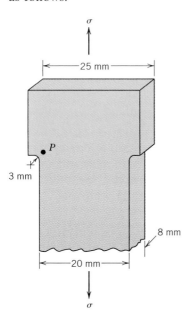

(a) Compute the magnitude of the stress at point P when the externally applied stress is 100 MPa (14,500 psi).

(b) How much will the radius of curvature at point P have to be increased to reduce this stress by 20%?

9.13* A cylindrical hole 25 mm (1.0 in.) in diameter passes entirely through the thickness of a steel plate 15 mm (0.6 in.) thick, 100 mm (4 in.) wide, and 400 mm (15.75 in.) long (see Figure 9.8a).

(a) Calculate the stress at the edge of this hole when a tensile stress of 50 MPa (7250 psi) is applied in a lengthwise direction.

(b) Calculate the stress at the hole edge when the same stress in part (a) is applied in a widthwise direction.

9.14* Cite the significant differences between the stress intensity factor, the plane stress fracture toughness, and the plane strain fracture toughness.

9.15* For each of the metal alloys listed in Table 9.1, compute the minimum component thickness for which the condition of plane strain is valid.

9.16 A specimen of a 4340 steel alloy having a plane strain fracture toughness of 45 MPa $\sqrt{m}$ (41 ksi $\sqrt{in.}$) is exposed to a stress of 1000 MPa (145,000 psi). Will this specimen experience fracture if it is known that the largest surface crack is 0.75 mm (0.03 in.) long? Why or why not? Assume that the parameter Y has a value of 1.0.

9.17 Some aircraft component is fabricated from an aluminum alloy that has a plane strain fracture toughness of 35 MPa $\sqrt{m}$ (31.9 ksi $\sqrt{in.}$). It has been determined that fracture results at a stress of 250 MPa (36,250 psi) when the maximum (or critical) internal crack length is 2.0 mm (0.08 in.). For this same component and alloy, will fracture occur at a stress level of 325 MPa (47,125 psi) when the maximum internal crack length is 1.0 mm (0.04 in.)? Why or why not?

9.18 Suppose that a wing component on an aircraft is fabricated from an aluminum alloy that has a plane strain fracture toughness of 40 MPa $\sqrt{m}$ (36.4 ksi $\sqrt{in.}$). It has been determined that fracture results at a stress of 365 MPa (53,000 psi) when the maximum internal crack length is 2.5 mm (0.10 in.). For this same component and alloy, compute the stress level at which fracture will occur for a critical internal crack length of 4.0 mm (0.16 in.).

9.19 A large plate is fabricated from a steel alloy that has a plane strain fracture toughness of 55 MPa $\sqrt{m}$ (50 ksi $\sqrt{in.}$). If, during service use, the plate is exposed to a tensile stress of 200 MPa (29,000 psi), determine the minimum length of a surface crack that will lead to fracture. Assume a value of 1.0 for Y.

9.20 Calculate the maximum internal crack length allowable for a 7075-T651 aluminum alloy (Table 9.1) component that is loaded to a stress one half of its yield strength. Assume that the value of Y is 1.35.

9.21 A structural component in the form of a wide plate is to be fabricated from a steel alloy that has a plane strain fracture toughness of 77 MPa $\sqrt{m}$ (70.1 ksi $\sqrt{in.}$) and a yield strength of 1400 MPa (205,000 psi). The flaw size resolution limit of the flaw detection apparatus is 4.0 mm (0.16 in.). If the design stress is one half of the yield strength and the value of Y is 1.0, determine whether or not a critical flaw for this plate is subject to detection.

9.22* A structural component in the shape of a flat plate 12.5 mm (0.5 in.) thick is to be fabricated from a metal alloy for which the yield strength and plane strain fracture toughness values are 350 MPa (50,750 psi) and 33 MPa $\sqrt{m}$ (30 ksi $\sqrt{in.}$), respectively; for this particular geometry, the value of Y is 1.75. Assuming a design stress of one half of the yield strength, is it possible to compute the critical length of a surface flaw? If so, determine its length; if this computation is not possible from the given data, then explain why.

9.23 After consultation of other references, write a brief report on one or two nondestructive test techniques that are used to detect and measure internal and/or surface flaws in metal alloys.

9.24 The fracture strength of glass may be increased by etching away a thin surface layer. It is believed that the etching may alter surface crack geometry (i.e., reduce crack length and increase the tip radius). Compute the ratio of the original and etched crack tip radii for an eightfold increase in fracture strength if two-thirds of the crack length is removed.

9.25 For thermoplastic polymers, cite five factors that favor brittle fracture.

9.26 Tabulated below are data that were gathered from a series of Charpy impact tests on a ductile cast iron:

Temperature (°C)	Impact Energy (J)
−25	124
−50	123
−75	115
−85	100
−100	73
−110	52
−125	26
−150	9
−175	6

(a) Plot the data as impact energy versus temperature.

(b) Determine a ductile-to-brittle transition temperature as that temperature corre-sponding to the average of the maximum and minimum impact energies.

(c) Determine a ductile-to-brittle transition temperature as that temperature at which the impact energy is 80 J.

9.27 Tabulated as follows are data that were gath-ered from a series of Charpy impact tests on a tempered 4140 steel alloy:

Temperature (°C)	Impact Energy (J)
100	89.3
75	88.6
50	87.6
25	85.4
0	82.9
−25	78.9
−50	73.1
−65	66.0
−75	59.3
−85	47.9
−100	34.3
−125	29.3
−150	27.1
−175	25.0

(a) Plot the data as impact energy versus temperature.

(b) Determine a ductile-to-brittle transition temperature as that temperature corre-sponding to the average of the maximum and minimum impact energies.

(c) Determine a ductile-to-brittle transition temperature as that temperature at which the impact energy is 70 J.

9.28 Briefly explain why BCC and HCP metal alloys may experience a ductile-to-brittle transition with decreasing temperature, whereas FCC alloys do not experience such a transition.

9.29 A fatigue test was conducted in which the mean stress was 50 MPa (7250 psi) and the stress amplitude was 225 MPa (32,625 psi).

(a) Compute the maximum and minimum stress levels.

(b) Compute the stress ratio.

(c) Compute the magnitude of the stress range.

9.30 A cylindrical 1045 steel bar (Figure 9.46) is subjected to repeated compression-tension stress cycling along its axis. If the load ampli-

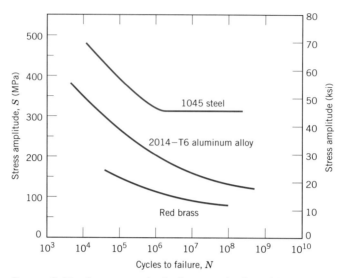

FIGURE 9.46 Stress magnitude S versus the logarithm of the number N of cycles to fatigue failure for red brass, an aluminum alloy, and a plain carbon steel. (Adapted from H. W. Hayden, W. G. Moffatt, and J. Wulff, *The Structure and Properties of Materials*, Vol. III, *Mechanical Behavior*, p. 15. Copyright © 1965 by John Wiley & Sons, New York. Reprinted by permission of John Wiley & Sons, Inc. Also adapted from *ASM Handbook*, Vol. 2, *Properties and Selection: Nonferrous Alloys and Special-Purpose Materials*, 1990. Reprinted by permission of ASM International.)

tude is 22,000 N (4950 lb$_f$), compute the minimum allowable bar diameter to ensure that fatigue failure will not occur. Assume a factor of safety of 2.0.

9.31 An 8.0 mm (0.31 in.) diameter cylindrical rod fabricated from a red brass alloy (Figure 9.46) is subjected to reversed tension-compression load cycling along its axis. If the maximum tensile and compressive loads are +7500 N (1700 lb$_f$) and −7500 N (−1700 lb$_f$), respectively, determine its fatigue life. Assume that the stress plotted in Figure 9.46 is stress amplitude.

9.32 A 12.5 mm (0.50 in.) diameter cylindrical rod fabricated from a 2014-T6 alloy (Figure 9.46) is subjected to a repeated tension-compression load cycling along its axis. Compute the maximum and minimum loads that will be applied to yield a fatigue life of 1.0×10^7 cycles. Assume that the stress plotted on the vertical axis is stress amplitude, and data were taken for a mean stress of 50 MPa (7250 psi).

9.33 The fatigue data for a brass alloy are given as follows:

Stress Amplitude (MPa)	Cycles to Failure
310	2×10^5
223	1×10^6
191	3×10^6
168	1×10^7
153	3×10^7
143	1×10^8
134	3×10^8
127	1×10^9

(a) Make an S–N plot (stress amplitude versus logarithm cycles to failure) using these data.

(b) Determine the fatigue strength at 5×10^5 cycles.

(c) Determine the fatigue life for 200 MPa.

9.34 Suppose that the fatigue data for the brass alloy in Problem 9.33 were taken from torsional tests, and that a shaft of this alloy is to be used for a coupling that is attached to an electric motor operating at 1500 rpm. Give the maximum torsional stress amplitude possible for each of the following lifetimes of the coupling: **(a)** 1 year, **(b)** 1 month, **(c)** 1 day, and **(d)** 2 hours.

9.35 The fatigue data for a ductile cast iron are given as follows:

Stress Amplitude [MPa (ksi)]	Cycles to Failure
248 (36.0)	1×10^5
236 (34.2)	3×10^5
224 (32.5)	1×10^6
213 (30.9)	3×10^6
201 (29.1)	1×10^7
193 (28.0)	3×10^7
193 (28.0)	1×10^8
193 (28.0)	3×10^8

(a) Make an S–N plot (stress amplitude versus logarithm cycles to failure) using the data.

(b) What is the fatigue limit for this alloy?

(c) Determine fatigue lifetimes at stress amplitudes of 230 MPa (33,500 psi) and 175 MPa (25,000 psi).

(d) Estimate fatigue strengths at 2×10^5 and 6×10^6 cycles.

9.36 Suppose that the fatigue data for the cast iron in Problem 9.35 were taken for bending-rotating tests, and that a rod of this alloy is to be used for an automobile axle that rotates at an average rotational velocity of 750 revolutions per minute. Give maximum lifetimes of continuous driving that are allowable for the following stress levels: **(a)** 250 MPa (36,250 psi), **(b)** 215 MPa (31,000 psi), **(c)** 200 MPa (29,000 psi), and **(d)** 150 MPa (21,750 psi).

9.37 Three identical fatigue specimens (denoted A, B, and C) are fabricated from a nonferrous alloy. Each is subjected to one of the maximum-minimum stress cycles listed below; the frequency is the same for all three tests.

Specimen	σ_{max} (MPa)	σ_{min} (MPa)
A	+450	−350
B	+400	−300
C	+340	−340

(a) Rank the fatigue lifetimes of these three specimens from the longest to the shortest.

(b) Now justify this ranking using a schematic S–N plot.

9.38 **(a)** Compare the fatigue limits for polystyrene (Figure 9.27) and the cast iron for which fatigue data are given in Problem 9.35.

(b) Compare the fatigue strengths at 10^6 cycles for polyethylene terephthalate (PET, Figure 9.27) and red brass (Figure 9.46).

9.39 Cite five factors that may lead to scatter in fatigue life data.

9.40 Make a schematic sketch of the fatigue behavior for some metal for which the stress ratio R has a value of $+1$.

9.41 Using Equations 9.23 and 9.24, demonstrate that increasing the value of the stress ratio R produces a decrease in stress amplitude σ_a.

9.42 Surfaces for some steel specimens that have failed by fatigue have a bright crystalline or grainy appearance. Laymen may explain the failure by saying that the metal crystallized while in service. Offer a criticism for this explanation.

9.43 Briefly explain the difference between fatigue striations and beachmarks both in terms of **(a)** size and **(b)** origin.

9.44 List four measures that may be taken to increase the resistance to fatigue of a metal alloy.

9.45 Give the approximate temperature at which creep deformation becomes an important consideration for each of the following metals: nickel, copper, iron, tungsten, lead, and aluminum.

9.46 Superimpose on the same strain-versus-time plot schematic creep curves for both constant tensile stress and constant load, and explain the differences in behavior.

9.47 The following creep data were taken on an aluminum alloy at 400°C (750°F) and a constant stress of 25 MPa (3660 psi). Plot the data as strain versus time, then determine the steady-state or minimum creep rate. *Note:* The initial and instantaneous strain is not included.

Time (min)	Strain	Time (min)	Strain
0	0.000	16	0.135
2	0.025	18	0.153
4	0.043	20	0.172
6	0.065	22	0.193
8	0.078	24	0.218
10	0.092	26	0.255
12	0.109	28	0.307
14	0.120	30	0.368

9.48 A specimen 750 mm (30 in.) long of a low carbon–nickel alloy (Figure 9.43) is to be exposed to a tensile stress of 40 MPa (5800 psi) at 538°C (1000°F). Determine its elongation after 5000 h. Assume that the total of both instantaneous and primary creep elongations is 1.5 mm (0.06 in.).

9.49 For a cylindrical low carbon–nickel alloy specimen (Figure 9.43) originally 10 mm (0.40 in.) in diameter and 500 mm (20 in.) long, what tensile load is necessary to produce a total elongation of 3.2 mm (0.13 in.) after 10,000 h at 427°C (800°F)? Assume that the sum of instantaneous and primary creep elongations is 0.8 mm (0.03 in.).

9.50 If a component fabricated from a low carbon–nickel alloy (Figure 9.42) is to be exposed to a tensile stress of 60 MPa (8700 psi) at 538°C (1000°F), estimate its rupture lifetime.

9.51 A cylindrical component constructed from a low carbon–nickel alloy (Figure 9.42) has a diameter of 12 mm (0.50 in.). Determine the maximum load that may be applied for it to survive 500 h at 649°C (1200°F).

9.52* From Equation 9.33, if the logarithm of $\dot{\epsilon}_s$ is plotted versus the logarithm of σ, then a straight line should result, the slope of which is the stress exponent n. Using Figure 9.43, determine the value of n for the low carbon–nickel alloy at each of the three temperatures.

9.53* **(a)** Estimate the activation energy for creep (i.e., Q_c in Equation 9.34) for the low carbon–nickel alloy having the steady-state creep behavior shown in Figure 9.43. Use data taken at a stress level of 55 MPa (8000 psi) and temperatures of 427°C and 538°C. Assume that the stress exponent n is inde-

pendent of temperature. **(b)** Estimate $\dot{\epsilon}_s$ at 649°C (922 K).

9.54* Steady-state creep rate data are given below for nickel at 1000°C (1273 K):

$\dot{\epsilon}_s$ (s^{-1})	σ[MPa (psi)]
10^{-4}	15 (2175)
10^{-6}	4.5 (650)

If it is known that the activation energy for creep is 272,000 J/mol, compute the steady-state creep rate at a temperature of 850°C (1123 K) and a stress level of 25 MPa (3625 psi).

9.55* Steady-state creep data taken for a stainless steel at a stress level of 70 MPa (10,000 psi) are given as follows:

$\dot{\epsilon}_s$ (s^{-1})	T (K)
1×10^{-5}	977
2.5×10^{-3}	1089

If it is known that the value of the stress exponent n for this alloy is 7.0, compute the steady-state creep rate at 1250 K and a stress level of 50 MPa (7250 psi).

9.56 Cite three metallurgical/processing techniques that are employed to enhance the creep resistance of metal alloys.

Design Problems

9.D1* Consider a flat plate of width 90 mm (3.5 in.) that contains a centrally positioned, through-thickness crack (Figure 9.12) of length (i.e., $2a$) 20 mm (0.8 in.). Determine the minimum plane strain fracture toughness necessary to ensure that fracture will not occur for a design stress of 375 MPa (54,400 psi). The $\pi a/W$ ratio is in radians.

9.D2* A flat plate of some metal alloy contains a centrally positioned, through-thickness crack (Figure 9.12). Determine the critical crack length if the plane strain fracture toughness of the alloy is 38 MPa $\sqrt{m}$ (34.6 ksi $\sqrt{in.}$), the plate width is 50 mm (2 in.), and the design stress is 300 MPa (43,500 psi). The $\pi a/W$ ratio is in radians.

9.D3* Consider a steel plate having a through-thickness edge crack similar to that shown in Figure 9.13a. If it is known that the minimum crack length subject to detection is 2 mm (0.08 in.), determine the minimum allowable plate width assuming a plane strain fracture toughness of 80 MPa $\sqrt{m}$ (72.8 ksi $\sqrt{in.}$), a yield strength of 825 MPa (125,000 psi), and that the plate is to be loaded to one half of its yield strength.

9.D4* Consider a steel plate having a through-thickness edge crack similar to that shown in Figure 9.13a; the plate width (W) is 75 mm (3 in.), and its thickness (B) is 12.0 mm (0.50 in.). Furthermore, plane strain fracture toughness and yield strength values for this material are 80 MPa $\sqrt{m}$ (72.8 ksi $\sqrt{in.}$) and 1200 MPa (175,000 psi), respectively. If the plate is to be loaded to a stress of 300 MPa (43,500 psi), would you expect failure to occur if the crack length a is 15 mm (0.60 in.)? Why or why not?

9.D5* A small and thin flat plate of a brittle material having a through-thickness surface crack is to be loaded in the manner of Figure 9.13c; the K_{Ic} value for this material is 0.45 MPa $\sqrt{m}$ (0.41 ksi $\sqrt{in.}$). For a crack length of 0.25 mm (0.01 in.), determine the maximum load that may be applied without failure for B = 4 mm (0.16 in.), S = 8 mm (0.31 in.), and W = 1 mm (0.04 in.). Assume that the crack is located at the $S/2$ position.

9.D6 **(a)** For the thin-walled spherical tank discussed in Design Example 9.1, on the basis of critical crack size criterion [as addressed in part (a)], rank the following polymers from longest to shortest critical crack length: nylon 6,6 (50% relative humidity), polycarbonate, polyethylene terephthalate, and polymethyl methacrylate. Comment on the magnitude range of the computed values used in the ranking relative to those tabulated for metal alloys as provided in Table 9.2. For these computations, use data contained in Tables B.4 and B.5 in Appendix B.

(b) Now rank these same four polymers relative to maximum allowable pressure according to the leak-before-break criterion, as described in the (b) portion of Design Example 9.1. As above, comment on

these values in relation to those for the metal alloys that are tabulated in Table 9.3.

9.D7* Consider a flat plate of some metal alloy that is to be exposed to repeated tensile-compressive cycling in which the mean stress is 25 MPa. If the initial and critical surface crack lengths are 0.15 and 4.5 mm, respectively, and the values of m and A are 3.5 and 2×10^{-14}, respectively (for $\Delta\sigma$ in MPa and a in m), estimate the maximum tensile stress to yield a fatigue life of 2.5×10^7 cycles. Assume the parameter Y has a value of 1.4, which is independent of crack length.

9.D8* Consider a large, flat plate of a titanium alloy which is to be exposed to reversed tensile-compressive cycles of stress amplitude 100 MPa. If initially the length of the largest surface crack in this specimen is 0.30 mm and the plane strain fracture toughness is 55 MPa $\sqrt{m}$, whereas the values of m and A are 3.0 and 2×10^{-11}, respectively (for $\Delta\sigma$ in MPa and a in m), estimate the fatigue life of this plate. Assume that the parameter Y has a value of 1.45 which is independent of crack length.

9.D9* Consider a metal component that is exposed to cyclic tensile-compressive stresses. If the fatigue lifetime must be a minimum of 1×10^7 cycles and it is known that the maximum initial surface crack length is 0.01 in. and the maximum tensile stress is 15,000 psi, compute the critical surface crack length. Assume that Y is independent of crack length and has a value of 1.75, and that m and A have values of 2.5 and 1.5×10^{-18}, respectively, for $\Delta\sigma$ and a in units of psi and in., respectively.

9.D10* Consider a thin metal plate 20 mm wide which contains a centrally positioned, through-thickness crack in the manner shown in Figure 9.12. This plate is to be exposed to reversed tensile-compressive cycles of stress amplitude 125 MPa. If the initial and critical crack lengths are 0.20 and 8.0 mm, respectively, and the values of m and A are 4 and 5×10^{-12}, respectively (for $\Delta\sigma$ in MPa and a in m), estimate the fatigue life of this plate.

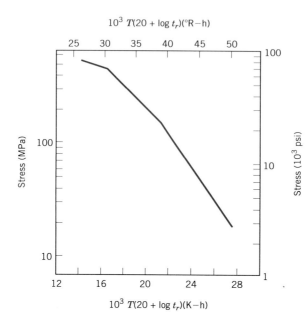

$10^3 \, T(20 + \log t_r)(°\text{R}-\text{h})$

FIGURE 9.47 Logarithm stress versus the Larson–Miller parameter for an 18-8 Mo stainless steel. (From F. R. Larson and J. Miller, *Trans. ASME*, **74**, 765, 1952. Reprinted by permission of ASME.)

$10^3 \, T(20 + \log t_r)(\text{K}-\text{h})$

9.D11* For an edge crack in a plate of finite width (Figure 9.13*a*), *Y* is a function of the crack length–specimen width ratio as

$$Y = \frac{1.1\left(1 - \dfrac{0.2a}{W}\right)}{\left(1 - \dfrac{a}{W}\right)^{3/2}} \qquad (9.36)$$

Now consider a 60 mm wide plate that is exposed to cyclic tensile-compressive stresses (reversed stress cycle) for which $\sigma_{min} = -135$ MPa. Estimate the fatigue life of this plate if the initial and critical crack lengths are 5 mm and 12 mm, respectively. Assume values of 3.5 and 1.5×10^{-12} for the *m* and *A* parameters, respectively, for σ in units of megapascals and *a* in meters.

9.D12* The spherical tank shown in Figure 9.15 is alternately pressurized and depressurized between atmospheric pressure and a positive pressure *p*; thus, fatigue failure is a possibility. Utilizing Equation 9.31, derive an expression for the fatigue life N_f in terms of *p*, the tank radius *r* and thickness *t*, and other parameters subject to the following assumptions: *Y* is independent of crack length, $m \neq 2$, and the original and critical crack lengths are variable parameters.

9.D13* An S-590 iron component (Figure 9.44) must have a creep rupture lifetime of at least 100 days at 500°C (773 K). Compute the maximum allowable stress level.

9.D14* Consider an S-590 iron component (Figure 9.44) that is subjected to a stress of 200 MPa (29,000 psi). At what temperature will the rupture lifetime be 500 h?

9.D15* For an 18-8 Mo stainless steel (Figure 9.47), predict the time to rupture for a component that is subjected to a stress of 80 MPa (11,600 psi) at 700°C (973 K).

9.D16* Consider an 18-8 Mo stainless steel component (Figure 9.47) that is exposed to a temperature of 500°C (773 K). What is the maximum allowable stress level for a rupture lifetime of 5 years? 20 years?

Chapter 10 / Phase Diagrams

A scanning electron micrograph which shows the microstructure of a plain carbon steel that contains 0.44 wt% C. The large dark areas are proeutectoid ferrite. Regions having the alternating light and dark lamellar structure are pearlite; the dark and light layers in the pearlite correspond, respectively, to ferrite and cementite phases. During etching of the surface prior to examination, the ferrite phase was preferentially dissolved; thus, the pearlite appears in topographical relief with cementite layers being elevated above the ferrite layers. 3000×. (Micrograph courtesy of Republic Steel Corporation.)

Why Study *Phase Diagrams?*

One reason why a knowledge and understanding of phase diagrams is important to the engineer relates to the design and control of heat treating procedures; some properties of materials are functions of their microstructures, and, consequently, of their thermal histories. Even though most phase diagrams represent stable (or equilibrium) states and microstructures, they are, nevertheless useful in un-

derstanding the development and preservation of nonequilibrium structures and their attendant properties; it is often the case that these properties are more desirable than those associated with the equilibrium state. This is aptly illustrated by the phenomenon of precipitation hardening (Sections 11.10 and 11.11).

After studying this chapter you should be able to do the following:

1. (a) Schematically sketch simple isomorphous and eutectic phase diagrams.
   (b) On these diagrams label the various phase regions.
   (c) Label liquidus, solidus, and solvus lines.
2. Given a binary phase diagram, the composition of an alloy, its temperature, and assuming that the alloy is at equilibrium, determine:
   (a) what phase(s) is (are) present;
   (b) the composition(s) of the phase(s); and
   (c) the mass fraction(s) of the phase(s).
3. For some given binary phase diagram, do the following:
   (a) locate the temperatures and compositions of all eutectic, eutectoid, peritectic, and congruent phase transformations; and
   (b) write reactions for all these transformations for either heating or cooling.
4. Given the composition of an iron–carbon alloy containing between 0.022 wt% C and 2.14 wt% C, be able to
   (a) specify whether the alloy is hypoeutectoid or hypereutectoid;
   (b) name the proeutectoid phase;
   (c) compute the mass fractions of proeutectoid phase and pearlite; and
   (d) make a schematic diagram of the microstructure at a temperature just below the eutectoid.

10.1 INTRODUCTION

The understanding of phase diagrams for alloy systems is extremely important because there is a strong correlation between microstructure and mechanical properties, and the development of microstructure of an alloy is related to the characteristics of its phase diagram. In addition, phase diagrams provide valuable information about melting, casting, crystallization, and other phenomena.

This chapter presents and discusses the following topics: (1) terminology associated with phase diagrams and phase transformations; (2) the interpretation of phase diagrams; (3) some of the common and relatively simple binary phase diagrams, including that for the iron–carbon system; and (4) the development of equilibrium microstructures, upon cooling, for several situations.

DEFINITIONS AND BASIC CONCEPTS

It is necessary to establish a foundation of definitions and basic concepts relating to alloys, phases, and equilibrium before delving into the interpretation and utilization of phase diagrams. The term **component** is frequently used in this discussion; components are pure metals and/or compounds of which an alloy is composed. For example, in a copper–zinc brass, the components are Cu and Zn. *Solute* and *solvent*, which are also common terms, were defined in Section 5.4. Another term used in this context is **system,** which has two meanings. First, "system" may refer to a specific body of material under consideration (e.g., a ladle of molten steel). Or, it may relate to the series of possible alloys consisting of the same components, but without regard to alloy composition (e.g., the iron–carbon system).

The concept of a solid solution was introduced in Section 5.4. By way of review, a solid solution consists of atoms of at least two different types; the solute atoms occupy either substitutional or interstitial positions in the solvent lattice, and the crystal structure of the solvent is maintained.

10.2 SOLUBILITY LIMIT

For many alloy systems and at some specific temperature, there is a maximum concentration of solute atoms that may dissolve in the solvent to form a solid solution; this is called a **solubility limit.** The addition of solute in excess of this solubility limit results in the formation of another solid solution or compound that has a distinctly different composition. To illustrate this concept, consider the sugar–water ($C_{12}H_{22}O_{11}$–H_2O) system. Initially, as sugar is added to water, a sugar–water solution or syrup forms. As more sugar is introduced, the solution becomes more concentrated, until the solubility limit is reached, or the solution becomes saturated with sugar. At this time the solution is not capable of dissolving any more sugar, and further additions simply settle to the bottom of the container. Thus, the system now consists of two separate substances: a sugar–water syrup liquid solution and solid crystals of undissolved sugar.

This solubility limit of sugar in water depends on the temperature of the water and may be represented in graphical form on a plot of temperature along the ordinate and composition (in weight percent sugar) along the abscissa, as shown in Figure 10.1. Along the composition axis, increasing sugar concentration is from left to right, and percentage of water is read from right to left. Since only two components are involved (sugar and water), the sum of the concentrations at any composition will equal 100 wt%. The solubility limit is represented as the nearly vertical line in the figure. For compositions and temperatures to the left of the solubility line, only the syrup liquid solution exists; to the right of the line, syrup and solid sugar coexist. The solubility limit at some temperature is the composition that corresponds to the intersection of the given temperature coordinate and the solubility limit line. For example, at 20°C the maximum solubility of sugar in water is 65 wt%. As Figure 10.1 indicates, the solubility limit increases slightly with rising temperature.

10.3 PHASES

Also critical to the understanding of phase diagrams is the concept of a **phase.** A phase may be defined as a homogeneous portion of a system that has uniform physical and chemical characteristics. Every pure material is considered to be a

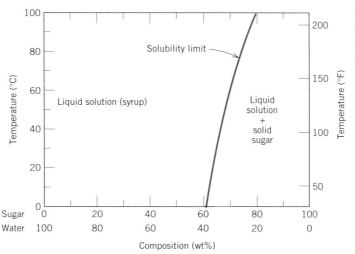

FIGURE 10.1 The solubility of sugar ($C_{12}H_{22}O_{11}$) in a sugar–water syrup.

phase; so also is every solid, liquid, and gaseous solution. For example, the sugar–water syrup solution just discussed is one phase, and solid sugar is another. Each has different physical properties (one is a liquid, the other is a solid); furthermore, each is different chemically (i.e., has a different chemical composition); one is virtually pure sugar, the other is a solution of H_2O and $C_{12}H_{22}O_{11}$. If more than one phase is present in a given system, each will have its own distinct properties, and a boundary separating the phases will exist across which there will be a discontinuous and abrupt change in physical and/or chemical characteristics. When two phases are present in a system, it is not necessary that there be a difference in both physical and chemical properties; a disparity in one or the other set of properties is sufficient. When water and ice are present in a container, two separate phases exist; they are physically dissimilar (one is a solid, the other is a liquid) but identical in chemical makeup. Also, when a substance can exist in two or more polymorphic forms (e.g., having both FCC and BCC structures), each of these structures is a separate phase because their respective physical characteristics differ.

Sometimes, a single-phase system is termed "homogeneous." Systems composed of two or more phases are termed "mixtures" or "heterogeneous systems." Most metallic alloys, and, for that matter, ceramic, polymeric, and composite systems are heterogeneous. Ordinarily, the phases interact in such a way that the property combination of the multiphase system is different from, and more attractive than, either of the individual phases.

10.4 MICROSTRUCTURE

Many times, the physical properties and, in particular, the mechanical behavior of a material depend on the microstructure. Microstructure is subject to direct microscopic observation, using optical or electron microscopes; {this topic was touched on in Section 5.12.} In metal alloys, microstructure is characterized by the number of phases present, their proportions, and the manner in which they are distributed or arranged. The microstructure of an alloy depends on such variables as the alloying elements present, their concentrations, and the heat treatment of the alloy (i.e., the temperature, the heating time at temperature, and the rate of cooling to room temperature).

{The procedure of specimen preparation for microscopic examination was briefly outlined in Section 5.12. After appropriate polishing and etching, the different phases may be distinguished by their appearance.} For example, for a two-phase alloy, one phase may appear light, and the other phase dark, as in the chapter-opening photograph for this chapter. When only a single phase or solid solution is present, the texture will be uniform, except for grain boundaries that may be revealed {(Figure 5.16b).}

10.5 PHASE EQUILIBRIA

Equilibrium is another essential concept. It is best described in terms of a thermodynamic quantity called the **free energy.** In brief, free energy is a function of the internal energy of a system, and also the randomness or disorder of the atoms or molecules (or entropy). A system is at equilibrium if its free energy is at a minimum under some specified combination of temperature, pressure, and composition. In a macroscopic sense, this means that the characteristics of the system do not change with time but persist indefinitely; that is, the system is stable. A change in temperature, pressure, and/or composition for a system in equilibrium will result in an increase in the free energy and in a possible spontaneous change to another state whereby the free energy is lowered.

The term **phase equilibrium,** often used in the context of this discussion, refers to equilibrium as it applies to systems in which more than one phase may exist. Phase equilibrium is reflected by a constancy with time in the phase characteristics of a system. Perhaps an example best illustrates this concept. Suppose that a sugar–water syrup is contained in a closed vessel and the solution is in contact with solid sugar at 20°C. If the system is at equilibrium, the composition of the syrup is 65 wt% $C_{12}H_{22}O_{11}$–35 wt% H_2O (Figure 10.1), and the amounts and compositions of the syrup and solid sugar will remain constant with time. If the temperature of the system is suddenly raised—say, to 100°C—this equilibrium or balance is temporarily upset in that the solubility limit has been increased to 80 wt% $C_{12}H_{22}O_{11}$ (Figure 10.1). Thus, some of the solid sugar will go into solution in the syrup. This will continue until the new equilibrium syrup concentration is established at the higher temperature.

This sugar–syrup example has illustrated the principle of phase equilibrium using a liquid–solid system. In many metallurgical and materials systems of interest, phase equilibrium involves just solid phases. In this regard the state of the system is reflected in the characteristics of the microstructure, which necessarily include not only the phases present and their compositions but, in addition, the relative phase amounts and their spatial arrangement or distribution.

Free energy considerations and diagrams similar to Figure 10.1 provide information about the equilibrium characteristics of a particular system, which is important; but they do not indicate the time period necessary for the attainment of a new equilibrium state. It is often the case, especially in solid systems, that a state of equilibrium is never completely achieved because the rate of approach to equilibrium is extremely slow; such a system is said to be in a nonequilibrium or **metastable** state. A metastable state or microstructure may persist indefinitely, experiencing only extremely slight and almost imperceptible changes as time progresses. Often, metastable structures are of more practical significance than equilibrium ones. For example, some steel and aluminum alloys rely for their strength on the development of metastable microstructures during carefully designed heat treatments (Sections 11.5 and 11.10).

Thus not only is an understanding of equilibrium states and structures important, but the speed or rate at which they are established and, in addition, the factors that affect the rate must be considered. This chapter is devoted almost exclusively to equilibrium structures; the treatment of reaction rates and nonequilibrium structures is deferred to Chapter 11.

EQUILIBRIUM PHASE DIAGRAMS

Much of the information about the control of microstructure or phase structure of a particular alloy system is conveniently and concisely displayed in what is called a **phase diagram,** also often termed an *equilibrium* or *constitutional diagram.* Many microstructures develop from phase transformations, the changes that occur between phases when the temperature is altered (ordinarily upon cooling). This may involve the transition from one phase to another, or the appearance or disappearance of a phase. Phase diagrams are helpful in predicting phase transformations and the resulting microstructures, which may have equilibrium or nonequilibrium character.

Equilibrium phase diagrams represent the relationships between temperature and the compositions and the quantities of phases at equilibrium. There are several different varieties; but in the present discussion, temperature and composition are the variable parameters, for binary alloys. A binary alloy is one that contains two components. If more than two components are present, phase diagrams become extremely complicated and difficult to represent. The principles of microstructural control with the aid of phase diagrams can be illustrated with binary alloys even though, in reality, most alloys contain more than two components. External pressure is also a parameter that influences the phase structure. However, in practicality, pressure remains virtually constant in most applications; thus, the phase diagrams presented here are for a constant pressure of one atmosphere (1 atm).

10.6 BINARY ISOMORPHOUS SYSTEMS

Possibly the easiest type of binary phase diagram to understand and interpret is that which is characterized by the copper–nickel system (Figure 10.2a). Temperature is plotted along the ordinate, and the abscissa represents the composition of the alloy, in weight percent (bottom) and atom percent (top) of nickel. The composition ranges from 0 wt% Ni (100 wt% Cu) on the left horizontal extremity to 100 wt% Ni (0 wt% Cu) on the right. Three different phase regions, or fields, appear on the diagram, an alpha (α) field, a liquid (L) field, and a two-phase $\alpha + L$ field. Each region is defined by the phase or phases that exist over the range of temperatures and compositions delimited by the phase boundary lines.

The liquid L is a homogeneous liquid solution composed of both copper and nickel. The α phase is a substitutional solid solution consisting of both Cu and Ni atoms, and having an FCC crystal structure. At temperatures below about 1080°C, copper and nickel are mutually soluble in each other in the solid state for all compositions. This complete solubility is explained by the fact that both Cu and Ni have the same crystal structure (FCC), nearly identical atomic radii and electronegativities, and similar valences, as discussed in Section 5.4. The copper–nickel system is termed **isomorphous** because of this complete liquid and solid solubility of the two components.

A couple of comments are in order regarding nomenclature. First, for metallic alloys, solid solutions are commonly designated by lowercase Greek letters (α, β, γ, etc.). Furthermore, with regard to phase boundaries, the line separating the L and $\alpha + L$ phase fields is termed the *liquidus line,* as indicated in Figure 10.2a; the liquid phase is present at all temperatures and compositions above this line. The *solidus line* is located between the α and $\alpha + L$ regions, below which only the solid α phase exists.

For Figure 10.2a, the solidus and liquidus lines intersect at the two composition extremities; these correspond to the melting temperatures of the pure components. For example, the melting temperatures of pure copper and nickel are 1085°C and 1453°C, respectively. Heating pure copper corresponds to moving vertically up the left-hand temperature axis. Copper remains solid until its melting temperature is reached. The solid-to-liquid transformation takes place at the melting temperature, and no further heating is possible until this transformation has been completed.

For any composition other than pure components, this melting phenomenon will occur over the range of temperatures between the solidus and liquidus lines;

FIGURE 10.2 (*a*) The copper–nickel phase diagram. (Adapted from *Phase Diagrams of Binary Nickel Alloys*, P. Nash, Editor, 1991. Reprinted by permission of ASM International, Materials Park, OH.) (*b*) A portion of the copper–nickel phase diagram for which compositions and phase amounts are determined at point *B*.

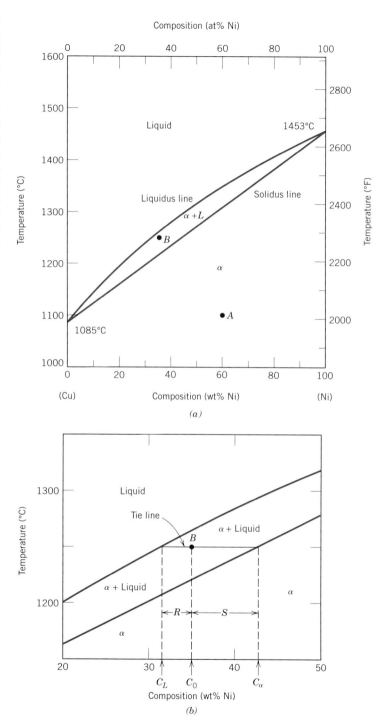

both solid α and liquid phases will be in equilibrium within this temperature range. For example, upon heating an alloy of composition 50 wt% Ni–50 wt% Cu (Figure 10.2*a*), melting begins at approximately 1280°C (2340°F); the amount of liquid phase continuously increases with temperature until about 1320°C (2410°F), at which the alloy is completely liquid.

10.7 INTERPRETATION OF PHASE DIAGRAMS

For a binary system of known composition and temperature that is at equilibrium, at least three kinds of information are available: (1) the phases that are present, (2) the compositions of these phases, and (3) the percentages or fractions of the phases. The procedures for making these determinations will be demonstrated using the copper–nickel system.

PHASES PRESENT

The establishment of what phases are present is relatively simple. One just locates the temperature–composition point on the diagram and notes the phase(s) with which the corresponding phase field is labeled. For example, an alloy of composition 60 wt% Ni–40 wt% Cu at 1100°C would be located at point A in Figure 10.2a; since this is within the α region, only the single α phase will be present. On the other hand, a 35 wt% Ni–65 wt% Cu alloy at 1250°C (point B) will consist of both α and liquid phases at equilibrium.

DETERMINATION OF PHASE COMPOSITIONS

The first step in the determination of phase compositions (in terms of the concentrations of the components) is to locate the temperature–composition point on the phase diagram. Different methods are used for single- and two-phase regions. If only one phase is present, the procedure is trivial: the composition of this phase is simply the same as the overall composition of the alloy. For example, consider the 60 wt% Ni–40 wt% Cu alloy at 1100°C (point A, Figure 10.2a). At this composition and temperature, only the α phase is present, having a composition of 60 wt% Ni–40 wt% Cu.

For an alloy having composition and temperature located in a two-phase region, the situation is more complicated. In all two-phase regions (and in two-phase regions only), one may imagine a series of horizontal lines, one at every temperature; each of these is known as a **tie line,** or sometimes as an isotherm. These tie lines extend across the two-phase region and terminate at the phase boundary lines on either side. To compute the equilibrium concentrations of the two phases, the following procedure is used:

1. A tie line is constructed across the two-phase region at the temperature of the alloy.
2. The intersections of the tie line and the phase boundaries on either side are noted.
3. Perpendiculars are dropped from these intersections to the horizontal composition axis, from which the composition of each of the respective phases is read.

For example, consider again the 35 wt% Ni–65 wt% Cu alloy at 1250°C, located at point B in Figure 10.2b and lying within the $\alpha + L$ region. Thus, the problem is to determine the composition (in wt% Ni and Cu) for both the α and liquid phases. The tie line has been constructed across the $\alpha + L$ phase region, as shown in Figure 10.2b. The perpendicular from the intersection of the tie line with the liquidus boundary meets the composition axis at 31.5 wt% Ni–68.5 wt% Cu, which is the composition of the liquid phase, C_L. Likewise, for the solidus–tie line intersection, we find a composition for the α solid-solution phase, C_α, of 42.5 wt% Ni–57.5 wt% Cu.

DETERMINATION OF PHASE AMOUNTS

The relative amounts (as fraction or as percentage) of the phases present at equilibrium may also be computed with the aid of phase diagrams. Again, the single- and two-phase situations must be treated separately. The solution is obvious in the single-phase region: Since only one phase is present, the alloy is composed entirely of that phase; that is, the phase fraction is 1.0 or, alternatively, the percentage is 100%. From the previous example for the 60 wt% Ni–40 wt% Cu alloy at 1100°C (point A in Figure 10.2a), only the α phase is present; hence, the alloy is completely or 100% α.

If the composition and temperature position is located within a two-phase region, things are more complex. The tie line must be utilized in conjunction with a procedure that is often called the **lever rule** (or the *inverse lever rule*), which is applied as follows:

1. The tie line is constructed across the two-phase region at the temperature of the alloy.
2. The overall alloy composition is located on the tie line.
3. The fraction of one phase is computed by taking the length of tie line from the overall alloy composition to the phase boundary for the *other* phase, and dividing by the total tie line length.
4. The fraction of the other phase is determined in the same manner.
5. If phase percentages are desired, each phase fraction is multiplied by 100. When the composition axis is scaled in weight percent, the phase fractions computed using the lever rule are mass fractions—the mass (or weight) of a specific phase divided by the total alloy mass (or weight). The mass of each phase is computed from the product of each phase fraction and the total alloy mass.

In the employment of the lever rule, tie line segment lengths may be determined either by direct measurement from the phase diagram using a linear scale, preferably graduated in millimeters, or by subtracting compositions as taken from the composition axis.

Consider again the example shown in Figure 10.2b, in which at 1250°C both α and liquid phases are present for a 35 wt% Ni–65 wt% Cu alloy. The problem is to compute the fraction of each of the α and liquid phases. The tie line has been constructed that was used for the determination of α and L phase compositions. Let the overall alloy composition be located along the tie line and denoted as C_0, and mass fractions be represented by W_L and W_α for the respective phases. From the lever rule, W_L may be computed according to

$$W_L = \frac{S}{R + S}$$

(10.1a)

or, by subtracting compositions,

$$W_L = \frac{C_\alpha - C_0}{C_\alpha - C_L}$$

(10.1b)

Composition need be specified in terms of only one of the constituents for a binary alloy; for the computation above, weight percent nickel will be used (i.e., $C_0 = 35$

wt% Ni, $C_\alpha = 42.5$ wt% Ni, and $C_L = 31.5$ wt% Ni), and

$$W_L = \frac{42.5 - 35}{42.5 - 31.5} = 0.68$$

Similarly, for the α phase,

$$W_\alpha = \frac{R}{R + S} \tag{10.2a}$$

$$= \frac{C_0 - C_L}{C_\alpha - C_L} \tag{10.2b}$$

$$= \frac{35 - 31.5}{42.5 - 31.5} = 0.32$$

Of course, identical answers are obtained if compositions are expressed in weight percent copper instead of nickel.

Thus, the lever rule may be employed to determine the relative amounts or fractions of phases in any two-phase region for a binary alloy if the temperature and composition are known and if equilibrium has been established. Its derivation is presented as an example problem.

It is easy to confuse the foregoing procedures for the determination of phase compositions and fractional phase amounts; thus, a brief summary is warranted. *Compositions* of phases are expressed in terms of weight percents of the components (e.g., wt% Cu, wt% Ni). For any alloy consisting of a single phase, the composition of that phase is the same as the total alloy composition. If two phases are present, the tie line must be employed, the extremities of which determine the compositions of the respective phases. With regard to *fractional phase amounts* (e.g., mass fraction of the α or liquid phase), when a single phase exists, the alloy is completely that phase. For a two-phase alloy, on the other hand, the lever rule is utilized, in which a ratio of tie line segment lengths is taken.

EXAMPLE PROBLEM 10.1

Derive the lever rule.

SOLUTION

Consider the phase diagram for copper and nickel (Figure 10.2*b*) and alloy of composition C_0 at 1250°C, and let C_α, C_L, W_α, and W_L represent the same parameters as above. This derivation is accomplished through two conservation-of-mass expressions. With the first, since only two phases are present, the sum of their mass fractions must be equal to unity, that is,

$$W_\alpha + W_L = 1 \tag{10.3}$$

For the second, the mass of one of the components (either Cu or Ni) that is present in both of the phases must be equal to the mass of that component in the total alloy, or

$$W_\alpha C_\alpha + W_L C_L = C_0 \tag{10.4}$$

Simultaneous solution of these two equations leads to the lever rule expressions for this particular situation, Equations 10.1b and 10.2b:

$$W_L = \frac{C_\alpha - C_0}{C_\alpha - C_L} \tag{10.1b}$$

$$W_\alpha = \frac{C_0 - C_L}{C_\alpha - C_L} \tag{10.2b}$$

For multiphase alloys, it is often more convenient to specify relative phase amount in terms of volume fraction rather than mass fraction. Phase volume fractions are preferred because they (rather than mass fractions) may be determined from examination of the microstructure; furthermore, the properties of a multiphase alloy may be estimated on the basis of volume fractions.

For an alloy consisting of α and β phases, the volume fraction of the α phase, V_α, is defined as

$$V_\alpha = \frac{v_\alpha}{v_\alpha + v_\beta} \tag{10.5}$$

where v_α and v_β denote the volumes of the respective phases in the alloy. Of course, an analogous expression exists for V_β; and, for an alloy consisting of just two phases, it is the case that $V_\alpha + V_\beta = 1$.

On occasion conversion from mass fraction to volume fraction (or vice versa) is desired. Equations that facilitate these conversions are as follows:

$$V_\alpha = \frac{\dfrac{W_\alpha}{\rho_\alpha}}{\dfrac{W_\alpha}{\rho_\alpha} + \dfrac{W_\beta}{\rho_\beta}} \tag{10.6a}$$

$$V_\beta = \frac{\dfrac{W_\beta}{\rho_\beta}}{\dfrac{W_\alpha}{\rho_\alpha} + \dfrac{W_\beta}{\rho_\beta}} \tag{10.6b}$$

and

$$W_\alpha = \frac{V_\alpha \rho_\alpha}{V_\alpha \rho_\alpha + V_\beta \rho_\beta} \tag{10.7a}$$

$$W_\beta = \frac{V_\beta \rho_\beta}{V_\alpha \rho_\alpha + V_\beta \rho_\beta} \tag{10.7b}$$

In these expressions, ρ_α and ρ_β are the densities of the respective phases; {these may be determined approximately using Equations 5.10a and 5.10b.}

When the densities of the phases in a two-phase alloy differ significantly, there will be quite a disparity between mass and volume fractions; conversely, if the phase densities are the same, mass and volume fractions are identical.

10.8 DEVELOPMENT OF MICROSTRUCTURE IN ISOMORPHOUS ALLOYS

EQUILIBRIUM COOLING (CD-ROM)

NONEQUILIBRIUM COOLING (CD-ROM)

10.9 MECHANICAL PROPERTIES OF ISOMORPHOUS ALLOYS

We shall now briefly explore how the mechanical properties of solid isomorphous alloys are affected by composition as other structural variables (e.g., grain size) are held constant. For all temperatures and compositions below the melting temperature of the lowest-melting component, only a single solid phase will exist. Therefore, each component will experience solid-solution strengthening (Section 8.10), or an increase in strength and hardness by additions of the other component. This effect is demonstrated in Figure 10.5a as tensile strength versus composition for the copper–nickel system at room temperature; at some intermediate composition, the curve necessarily passes through a maximum. Plotted in Figure 10.5b is the ductility (%EL)–composition behavior, which is just the opposite of tensile strength; that is, ductility decreases with additions of the second component, and the curve exhibits a minimum.

10.10 BINARY EUTECTIC SYSTEMS

Another type of common and relatively simple phase diagram found for binary alloys is shown in Figure 10.6 for the copper–silver system; this is known as a binary eutectic phase diagram. A number of features of this phase diagram are important and worth noting. First of all, three single-phase regions are found on the diagram: α, β, and liquid. The α phase is a solid solution rich in copper; it has silver as the solute component and an FCC crystal structure. The β phase solid solution also

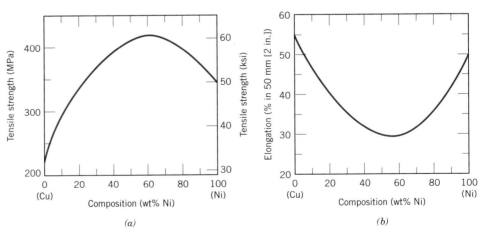

FIGURE 10.5 For the copper–nickel system, (a) tensile strength versus composition, and (b) ductility (%EL) versus composition at room temperature. A solid solution exists over all compositions for this system.

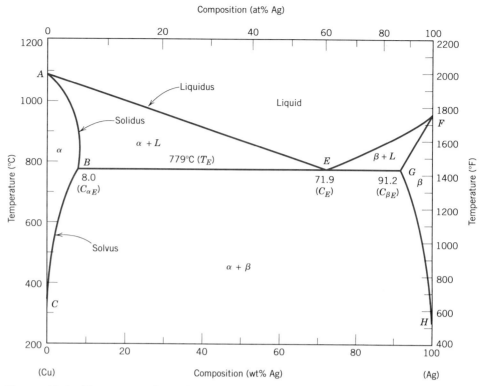

FIGURE 10.6   The copper–silver phase diagram. (Adapted from *Binary Alloy Phase Diagrams*, 2nd edition, Vol. 1, T. B. Massalski, Editor-in-Chief, 1990. Reprinted by permission of ASM International, Materials Park, OH.)

has an FCC structure, but copper is the solute. The α and β phases are considered to include pure copper and pure silver, respectively.

Thus, the solubility in each of these solid phases is limited, in that at any temperature below line *BEG* only a limited concentration of silver will dissolve in copper (for the α phase), and similarly for copper in silver (for the β phase). The solubility limit for the α phase corresponds to the boundary line, labeled *CBA*, between the $\alpha/(\alpha + \beta)$ and $\alpha/(\alpha + L)$ phase regions; it increases with temperature to a maximum [8.0 wt% Ag at 779°C (1434°F)] at point *B*, and decreases back to zero at the melting temperature of pure copper, point *A* [1085°C (1985°F)]. At temperatures below 779°C (1434°F), the solid solubility limit line separating the α and $\alpha + \beta$ phase regions is termed a **solvus line;** the boundary *AB* between the α and $\alpha + L$ fields is the **solidus line,** as indicated in Figure 10.6. For the β phase, both solvus and solidus lines also exist, *HG* and *GF*, respectively, as shown. The maximum solubility of copper in the β phase, point *G* (8.8 wt% Cu), also occurs at 779°C (1434°F). This horizontal line *BEG*, which is parallel to the composition axis and extends between these maximum solubility positions, may also be considered to be a solidus line; it represents the lowest temperature at which a liquid phase may exist for any copper–silver alloy that is at equilibrium.

There are also three two-phase regions found for the copper–silver system (Figure 10.6): $\alpha + L$, $\beta + L$, and $\alpha + \beta$. The α and β phase solid solutions coexist for all compositions and temperatures within the $\alpha + \beta$ phase field; the α + liquid and β + liquid phases also coexist in their respective phase regions. Furthermore, compositions and relative amounts for the phases may be determined using tie lines and the lever rule as outlined previously.

As silver is added to copper, the temperature at which the alloys become totally liquid decreases along the liquidus line, line AE; thus, the melting temperature of copper is lowered by silver additions. The same may be said for silver: the introduction of copper reduces the temperature of complete melting along the other liquidus line, FE. These liquidus lines meet at the point E on the phase diagram, through which also passes the horizontal isotherm line BEG. Point E is called an **invariant point,** which is designated by the composition C_E and temperature T_E; for the copper–silver system, the values of C_E and T_E are 71.9 wt% Ag and 779°C (1434°F), respectively.

An important reaction occurs for an alloy of composition C_E as it changes temperature in passing through T_E; this reaction may be written as follows:

$$L(C_E) \underset{\text{heating}}{\overset{\text{cooling}}{\rightleftharpoons}} \alpha(C_{\alpha E}) + \beta(C_{\beta E}) \tag{10.8}$$

Or, upon cooling, a liquid phase is transformed into the two solid α and β phases at the temperature T_E; the opposite reaction occurs upon heating. This is called a **eutectic reaction** (eutectic means easily melted), and C_E and T_E represent the eutectic composition and temperature, respectively; $C_{\alpha E}$ and $C_{\beta E}$ are the respective compositions of the α and β phases at T_E. Thus, for the copper–silver system, the eutectic reaction, Equation 10.8, may be written as follows:

$$L(71.9\,\text{wt\% Ag}) \underset{\text{heating}}{\overset{\text{cooling}}{\rightleftharpoons}} \alpha(8.0\,\text{wt\% Ag}) + \beta(91.2\,\text{wt\% Ag})$$

Often, the horizontal solidus line at T_E is called the *eutectic isotherm.*

The eutectic reaction, upon cooling, is similar to solidification for pure components in that the reaction proceeds to completion at a constant temperature, or isothermally, at T_E. However, the solid product of eutectic solidification is always two solid phases, whereas for a pure component only a single phase forms. Because of this eutectic reaction, phase diagrams similar to that in Figure 10.6 are termed eutectic phase diagrams; components exhibiting this behavior comprise a eutectic system.

In the construction of binary phase diagrams, it is important to understand that one or at most two phases may be in equilibrium within a phase field. This holds true for the phase diagrams in Figures 10.2a and 10.6. For a eutectic system, three phases (α, β, and L) may be in equilibrium, but only at points along the eutectic isotherm. Another general rule is that single-phase regions are always separated from each other by a two-phase region that consists of the two single phases that it separates. For example, the $\alpha + \beta$ field is situated between the α and β single-phase regions in Figure 10.6.

Another common eutectic system is that for lead and tin; the phase diagram (Figure 10.7) has a general shape similar to that for copper–silver. For the lead–tin system the solid solution phases are also designated by α and β; in this case, α represents a solid solution of tin in lead, and for β, tin is the solvent and lead is the solute. The eutectic invariant point is located at 61.9 wt% Sn and 183°C (361°F). Of course, maximum solid solubility compositions as well as component melting temperatures will be different for the copper–silver and lead–tin systems, as may be observed by comparing their phase diagrams.

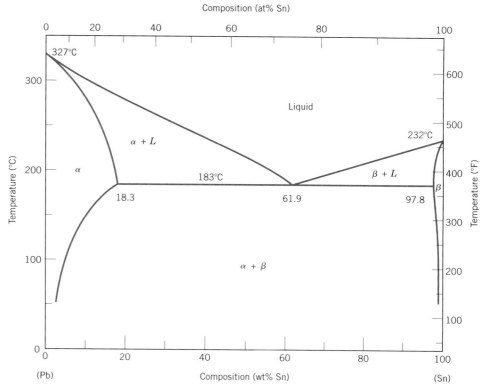

FIGURE 10.7 The lead–tin phase diagram. (Adapted from *Binary Alloy Phase Diagrams,* 2nd edition, Vol. 3, T. B. Massalski, Editor-in-Chief, 1990. Reprinted by permission of ASM International, Materials Park, OH.)

On occasion, low-melting-temperature alloys are prepared having near-eutectic compositions. A familiar example is the 60–40 solder, containing 60 wt% Sn and 40 wt% Pb. Figure 10.7 indicates that an alloy of this composition is completely molten at about 185°C (365°F), which makes this material especially attractive as a low-temperature solder, since it is easily melted.

EXAMPLE PROBLEM 10.2

For a 40 wt% Sn–60 wt% Pb alloy at 150°C (300°F), (a) What phase(s) is (are) present? (b) What is (are) the composition(s) of the phase(s)?

SOLUTION

(a) Locate this temperature–composition point on the phase diagram (point B in Figure 10.8). Inasmuch as it is within the $\alpha + \beta$ region, both α and β phases will coexist.

(b) Since two phases are present, it becomes necessary to construct a tie line across the $\alpha + \beta$ phase field at 150°C, as indicated in Figure 10.8. The composition of the α phase corresponds to the tie line intersection with the $\alpha/(\alpha + \beta)$ solvus phase boundary—about 10 wt% Sn–90 wt% Pb, denoted as C_α. Similarly for the β phase, which will have a composition approximately 98 wt% Sn–2 wt% Pb (C_β).

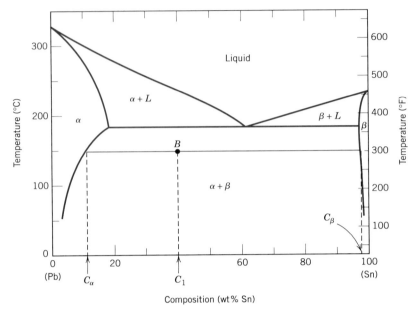

FIGURE 10.8   The lead–tin phase diagram. For a 40 wt% Sn–60 wt% Pb alloy at 150°C (point B), phase compositions and relative amounts are computed in Example Problems 10.2 and 10.3.

EXAMPLE PROBLEM 10.3

For the lead–tin alloy in Example Problem 10.2, calculate the relative amount of each phase present in terms of (a) mass fraction and (b) volume fraction. At 150°C take the densities of Pb and Sn to be 11.23 and 7.24 g/cm³, respectively.

SOLUTION

(a) Since the alloy consists of two phases, it is necessary to employ the lever rule. If C_1 denotes the overall alloy composition, mass fractions may be computed by subtracting compositions, in terms of weight percent tin, as follows:

$$W_\alpha = \frac{C_\beta - C_1}{C_\beta - C_\alpha} = \frac{98 - 40}{98 - 10} = 0.66$$

$$W_\beta = \frac{C_1 - C_\alpha}{C_\beta - C_\alpha} = \frac{40 - 10}{98 - 10} = 0.34$$

(b) To compute volume fractions it is first necessary to determine the density of each phase {using Equation 5.10a.} Thus

$$\rho_\alpha = \frac{100}{\dfrac{C_{Sn(\alpha)}}{\rho_{Sn}} + \dfrac{C_{Pb(\alpha)}}{\rho_{Pb}}}$$

where $C_{Sn(\alpha)}$ and $C_{Pb(\alpha)}$ denote the concentrations in weight percent of tin and lead, respectively, in the α phase. From Example Problem 10.2, these values are 10 wt% and 90 wt%. Incorporation of these values along with the densities of the two components lead to

$$\rho_\alpha = \frac{100}{\dfrac{10}{7.24 \text{ g/cm}^3} + \dfrac{90}{11.23 \text{ g/cm}^3}} = 10.64 \text{ g/cm}^3$$

Similarly for the β phase

$$\rho_\beta = \frac{100}{\dfrac{C_{Sn(\beta)}}{\rho_{Sn}} + \dfrac{C_{Pb(\beta)}}{\rho_{Pb}}}$$

$$= \frac{100}{\dfrac{98}{7.24 \text{ g/cm}^3} + \dfrac{2}{11.23 \text{ g/cm}^3}} = 7.29 \text{ g/cm}^3$$

Now it becomes necessary to employ Equations 10.6a and 10.6b to determine V_α and V_β as

$$V_\alpha = \frac{\dfrac{W_\alpha}{\rho_\alpha}}{\dfrac{W_\alpha}{\rho_\alpha} + \dfrac{W_\beta}{\rho_\beta}}$$

$$= \frac{\dfrac{0.66}{10.64 \text{ g/cm}^3}}{\dfrac{0.66}{10.64 \text{ g/cm}^3} + \dfrac{0.34}{7.29 \text{ g/cm}^3}} = 0.57$$

$$V_\beta = \frac{\dfrac{W_\beta}{\rho_\beta}}{\dfrac{W_\alpha}{\rho_\alpha} + \dfrac{W_\beta}{\rho_\beta}}$$

$$= \frac{\dfrac{0.34}{7.29 \text{ g/cm}^3}}{\dfrac{0.66}{10.64 \text{ g/cm}^3} + \dfrac{0.34}{7.29 \text{ g/cm}^3}} = 0.43$$

10.11 DEVELOPMENT OF MICROSTRUCTURE IN EUTECTIC ALLOYS (CD-ROM)

10.12 EQUILIBRIUM DIAGRAMS HAVING INTERMEDIATE PHASES OR COMPOUNDS

The isomorphous and eutectic phase diagrams discussed thus far are relatively simple, but those for many binary alloy systems are much more complex. The eutectic copper–silver and lead–tin phase diagrams (Figures 10.6 and 10.7) have only two solid phases, α and β; these are sometimes termed **terminal solid solutions,** because they exist over composition ranges near the concentration extremities of the phase diagram. For other alloy systems, **intermediate solid solutions** (or *intermediate phases*) may be found at other than the two composition extremes. Such is the case for the copper–zinc system. Its phase diagram (Figure 10.17) may at first appear formidable because there are some invariant points and reactions similar to the eutectic that have not yet been discussed. In addition, there are six different solid solutions—two terminal (α and η) and four intermediate (β, γ, δ, and ϵ). (The β' phase is termed an ordered solid solution, one in which the copper and zinc atoms are situated in a specific and ordered arrangement within each unit cell.) Some phase boundary lines near the bottom of Figure 10.17 are dashed to indicate that their positions have not been exactly determined. The reason for this is that at low temperatures, diffusion rates are very slow and inordinately long times are required for the attainment of equilibrium. Again, only single- and two-phase regions are found on the diagram, and the same rules outlined in Section 10.7 are utilized for computing phase compositions and relative amounts. The commercial brasses are copper-rich copper–zinc alloys; for example, cartridge brass has a composition of 70 wt% Cu–30 wt% Zn and a microstructure consisting of a single α phase.

For some systems, discrete intermediate compounds rather than solid solutions may be found on the phase diagram, and these compounds have distinct chemical formulas; for metal–metal systems, they are called **intermetallic compounds.** For example, consider the magnesium–lead system (Figure 10.18). The compound Mg_2Pb has a composition of 19 wt% Mg–81 wt% Pb (33 at% Pb), and is represented as a vertical line on the diagram, rather than as a phase region of finite width; hence, Mg_2Pb can exist by itself only at this precise composition.

Several other characteristics are worth noting for this magnesium–lead system. First, the compound Mg_2Pb melts at approximately 550°C (1020°F), as indicated by point M in Figure 10.18. Also, the solubility of lead in magnesium is rather extensive, as indicated by the relatively large composition span for the α phase field. On the other hand, the solubility of magnesium in lead is extremely limited. This is evident from the very narrow β terminal solid-solution region on the right or lead-rich side of the diagram. Finally, this phase diagram may be thought of as two simple eutectic diagrams joined back to back, one for the Mg–Mg_2Pb system, the other for Mg_2Pb–Pb; as such, the compound Mg_2Pb is really considered to be a component. This separation of complex phase diagrams into smaller-component units may simplify them and, furthermore, expedite their interpretation.

10.13 EUTECTOID AND PERITECTIC REACTIONS

In addition to the eutectic, other invariant points involving three different phases are found for some alloy systems. One of these occurs for the copper–zinc system (Figure 10.17) at 560°C (1040°F) and 74 wt% Zn–26 wt% Cu. A portion of the phase diagram in this vicinity appears enlarged in Figure 10.19. Upon cooling, a solid δ phase transforms into two other solid phases (γ and ϵ) according to the reaction

$$\delta \underset{\text{heating}}{\overset{\text{cooling}}{\rightleftharpoons}} \gamma + \epsilon \qquad (10.14)$$

The reverse reaction occurs upon heating. It is called a **eutectoid** (or eutecticlike) **reaction,** and the invariant point (point E, Figure 10.19) and the horizontal tie line

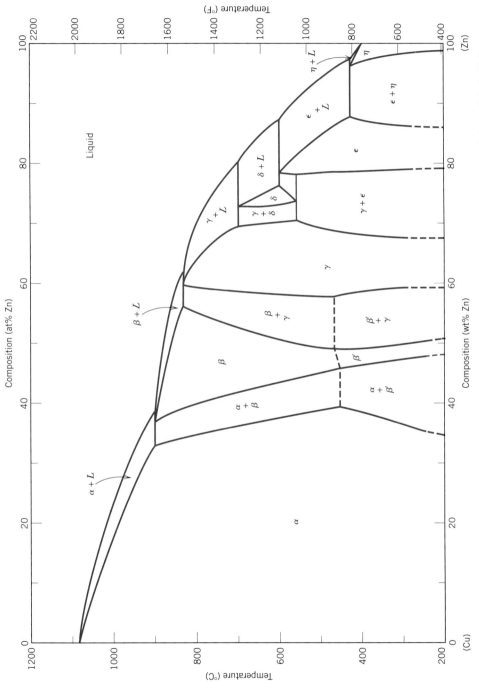

FIGURE 10.17 The copper–zinc phase diagram. (Adapted from *Binary Alloy Phase Diagrams*, 2nd edition, Vol. 2, T. B. Massalski, Editor-in-Chief, 1990. Reprinted by permission of ASM International, Materials Park, OH.)

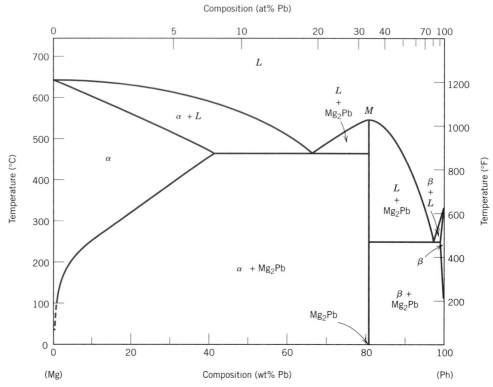

FIGURE 10.18 The magnesium–lead phase diagram. (Adapted from *Phase Diagrams of Binary Magnesium Alloys,* A. A. Nayeb-Hashemi and J. B. Clark, Editors, 1988. Reprinted by permission of ASM International, Materials Park, OH.)

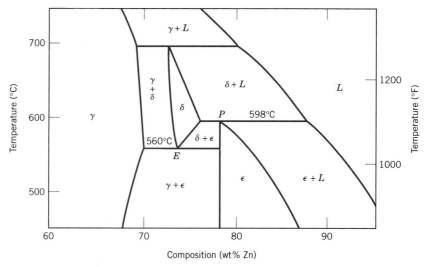

FIGURE 10.19 A region of the copper–zinc phase diagram that has been enlarged to show eutectoid and peritectic invariant points, labeled *E* (560°C, 74 wt% Zn) and *P* (598°C, 78.6 wt% Zn), respectively. (Adapted from *Binary Alloy Phase Diagrams,* 2nd edition, Vol. 2, T. B. Massalski, Editor-in-Chief, 1990. Reprinted by permission of ASM International, Materials Park, OH.)

at 560°C are termed the *eutectoid* and *eutectoid isotherm,* respectively. The feature distinguishing "eutectoid" from "eutectic" is that one solid phase instead of a liquid transforms into two other solid phases at a single temperature. A eutectoid reaction is found in the iron–carbon system (Section 10.18) that is very important in the heat treating of steels.

The **peritectic reaction** is yet another invariant reaction involving three phases at equilibrium. With this reaction, upon heating, one solid phase transforms into a liquid phase and another solid phase. A peritectic exists for the copper–zinc system (Figure 10.19, point *P*) at 598°C (1108°F) and 78.6 wt% Zn–21.4 wt% Cu; this reaction is as follows:

$$\delta + L \underset{\text{heating}}{\overset{\text{cooling}}{\rightleftharpoons}} \epsilon \qquad (10.15)$$

The low-temperature solid phase may be an intermediate solid solution (e.g., ϵ in the above reaction), or it may be a terminal solid solution. One of the latter peritectics exists at about 97 wt% Zn and 435°C (815°F) (see Figure 10.17), wherein the η phase, when heated, transforms to ϵ and liquid phases. Three other peritectics are found for the Cu–Zn system, the reactions of which involve β, δ, and γ intermediate solid solutions as the low-temperature phases that transform upon heating.

10.14 CONGRUENT PHASE TRANSFORMATIONS

Phase transformations may be classified according to whether or not there is any change in composition for the phases involved. Those for which there are no compositional alterations are said to be **congruent transformations.** Conversely, for *incongruent transformations,* at least one of the phases will experience a change in composition. Examples of congruent transformations include allotropic transformations (Section 3.10) and melting of pure materials. Eutectic and eutectoid reactions, as well as the melting of an alloy that belongs to an isomorphous system, all represent incongruent transformations.

Intermediate phases are sometimes classified on the basis of whether they melt congruently or incongruently. The intermetallic compound Mg_2Pb melts congruently at the point designated *M* on the magnesium–lead phase diagram, Figure 10.18. Also, for the nickel–titanium system, Figure 10.20, there is a congruent melting point for the γ solid solution that corresponds to the point of tangency for the pairs of liquidus and solidus lines, at 1310°C and 44.9 wt% Ti. Furthermore, the peritectic reaction is an example of incongruent melting for an intermediate phase.

10.15 CERAMIC PHASE DIAGRAMS (CD-ROM)

10.16 TERNARY PHASE DIAGRAMS

Phase diagrams have also been determined for metallic (as well as ceramic) systems containing more than two components; however, their representation and interpretation may be exceedingly complex. For example, a ternary, or three-component, composition–temperature phase diagram in its entirety is depicted by a three-dimensional model. Portrayal of features of the diagram or model in two dimensions is possible but somewhat difficult.

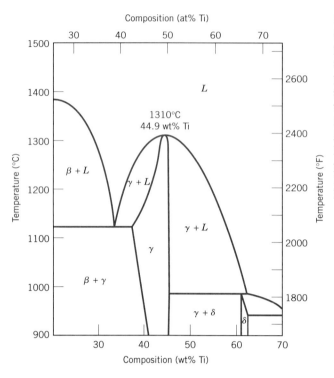

FIGURE 10.20 A portion of the nickel–titanium phase diagram on which is shown a congruent melting point for the γ phase solid solution at 1310°C and 44.9 wt% Ti. (Adapted from *Phase Diagrams of Binary Nickel Alloys,* P. Nash, Editor, 1991. Reprinted by permission of ASM International, Materials Park, OH.)

10.17 THE GIBBS PHASE RULE (CD-ROM)

THE IRON–CARBON SYSTEM

Of all binary alloy systems, the one that is possibly the most important is that for iron and carbon. Both steels and cast irons, primary structural materials in every technologically advanced culture, are essentially iron–carbon alloys. This section is devoted to a study of the phase diagram for this system and the development of several of the possible microstructures. The relationships between heat treatment, microstructure, and mechanical properties are explored in Chapter 11.

10.18 THE IRON–IRON CARBIDE (Fe–Fe₃C) PHASE DIAGRAM

A portion of the iron–carbon phase diagram is presented in Figure 10.26. Pure iron, upon heating, experiences two changes in crystal structure before it melts. At room temperature the stable form, called **ferrite,** or α iron, has a BCC crystal structure. Ferrite experiences a polymorphic transformation to FCC **austenite,** or γ iron, at 912°C (1674°F). This austenite persists to 1394°C (2541°F), at which temperature the FCC austenite reverts back to a BCC phase known as δ ferrite, which finally melts at 1538°C (2800°F). All these changes are apparent along the left vertical axis of the phase diagram.

The composition axis in Figure 10.26 extends only to 6.70 wt% C; at this concentration the intermediate compound iron carbide, or **cementite** (Fe₃C), is

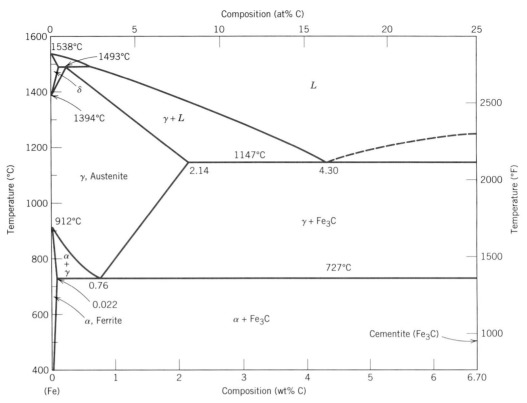

FIGURE 10.26   The iron–iron carbide phase diagram. (Adapted from *Binary Alloy Phase Diagrams*, 2nd edition, Vol. 1, T. B. Massalski, Editor-in-Chief, 1990. Reprinted by permission of ASM International, Materials Park, OH.)

formed, which is represented by a vertical line on the phase diagram. Thus, the iron–carbon system may be divided into two parts: an iron-rich portion, as in Figure 10.26; and the other (not shown) for compositions between 6.70 and 100 wt% C (pure graphite). In practice, all steels and cast irons have carbon contents less than 6.70 wt% C; therefore, we consider only the iron–iron carbide system. Figure 10.26 would be more appropriately labeled the Fe–Fe₃C phase diagram, since Fe₃C is now considered to be a component. Convention and convenience dictate that composition still be expressed in "wt% C" rather than "wt% Fe₃C"; 6.70 wt% C corresponds to 100 wt% Fe₃C.

Carbon is an interstitial impurity in iron and forms a solid solution with each of α and δ ferrites, and also with austenite, as indicated by the α, δ, and γ single-phase fields in Figure 10.26. In the BCC α ferrite, only small concentrations of carbon are soluble; the maximum solubility is 0.022 wt% at 727°C (1341°F). The limited solubility is explained by the shape and size of the BCC interstitial positions, which make it difficult to accommodate the carbon atoms. Even though present in relatively low concentrations, carbon significantly influences the mechanical properties of ferrite. This particular iron–carbon phase is relatively soft, may be made magnetic at temperatures below 768°C (1414°F), and has a density of 7.88 g/cm³. Figure 10.27a is a photomicrograph of α ferrite.

The austenite, or γ phase of iron, when alloyed with just carbon, is not stable below 727°C (1341°F), as indicated in Figure 10.26. The maximum solubility of

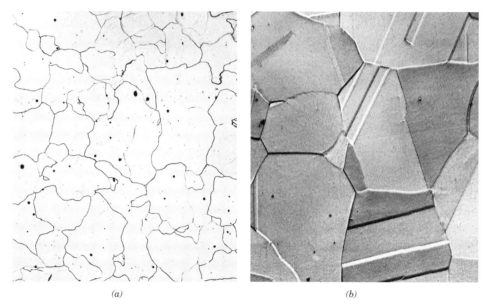

FIGURE 10.27 Photomicrographs of (a) α ferrite (90×) and (b) austenite (325×). (Copyright 1971 by United States Steel Corporation.)

carbon in austenite, 2.14 wt%, occurs at 1147°C (2097°F). This solubility is approximately 100 times greater than the maximum for BCC ferrite, since the FCC interstitial positions are larger (see the results of Problem 5.9), and, therefore, the strains imposed on the surrounding iron atoms are much lower. As the discussions that follow demonstrate, phase transformations involving austenite are very important in the heat treating of steels. In passing, it should be mentioned that austenite is nonmagnetic. Figure 10.27b shows a photomicrograph of this austenite phase.

The δ ferrite is virtually the same as α ferrite, except for the range of temperatures over which each exists. Since the δ ferrite is stable only at relatively high temperatures, it is of no technological importance and is not discussed further.

Cementite (Fe_3C) forms when the solubility limit of carbon in α ferrite is exceeded below 727°C (1341°F) (for compositions within the α + Fe_3C phase region). As indicated in Figure 10.26, Fe_3C will also coexist with the γ phase between 727 and 1147°C (1341 and 2097°F). Mechanically, cementite is very hard and brittle; the strength of some steels is greatly enhanced by its presence.

Strictly speaking, cementite is only metastable; that is, it will remain as a compound indefinitely at room temperature. But if heated to between 650 and 700°C (1200 and 1300°F) for several years, it will gradually change or transform into α iron and carbon, in the form of graphite, which will remain upon subsequent cooling to room temperature. Thus, the phase diagram in Figure 10.26 is not a true equilibrium one because cementite is not an equilibrium compound. However, inasmuch as the decomposition rate of cementite is extremely sluggish, virtually all the carbon in steel will be as Fe_3C instead of graphite, and the iron–iron carbide phase diagram is, for all practical purposes, valid. As will be seen in Section 13.2, addition of silicon to cast irons greatly accelerates this cementite decomposition reaction to form graphite.

The two-phase regions are labeled in Figure 10.26. It may be noted that one eutectic exists for the iron–iron carbide system, at 4.30 wt% C and 1147°C (2097°F);

for this eutectic reaction,

$$L \underset{\text{heating}}{\overset{\text{cooling}}{\rightleftharpoons}} \gamma + \text{Fe}_3\text{C} \tag{10.18}$$

the liquid solidifies to form austenite and cementite phases. Of course, subsequent cooling to room temperature will promote additional phase changes.

It may be noted that a eutectoid invariant point exists at a composition of 0.76 wt% C and a temperature of 727°C (1341°F). This eutectoid reaction may be represented by

$$\gamma(0.76 \text{ wt\% C}) \underset{\text{heating}}{\overset{\text{cooling}}{\rightleftharpoons}} \alpha(0.022 \text{ wt\% C}) + \text{Fe}_3\text{C} (6.7 \text{ wt\% C}) \tag{10.19}$$

or, upon cooling, the solid γ phase is transformed into α iron and cementite. (Eutectoid phase transformations were addressed in Section 10.13.) The eutectoid phase changes described by Equation 10.19 are very important, being fundamental to the heat treatment of steels, as explained in subsequent discussions.

Ferrous alloys are those in which iron is the prime component, but carbon as well as other alloying elements may be present. In the classification scheme of ferrous alloys based on carbon content, there are three types: iron, steel, and cast iron. Commercially pure iron contains less than 0.008 wt% C and, from the phase diagram, is composed almost exclusively of the ferrite phase at room temperature. The iron–carbon alloys that contain between 0.008 and 2.14 wt% C are classified as steels. In most steels the microstructure consists of both α and Fe_3C phases. Upon cooling to room temperature, an alloy within this composition range must pass through at least a portion of the γ phase field; distinctive microstructures are subsequently produced, as discussed below. Although a steel alloy may contain as much as 2.14 wt% C, in practice, carbon concentrations rarely exceed 1.0 wt%. The properties and various classifications of steels are treated in Section 13.2. Cast irons are classified as ferrous alloys that contain between 2.14 and 6.70 wt% C. However, commercial cast irons normally contain less than 4.5 wt% C. These alloys are also discussed in Section 13.2.

10.19 DEVELOPMENT OF MICROSTRUCTURES IN IRON–CARBON ALLOYS

Several of the various microstructures that may be produced in steel alloys and their relationships to the iron–iron carbon phase diagram are now discussed, and it is shown that the microstructure that develops depends on both the carbon content and heat treatment. The discussion is confined to very slow cooling of steel alloys, in which equilibrium is continuously maintained. A more detailed exploration of the influence of heat treatment on microstructure, and ultimately on the mechanical properties of steels, is contained in Chapter 11.

Phase changes that occur upon passing from the γ region into the $\alpha + \text{Fe}_3\text{C}$ phase field (Figure 10.26) are relatively complex {and similar to those described for the eutectic systems in Section 10.11.} Consider, for example, an alloy of eutectoid composition (0.76 wt% C) as it is cooled from a temperature within the γ phase region, say, 800°C, that is, beginning at point a in Figure 10.28 and moving down the vertical line xx'. Initially, the alloy is composed entirely of the austenite phase

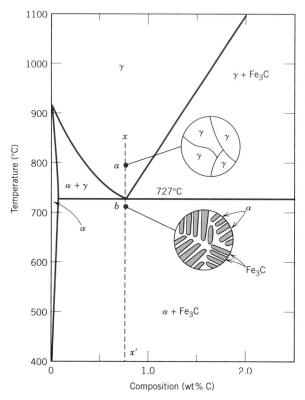

FIGURE 10.28 Schematic representations of the microstructures for an iron–carbon alloy of eutectoid composition (0.76 wt% C) above and below the eutectoid temperature.

having a composition of 0.76 wt% C and corresponding microstructure, also indicated in Figure 10.28. As the alloy is cooled, there will occur no changes until the eutectoid temperature (727°C) is reached. Upon crossing this temperature to point *b*, the austenite transforms according to Equation 10.19.

The microstructure for this eutectoid steel that is slowly cooled through the eutectoid temperature consists of alternating layers or lamellae of the two phases (α and Fe_3C) that form simultaneously during the transformation. In this case, the relative layer thickness is approximately 8 to 1. This microstructure, represented schematically in Figure 10.28, point *b*, is called **pearlite** because it has the appearance of mother of pearl when viewed under the microscope at low magnifications. Figure 10.29 is a photomicrograph of a eutectoid steel showing the pearlite. The pearlite exists as grains, often termed "colonies"; within each colony the layers are oriented in essentially the same direction, which varies from one colony to another. The thick light layers are the ferrite phase, and the cementite phase appears as thin lamellae most of which appear dark. Many cementite layers are so thin that adjacent phase boundaries are indistinguishable, which layers appear dark at this magnification. Mechanically, pearlite has properties intermediate between the soft, ductile ferrite and the hard, brittle cementite.

The alternating α and Fe_3C layers in pearlite form {as such for the same reason that the eutectic structure (Figures 10.11 and 10.12) forms}—because the composition of the parent phase [in this case austenite (0.76 wt% C)] is different from either of the product phases [ferrite (0.022 wt% C) and cementite (6.7 wt% C)], and the phase transformation requires that there be a redistribution of the carbon by diffusion. Figure 10.30 illustrates schematically microstructural changes that accompany this eutectoid reaction; here the directions of carbon diffusion are

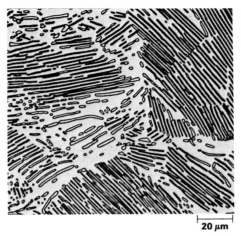

FIGURE 10.29 Photomicrograph of a eutectoid steel showing the pearlite microstructure consisting of alternating layers of α ferrite (the light phase) and Fe_3C (thin layers most of which appear dark). 500×. (Reproduced with permission from *Metals Handbook,* Vol. 9, 9th edition, *Metallography and Microstructures,* American Society for Metals, Materials Park, OH, 1985.)

20 μm

indicated by arrows. Carbon atoms diffuse away from the 0.022 wt% ferrite regions and to the 6.7 wt% cementite layers, as the pearlite extends from the grain boundary into the unreacted austenite grain. The layered pearlite forms because carbon atoms need diffuse only minimal distances with the formation of this structure.

Furthermore, subsequent cooling of the pearlite from point b in Figure 10.28 will produce relatively insignificant microstructural changes.

HYPOEUTECTOID ALLOYS

Microstructures for iron–iron carbide alloys having other than the eutectoid composition are now explored; {these are analogous to the fourth case described in Section 10.11 and illustrated in Figure 10.14 for the eutectic system.} Consider a composition C_0 to the left of the eutectoid, between 0.022 and 0.76 wt% C; this is termed a **hypoeutectoid** (less than eutectoid) **alloy.** Cooling an alloy of this composition is represented by moving down the vertical line yy' in Figure 10.31. At about 875°C, point c, the microstructure will consist entirely of grains of the γ phase, as shown schematically in the figure. In cooling to point d, about 775°C, which is within the $\alpha + \gamma$ phase region, both these phases will coexist as in the schematic microstructure. Most of the small α particles will form along the original γ grain boundaries. The

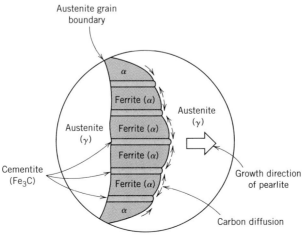

Austenite grain boundary

α

Ferrite (α)

Austenite (γ)

Ferrite (α)

Austenite (γ)

Ferrite (α)

Ferrite (α)

Cementite (Fe_3C)

Ferrite (α)

α

Growth direction of pearlite

Carbon diffusion

FIGURE 10.30 Schematic representation of the formation of pearlite from austenite; direction of carbon diffusion indicated by arrows.

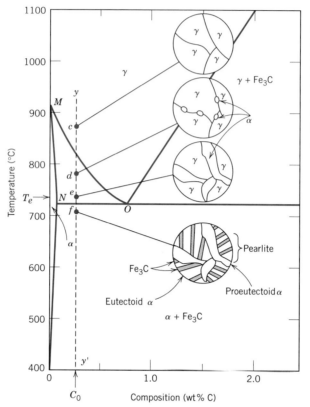

FIGURE 10.31 Schematic representations of the microstructures for an iron–carbon alloy of hypoeutectoid composition C_0 (containing less than 0.76 wt% C) as it is cooled from within the austenite phase region to below the eutectoid temperature.

compositions of both α and γ phases may be determined using the appropriate tie line; these compositions correspond, respectively, to about 0.020 and 0.40 wt% C.

While cooling an alloy through the $\alpha + \gamma$ phase region, the composition of the ferrite phase changes with temperature along the $\alpha - (\alpha + \gamma)$ phase boundary, line *MN*, becoming slightly richer in carbon. On the other hand, the change in composition of the austenite is more dramatic, proceeding along the $(\alpha + \gamma) - \gamma$ boundary, line *MO*, as the temperature is reduced.

Cooling from point *d* to *e*, just above the eutectoid but still in the $\alpha + \gamma$ region, will produce an increased fraction of the α phase and a microstructure similar to that also shown: the α particles will have grown larger. At this point, the compositions of the α and γ phases are determined by constructing a tie line at the temperature T_e; the α phase will contain 0.022 wt% C, while the γ phase will be of the eutectoid composition, 0.76 wt% C.

As the temperature is lowered just below the eutectoid, to point *f*, all the γ phase that was present at temperature T_e (and having the eutectoid composition) will transform to pearlite, according to the reaction in Equation 10.19. There will be virtually no change in the α phase that existed at point *e* in crossing the eutectoid temperature—it will normally be present as a continuous matrix phase surrounding the isolated pearlite colonies. The microstructure at point *f* will appear as the corresponding schematic inset of Figure 10.31. Thus the ferrite phase will be present both in the pearlite and also as the phase that formed while cooling through the $\alpha + \gamma$ phase region. The ferrite that is present in the pearlite is called *eutectoid ferrite*, whereas the other, that formed above T_e, is termed **proeutectoid** (meaning pre- or before eutectoid) **ferrite**, as labeled in Figure 10.31. Figure 10.32 is a photomicrograph of a 0.38 wt% C steel; large, white regions correspond to the proeutectoid

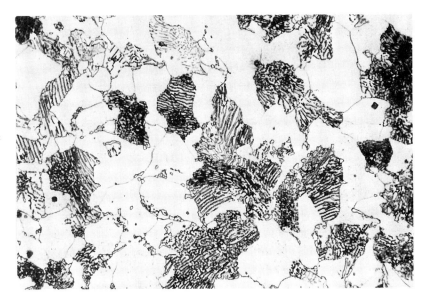

FIGURE 10.32
Photomicrograph of a 0.38 wt% C steel having a microstructure consisting of pearlite and proeutectoid ferrite. 635×. (Photomicrograph courtesy of Republic Steel Corporation.)

ferrite. For pearlite, the spacing between the α and Fe_3C layers varies from grain to grain; some of the pearlite appears dark because the many close-spaced layers are unresolved at the magnification of the photomicrograph. The chapter-opening photograph for this chapter is a scanning electron micrograph of a hypoeutectoid (0.44 wt% C) steel in which may also be seen both pearlite and proeutectoid ferrite, only at a higher magnification. It should also be noted that two microconstituents are present in these micrographs—proeutectoid ferrite and pearlite—which will appear in all hypoeutectoid iron–carbon alloys that are slowly cooled to a temperature below the eutectoid.

The relative amounts of the proeutectoid α and pearlite may be determined {in a manner similar to that described in Section 10.11 for primary and eutectic microconstituents.} We use the lever rule in conjunction with a tie line that extends from the $\alpha - (\alpha + Fe_3C)$ phase boundary (0.022 wt% C) to the eutectoid composition (0.76 wt% C), inasmuch as pearlite is the transformation product of austenite having this composition. For example, let us consider an alloy of composition C_0' in Figure 10.33. Thus, the fraction of pearlite, W_p, may be determined according to

$$W_p = \frac{T}{T + U}$$

$$= \frac{C_0' - 0.022}{0.76 - 0.022} = \frac{C_0' - 0.022}{0.74} \tag{10.20}$$

Furthermore, the fraction of proeutectoid α, $W_{\alpha'}$, is computed as follows:

$$W_{\alpha'} = \frac{U}{T + U}$$

$$= \frac{0.76 - C_0'}{0.76 - 0.022} = \frac{0.76 - C_0'}{0.74} \tag{10.21}$$

FIGURE 10.33 A portion of the Fe–Fe$_3$C phase diagram used in computations for relative amounts of proeutectoid and pearlite microconstituents for hypoeutectoid (C_0') and hypereutectoid (C_1') compositions.

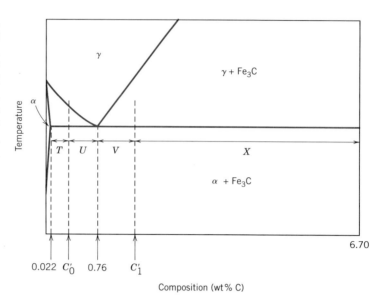

Of course, fractions of both total α (eutectoid and proeutectoid) and cementite are determined using the lever rule and a tie line that extends across the entirety of the α + Fe$_3$C phase region, from 0.022 to 6.7 wt% C.

HYPEREUTECTOID ALLOYS

Analogous transformations and microstructures result for **hypereutectoid alloys,** those containing between 0.76 and 2.14 wt% C, which are cooled from temperatures within the γ phase field. Consider an alloy of composition C_1 in Figure 10.34 which, upon cooling, moves down the line zz'. At point g only the γ phase will be present with a composition of C_1; the microstructure will appear as shown, having only γ grains. Upon cooling into the γ + Fe$_3$C phase field, say, to point h, the cementite phase will begin to form along the initial γ grain boundaries, similar to the α phase in Figure 10.31, point d. This cementite is called **proeutectoid cementite**—that which forms before the eutectoid reaction. Of course, the cementite composition remains constant (6.70 wt% C) as the temperature changes. However, the composition of the austenite phase will move along line PO toward the eutectoid. As the temperature is lowered through the eutectoid to point i, all remaining austenite of eutectoid composition is converted into pearlite; thus, the resulting microstructure consists of pearlite and proeutectoid cementite as microconstituents (Figure 10.34). In the photomicrograph of a 1.4 wt% C steel (Figure 10.35), note that the proeutectoid cementite appears light. Since it has much the same appearance as proeutectoid ferrite (Figure 10.32), there is some difficulty in distinguishing between hypoeutectoid and hypereutectoid steels on the basis of microstructure.

Relative amounts of both pearlite and proeutectoid Fe$_3$C microconstituents may be computed for hypereutectoid steel alloys in a manner analogous to that for hypoeutectoid materials; the appropriate tie line extends between 0.76 and 6.70 wt% C. Thus, for an alloy having composition C_1' in Figure 10.33, fractions of pearlite W_p and proeutectoid cementite $W_{Fe_3C'}$ are determined from the following lever rule expressions:

$$W_p = \frac{X}{V + X} = \frac{6.70 - C_1'}{6.70 - 0.76} = \frac{6.70 - C_1'}{5.94} \tag{10.22}$$

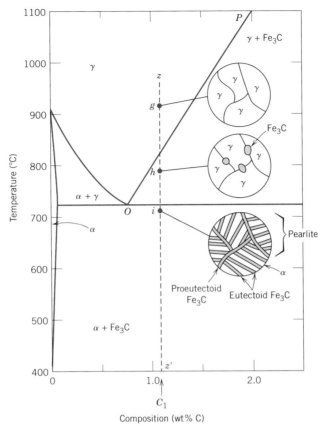

FIGURE 10.34 Schematic representations of the microstructures for an iron–carbon alloy of hypereutectoid composition C_1 (containing between 0.76 and 2.14 wt% C), as it is cooled from within the austenite phase region to below the eutectoid temperature.

FIGURE 10.35 Photomicrograph of a 1.4 wt% C steel having a microstructure consisting of a white proeutectoid cementite network surrounding the pearlite colonies. 1000×. (Copyright 1971 by United States Steel Corporation.)

and

$$W_{Fe_3C'} = \frac{V}{V+X} = \frac{C_1' - 0.76}{6.70 - 0.76} = \frac{C_1' - 0.76}{5.94} \tag{10.23}$$

EXAMPLE PROBLEM 10.4

For a 99.65 wt% Fe–0.35 wt% C alloy at a temperature just below the eutectoid, determine the following:

(a) The fractions of total ferrite and cementite phases.

(b) The fractions of the proeutectoid ferrite and pearlite.

(c) The fraction of eutectoid ferrite.

SOLUTION

(a) This part of the problem is solved by application of the lever rule expressions employing a tie line that extends all the way across the α + Fe_3C phase field. Thus, C_0' is 0.35 wt% C, and

$$W_\alpha = \frac{6.70 - 0.35}{6.70 - 0.022} = 0.95$$

and

$$W_{Fe_3C} = \frac{0.35 - 0.022}{6.70 - 0.022} = 0.05$$

(b) The fractions of proeutectoid ferrite and pearlite are determined by using the lever rule, and a tie line that extends only to the eutectoid composition (i.e., Equations 10.20 and 10.21). Or

$$W_p = \frac{0.35 - 0.022}{0.76 - 0.022} = 0.44$$

and

$$W_{\alpha'} = \frac{0.76 - 0.35}{0.76 - 0.022} = 0.56$$

(c) All ferrite is either as proeutectoid or eutectoid (in the pearlite). Therefore, the sum of these two ferrite fractions will equal the fraction of total ferrite, that is,

$$W_{\alpha'} + W_{\alpha e} = W_\alpha$$

where $W_{\alpha e}$ denotes the fraction of the total alloy that is eutectoid ferrite. Values for W_α and $W_{\alpha'}$ were determined in parts (a) and (b) as 0.95 and 0.56, respectively. Therefore,

$$W_{\alpha e} = W_\alpha - W_{\alpha'} = 0.95 - 0.56 = 0.39$$

NONEQUILIBRIUM COOLING

In this discussion on the microstructural development of iron–carbon alloys it has been assumed that, upon cooling, conditions of metastable equilibrium[1] have been

---

[1] The term "metastable equilibrium" is used in this discussion inasmuch as Fe_3C is only a metastable compound.

continuously maintained; that is, sufficient time has been allowed at each new temperature for any necessary adjustment in phase compositions and relative amounts as predicted from the Fe–Fe$_3$C phase diagram. In most situations these cooling rates are impractically slow and really unnecessary; in fact, on many occasions nonequilibrium conditions are desirable. Two nonequilibrium effects of practical importance are (1) the occurrence of phase changes or transformations at temperatures other than those predicted by phase boundary lines on the phase diagram, and (2) the existence at room temperature of nonequilibrium phases that do not appear on the phase diagram. Both are discussed in the next chapter.

10.20 THE INFLUENCE OF OTHER ALLOYING ELEMENTS (CD-ROM)

SUMMARY

Equilibrium phase diagrams are a convenient and concise way of representing the most stable relationships between phases in alloy systems. This discussion considered binary phase diagrams for which temperature and composition are variables. Areas, or phase regions, are defined on these temperature-versus-composition plots within which either one or two phases exist. For an alloy of specified composition and at a known temperature, the phases present, their compositions, and relative amounts under equilibrium conditions may be determined. Within two-phase regions, tie lines and the lever rule must be used for phase composition and mass fraction computations, respectively.

Several different kinds of phase diagram were discussed for metallic systems. Isomorphous diagrams are those for which there is complete solubility in the solid phase; the copper–nickel system displays this behavior. {Also discussed for alloys belonging to isomorphous systems were the development of microstructure for both cases of equilibrium and nonequilibrium cooling} and the dependence of mechanical characteristics on composition.

In a eutectic reaction, as found in some alloy systems, a liquid phase transforms isothermally to two different solid phases upon cooling. Such a reaction is noted on the copper–silver and lead–tin phase diagrams. Complete solid solubility for all compositions does not exist; instead, solid solutions are terminal—there is only a limited solubility of each component in the other. {Four different kinds of microstructures that may develop for the equilibrium cooling of alloys belonging to eutectic systems were discussed.}

Other equilibrium phase diagrams are more complex, having intermediate compounds and/or phases, possibly more than a single eutectic, and other reactions including eutectoid, peritectic, and congruent phase transformations. These are found for copper–zinc and magnesium–lead systems.

{Phase diagrams for the Al$_2$O$_3$–Cr$_2$O$_3$, MgO–Al$_2$O$_3$, ZrO$_2$–CaO, and SiO$_2$–Al$_2$O$_3$ systems were discussed. These diagrams are especially useful in assessing the high-temperature performance of ceramic materials.}

{The Gibbs phase rule was introduced; it is a simple equation that relates the number of phases present in a system at equilibrium with the number of degrees of freedom, the number of components, and the number of noncompositional variables.}

Considerable attention was given to the iron–carbon system, and specifically, the iron–iron carbide phase diagram, which technologically is one of the most important. The development of microstructure in many iron–carbon alloys and steels depends on the eutectoid reaction in which the FCC austenite phase of composition 0.76 wt% C transforms isothermally to the BCC α ferrite phase (0.022 wt% C) and the intermetallic compound, cementite (Fe_3C). The microstructural product of an iron–carbon alloy of eutectoid composition is pearlite, a microconstituent consisting of alternating layers of ferrite and cementite. The microstructures of alloys having carbon contents less than the eutectoid (hypoeutectoid) are comprised of a proeutectoid ferrite phase in addition to pearlite. On the other hand, pearlite and proeutectoid cementite constitute the microconstituents for hypereutectoid alloys—those with carbon contents in excess of the eutectoid composition.

IMPORTANT TERMS AND CONCEPTS

Austenite
Cementite
Component
Congruent transformation
Equilibrium
Eutectic phase
Eutectic reaction
Eutectic structure
Eutectoid reaction
Ferrite
Free energy
Gibbs phase rule

Hypereutectoid alloy
Hypoeutectoid alloy
Intermediate solid solution
Intermetallic compound
Invariant point
Isomorphous
Lever rule
Liquidus line
Metastable
Microconstituent
Pearlite
Peritectic reaction

Phase
Phase diagram
Phase equilibrium
Primary phase
Proeutectoid cementite
Proeutectoid ferrite
Solidus line
Solubility limit
Solvus line
System
Terminal solid solution
Tie line

REFERENCES

ASM Handbook, Vol. 3, *Alloy Phase Diagrams,* ASM International, Materials Park, OH, 1992.

ASM Handbook, Vol. 9, *Metallography and Microstructures,* ASM International, Materials Park, OH, 1985.

Bergeron, C. G. and S. H. Risbud, *Introduction to Phase Equilibria in Ceramics,* American Ceramic Society, Columbus, OH, 1984.

Cook, L. P. and H. F. McMurdie (Editors), *Phase Diagrams for Ceramists,* Vol. VII, American Ceramic Society, Columbus, OH, 1989.

Gordon, P., *Principles of Phase Diagrams in Materials Systems,* McGraw-Hill Book Company, New York, 1968. Reprinted by Krieger Publishing Company, Melbourne, FL, 1983.

Hansen, M. and K. Anderko, *Constitution of Binary Alloys,* 2nd edition, McGraw-Hill Book Company, New York, 1958. *First Supplement* (R. P. Elliott), 1965. *Second Supplement* (F. A.

Shunk), 1969. Reprinted by Genium Publishing Corp., Schenectady, NY.

Kingery, W. D., H. K. Bowen, and D. R. Uhlmann, *Introduction to Ceramics,* 2nd edition, John Wiley & Sons, New York, 1976. Chapter 7.

Levin, E. M., C. R. Robbins, and H. F. McMurdie (Editors), *Phase Diagrams for Ceramists,* Vol. I, American Ceramic Society, Columbus, OH, 1964. Also supplementary Volumes II, III, IV, V and VI, published in 1969, 1973, 1981, 1983 and 1987, respectively.

Massalski, T. B. (Editor), *Binary Phase Diagrams,* 2nd edition, ASM International, Materials Park, OH, 1990. Three volumes. On CD-ROM with updates.

Mysen, B. O. (Editor), *Phase Diagrams for Ceramists,* Vol. VIII, American Ceramic Society, Columbus, OH, 1990.

Petzow, G. and G. Effenberg, *Ternary Alloys, A*

Comprehensive Compendium of Evaluated Constitutional Data and Phase Diagrams, VCH Publishers, New York, 1988. Eight volumes.

Rhines, F. N., *Phase Diagrams in Metallurgy— Their Development and Application,* McGraw-Hill Book Company, Inc., New York, 1956.

QUESTIONS AND PROBLEMS

Note: To solve those problems having an asterisk (*) by their numbers, consultation of supplementary topics [appearing only on the CD-ROM (and not in print)] will probably be necessary.

10.1 Cite three variables that determine the microstructure of an alloy.

10.2 What thermodynamic condition must be met for a state of equilibrium to exist?

10.3 For metal alloys, the development of microstructure depends on the phenomenon of diffusion (Figures {10.13} and 10.30). It was noted in Section 6.3 that the driving force for steady-state diffusion is a concentration gradient. However, concentration gradients are normally absent in regions where diffusion is occurring, as represented in Figures {10.13} and 10.30; for these situations, what is the driving force?

10.4 What is the difference between the states of phase equilibrium and metastability?

10.5 Cite the phases that are present and the phase compositions for the following alloys:

(a) 90 wt% Zn–10 wt% Cu at 400°C (750°F).

(b) 75 wt% Sn–25 wt% Pb at 175°C (345°F).

(c) 55 wt% Ag–45 wt% Cu at 900°C (1650°F).

(d) 30 wt% Pb–70 wt% Mg at 425°C (795°F).

(e) 2.12 kg Zn and 1.88 kg Cu at 500°C (930°F).

(f) 37 lb$_m$ Pb and 6.5 lb$_m$ Mg at 400°C (750°F).

(g) 8.2 mol Ni and 4.3 mol Cu at 1250°C (2280°F).

(h) 4.5 mol Sn and 0.45 mol Pb at 200°C (390°F).

10.6 For an alloy of composition 74 wt% Zn–26 wt% Cu, cite the phases present and their compositions at the following temperatures: 850°C, 750°C, 680°C, 600°C, and 500°C.

10.7 Determine the relative amounts (in terms of mass fractions) of the phases for the alloys and temperatures given in Problem 10.5.

10.8 Derive Equations 10.6a and 10.7a, which may be used to convert mass fraction to volume fraction, and vice versa.

10.9 Determine the relative amounts (in terms of volume fractions) of the phases for the alloys and temperatures given in Problem 10.5a, b, and c. Below are given the approximate densities of the various metals at the alloy temperatures:

Metal	Temperature (°C)	Density (g/cm³)
Ag	900	9.97
Cu	400	8.77
Cu	900	8.56
Pb	175	11.20
Sn	175	7.22
Zn	400	6.83

10.10 Below is a portion of the H_2O–NaCl phase diagram:

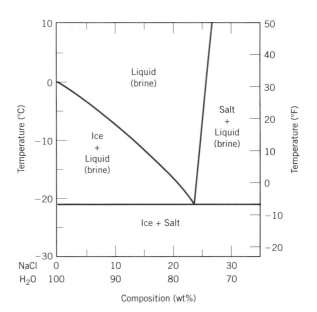

(a) Using this diagram, briefly explain how spreading salt on ice that is at a temperature below 0°C (32°F) can cause the ice to melt.

(b) What concentration of salt is necessary to have a 50% ice–50% liquid brine at −10°C (14°F)?

10.11 A 1.5-kg specimen of a 90 wt% Pb–10 wt% Sn alloy is heated to 250°C (480°F), at which temperature it is entirely an α-phase solid solution (Figure 10.7). The alloy is to be melted to the extent that 50% of the specimen is liquid, the remainder being the α phase. This may be accomplished either by heating the alloy or changing its composition while holding the temperature constant.

(a) To what temperature must the specimen be heated?

(b) How much tin must be added to the 1.5-kg specimen at 250°C to achieve this state?

10.12 Consider the sugar–water phase diagram of Figure 10.1.

(a) How much sugar will dissolve in 1500 g water at 90°C (194°F)?

(b) If the saturated liquid solution in part (a) is cooled to 20°C (68°F), some of the sugar will precipitate out as a solid. What will be the composition of the saturated liquid solution (in wt% sugar) at 20°C?

(c) How much of the solid sugar will come out of solution upon cooling to 20°C?

10.13 Consider a specimen of ice I which is at −10°C and 1 atm pressure. Using Figure 10.38, the pressure–temperature phase diagram for H_2O, determine the pressure to which the specimen must be raised or lowered to cause it **(a)** to melt, and **(b)** to sublime.

10.14 At a pressure of 0.01 atm, determine **(a)** the melting temperature for ice I, and **(b)** the boiling temperature for water.

10.15 A magnesium–lead alloy of mass 5.5 kg consists of a solid α phase that has a composition that is just slightly below the solubility limit at 200°C (390°F).

(a) What mass of lead is in the alloy?

(b) If the alloy is heated to 350°C (660°F), how much more lead may be dissolved in the α phase without exceeding the solubility limit of this phase?

10.16* **(a)** Briefly describe the phenomenon of coring and why it occurs.

(b) Cite one undesirable consequence of coring.

10.17 It is desired to produce a copper–nickel alloy that has a minimum noncold-worked tensile strength of 350 MPa (50,750 psi) and a ductility of at least 48%EL. Is such an alloy possible? If so, what must be its composition? If this is not possible, then explain why.

10.18 Is it possible to have a copper–silver alloy that, at equilibrium, consists of a β phase

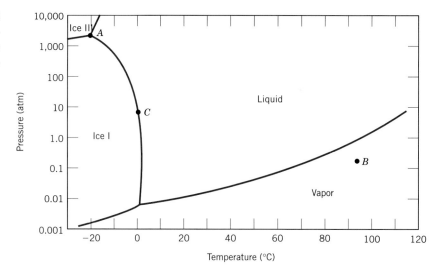

FIGURE 10.38 Logarithm pressure-versus-temperature phase diagram for H_2O.

of composition 92 wt% Ag–8 wt% Cu, and also a liquid phase of composition 76 wt% Ag–24 wt% Cu? If so, what will be the approximate temperature of the alloy? If this is not possible, explain why.

10.19 Is it possible to have a copper–zinc alloy that, at equilibrium, consists of an ϵ phase of composition 80 wt% Zn–20 wt% Cu, and also a liquid phase of composition 95 wt% Zn–5 wt% Cu? If so, what will be the approximate temperature of the alloy? If this is not possible, explain why.

10.20 A copper–nickel alloy of composition 70 wt% Ni–30 wt% Cu is slowly heated from a temperature of 1300°C (2370°F).

(a) At what temperature does the first liquid phase form?

(b) What is the composition of this liquid phase?

(c) At what temperature does complete melting of the alloy occur?

(d) What is the composition of the last solid remaining prior to complete melting?

10.21 A 50 wt% Pb–50 wt% Mg alloy is slowly cooled from 700°C (1290°F) to 400°C (750°F).

(a) At what temperature does the first solid phase form?

(b) What is the composition of this solid phase?

(c) At what temperature does the liquid solidify?

(d) What is the composition of this last remaining liquid phase?

10.22 A 90 wt% Ag–10 wt% Cu alloy is heated to a temperature within the β + liquid phase region. If the composition of the liquid phase is 85 wt% Ag, determine **(a)** the temperature of the alloy, **(b)** the composition of the β phase, and **(c)** the mass fractions of both phases.

10.23 Below are given the solidus and liquidus temperatures for the germanium–silicon system. Construct the phase diagram for this system and label each region.

Composition (wt% Si)	Solidus Temperature (°C)	Liquidus Temperature (°C)
0	938	938
10	1005	1147
20	1065	1226
30	1123	1278
40	1178	1315
50	1232	1346
60	1282	1367
70	1326	1385
80	1359	1397
90	1390	1408
100	1414	1414

10.24 A 30 wt% Sn–70 wt% Pb alloy is heated to a temperature within the α + liquid phase region. If the mass fraction of each phase is 0.5, estimate **(a)** the temperature of the alloy, and **(b)** the compositions of the two phases.

10.25* When kaolinite clay $[Al_2(Si_2O_5)(OH)_4]$ is heated to a sufficiently high temperature, chemical water is driven off.

(a) Under these circumstances what is the composition of the remaining product?

(b) What are the liquidus and solidus temperatures of this material?

10.26 For alloys of two hypothetical metals A and B, there exist an α, A-rich phase and a β, B-rich phase. From the mass fractions of both phases for two different alloys, which are at the same temperature, determine the composition of the phase boundary (or solubility limit) for both α and β phases at this temperature.

Alloy Composition	Fraction α Phase	Fraction β Phase
60 wt% A–40 wt% B	0.57	0.43
30 wt% A–70 wt% B	0.14	0.86

10.27 A hypothetical A–B alloy of composition 55 wt% B–45 wt% A at some temperature is found to consist of mass fractions of 0.5 for both α and β phases. If the composition of the β phase is 90 wt% B–10 wt% A, what is the composition of the α phase?

10.28 Is it possible to have a copper–silver alloy of composition 50 wt% Ag–50 wt% Cu, which, at equilibrium, consists of α and β phases having mass fractions $W_\alpha = 0.60$ and $W_\beta = 0.40$? If so, what will be the approximate temperature of the alloy? If such an alloy is not possible, explain why.

10.29 For 11.20 kg of a magnesium–lead alloy of composition 30 wt% Pb–70 wt% Mg, is it possible, at equilibrium, to have α and Mg_2Pb phases having respective masses of 7.39 kg and 3.81 kg? If so, what will be the approximate temperature of the alloy? If such an alloy is not possible, explain why.

10.30 At 700°C (1290°F), what is the maximum solubility **(a)** of Cu in Ag? **(b)** Of Ag in Cu?

10.31* A 45 wt% Pb–55 wt% Mg alloy is rapidly quenched to room temperature from an elevated temperature in such a way that the high-temperature microstructure is preserved. This microstructure is found to consist of the α phase and Mg_2Pb, having respective mass fractions of 0.65 and 0.35. Determine the approximate temperature from which the alloy was quenched.

10.32* Is it possible to have a copper–silver alloy in which the mass fractions of primary β and total β are 0.68 and 0.925, respectively, at 775°C (1425°F)? Why or why not?

10.33* For 6.70 kg of a magnesium–lead alloy, is it possible to have the masses of primary α and total α of 4.23 kg and 6.00 kg, respectively, at 460°C (860°F)? Why or why not?

10.34* For a copper–silver alloy of composition 25 wt% Ag–75 wt% Cu and at 775°C (1425°F) do the following:

(a) Determine the mass fractions of the α and β phases.

(b) Determine the mass fractions of primary α and eutectic microconstituents.

(c) Determine the mass fraction of eutectic α.

10.35* The microstructure of a lead–tin alloy at 180°C (355°F) consists of primary β and eutectic structures. If the mass fractions of these two microconstituents are 0.57 and 0.43, respectively, determine the composition of the alloy.

10.36* Consider the hypothetical eutectic phase diagram for metals A and B, which is similar to that for the lead–tin system, Figure 10.7. Assume that (1) α and β phases exist at the A and B extremities of the phase diagram, respectively; (2) the eutectic composition is 47 wt% B–53 wt% A; and (3) the composition of the β phase at the eutectic temperature is 92.6 wt% B–7.4 wt% A. Determine the composition of an alloy that will yield primary α and total α mass fractions of 0.356 and 0.693, respectively.

10.37* Briefly explain why, upon solidification, an alloy of eutectic composition forms a microstructure consisting of alternating layers of the two solid phases.

10.38* For an 85 wt% Pb–15 wt% Mg alloy, make schematic sketches of the microstructure that would be observed for conditions of very slow cooling at the following temperatures: 600°C (1110°F), 500°C (930°F), 270°C (520°F), and 200°C (390°F). Label all phases and indicate their approximate compositions.

10.39* For a 68 wt% Zn–32 wt% Cu alloy, make schematic sketches of the microstructure that would be observed for conditions of very slow cooling at the following temperatures: 1000°C (1830°F), 760°C (1400°F), 600°C (1110°F), and 400°C (750°F). Label all phases and indicate their approximate compositions.

10.40* For a 30 wt% Zn–70 wt% Cu alloy, make schematic sketches of the microstructure that would be observed for conditions of very slow cooling at the following temperatures: 1100°C (2010°F), 950°C (1740°F), 900°C (1650°F), and 700°C (1290°F). Label all phases and indicate their approximate compositions.

10.41 What is the principal difference between congruent and incongruent phase transformations?

10.42 Figure 10.39 is the aluminum–neodymium phase diagram, for which only single-phase regions are labeled. Specify temperature–composition points at which all eutectics, eutectoids, peritectics, and congruent phase transformations occur. Also, for each, write the reaction upon cooling.

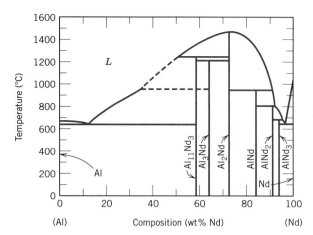

FIGURE 10.39 The aluminum–neodymium phase diagram. (Adapted from *ASM Handbook,* Vol. 3, *Alloy Phase Diagrams,* H. Baker, Editor, 1992. Reprinted by permission of ASM International, Materials Park, OH.)

10.43 Figure 10.40 is a portion of the titanium–copper phase diagram for which only single-phase regions are labeled. Specify all temperature–composition points at which eutectics, eutectoids, peritectics, and congruent phase transformations occur. Also, for each, write the reaction upon cooling.

10.44* For the ZrO_2–CaO system (Figure 10.23), write all eutectic and eutectoid reactions for cooling.

10.45* From Figure 10.22, the phase diagram for the MgO–Al_2O_3 system, it may be noted that the spinel solid solution exists over a range of compositions, which means that it is nonstoichiometric at compositions other than 50 mol% MgO–50 mol% Al_2O_3.

(a) The maximum nonstoichiometry on the Al_2O_3-rich side of the spinel phase field exists at about 2000°C (3630°F) corresponding to approximately 82 mol% (92 wt%) Al_2O_3.

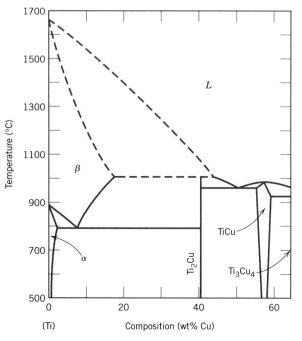

FIGURE 10.40 The titanium–copper phase diagram. (Adapted from *Phase Diagrams of Binary Titanium Alloys,* J. L. Murray, Editor, 1987. Reprinted by permission of ASM International, Materials Park, OH.)

Determine the type of vacancy defect that is produced and the percentage of vacancies that exist at this composition.

(b) The maximum nonstoichiometry on the MgO-rich side of the spinel phase field exists at about 2000°C (3630°F) corresponding to approximately 39 mol% (62 wt%) Al_2O_3. Determine the type of vacancy defect that is produced and the percentage of vacancies that exist at this composition.

10.46* For a ternary system, three components are present; temperature is also a variable. Compute the maximum number of phases that may be present for a ternary system, assuming that pressure is held constant.

10.47* In Figure 10.38 is shown the pressure–temperature phase diagram for H_2O. Apply the Gibbs phase rule at points A, B, and C; that is, specify the number of degrees of freedom at each of the points, that is, the number of externally controllable variables that need be specified to completely define the system.

10.48 Construct the hypothetical phase diagram for metals A and B between temperatures of 600°C and 1000°C given the following information:

- The melting temperature of metal A is 940°C.
- The solubility of B in A is negligible at all temperatures.
- The melting temperature of metal B is 830°C.
- The maximum solubility of A in B is 12 wt% A, which occurs at 700°C.
- At 600°C, the solubility of A in B is 8 wt% A.
- One eutectic occurs at 700°C and 75 wt% B–25 wt% A.
- A second eutectic occurs at 730°C and 60 wt% B–40 wt% A.
- A third eutectic occurs at 755°C and 40 wt% B–60 wt% A.
- One congruent melting point occurs at 780°C and 51 wt% B–49 wt% A.
- A second congruent melting point occurs at 755°C and 67 wt% B–33 wt% A.
- The intermetallic compound AB exists at 51 wt% B–49 wt% A.

- The intermetallic compound AB_2 exists at 67 wt% B–33 wt% A.

10.49* Two intermetallic compounds, AB and AB_2, exist for elements A and B. If the compositions for AB and AB_2 are 34.3 wt% A–65.7 wt% B and 20.7 wt% A–79.3 wt% B, respectively, and element A is potassium, identify element B.

10.50 Compute the mass fractions of α ferrite and cementite in pearlite.

10.51 What is the difference between a phase and a microconstituent?

10.52 **(a)** What is the distinction between hypoeutectoid and hypereutectoid steels?

(b) In a hypoeutectoid steel, both eutectoid and proeutectoid ferrite exist. Explain the difference between them. What will be the carbon concentration in each?

10.53 Briefly explain why a proeutectoid phase forms along austenite grain boundaries. *Hint:* Consult Section 5.8.

10.54 What is the carbon concentration of an iron–carbon alloy for which the fraction of total ferrite is 0.94?

10.55 What is the proeutectoid phase for an iron–carbon alloy in which the mass fractions of total ferrite and total cementite are 0.92 and 0.08, respectively? Why?

10.56 Consider 1.0 kg of austenite containing 1.15 wt% C, cooled to below 727°C (1341°F).

(a) What is the proeutectoid phase?

(b) How many kilograms each of total ferrite and cementite form?

(c) How many kilograms each of pearlite and the proeutectoid phase form?

(d) Schematically sketch and label the resulting microstructure.

10.57 Consider 2.5 kg of austenite containing 0.65 wt% C, cooled to below 727°C (1341°F).

(a) What is the proeutectoid phase?

(b) How many kilograms each of total ferrite and cementite form?

(c) How many kilograms each of pearlite and the proeutectoid phase form?

(d) Schematically sketch and label the resulting microstructure.

10.58 Compute the mass fractions of proeutectoid ferrite and pearlite that form in an iron–carbon alloy containing 0.25 wt% C.

10.59 The microstructure of an iron–carbon alloy consists of proeutectoid ferrite and pearlite; the mass fractions of these two microconstituents are 0.286 and 0.714, respectively. Determine the concentration of carbon in this alloy.

10.60 The mass fractions of total ferrite and total cementite in an iron–carbon alloy are 0.88 and 0.12, respectively. Is this a hypoeutectoid or hypereutectoid alloy? Why?

10.61 The microstructure of an iron–carbon alloy consists of proeutectoid ferrite and pearlite; the mass fractions of these microconstituents are 0.20 and 0.80, respectively. Determine the concentration of carbon in this alloy.

10.62 Consider 2.0 kg of a 99.6 wt% Fe–0.4 wt% C alloy that is cooled to a temperature just below the eutectoid.

(a) How many kilograms of proeutectoid ferrite form?

(b) How many kilograms of eutectoid ferrite form?

(c) How many kilograms of cementite form?

10.63 Compute the maximum mass fraction of proeutectoid cementite possible for a hypereutectoid iron–carbon alloy.

10.64 Is it possible to have an iron–carbon alloy for which the mass fractions of total ferrite and proeutectoid cementite are 0.846 and 0.049, respectively? Why or why not?

10.65 Is it possible to have an iron–carbon alloy for which the mass fractions of total cementite and pearlite are 0.039 and 0.417, respectively? Why or why not?

10.66 Compute the mass fraction of eutectoid ferrite in an iron–carbon alloy that contains 0.43 wt% C.

10.67 The mass fraction of *eutectoid* cementite in an iron–carbon alloy is 0.104. On the basis of this information, is it possible to determine the composition of the alloy? If so, what is its composition? If this is not possible, explain why.

10.68 The mass fraction of *eutectoid* ferrite in an iron–carbon alloy is 0.82. On the basis of this information, is it possible to determine the composition of the alloy? If so, what is its composition? If this is not possible, explain why.

10.69 For an iron–carbon alloy of composition 5 wt% C–95 wt% Fe, make schematic sketches of the microstructure that would be observed for conditions of very slow cooling at the following temperatures: 1175°C (2150°F), 1145°C (2095°F), and 700°C (1290°F). Label the phases and indicate their compositions (approximate).

10.70 Often, the properties of multiphase alloys may be approximated by the relationship

$$E \text{ (alloy)} = E_\alpha V_\alpha + E_\beta V_\beta$$

where E represents a specific property (modulus of elasticity, hardness, etc.), and V is the volume fraction. The subscripts α and β denote the existing phases or microconstituents. Employ the relationship above to determine the approximate Brinell hardness of a 99.80 wt% Fe–0.20 wt% C alloy. Assume Brinell hardnesses of 80 and 280 for ferrite and pearlite, respectively, and that volume fractions may be approximated by mass fractions.

10.71* On the basis of the photomicrograph (i.e., the relative amounts of the microconstituents) for the lead–tin alloy shown in Figure 10.15 and the Pb–Sn phase diagram (Figure 10.7), estimate the composition of the alloy, and then compare this estimate with the composition given in the figure legend of Figure 10.15. Make the following assumptions: (1) the area fraction of each phase and microconstituent in the photomicrograph is equal to its volume fraction; (2) the densities of the α and β phases as well as the eutectic structure are 11.2, 7.3, and 8.7 g/cm^3, respectively; and (3) this photomicrograph represents the equilibrium microstructure at 180°C (356°F).

10.72* A steel alloy contains 97.5 wt% Fe, 2.0 wt% Mo, and 0.5 wt% C.

(a) What is the eutectoid temperature of this alloy?

(b) What is the eutectoid composition?

(c) What is the proeutectoid phase?

Assume that there are no changes in the positions of other phase boundaries with the addition of Mo.

10.73* A steel alloy is known to contain 93.8 wt% Fe, 6.0 wt% Ni, and 0.2 wt% C.

(a) What is the approximate eutectoid temperature of this alloy?

(b) What is the proeutectoid phase when this alloy is cooled to a temperature just below the eutectoid?

(c) Compute the relative amounts of the proeutectoid phase and pearlite. Assume that there are no alterations in the positions of other phase boundaries with the addition of Ni.

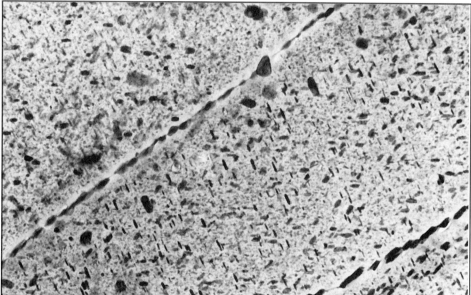

*T*op: A Boeing 767 airplane in flight. (Photograph courtesy of the Boeing Commercial Airplane Company.) Bottom: A transmission electron micrograph showing the microstructure of the aluminum alloy that is used for the upper wing skins, parts of the internal wing structures, and selected areas of the fuselage of the Boeing 767 above. This is a 7150–T651 alloy (6.2Zn, 2.3Cu, 2.3Mg, 0.12Zr, the balance Al) that has been precipitation hardened. The light matrix phase in the micrograph is an aluminum solid solution. The majority of the small plate-shaped dark precipitate particles are a transition η' phase, the remainder being the equilibrium η ($MgZn_2$) phase. Note that grain boundaries are "decorated" by some of these particles. 80,475×. (Electron micrograph courtesy of G. H. Narayanan and A. G. Miller, Boeing Commercial Airplane Company.)

Why Study *Phase Transformations?*

The development of a set of desirable mechanical characteristics for a material often results from a phase transformation, which is wrought by a heat treatment. The time and temperature dependencies of some phase transformations are conveniently represented on modified phase diagrams. It is important to know how to use these diagrams in order to

design a heat treatment for some alloy that will yield the desired room-temperature mechanical properties. For example, the tensile strength of an iron–carbon alloy of eutectoid composition (0.76 wt% C) can be varied between approximately 700 MPa (100,000 psi) and 2000 MPa (300,000 psi) depending on the heat treatment employed.

323

After careful study of this chapter you should be able to do the following:

1. Make a schematic fraction transformation-versus-logarithm of time plot for a typical solid–solid transformation; cite the equation that describes this behavior.
2. Briefly describe the microstructure for each of the following microconstituents that are found in steel alloys: fine pearlite, coarse pearlite, spheroidite, bainite, martensite, and tempered martensite.
3. Cite the general mechanical characteristics for each of the following microconstituents: fine pearlite, coarse pearlite, spheroidite, bainite, martensite, and tempered martensite. Now, in terms of microstructure (or crystal structure), briefly explain these behaviors.
4. Given the isothermal transformation (or continuous cooling transformation) diagram for some iron–carbon alloy, design a heat treatment that will produce a specified microstructure.
5. Using a phase diagram, describe and explain the two heat treatments that are used to precipitation-harden a metal alloy.
6. Make a schematic plot of room-temperature strength (or hardness) versus the logarithm of time for a precipitation heat treatment at constant temperature. Explain the shape of this curve in terms of the mechanism of precipitation hardening.
7. Schematically plot specific volume versus temperature for crystalline, semicrystalline, and amorphous polymers, noting glass transition and melting temperatures.
{8. List four characteristics or structural components of a polymer that affect both its melting and glass transition temperatures.}

11.1 INTRODUCTION

Mechanical and other properties of many materials depend on their microstructures, which are often produced as a result of phase transformations. In the first portion of this chapter we discuss the basic principles of phase transformations. Next, we address the role these transformations play in the development of microstructure for iron–carbon, as well as other alloys, and how the mechanical properties are affected by these microstructural changes. Finally, we treat crystallization, melting, and glass transition transformations in polymers.

PHASE TRANSFORMATIONS IN METALS

One reason for the versatility of metallic materials lies in the wide range of mechanical properties they possess, which are accessible to management by various means. Three strengthening mechanisms were discussed in Chapter 8, namely, grain size refinement, solid-solution strengthening, and strain hardening. Additional techniques are available wherein the mechanical properties are reliant on the characteristics of the microstructure.

The development of microstructure in both single- and two-phase alloys ordinarily involves some type of phase transformation—an alteration in the number and/or character of the phases. The first portion of this chapter is devoted to a brief discussion of some of the basic principles relating to transformations involving solid phases. Inasmuch as most phase transformations do not occur instantaneously, consideration is given to the dependence of reaction progress on time, or the **transformation rate.** This is followed by a discussion of the development of two-phase microstructures for iron–carbon alloys. Modified phase diagrams are introduced which permit determination of the microstructure that results from a specific heat treatment. Finally, other microconstituents in addition to pearlite are presented, and, for each, the mechanical properties are discussed.

11.2 BASIC CONCEPTS

A variety of **phase transformations** are important in the processing of materials, and usually they involve some alteration of the microstructure. For purposes of this discussion, these transformations are divided into three classifications. In one group are simple diffusion-dependent transformations in which there is no change in either the number or composition of the phases present. These include solidification of a pure metal, allotropic transformations, and, recrystallization and grain growth (see Sections 8.13 and 8.14).

In another type of diffusion-dependent transformation, there is some alteration in phase compositions and often in the number of phases present; the final microstructure ordinarily consists of two phases. The eutectoid reaction, described by Equation 10.19, is of this type; it receives further attention in Section 11.5.

The third kind of transformation is diffusionless, wherein a metastable phase is produced. As discussed in Section 11.5, a martensitic transformation, which may be induced in some steel alloys, falls into this category.

11.3 THE KINETICS OF SOLID-STATE REACTIONS

Most solid-state transformations do not occur instantaneously because obstacles impede the course of the reaction and make it dependent on time. For example, since most transformations involve the formation of at least one new phase that has a composition and/or crystal structure different from that of the parent one, some atomic rearrangements via diffusion are required. Diffusion is a time-dependent phenomenon, as discussed in Section 6.4. A second impediment to the formation of a new phase is the increase in energy associated with the phase boundaries that are created between parent and product phases.

From a microstructural standpoint, the first process to accompany a phase transformation is **nucleation**—the formation of very small (often submicroscopic) particles, or nuclei, of the new phase, which are capable of growing. Favorable positions for the formation of these nuclei are imperfection sites, especially grain boundaries. The second stage is *growth*, in which the nuclei increase in size; during this process, of course, some volume of the parent phase disappears. The transformation reaches completion if growth of these new phase particles is allowed to proceed until the equilibrium fraction is attained.

As would be expected, the time dependence of the transformation rate (which is often termed the **kinetics** of a transformation) is an important consideration in the heat treatment of materials. With many kinetic investigations, the fraction of reaction that has occurred is measured as a function of time, while the temperature is maintained constant. Transformation progress is usually ascertained by either microscopic examination or measurement of some physical property (such as electrical conductivity) the magnitude of which is distinctive of the new phase. Data are plotted as the fraction of transformed material versus the logarithm of time; an S-shaped curve similar to that in Figure 11.1 represents the typical kinetic behavior for most solid-state reactions. Nucleation and growth stages are indicated in the figure.

For solid-state transformations displaying the kinetic behavior in Figure 11.1, the fraction of transformation y is a function of time t as follows:

$$y = 1 - \exp(-kt^n) \qquad (11.1)$$

where k and n are time-independent constants for the particular reaction. The above expression is often referred to as the *Avrami equation*.

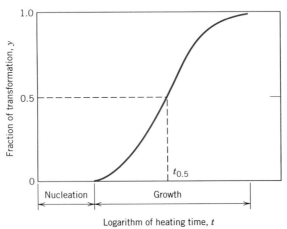

FIGURE 11.1 Plot of fraction reacted versus the logarithm of time typical of many solid-state transformations in which temperature is held constant.

By convention, the rate of a transformation r is taken as the reciprocal of time required for the transformation to proceed halfway to completion, $t_{0.5}$, or

$$r = \frac{1}{t_{0.5}} \qquad (11.2)$$

This $t_{0.5}$ is also noted in Figure 11.1.

Temperature is one variable in a heat treatment process that is subject to control, and it may have a profound influence on the kinetics and thus on the rate of a transformation. This is demonstrated in Figure 11.2, where y-versus-log t S-shaped curves at several temperatures for the recrystallization of copper are shown.

For most reactions and over specific temperature ranges, rate increases with temperature according to

$$r = Ae^{-Q/RT} \qquad (11.3)$$

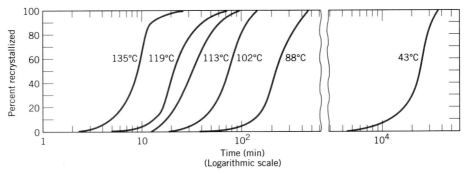

FIGURE 11.2 Percent recrystallization as a function of time and at constant temperature for pure copper. (Reprinted with permission from *Metallurgical Transactions*, Vol. 188, 1950, a publication of The Metallurgical Society of AIME, Warrendale, Pennsylvania. Adapted from B. F. Decker and D. Harker, "Recrystallization in Rolled Copper," *Trans. AIME,* **188,** 1950, p. 888.)

where

R = the gas constant

T = absolute temperature

A = a temperature-independent constant

Q = an activation energy for the particular reaction

It may be recalled that the diffusion coefficient has the same temperature dependence (Equation 6.8). Processes the rates of which exhibit this relationship with temperature are sometimes termed **thermally activated.**

11.4 MULTIPHASE TRANSFORMATIONS

Phase transformations may be wrought in metal alloy systems by varying temperature, composition, and the external pressure; however, temperature changes by means of heat treatments are most conveniently utilized to induce phase transformations. This corresponds to crossing a phase boundary on the composition–temperature phase diagram as an alloy of given composition is heated or cooled.

During a phase transformation, an alloy proceeds toward an equilibrium state that is characterized by the phase diagram in terms of the product phases, their compositions, and relative amounts. Most phase transformations require some finite time to go to completion, and the speed or rate is often important in the relationship between the heat treatment and the development of microstructure. One limitation of phase diagrams is their inability to indicate the time period required for the attainment of equilibrium.

The rate of approach to equilibrium for solid systems is so slow that true equilibrium structures are rarely achieved. Equilibrium conditions are maintained only if heating or cooling is carried out at extremely slow and unpractical rates. For other than equilibrium cooling, transformations are shifted to lower temperatures than indicated by the phase diagram; for heating, the shift is to higher temperatures. These phenomena are termed **supercooling** and **superheating,** respectively. The degree of each depends on the rate of temperature change; the more rapid the cooling or heating, the greater the supercooling or superheating. For example, for normal cooling rates the iron–carbon eutectoid reaction is typically displaced 10 to 20°C (18 to 36°F) below the equilibrium transformation temperature.

For many technologically important alloys, the preferred state or microstructure is a metastable one, intermediate between the initial and equilibrium states; on occasion, a structure far removed from the equilibrium one is desired. It thus becomes imperative to investigate the influence of time on phase transformations. This kinetic information is, in many instances, of greater value than a knowledge of the final equilibrium state.

MICROSTRUCTURAL AND PROPERTY CHANGES IN IRON–CARBON ALLOYS

Some of the basic kinetic principles of solid-state transformations are now extended and applied specifically to iron–carbon alloys in terms of the relationships between heat treatment, the development of microstructure, and mechanical properties.

This system has been chosen because it is familiar and because a wide variety of microstructures and mechanical properties are possible for iron–carbon (or steel) alloys.

11.5 ISOTHERMAL TRANSFORMATION DIAGRAMS

PEARLITE

Consider again the iron–iron carbide eutectoid reaction

$$\gamma(0.76 \text{ wt\% C}) \xrightleftharpoons[\text{heating}]{\text{cooling}} \alpha(0.022 \text{ wt\% C}) + Fe_3C(6.70 \text{ wt\% C}) \qquad (10.19)$$

which is fundamental to the development of microstructure in steel alloys. Upon cooling, austenite, having an intermediate carbon concentration, transforms to a ferrite phase, having a much lower carbon content, and also cementite, with a much higher carbon concentration. Pearlite is one microstructural product of this transformation (Figure 10.29), and the mechanism of pearlite formation was discussed previously (Section 10.19) and demonstrated in Figure 10.30.

Temperature plays an important role in the rate of the austenite-to-pearlite transformation. The temperature dependence for an iron–carbon alloy of eutectoid composition is indicated in Figure 11.3, which plots S-shaped curves of the percentage transformation versus the logarithm of time at three different temperatures. For each curve, data were collected after rapidly cooling a specimen composed of 100% austenite to the temperature indicated; that temperature was maintained constant throughout the course of the reaction.

A more convenient way of representing both the time and temperature dependence of this transformation is in the bottom portion of Figure 11.4. Here, the vertical and horizontal axes are, respectively, temperature and the logarithm of time. Two solid curves are plotted; one represents the time required at each temperature for the initiation or start of the transformation; the other is for the transformation conclusion. The dashed curve corresponds to 50% of transformation completion. These curves were generated from a series of plots of the percentage transformation versus the logarithm of time taken over a range of temperatures. The S-shaped curve [for 675°C (1247°F)], in the upper portion of Figure 11.4, illustrates how the data transfer is made.

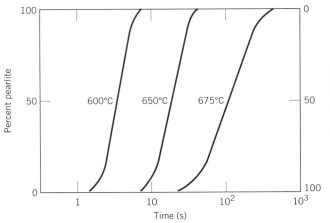

FIGURE 11.3 For an iron–carbon alloy of eutectoid composition (0.76 wt% C), isothermal fraction reacted versus the logarithm of time for the austenite-to-pearlite transformation.

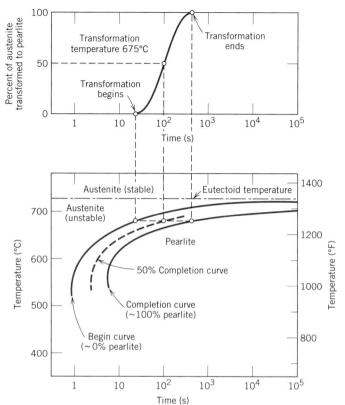

FIGURE 11.4
Demonstration of how
an isothermal transfor-
mation diagram (bot-
tom) is generated from
percent transformation-
versus-logarithm of time
measurements (top).
(Adapted from H. Boyer,
Editor, *Atlas of Isothermal Transformation and
Cooling Transformation
Diagrams*, American So-
ciety for Metals, 1977,
p. 369.)

In interpreting this diagram, note first that the eutectoid temperature [727°C (1341°F)] is indicated by a horizontal line; at temperatures above the eutectoid and for all times, only austenite will exist, as indicated in the figure. The austenite-to-pearlite transformation will occur only if an alloy is supercooled to below the eutectoid; as indicated by the curves, the time necessary for the transformation to begin and then end depends on temperature. The start and finish curves are nearly parallel, and they approach the eutectoid line asymptotically. To the left of the transformation start curve, only austenite (which is unstable) will be present, whereas to the right of the finish curve, only pearlite will exist. In between, the austenite is in the process of transforming to pearlite, and thus both microconstituents will be present.

According to Equation 11.2, the transformation rate at some particular temperature is inversely proportional to the time required for the reaction to proceed to 50% completion (to the dashed curve in Figure 11.4). That is, the shorter this time, the higher the rate. Thus, from Figure 11.4, at temperatures just below the eutectoid (corresponding to just a slight degree of undercooling) very long times (on the order of 10^5 s) are required for the 50% transformation, and therefore the reaction rate is very slow. The transformation rate increases with decreasing temperature such that at 540°C (1000°F) only about 3 s is required for the reaction to go to 50% completion.

This rate–temperature behavior is in apparent contradiction of Equation 11.3, which stipulates that rate *increases* with increasing temperature. The reason for this disparity is that over this range of temperatures (i.e., 540 to 727°C), the transformation rate is controlled by the rate of pearlite nucleation, and nucleation rate decreases with rising temperature (i.e., less supercooling). This behavior may be explained by Equation 11.3, wherein the activation energy Q for nucleation is a function of, and increases with, increasing temperature. We shall find that at lower temperatures, the austenite decomposition transformation is diffusion-controlled and that the rate behavior is as predicted by Equation 11.3, with a temperature-independent activation energy for diffusion.

Several constraints are imposed on using diagrams like Figure 11.4. First, this particular plot is valid only for an iron–carbon alloy of eutectoid composition; for other compositions, the curves will have different configurations. In addition, these plots are accurate only for transformations in which the temperature of the alloy is held constant throughout the duration of the reaction. Conditions of constant temperature are termed *isothermal;* thus, plots such as Figure 11.4 are referred to as **isothermal transformation diagrams,** or sometimes as *time–temperature–transformation* (or *T–T–T*) plots.

An actual isothermal heat treatment curve (*ABCD*) is superimposed on the isothermal transformation diagram for a eutectoid iron–carbon alloy in Figure 11.5. Very rapid cooling of austenite to a temperature is indicated by the near-vertical line *AB*, and the isothermal treatment at this temperature is represented by the horizontal segment *BCD*. Of course, time increases from left to right along this line. The transformation of austenite to pearlite begins at the intersection, point *C* (after approximately 3.5 s), and has reached completion by about 15 s, corresponding

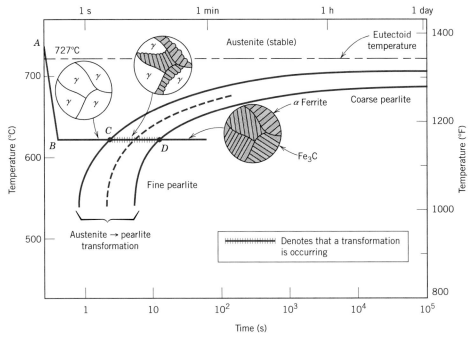

FIGURE 11.5   Isothermal transformation diagram for a eutectoid iron–carbon alloy, with superimposed isothermal heat treatment curve (*ABCD*). Microstructures before, during, and after the austenite-to-pearlite transformation are shown. (Adapted from H. Boyer, Editor, *Atlas of Isothermal Transformation and Cooling Transformation Diagrams,* American Society for Metals, 1977, p. 28.)

to point *D*. Figure 11.5 also shows schematic microstructures at various times during the progression of the reaction.

The thickness ratio of the ferrite and cementite layers in pearlite is approximately 8 to 1. However, the absolute layer thickness depends on the temperature at which the isothermal transformation is allowed to occur. At temperatures just below the eutectoid, relatively thick layers of both the α-ferrite and Fe₃C phases are produced; this microstructure is called **coarse pearlite,** and the region at which it forms is indicated to the right of the completion curve on Figure 11.5. At these temperatures, diffusion rates are relatively high, such that during the transformation illustrated in Figure 10.30 carbon atoms can diffuse relatively long distances, which results in the formation of thick lamellae. With decreasing temperature, the carbon diffusion rate decreases, and the layers become progressively thinner. The thin-layered structure produced in the vicinity of 540°C is termed **fine pearlite;** this is also indicated in Figure 11.5. To be discussed in Section 11.7 is the dependence of mechanical properties on lamellar thickness. Photomicrographs of coarse and fine pearlite for a eutectoid composition are shown in Figure 11.6.

For iron–carbon alloys of other compositions, a proeutectoid phase (either ferrite or cementite) will coexist with pearlite, as discussed in Section 10.19. Thus additional curves corresponding to a proeutectoid transformation also must be included on the isothermal transformation diagram. A portion of one such diagram for a 1.13 wt% C alloy is shown in Figure 11.7.

FIGURE 11.6 Photomicrographs of (*a*) coarse pearlite and (*b*) fine pearlite. 3000×. (From K. M. Ralls, et al., *An Introduction to Materials Science and Engineering,* p. 361. Copyright © 1976 by John Wiley & Sons, New York. Reprinted by permission of John Wiley & Sons, Inc.)

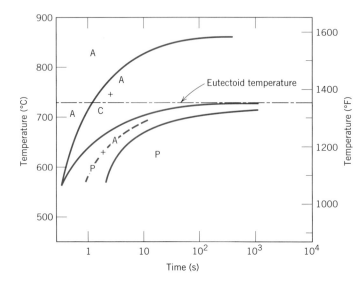

FIGURE 11.7
Isothermal transformation diagram for a 1.13 wt% C iron–carbon alloy: A, austenite; C, proeutectoid cementite; P, pearlite. (Adapted from H. Boyer, Editor, *Atlas of Isothermal Transformation and Cooling Transformation Diagrams*, American Society for Metals, 1977, p. 33.)

BAINITE

In addition to pearlite, other microconstituents that are products of the austenitic transformation exist; one of these is called **bainite.** The microstructure of bainite consists of ferrite and cementite phases, and thus diffusional processes are involved in its formation. Bainite forms as needles or plates, depending on the temperature of the transformation; the microstructural details of bainite are so fine that their resolution is possible only using electron microscopy. Figure 11.8 is an electron micrograph that shows a grain of bainite (positioned diagonally from lower left to upper right); it is composed of needles of ferrite that are separated by elongated particles of the Fe_3C phase; the various phases in this micrograph have been labeled. In addition, the phase that surrounds the needle is martensite, the topic to which a subsequent section is addressed. Furthermore, no proeutectoid phase forms with bainite.

The time–temperature dependence of the bainite transformation may also be represented on the isothermal transformation diagram. It occurs at temperatures below those at which pearlite forms; begin-, end-, and half-reaction curves are just extensions of those for the pearlitic transformation, as shown in Figure 11.9, the isothermal transformation diagram for an iron–carbon alloy of eutectoid composition that has been extended to lower temperatures. All three curves are C-shaped and have a "nose" at point *N*, where the rate of transformation is a maximum. As may be noted, whereas pearlite forms above the nose—that is, over the temperature range of about 540 to 727°C (1000 to 1341°F)—for isothermal treatments at temperatures between about 215 and 540°C (420 and 1000°F), bainite is the transformation product.

It should also be noted that pearlitic and bainitic transformations are really competitive with each other, and once some portion of an alloy has transformed to either pearlite or bainite, transformation to the other microconstituent is not possible without reheating to form austenite.

In passing, it should be mentioned that the kinetics of the bainite transformation (below the nose in Figure 11.9) obey Equation 11.3; that is, rate ($1/t_{0.5}$, Equation 11.2) increases exponentially with rising temperature. Furthermore, the kinetics of

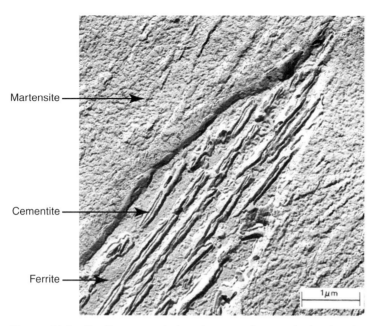

FIGURE 11.8   Replica transmission electron micrograph showing the structure of bainite. A grain of bainite passes from lower left to upper right-hand corners, which consists of elongated and needle-shaped particles of Fe₃C within a ferrite matrix. The phase surrounding the bainite is martensite. (Reproduced with permission from *Metals Handbook*, Vol. 8, 8th edition, *Metallography, Structures and Phase Diagrams*, American Society for Metals, Materials Park, OH, 1973.)

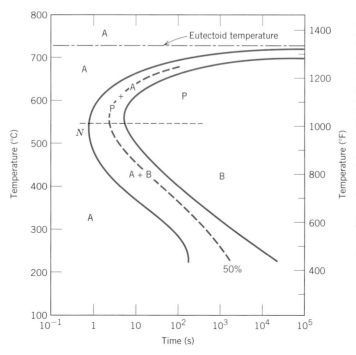

FIGURE 11.9 Isothermal transformation diagram for an iron–carbon alloy of eutectoid composition, including austenite-to-pearlite (A–P) and austenite-to-bainite (A–B) transformations. (Adapted from H. Boyer, Editor, *Atlas of Isothermal Transformation and Cooling Transformation Diagrams*, American Society for Metals, 1977, p. 28.)

many solid-state transformations are represented by this characteristic C-shaped curve (Figure 11.9).

SPHEROIDITE

If a steel alloy having either pearlitic or bainitic microstructures is heated to, and left at, a temperature below the eutectoid for a sufficiently long period of time—for example, at about 700°C (1300°F) for between 18 and 24 h—yet another microstructure will form. It is called **spheroidite** (Figure 11.10). Instead of the alternating ferrite and cementite lamellae (pearlite), or the microstructures observed for bainite, the Fe_3C phase appears as spherelike particles embedded in a continuous α phase matrix. This transformation has occurred by additional carbon diffusion with no change in the compositions or relative amounts of ferrite and cementite phases. The photomicrograph in Figure 11.11 shows a pearlitic steel that has partially transformed to spheroidite. The driving force for this transformation is the reduction in α–Fe_3C phase boundary area. The kinetics of spheroidite formation are not included on isothermal transformation diagrams.

MARTENSITE

Yet another microconstituent or phase called **martensite** is formed when austenitized iron–carbon alloys are rapidly cooled (or quenched) to a relatively low temperature (in the vicinity of the ambient). Martensite is a nonequilibrium single-phase structure that results from a diffusionless transformation of austenite. It may be thought of as a transformation product that is competitive with pearlite and bainite. The martensitic transformation occurs when the quenching rate is rapid enough to prevent carbon diffusion. Any diffusion whatsoever will result in the formation of ferrite and cementite phases.

The martensitic transformation is not well understood. However, large numbers of atoms experience cooperative movements, in that there is only a slight displacement of each atom relative to its neighbors. This occurs in such a way that the FCC austenite experiences a polymorphic transformation to a body-centered tetragonal (BCT) martensite. A unit cell of this crystal structure (Figure

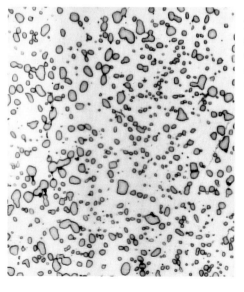

FIGURE 11.10 Photomicrograph of a steel having a spheroidite microstructure. The small particles are cementite; the continuous phase is α ferrite. 1000×. (Copyright 1971 by United States Steel Corporation.)

FIGURE 11.11 A photomicrograph of a pearlitic steel that has partially transformed to spheroidite. 1400×. (Courtesy of United States Steel Corporation.)

11.12) is simply a body-centered cube that has been elongated along one of its dimensions; this structure is distinctly different from that for BCC ferrite. All the carbon atoms remain as interstitial impurities in martensite; as such, they constitute a supersaturated solid solution that is capable of rapidly transforming to other structures if heated to temperatures at which diffusion rates become appreciable. Many steels, however, retain their martensitic structure almost indefinitely at room temperature.

The martensitic transformation is not, however, unique to iron–carbon alloys. It is found in other systems and is characterized, in part, by the diffusionless transformation.

Since the martensitic transformation does not involve diffusion, it occurs almost instantaneously; the martensite grains nucleate and grow at a very rapid rate—the velocity of sound within the austenite matrix. Thus the martensitic transformation rate, for all practical purposes, is time independent.

Martensite grains take on a platelike or needlelike appearance, as indicated in Figure 11.13. The white phase in the micrograph is austenite (retained austenite)

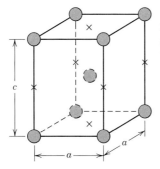

FIGURE 11.12 The body-centered tetragonal unit cell for martensitic steel showing iron atoms (circles) and sites that may be occupied by carbon atoms (crosses). For this tetragonal unit cell, $c > a$.

FIGURE 11.13 Photomicrograph showing the martensitic microstructure. The needle-shaped grains are the martensite phase, and the white regions are austenite that failed to transform during the rapid quench. 1220×. (Photomicrograph courtesy of United States Steel Corporation.)

that did not transform during the rapid quench. As has already been mentioned, martensite as well as other microconstituents (e.g., pearlite) can coexist.

Being a nonequilibrium phase, martensite does not appear on the iron–iron carbide phase diagram (Figure 10.26). The austenite-to-martensite transformation is, however, represented on the isothermal transformation diagram. Since the mar-

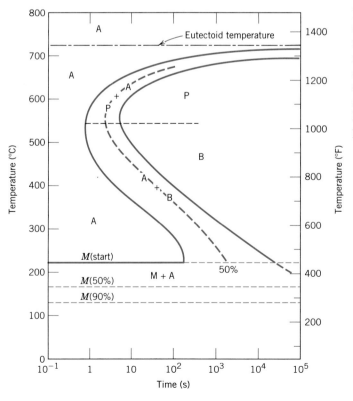

FIGURE 11.14 The complete isothermal transformation diagram for an iron–carbon alloy of eutectoid composition: A, austenite; B, bainite; M, martensite; P, pearlite.

tensitic transformation is diffusionless and instantaneous, it is not depicted in this diagram like the pearlitic and bainitic reactions. The beginning of this transformation is represented by a horizontal line designated M(start) (Figure 11.14). Two other horizontal and dashed lines, labeled M(50%) and M(90%), indicate percentages of the austenite-to-martensite transformation. The temperatures at which these lines are located vary with alloy composition but, nevertheless, must be relatively low because carbon diffusion must be virtually nonexistent. The horizontal and linear character of these lines indicates that the martensitic transformation is independent of time; it is a function only of the temperature to which the alloy is quenched or rapidly cooled. A transformation of this type is termed an **athermal transformation.**

Consider an alloy of eutectoid composition that is very rapidly cooled from a temperature above 727°C (1341°F) to, say, 165°C (330°F). From the isothermal transformation diagram (Figure 11.14) it may be noted that 50% of the austenite will immediately transform to martensite; and as long as this temperature is maintained, there will be no further transformation.

The presence of alloying elements other than carbon (e.g., Cr, Ni, Mo, and W) may cause significant changes in the positions and shapes of the curves in the isothermal transformation diagrams. These include (1) shifting to longer times the nose of the austenite-to-pearlite transformation (and also a proeutectoid phase nose, if such exists), and (2) the formation of a separate bainite nose. These alterations may be observed by comparing Figures 11.14 and 11.15, which are isothermal transformation diagrams for carbon and alloy steels, respectively.

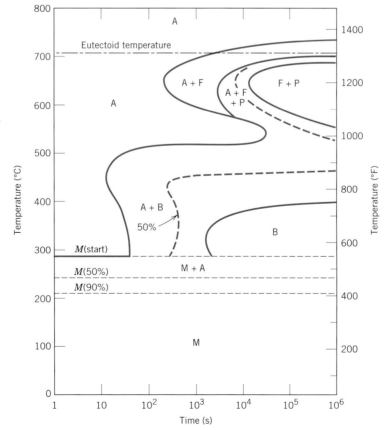

FIGURE 11.15 Isothermal transformation diagram for an alloy steel (type 4340): A, austenite; B, bainite; P, pearlite; M, martensite; F, proeutectoid ferrite. (Adapted from H. Boyer, Editor, *Atlas of Isothermal Transformation and Cooling Transformation Diagrams*, American Society for Metals, 1977, p. 181.)

Steels in which carbon is the prime alloying element are termed **plain carbon steels,** whereas **alloy steels** contain appreciable concentrations of other elements, including those cited in the preceding paragraph. Chapter 13 tells more about the classification and properties of ferrous alloys.

EXAMPLE PROBLEM 11.1

Using the isothermal transformation diagram for an iron–carbon alloy of eutectoid composition (Figure 11.14), specify the nature of the final microstructure (in terms of microconstituents present and approximate percentages) of a small specimen that has been subjected to the following time–temperature treatments. In each case assume that the specimen begins at 760°C (1400°F) and that it has been held at this temperature long enough to have achieved a complete and homogeneous austenitic structure.

(a) Rapidly cool to 350°C (660°F), hold for 10^4 s, and quench to room temperature.

(b) Rapidly cool to 250°C (480°F), hold for 100 s, and quench to room temperature.

(c) Rapidly cool to 650°C (1200°F), hold for 20 s, rapidly cool to 400°C (750°F), hold for 10^3 s, and quench to room temperature.

SOLUTION

The time–temperature paths for all three treatments are shown in Figure 11.16. In each case the initial cooling is rapid enough to prevent any transformation from occurring.

(a) At 350°C austenite isothermally transforms to bainite; this reaction begins after about 10 s and reaches completion at about 500 s elapsed time. Therefore, by 10^4 s, as stipulated in this problem, 100% of the specimen is bainite, and no further transformation is possible, even though the final quenching line passes through the martensite region of the diagram.

(b) In this case it takes about 150 s at 250°C for the bainite transformation to begin, so that at 100 s the specimen is still 100% austenite. As the specimen is cooled through the martensite region, beginning at about 215°C, progressively more of the austenite instantaneously transforms to martensite. This transformation is complete by the time room temperature is reached, such that the final microstructure is 100% martensite.

(c) For the isothermal line at 650°C, pearlite begins to form after about 7 s; by the time 20 s has elapsed, only approximately 50% of the specimen has transformed to pearlite. The rapid cool to 400°C is indicated by the vertical line; during this cooling, very little, if any, remaining austenite will transform to either pearlite or bainite, even though the cooling line passes through pearlite and bainite regions of the diagram. At 400°C, we begin timing at essentially zero time (as indicated in Figure 11.16); thus, by the time 10^3 s has elapsed, all of the remaining 50% austenite will have completely transformed to bainite. Upon quenching to room temperature, any further transformation is not possible inasmuch as no austenite remains; and so the final microstructure at room temperature consists of 50% pearlite and 50% bainite.

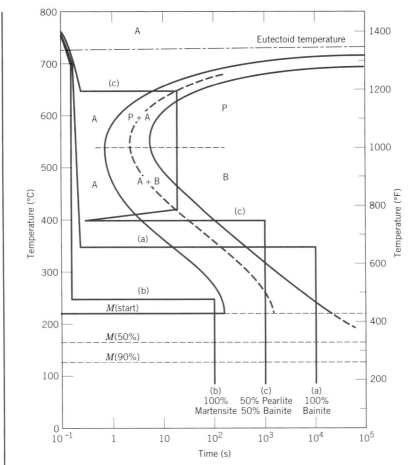

FIGURE 11.16 Isothermal transformation diagram for an iron–carbon alloy of eutectoid composition and the isothermal heat treatments (a), (b), and (c) in Example Problem 11.1.

11.6 CONTINUOUS COOLING TRANSFORMATION DIAGRAMS (CD-ROM)

11.7 MECHANICAL BEHAVIOR OF IRON–CARBON ALLOYS

We shall now discuss the mechanical behavior of iron–carbon alloys having the microstructures discussed heretofore, namely, fine and coarse pearlite, spheroidite, bainite, and martensite. For all but martensite, two phases are present (i.e., ferrite and cementite); and so an opportunity is provided to explore several mechanical property–microstructure relationships that exist for these alloys.

PEARLITE

Cementite is much harder but more brittle than ferrite. Thus, increasing the fraction of Fe_3C in a steel alloy while holding other microstructural elements constant will result in a harder and stronger material. This is demonstrated in Figure 11.21a, in which the tensile and yield strengths as well as the Brinell hardness number are plotted as a function of the weight percent carbon (or equivalently as the percent of Fe_3C) for steels that are composed of fine pearlite. All three parameters increase with increasing carbon concentration. Inasmuch as cementite is more brittle, increasing its content will result in a decrease in both ductility and toughness (or impact energy). These effects are shown in Figure 11.21b for the same fine pearlitic steels.

The layer thickness of each of the ferrite and cementite phases in the microstructure also influences the mechanical behavior of the material. Fine pearlite is harder and stronger than coarse pearlite, as demonstrated in Figure 11.22a, which plots hardness versus the carbon concentration.

The reasons for this behavior relate to phenomena that occur at the α–Fe_3C phase boundaries. First, there is a large degree of adherence between the two phases across a boundary. Therefore, the strong and rigid cementite phase severely restricts deformation of the softer ferrite phase in the regions adjacent to the boundary; thus the cementite may be said to reinforce the ferrite. The degree of this reinforcement is substantially higher in fine pearlite because of the greater phase boundary area per unit volume of material. In addition, phase boundaries serve as barriers to dislocation motion in much the same way as grain boundaries (Section 8.9). For fine pearlite there are more boundaries through which a dislocation must pass during plastic deformation. Thus, the greater reinforcement and restriction of dislocation motion in fine pearlite account for its greater hardness and strength.

Coarse pearlite is more ductile than fine pearlite, as illustrated in Figure 11.22b, which plots percent reduction in area versus carbon concentration for both microstructure types. This behavior results from the greater restriction to plastic deformation of the fine pearlite.

SPHEROIDITE

Other elements of the microstructure relate to the shape and distribution of the phases. In this respect, the cementite phase has distinctly different shapes and arrangements in the pearlite and spheroidite microstructures (Figures 11.6 and 11.10). Alloys containing pearlitic microstructures have greater strength and hardness than do those with spheroidite. This is demonstrated in Figure 11.22a, which compares the hardness as a function of the weight percent carbon for spheroidite with both the other pearlite structure types. This behavior is again explained in terms of reinforcement at, and impedance to, dislocation motion across the ferrite–cementite boundaries as discussed above. There is less boundary area per unit volume in spheroidite, and consequently plastic deformation is not nearly as constrained, which gives rise to a relatively soft and weak material. In fact, of all steel alloys, those that are softest and weakest have a spheroidite microstructure.

As would be expected, spheroidized steels are extremely ductile, much more than either fine or coarse pearlite (Figure 11.22b). In addition, they are notably tough because any crack can encounter only a very small fraction of the brittle cementite particles as it propagates through the ductile ferrite matrix.

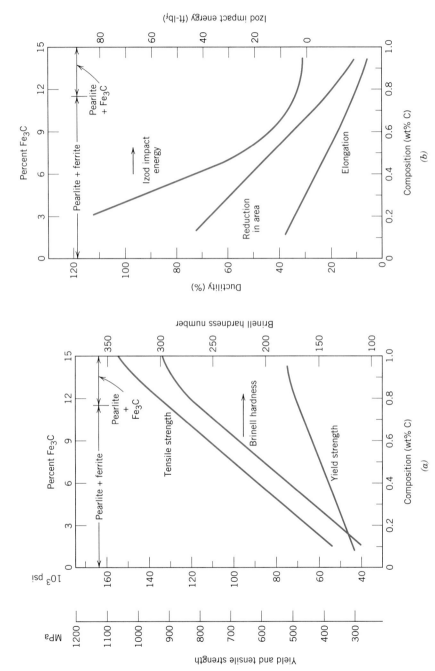

FIGURE 11.21 (a) Yield strength, tensile strength, and Brinell hardness versus carbon concentration for plain carbon steels having microstructures consisting of fine pearlite. (b) Ductility (%EL and %RA) and Izod impact energy versus carbon concentration for plain carbon steels having microstructures consisting of fine pearlite. (Data taken from Metals Handbook: Heat Treating, Vol. 4, 9th edition, V. Masseria, Managing Editor, American Society for Metals, 1981, p. 9.)

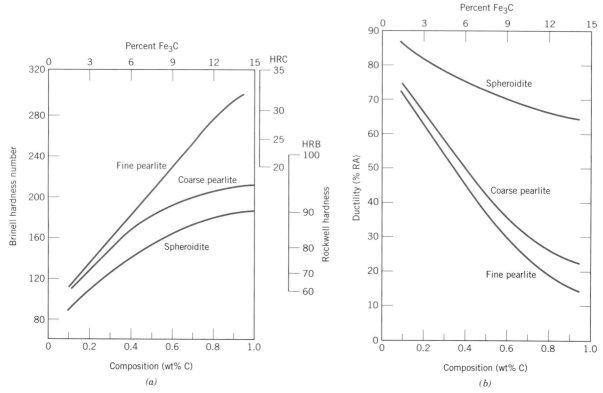

FIGURE 11.22   (a) Brinell and Rockwell hardness as a function of carbon concentration for plain carbon steels having fine and coarse pearlite as well as spheroidite microstructures. (b) Ductility (%RA) as a function of carbon concentration for plain carbon steels having fine and coarse pearlite as well as spheroidite microstructures. (Data taken from *Metals Handbook: Heat Treating,* Vol. 4, 9th edition, V. Masseria, Managing Editor, American Society for Metals, 1981, pp. 9 and 17.)

BAINITE

Because bainitic steels have a finer structure (i.e., smaller α-ferrite and Fe_3C particles), they are generally stronger and harder than pearlitic ones; yet they exhibit a desirable combination of strength and ductility. Figure 11.23 shows the influence of transformation temperature on the tensile strength and hardness for an iron–carbon alloy of eutectoid composition; temperature ranges over which pearlite and bainite form (consistent with the isothermal transformation diagram for this alloy, Figure 11.9) are noted at the top of Figure 11.23.

MARTENSITE

Of the various microstructures that may be produced for a given steel alloy, martensite is the hardest and strongest and, in addition, the most brittle; it has, in fact, negligible ductility. Its hardness is dependent on the carbon content, up to about 0.6 wt% as demonstrated in Figure 11.24, which plots the hardness of martensite and fine pearlite as a function of weight percent carbon. In contrast to pearlitic steels, strength and hardness of martensite are not thought to be related to microstructure.

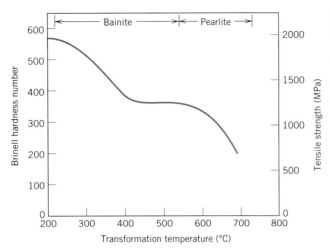

FIGURE 11.23
Brinell hardness and tensile strength as a function of isothermal transformation temperature for an iron–carbon alloy of eutectoid composition, taken over the temperature range at which bainitic and pearlitic microstructures form. (Adapted from E. S. Davenport, "Isothermal Transformation in Steels," *Trans. ASM*, **27**, 1939, p. 847. Reprinted by permission of ASM International.)

Rather, these properties are attributed to the effectiveness of the interstitial carbon atoms in hindering dislocation motion (as a solid-solution effect, Section 8.10), and to the relatively few slip systems (along which dislocations move) for the BCT structure.

Austenite is slightly denser than martensite, and therefore, during the phase transformation upon quenching, there is a net volume increase. Consequently, relatively large pieces that are rapidly quenched may crack as a result of internal stresses; this becomes a problem especially when the carbon content is greater than about 0.5 wt%.

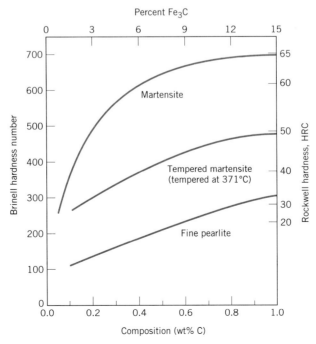

FIGURE 11.24 Hardness as a function of carbon concentration for plain carbon martensitic, tempered martensitic [tempered at 371°C (700°F)], and pearlitic steels. (Adapted from Edgar C. Bain, *Functions of the Alloying Elements in Steel*, American Society for Metals, 1939, p. 36; and R. A. Grange, C. R. Hribal, and L. F. Porter: *Metall. Trans. A*, Vol. 8A, p. 1776.)

11.8 TEMPERED MARTENSITE

In the as-quenched state, martensite, in addition to being very hard, is so brittle that it cannot be used for most applications; also, any internal stresses that may have been introduced during quenching have a weakening effect. The ductility and toughness of martensite may be enhanced and these internal stresses relieved by a heat treatment known as *tempering*.

Tempering is accomplished by heating a martensitic steel to a temperature below the eutectoid for a specified time period. Normally, tempering is carried out at temperatures between 250 and 650°C (480 and 1200°F); internal stresses, however, may be relieved at temperatures as low as 200°C (390°F). This tempering heat treatment allows, by diffusional processes, the formation of **tempered martensite,** according to the reaction

$$\text{martensite (BCT, single phase)} \rightarrow \text{tempered martensite } (\alpha + \text{Fe}_3\text{C phases})$$

(11.4)

where the single-phase BCT martensite, which is supersaturated with carbon, transforms to the tempered martensite, composed of the stable ferrite and cementite phases, as indicated on the iron–iron carbide phase diagram.

The microstructure of tempered martensite consists of extremely small and uniformly dispersed cementite particles embedded within a continuous ferrite matrix. This is similar to the microstructure of spheroidite except that the cementite particles are much, much smaller. An electron micrograph showing the microstructure of tempered martensite at a very high magnification is presented in Figure 11.25.

Tempered martensite may be nearly as hard and strong as martensite, but with substantially enhanced ductility and toughness. For example, on the hardness-versus-weight percent carbon plot of Figure 11.24 is included a curve for tempered martensite. The hardness and strength may be explained by the large ferrite–cementite phase boundary area per unit volume that exists for the very fine and numerous cementite particles. Again, the hard cementite phase reinforces the ferrite matrix along the boundaries, and these boundaries also act as barriers to dislocation motion during plastic deformation. The continuous ferrite phase is also very ductile and relatively tough, which accounts for the improvement of these two properties for tempered martensite.

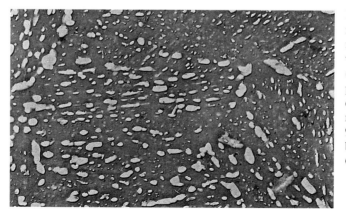

FIGURE 11.25 Electron micrograph of tempered martensite. Tempering was carried out at 594°C (1100°F). The small particles are the cementite phase; the matrix phase is α ferrite. 9300×. (Copyright 1971 by United States Steel Corporation.)

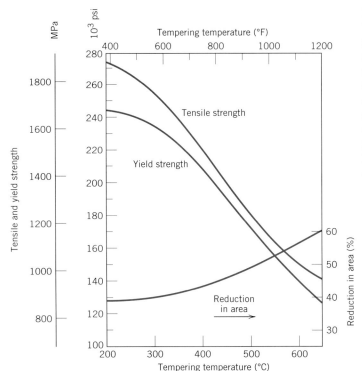

FIGURE 11.26
Tensile and yield
strengths and ductility
(%RA) versus
tempering
temperature for an
oil-quenched alloy
steel (type 4340).
(Adapted from figure
furnished courtesy
Republic Steel
Corporation.)

The size of the cementite particles influences the mechanical behavior of tempered martensite; increasing the particle size decreases the ferrite–cementite phase boundary area and, consequently, results in a softer and weaker material yet one that is tougher and more ductile. Furthermore, the tempering heat treatment determines the size of the cementite particles. Heat treatment variables are temperature and time, and most treatments are constant-temperature processes. Since carbon diffusion is involved in the martensite-tempered martensite transformation, increasing the temperature will accelerate diffusion, the rate of cementite particle growth, and, subsequently, the rate of softening. The dependence of tensile and yield strength and ductility on tempering temperature for an alloy steel is shown in Figure 11.26. Before tempering, the material was quenched in oil to produce the martensitic structure; the tempering time at each temperature was 1 h. This type of tempering data is ordinarily provided by the steel manufacturer.

The time dependence of hardness at several different temperatures is presented in Figure 11.27 for a water-quenched steel of eutectoid composition; the time scale is logarithmic. With increasing time the hardness decreases, which corresponds to the growth and coalescence of the cementite particles. At temperatures approaching the eutectoid [700°C (1300°F)] and after several hours, the microstructure will have become spheroiditic (Figure 11.10), with large cementite spheroids embedded within the continuous ferrite phase. Correspondingly, overtempered martensite is relatively soft and ductile.

TEMPER EMBRITTLEMENT

The tempering of some steels may result in a reduction of toughness as measured by impact tests (Section 9.8); this is termed *temper embrittlement*. The phenomenon

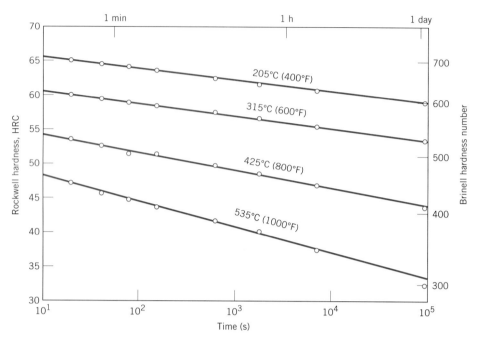

FIGURE 11.27 Hardness versus tempering time for a water-quenched eutectoid plain carbon (1080) steel. (Adapted from Edgar C. Bain, *Functions of the Alloying Elements in Steel*, American Society for Metals, 1939, p. 233.)

occurs when the steel is tempered at a temperature above about 575°C (1070°F) followed by slow cooling to room temperature, or when tempering is carried out at between approximately 375 and 575°C (700 and 1070°F). Steel alloys that are susceptible to temper embrittlement have been found to contain appreciable concentrations of the alloying elements manganese, nickel, or chromium and, in addition, one or more of antimony, phosphorus, arsenic, and tin as impurities in relatively low concentrations. The presence of these alloying elements and impurities shifts the ductile-to-brittle transition to significantly higher temperatures; the ambient temperature thus lies below this transition in the brittle regime. It has been observed that crack propagation of these embrittled materials is intergranular; that is, the fracture path is along the grain boundaries of the precursor austenite phase. Furthermore, alloy and impurity elements have been found to preferentially segregate in these regions.

Temper embrittlement may be avoided by (1) compositional control; and/or (2) tempering above 575°C or below 375°C, followed by quenching to room temperature. Furthermore, the toughness of steels that have been embrittled may be improved significantly by heating to about 600°C (1100°F) and then rapidly cooling to below 300°C (570°F).

11.9 REVIEW OF PHASE TRANSFORMATIONS FOR IRON–CARBON ALLOYS

In this chapter several different microstructures that may be produced in iron–carbon alloys depending on heat treatment have been discussed. Figure 11.28 sum-

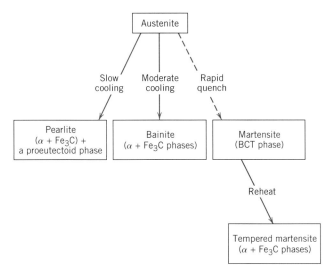

FIGURE 11.28 Possible transformations involving the decomposition of austenite. Solid arrows, transformations involving diffusion; dashed arrow, diffusionless transformation.

marizes the transformation paths that produce these various microstructures. Here, it is assumed that pearlite, bainite, and martensite result from continuous cooling treatments; furthermore, the formation of bainite is only possible for alloy steels (not plain carbon ones) as outlined above.

PRECIPITATION HARDENING

The strength and hardness of some metal alloys may be enhanced by the formation of extremely small uniformly dispersed particles of a second phase within the original phase matrix; this must be accomplished by phase transformations that are induced by appropriate heat treatments. The process is called **precipitation hardening** because the small particles of the new phase are termed "precipitates." "Age hardening" is also used to designate this procedure because the strength develops with time, or as the alloy ages. Examples of alloys that are hardened by precipitation treatments include aluminum–copper, copper–beryllium, copper–tin, and magnesium–aluminum; some ferrous alloys are also precipitation hardenable.

Precipitation hardening and the treating of steel to form tempered martensite are totally different phenomena, even though the heat treatment procedures are similar; therefore, the processes should not be confused. The principal difference lies in the mechanisms by which hardening and strengthening are achieved. These should become apparent as precipitation hardening is explained.

11.10 HEAT TREATMENTS

Inasmuch as precipitation hardening results from the development of particles of a new phase, an explanation of the heat treatment procedure is facilitated by use of a phase diagram. Even though, in practice, many precipitation-hardenable alloys contain two or more alloying elements, the discussion is simplified by reference to a binary system. The phase diagram must be of the form shown for the hypothetical A–B system in Figure 11.29.

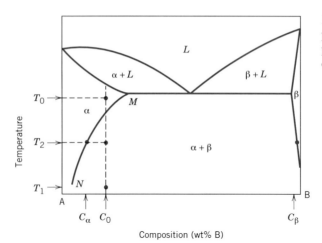

FIGURE 11.29 Hypothetical phase diagram for a precipitation hardenable alloy of composition C_0.

Two requisite features must be displayed by the phase diagrams of alloy systems for precipitation hardening: an appreciable maximum solubility of one component in the other, on the order of several percent; and a solubility limit that rapidly decreases in concentration of the major component with temperature reduction. Both these conditions are satisfied by this hypothetical phase diagram (Figure 11.29). The maximum solubility corresponds to the composition at point M. In addition, the solubility limit boundary between the α and $\alpha + \beta$ phase fields diminishes from this maximum concentration to a very low B content in A at point N. Furthermore, the composition of a precipitation-hardenable alloy must be less than the maximum solubility. These conditions are necessary but *not* sufficient for precipitation hardening to occur in an alloy system. An additional requirement is discussed below.

SOLUTION HEAT TREATING

Precipitation hardening is accomplished by two different heat treatments. The first is a **solution heat treatment** in which all solute atoms are dissolved to form a single-phase solid solution. Consider an alloy of composition C_0 in Figure 11.29. The treatment consists of heating the alloy to a temperature within the α phase field—say, T_0—and waiting until all the β phase that may have been present is completely dissolved. At this point, the alloy consists only of an α phase of composition C_0. This procedure is followed by rapid cooling or quenching to temperature T_1, which for many alloys is room temperature, to the extent that any diffusion and the accompanying formation of any of the β phase is prevented. Thus, a nonequilibrium situation exists in which only the α phase solid solution supersaturated with B atoms is present at T_1; in this state the alloy is relatively soft and weak. Furthermore, for most alloys diffusion rates at T_1 are extremely slow, such that the single α phase is retained at this temperature for relatively long periods.

PRECIPITATION HEAT TREATING

For the second or **precipitation heat treatment,** the supersaturated α solid solution is ordinarily heated to an intermediate temperature T_2 (Figure 11.29) within the $\alpha + \beta$ two-phase region, at which temperature diffusion rates become appreciable. The β precipitate phase begins to form as finely dispersed particles of composition C_β, which process is sometimes termed "aging." After the appropriate aging time

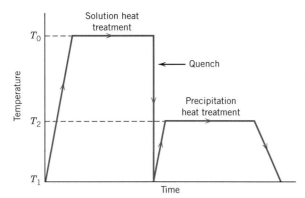

FIGURE 11.30 Schematic temperature-versus-time plot showing both solution and precipitation heat treatments for precipitation hardening.

at T_2, the alloy is cooled to room temperature; normally, this cooling rate is not an important consideration. Both solution and precipitation heat treatments are represented on the temperature-versus-time plot, Figure 11.30. The character of these β particles, and subsequently the strength and hardness of the alloy, depend on both the precipitation temperature T_2 and the aging time at this temperature. For some alloys, aging occurs spontaneously at room temperature over extended time periods.

The dependence of the growth of the precipitate β particles on time and temperature under isothermal heat treatment conditions may be represented by C-shaped curves similar to those in Figure 11.9 for the eutectoid transformation in steels. However, it is more useful and convenient to present the data as tensile strength, yield strength, or hardness at room temperature as a function of the logarithm of aging time, at constant temperature T_2. The behavior for a typical precipitation-hardenable alloy is represented schematically in Figure 11.31. With increasing time, the strength or hardness increases, reaches a maximum, and finally diminishes. This reduction in strength and hardness that occurs after long time periods is known as **overaging.** The influence of temperature is incorporated by the superposition, on a single plot, of curves at a variety of temperatures.

11.11 MECHANISM OF HARDENING

Precipitation hardening is commonly employed with high-strength aluminum alloys. Although a large number of these alloys have different proportions and combina-

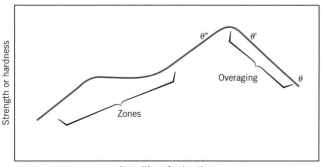

FIGURE 11.31 Schematic diagram showing strength and hardness as a function of the logarithm of aging time at constant temperature during the precipitation heat treatment.

FIGURE 11.32   The aluminum-rich side of the aluminum–copper phase diagram. (Adapted from J. L. Murray, *International Metals Review*, **30**, 5, 1985. Reprinted by permission of ASM International.)

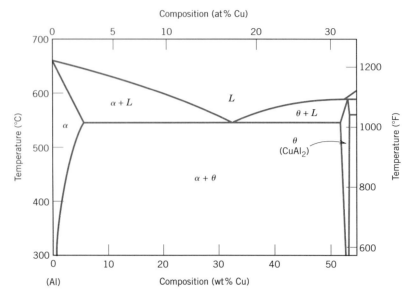

tions of alloying elements, the mechanism of hardening has perhaps been studied most extensively for the aluminum–copper alloys. Figure 11.32 presents the aluminum-rich portion of the aluminum–copper phase diagram. The α phase is a substitutional solid solution of copper in aluminum, whereas the intermetallic compound $CuAl_2$ is designated the θ phase. For an aluminum–copper alloy of, say, composition 96 wt% Al–4 wt% Cu, in the development of this equilibrium θ phase during the precipitation heat treatment, several transition phases are first formed in a specific sequence. The mechanical properties are influenced by the character of the particles of these transition phases. During the initial hardening stage (at short times, Figure 11.31), copper atoms cluster together in very small and thin discs that are only one or two atoms thick and approximately 25 atoms in diameter; these form at countless positions within the α phase. The clusters, sometimes called zones, are so small that they are really not regarded as distinct precipitate particles. However, with time and the subsequent diffusion of copper atoms, zones become particles as they increase in size. These precipitate particles then pass through two transition phases (denoted as θ'' and θ'), before the formation of the equilibrium θ phase (Figure 11.33c). Transition phase particles for a precipitation-hardened 7150 aluminum alloy are shown in the electron micrograph of the chapter-opening photograph for this chapter.

The strengthening and hardening effects shown in Figure 11.31 result from the innumerable particles of these transition and metastable phases. As noted in the figure, maximum strength coincides with the formation of the θ'' phase, which may be preserved upon cooling the alloy to room temperature. Overaging results from continued particle growth and the development of θ' and θ phases.

The strengthening process is accelerated as the temperature is increased. This is demonstrated in Figure 11.34a, a plot of tensile strength versus the logarithm of time for a 2014 aluminum alloy at several different precipitation temperatures. Ideally, temperature and time for the precipitation heat treatment should be designed to produce a hardness or strength in the vicinity of the maximum. Associated with an increase in strength is a reduction in ductility. This is demonstrated in Figure 11.34b for the same 2014 aluminum alloy at the several temperatures.

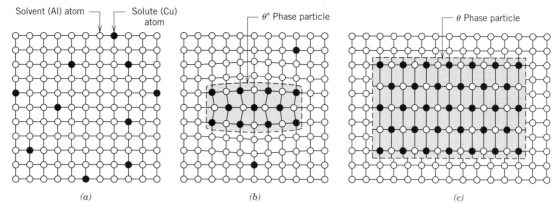

FIGURE 11.33 Schematic depiction of several stages in the formation of the equilibrium precipitate (θ) phase. (a) A supersaturated α solid solution. (b) A transition, θ'', precipitate phase. (c) The equilibrium θ phase, within the α matrix phase. Actual phase particle sizes are much larger than shown here.

Not all alloys that satisfy the aforementioned conditions relative to composition and phase diagram configuration are amenable to precipitation hardening. In addition, lattice strains must be established at the precipitate–matrix interface. For aluminum–copper alloys, there is a distortion of the crystal lattice structure around and within the vicinity of particles of these transition phases (Figure 11.33b). During plastic deformation, dislocation motions are effectively impeded as a result of these distortions, and, consequently, the alloy becomes harder and stronger. As the θ phase forms, the resultant overaging (softening and weakening) is explained by a reduction in the resistance to slip that is offered by these precipitate particles.

Alloys that experience appreciable precipitation hardening at room temperature and after relatively short time periods must be quenched to and stored under refrigerated conditions. Several aluminum alloys that are used for rivets exhibit this behavior. They are driven while still soft, then allowed to age harden at the normal ambient temperature. This is termed **natural aging**; **artificial aging** is carried out at elevated temperatures.

11.12 MISCELLANEOUS CONSIDERATIONS

The combined effects of strain hardening and precipitation hardening may be employed in high-strength alloys. The order of these hardening procedures is important in the production of alloys having the optimum combination of mechanical properties. Normally, the alloy is solution heat treated and then quenched. This is followed by cold working and finally by the precipitation hardening heat treatment. In the final treatment, little strength loss is sustained as a result of recrystallization. If the alloy is precipitation hardened before cold working, more energy must be expended in its deformation; in addition, cracking may also result because of the reduction in ductility that accompanies the precipitation hardening.

Most precipitation-hardened alloys are limited in their maximum service temperatures. Exposure to temperatures at which aging occurs may lead to a loss of strength due to overaging.

FIGURE 11.34 The precipitation hardening characteristics of a 2014 aluminum alloy (0.9 wt% Si, 4.4 wt% Cu, 0.8 wt% Mn, 0.5 wt% Mg) at four different aging temperatures: (a) tensile strength, and (b) ductility (%EL). (Adapted from *Metals Handbook: Properties and Selection: Nonferrous Alloys and Pure Metals,* Vol. 2, 9th edition, H. Baker, Managing Editor, American Society for Metals, 1979, p. 41.)

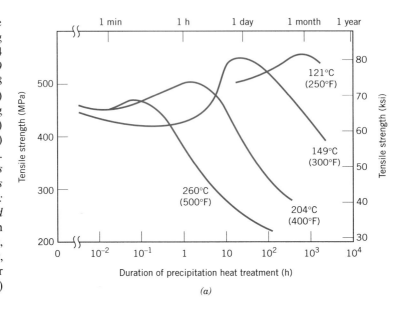

(a)

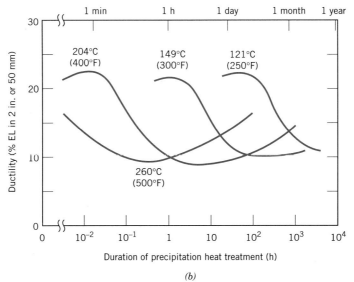

(b)

CRYSTALLIZATION, MELTING, AND GLASS TRANSITION PHENOMENA IN POLYMERS

Phase transformation phenomena are important with respect to the design and processing of polymeric materials. In the succeeding sections we discuss three of these phenomena—viz., crystallization, melting, and the glass transition.

Crystallization is the process by which, upon cooling, an ordered (i.e., crystalline) solid phase is produced from a liquid melt having a highly random molecular structure. The melting transformation is the reverse process that occurs when a polymer is heated. The glass-transition phenomenon occurs with amorphous or noncrystallizable polymers which, when cooled from a liquid melt, become rigid solids yet retain the disordered molecular structure that is characteristic of the liquid state. Of course, alterations of physical and mechanical properties attend crystallization, melting, and the glass transition. Furthermore, for semicrystalline polymers, crystalline regions will experience melting (and crystallization), while noncrystalline areas pass through the glass transition.

11.13 CRYSTALLIZATION

An understanding of the mechanism and kinetics of polymer crystallization is important inasmuch as the degree of crystallinity influences the mechanical and thermal properties of these materials. The crystallization of a molten polymer occurs by nucleation and growth processes, topics discussed in the context of phase transformations for metals in Section 11.3. For polymers, upon cooling through the melting temperature, nuclei form wherein small regions of the tangled and random molecules become ordered and aligned in the manner of chain-folded layers, Figure 4.13. At temperatures in excess of the melting temperature, these nuclei are unstable due to the thermal atomic vibrations that tend to disrupt the ordered molecular arrangements. Subsequent to nucleation and during the crystallization growth stage, nuclei grow by the continued ordering and alignment of additional molecular chain segments; that is, the chain-folded layers increase in lateral dimensions, or, for spherulitic structures (Figure 4.14) there is an increase in spherulite radius.

The time dependence of crystallization is the same as for many solid-state transformations—Figure 11.1; that is, a sigmoidal-shaped curve results when fraction transformation (i.e., fraction crystallized) is plotted versus the logarithm of time (at constant temperature). Such a plot is presented in Figure 11.35 for the crystallization of polypropylene at three temperatures. Mathematically, fraction crystallized y is a function of time t according to the Avrami equation, Equation 11.1, as

$$y = 1 - \exp(-kt^n) \tag{11.1}$$

where k and n are time-independent constants, which values depend on the crystallizing system. Normally, the extent of crystallization is measured by specimen volume changes since there will be a difference in volume for liquid and crystallized phases. Rate of crystallization may be specified in the same manner as for the

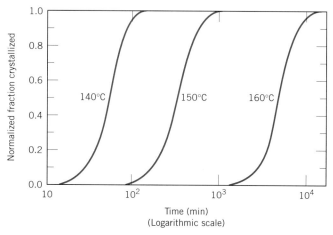

FIGURE 11.35 Plot of normalized fraction crystallized versus the logarithm of time for polypropylene at constant temperatures of 140°C, 150°C, and 160°C. (Adapted from P. Parrini and G. Corrieri, *Makromol. Chem.*, **62**, 83, 1963. Reprinted by permission of Hüthig & Wepf Publishers, Zug, Switzerland.)

transformations discussed in Section 11.3, and according to Equation 11.2; that is, rate is equal to the reciprocal of time required for crystallization to proceed to 50% completion. This rate is dependent on crystallization temperature (Figure 11.35) and also on the molecular weight of the polymer; rate decreases with increasing molecular weight.

For polypropylene, the attainment of 100% crystallinity is not possible. Therefore, in Figure 11.35, the vertical axis is scaled as "normalized fraction crystallized." A value of 1.0 for this parameter corresponds to the highest level of crystallization that is achieved during the tests, which, in reality, is less than complete crystallization.

11.14 MELTING

The melting of a polymer crystal corresponds to the transformation of a solid material, having an ordered structure of aligned molecular chains, to a viscous liquid in which the structure is highly random; this phenomenon occurs, upon heating, at the **melting temperature,** T_m. There are several features distinctive to the melting of polymers that are not normally observed with metals and ceramics; these are consequences of the polymer molecular structures and lamellar crystalline morphology. First of all, melting of polymers takes place over a range of temperatures; this phenomenon is discussed in more detail below. In addition, the melting behavior depends on the history of the specimen, in particular the temperature at which it crystallized. The thickness of chain-folded lamellae will depend on crystallization temperature; the thicker the lamellae, the higher the melting temperature. And finally, the apparent melting behavior is a function of the rate of heating; increasing this rate results in an elevation of the melting temperature.

{As Section 8.18 notes, polymeric materials are responsive to heat treatments that produce structural and property alterations. An increase in lamellar thickness may be induced by annealing just below the melting temperature. Annealing also raises the melting temperature of the polymer.}

11.15 THE GLASS TRANSITION

The glass transition occurs in amorphous (or glassy) and semicrystalline polymers, and is due to a reduction in motion of large segments of molecular chains with decreasing temperature. Upon cooling, the glass transition corresponds to the gradual transformation from a liquid to a rubbery material, and finally, to a rigid solid. The temperature at which the polymer experiences the transition from rubbery to rigid states is termed the **glass transition temperature,** T_g. Of course, this sequence of events occurs in the reverse order when a rigid glass at a temperature below T_g is heated. In addition, abrupt changes in other physical properties accompany this glass transition: e.g., stiffness {(Figure 7.28),} heat capacity, and coefficient of thermal expansion.

11.16 MELTING AND GLASS TRANSITION TEMPERATURES

Melting and glass transition temperatures are important parameters relative to in-service applications of polymers. They define, respectively, the upper and lower temperature limits for numerous applications, especially for semicrystalline poly-

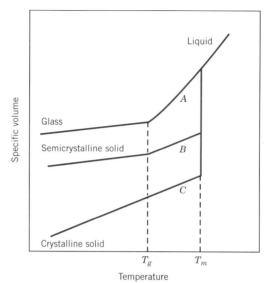

FIGURE 11.36 Specific volume versus temperature, upon cooling from the liquid melt, for totally amorphous (curve *A*), semicrystalline (curve *B*), and crystalline (curve *C*) polymers.

mers. The glass transition temperature may also define the upper use temperature for glassy amorphous materials. Furthermore, T_m and T_g also influence the fabrication and processing procedures for polymers and polymer-matrix composites. {These issues are discussed in other chapters.}

The temperatures at which melting and/or the glass transition occur for a polymer are determined in the same manner as for ceramic materials—from a plot of specific volume (the reciprocal of density) versus temperature. Figure 11.36 is such a plot, wherein curves *A* and *C*, for amorphous and crystalline polymers, respectively.[1] For the crystalline material, there is a discontinuous change in specific volume at the melting temperature T_m. The curve for the totally amorphous material is continuous but it experiences a slight decrease in slope at the glass transition temperature, T_g. The behavior is intermediate between these extremes for a semicrystalline polymer (curve *B*), in that both melting and glass transition phenomena are observed; T_m and T_g are properties of the respective crystalline and amorphous phases in this semicrystalline material. As discussed above, the behaviors represented in Figure 11.36 will depend on the rate of cooling or heating. Representative melting and glass transition temperatures of a number of polymers are contained in Table 11.1 and Appendix E.

11.17 FACTORS THAT INFLUENCE MELTING AND GLASS TRANSITION TEMPERATURES (CD-ROM)

---

[1] It should be noted that no engineering polymer is 100% crystalline; curve *C* is included in Figure 11.36 to illustrate the extreme behavior that would be displayed by a totally crystalline material.

Table 11.1 Melting and Glass Transition Temperatures for Some of the More Common Polymeric Materials

Material	Glass Transition Temperature [°C (°F)]	Melting Temperature [°C (°F)]
Polyethylene (low density)	−110 (−165)	115 (240)
Polytetrafluoroethylene	−97 (−140)	327 (620)
Polyethylene (high density)	−90 (−130)	137 (279)
Polypropylene	−18 (0)	175 (347)
Nylon 6,6	57 (135)	265 (510)
Polyester (PET)	69 (155)	265 (510)
Polyvinyl chloride	87 (190)	212 (415)
Polystyrene	100 (212)	240 (465)
Polycarbonate	150 (300)	265 (510)

S U M M A R Y

The first set of discussion topics for this chapter has included phase transformations in metals—modifications in the phase structure or microstructure—and how they affect mechanical properties. Some transformations involve diffusional phenomena, which means that their progress is time dependent. For these, some of the basic kinetic concepts were explored, including the relation between degree of reaction completion and time, the notion of transformation rate, and how rate depends on temperature.

As a practical matter, phase diagrams are severely restricted relative to transformations in multiphase alloys, because they provide no information as to phase transformation rates. The element of time is incorporated into both isothermal transformation and continuous cooling transformation diagrams; transformation progress as a function of temperature and elapsed time is expressed for a specific alloy at constant temperature {and for continuous cooling} treatments, respectively. Diagrams of both types were presented for iron–carbon steel alloys, and their utility with regard to the prediction of microstructural products was discussed.

Several microconstituents are possible for steels, the formation of which depends on composition and heat treatment. These microconstituents include fine and coarse pearlite, and bainite, which are composed of ferrite and cementite phases and result from the decomposition of austenite via diffusional processes. A spheroidite microstructure (also consisting of ferrite and cementite phases) may be produced when a steel specimen composed of any of the preceding microstructures is heat treated at a temperature just below the eutectoid. The mechanical characteristics of pearlitic, bainitic, and spheroiditic steels were compared and also explained in terms of their microconstituents.

Martensite, yet another transformation product in steels, results when austenite is cooled very rapidly. It is a metastable and single-phase structure that may be produced in steels by a diffusionless and almost instantaneous transformation of austenite. Transformation progress is dependent on temperature rather than time, and may be represented on both isothermal {and continuous cooling} transformation diagrams. Furthermore, alloying element additions retard the formation rate of pearlite and bainite, thus rendering the martensitic transformation more competitive. Mechanically, martensite is extremely hard; applicability, however, is limited by its brittleness. A tempering heat treatment increases the ductility at some sacrifice of strength and hardness. During tempering, martensite transforms to tempered martensite, which consists of the equilibrium ferrite and cementite phases. Embrittlement of some steel alloys results when specific alloying and impurity elements are present, and upon tempering within a definite temperature range.

Some alloys are amenable to precipitation hardening, that is, to strengthening by the formation of very small particles of a second, or precipitate, phase. Control of particle size, and subsequently the strength, is accomplished by two heat treatments. For the second or precipitation treatment at constant temperature, strength increases with time to a maximum and decreases during overaging. This process is accelerated with rising temperature. The strengthening phenomenon is explained in terms of an increased resistance to dislocation motion by lattice strains, which are established in the vicinity of these microscopically small precipitate particles.

Relative to polymeric materials, the molecular mechanics of crystallization, melting, and the glass transition were discussed. The manner in which melting and glass transition temperatures are determined was outlined; these parameters are important relative to the temperature range over which a particular polymer may be utilized and processed. {The magnitudes of T_m and T_g increase with increasing chain stiffness; stiffness is enhanced by the presence of chain double bonds and side groups that are either bulky or polar. Molecular weight and degree of branching also affect T_m and T_g.}

IMPORTANT TERMS AND CONCEPTS

Alloy steel
Artificial aging
Athermal transformation
Bainite
Coarse pearlite
Continuous cooling transformation diagram
Fine pearlite
Glass transition temperature

Isothermal transformation diagram
Kinetics
Martensite
Melting temperature
Natural aging
Nucleation
Overaging
Phase transformation
Plain carbon steel

Precipitation hardening
Precipitation heat treatment
Solution heat treatment
Spheroidite
Supercooling
Superheating
Tempered martensite
Thermally activated transformation
Transformation rate

REFERENCES

Atkins, M., *Atlas of Continuous Cooling Transformation Diagrams for Engineering Steels,* British Steel Corporation, Sheffield, England, 1980.

Atlas of Isothermal Transformation and Cooling Transformation Diagrams, American Society for Metals, Metals Park, OH, 1977.

Billmeyer, F. W., Jr., *Textbook of Polymer Science,* 3rd edition, Wiley-Interscience, New York, 1984. Chapter 10.

Brooks, C. R., *Principles of the Heat Treatment of Plain Carbon and Low Alloy Steels,* ASM International, Materials Park, OH, 1996.

Brophy, J. H., R. M. Rose, and J. Wulff, *The Structure and Properties of Materials,* Vol. II, *Thermodynamics of Structure,* John Wiley & Sons, New York, 1964. Reprinted by Books on Demand, Ann Arbor, MI.

Porter, D. A. and K. E. Easterling, *Phase Transformations in Metals and Alloys,* Van Nos-

358 • Chapter 11 / Phase Transformations

trand Reinhold (International) Co. Ltd., Workingham, Berkshire, England, 1981. Reprinted by Chapman and Hall, New York, 1992.

Shewmon, P. G., *Transformations in Metals,* McGraw-Hill Book Company, New York, 1969. Reprinted by Williams Book Company, Tulsa, OK.

Vander Voort, G. (Editor), *Atlas of Time–*

Temperature Diagrams for Irons and Steels, ASM International, Materials Park, OH, 1991.

Vander Voort, G. (Editor), *Atlas of Time–Temperature Diagrams for Nonferrous Alloys,* ASM International, Materials Park, OH, 1991.

Young, R. J. and P. Lovell, *Introduction to Polymers,* 2nd edition, Chapman and Hall, London, 1991.

QUESTIONS AND PROBLEMS

Note: To solve those problems having an asterisk (*) by their numbers, consultation of supplementary topics [appearing only on the CD-ROM (and not in print)] will probably be necessary.

11.1 Name the two stages involved in the formation of particles of a new phase. Briefly describe each.

11.2 For some transformation having kinetics that obey the Avrami equation (Equation 11.1), the parameter n is known to have a value of 1.7. If, after 100 s, the reaction is 50% complete, how long (total time) will it take the transformation to go to 99% completion?

11.3 Compute the rate of some reaction that obeys Avrami kinetics, assuming that the constants n and k have values of 3.0 and 7×10^{-3}, respectively, for time expressed in seconds.

11.4 It is known that the kinetics of recrystallization for some alloy obey the Avrami equation and that the value of n in the exponential is 2.5. If, at some temperature, the fraction recrystallized is 0.40 after 200 min, determine the rate of recrystallization at this temperature.

11.5 The kinetics of the austenite-to-pearlite transformation obey the Avrami relationship. Using the fraction transformed–time data given below, determine the total time required for 95% of the austenite to transform to pearlite:

Fraction Transformed	Time (s)
0.2	12.6
0.8	28.2

11.6 Below, the fraction recrystallized–time data for the recrystallization at 600°C of a previously deformed steel are tabulated. Assuming that the kinetics of this process obey the Avrami relationship, determine the fraction recrystallized after a total time of 22.8 min.

Fraction Recrystallized	Time (min)
0.20	13.1
0.70	29.1

11.7 **(a)** From the curves shown in Figure 11.2 and using Equation 11.2, determine the rate of recrystallization for pure copper at the several temperatures.

(b) Make a plot of ln(rate) versus the reciprocal of temperature (in K^{-1}), and determine the activation energy for this recrystallization process. (See Section 6.5.)

(c) By extrapolation, estimate the length of time required for 50% recrystallization at room temperature, 20°C (293 K).

11.8 In terms of heat treatment and the development of microstructure, what are two major limitations of the iron–iron carbide phase diagram?

11.9 **(a)** Briefly describe the phenomena of superheating and supercooling.

(b) Why do they occur?

11.10 Suppose that a steel of eutectoid composition is cooled to 550°C (1020°F) from 760°C

(1400°F) in less than 0.5 s and held at this temperature.

(a) How long will it take for the austenite-to-pearlite reaction to go to 50% completion? To 100% completion?

(b) Estimate the hardness of the alloy that has completely transformed to pearlite.

11.11 Briefly explain why the reaction rate for the austenite-to-pearlite transformation, as determined from Figure 11.5 and utilizing Equation 11.2, decreases with increasing temperature, in apparent contradiction with Equation 11.3.

11.12 Briefly cite the differences between pearlite, bainite, and spheroidite relative to microstructure and mechanical properties.

11.13 What is the driving force for the formation of spheroidite?

11.14 Using the isothermal transformation diagram for an iron–carbon alloy of eutectoid composition (Figure 11.14), specify the nature of the final microstructure (in terms of microconstituents present and approximate percentages of each) of a small specimen that has been subjected to the following time–temperature treatments. In each case assume that the specimen begins at 760°C (1400°F) and that it has been held at this temperature long enough to have achieved a complete and homogeneous austenitic structure.

(a) Cool rapidly to 700°C (1290°F), hold for 10^4 s, then quench to room temperature.

(b) Reheat the specimen in part a to 700°C (1290°F) for 20 h.

(c) Rapidly cool to 600°C (1110°F), hold for 4 s, rapidly cool to 450°C (840°F), hold for 10 s, then quench to room temperature.

(d) Cool rapidly to 400°C (750°F), hold for 2 s, then quench to room temperature.

(e) Cool rapidly to 400°C (750°F), hold for 20 s, then quench to room temperature.

(f) Cool rapidly to 400°C (750°F), hold for 200 s, then quench to room temperature.

(g) Rapidly cool to 575°C (1065°F), hold for 20 s, rapidly cool to 350°C (660°F), hold for 100 s, then quench to room temperature.

(h) Rapidly cool to 250°C (480°F), hold for 100 s, then quench to room temperature in water. Reheat to 315°C (600°F) for 1 h and slowly cool to room temperature.

11.15 Make a copy of the isothermal transformation diagram for an iron–carbon alloy of eutectoid composition (Figure 11.14) and then sketch and label on this diagram time–temperature paths to produce the following microstructures:

(a) 100% coarse pearlite.

(b) 100% tempered martensite.

(c) 50% coarse pearlite, 25% bainite, and 25% martensite.

11.16 Using the isothermal transformation diagram for a 0.45 wt% C steel alloy (Figure 11.38), determine the final microstructure (in terms of just the microconstituents present) of a small specimen that has been subjected to the following time–temperature treatments. In each case assume that the specimen begins at 845°C (1550°F), and that it has been held at this temperature long enough to have achieved a complete and homogeneous austenitic structure.

(a) Rapidly cool to 250°C (480°F), hold for 10^3 s, then quench to room temperature.

(b) Rapidly cool to 700°C (1290°F), hold for 30 s, then quench to room temperature.

(c) Rapidly cool to 400°C (750°F), hold for 500 s, then quench to room temperature.

(d) Rapidly cool to 700°C (1290°F), hold at this temperature for 10^5 s, then quench to room temperature.

(e) Rapidly cool to 650°C (1200°F), hold at this temperature for 3 s, rapidly cool to 400°C (750°F), hold for 10 s, then quench to room temperature.

(f) Rapidly cool to 450°C (840°F), hold for 10 s, then quench to room temperature.

(g) Rapidly cool to 625°C (1155°F), hold for 1 s, then quench to room temperature.

(h) Rapidly cool to 625°C (1155°F), hold at this temperature for 10 s, rapidly cool to 400°C (750°F), hold at this temperature for 5 s, then quench to room temperature.

FIGURE 11.38 Isothermal transformation diagram for a 0.45 wt% C iron–carbon alloy: A, austenite; B, bainite; F, proeutectoid ferrite; M, martensite; P, pearlite. (Adapted from *Atlas of Time-Temperature Diagrams for Irons and Steels,* G. F. Vander Voort, Editor, 1991. Reprinted by permission of ASM International, Materials Park, OH.)

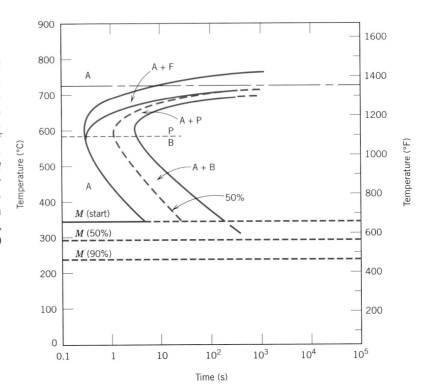

11.17 For parts a, c, d, f, and h of Problem 11.16, determine the approximate percentages of the microconstituents that form.

11.18 Make a copy of the isothermal transformation diagram for a 0.45 wt% C iron–carbon alloy (Figure 11.38), and then sketch and label on this diagram the time–temperature paths to produce the following microstructures:

(a) 42% proeutectoid ferrite and 58% coarse pearlite.

(b) 50% fine pearlite and 50% bainite.

(c) 100% martensite.

(d) 50% martensite and 50% austenite.

11.19* Name the microstructural products of eutectoid iron–carbon alloy (0.76 wt% C) specimens that are first completely transformed to austenite, then cooled to room temperature at the following rates: **(a)** 200°C/s, **(b)** 100°C/s, and **(c)** 20°C/s.

11.20* Figure 11.39 shows the continuous cooling transformation diagram for a 1.13 wt% C iron–carbon alloy. Make a copy of this figure and then sketch and label continuous cooling curves to yield the following microstructures:

(a) Fine pearlite and proeutectoid cementite.

(b) Martensite.

(c) Martensite and proeutectoid cementite.

(d) Coarse pearlite and proeutectoid cementite.

(e) Martensite, fine pearlite, and proeutectoid cementite.

11.21 Cite two major differences between martensitic and pearlitic transformations.

11.22* Cite two important differences between continuous cooling transformation diagrams for plain carbon and alloy steels.

11.23* Briefly explain why there is no bainite transformation region on the continuous cooling transformation diagram for an iron–carbon alloy of eutectoid composition.

11.24* Name the microstructural products of 4340 alloy steel specimens that are first completely transformed to austenite, then

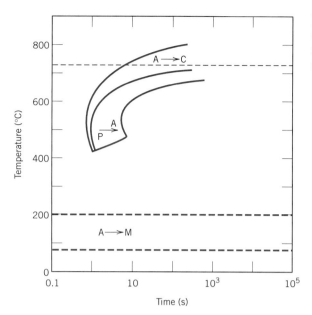

Figure 11.39 Continuous cooling transformation diagram for a 1.13 wt% C iron–carbon alloy.

cooled to room temperature at the following rates: **(a)** 10°C/s, **(b)** 1°C/s, **(c)** 0.1°C/s, and **(d)** 0.01°C/s.

11.25* Briefly describe the simplest continuous cooling heat treatment procedure that would be used in converting a 4340 steel from one microstructure to another.

(a) (Martensite + bainite) to (ferrite + pearlite).

(b) (Martensite + bainite) to spheroidite.

(c) (Martensite + bainite) to (martensite + bainite + ferrite).

11.26* On the basis of diffusion considerations, explain why fine pearlite forms for the moderate cooling of austenite through the eutectoid temperature, whereas coarse pearlite is the product for relatively slow cooling rates.

11.27 **(a)** Which is the more stable, the pearlitic or the spheroiditic microstructure?

(b) Why?

11.28 Briefly explain why fine pearlite is harder and stronger than coarse pearlite, which in turn is harder and stronger than spheroidite.

11.29 Cite two reasons why martensite is so hard and brittle.

11.30 Rank the following iron–carbon alloys and associated microstructures from the highest to the lowest tensile strength: **(a)** 0.25 wt%C with spheroidite, **(b)** 0.25 wt%C with coarse pearlite, **(c)** 0.6 wt%C with fine pearlite, and **(d)** 0.6 wt%C with coarse pearlite. Justify this ranking.

11.31 Briefly explain why the hardness of tempered martensite diminishes with tempering time (at constant temperature) and with increasing temperature (at constant tempering time).

11.32* Briefly describe the simplest heat treatment procedure that would be used in converting a 0.76 wt% C steel from one microstructure to the other, as follows:

(a) Spheroidite to tempered martensite.

(b) Tempered martensite to pearlite.

(c) Bainite to martensite.

(d) Martensite to pearlite.

(e) Pearlite to tempered martensite.

(f) Tempered martensite to pearlite.

(g) Bainite to tempered martensite.

(h) Tempered martensite to spheroidite.

11.33 **(a)** Briefly describe the microstructural difference between spheroidite and tempered martensite.

(b) Explain why tempered martensite is much harder and stronger.

11.34 Estimate the Rockwell hardnesses for specimens of an iron–carbon alloy of eutectoid composition that have been subjected to the heat treatments described in parts b, d, f, g, and h of Problem 11.14.

11.35 Estimate the Brinell hardnesses for specimens of a 0.45 wt% C iron–carbon alloy that have been subjected to the heat treatments described in parts a, d, and h of Problem 11.16.

11.36* Determine the approximate tensile strengths for specimens of a eutectoid iron–carbon alloy that have experienced the heat treatments described in parts a and c of Problem 11.19.

11.37 For a eutectoid steel, describe isothermal heat treatments that would be required to yield specimens having the following Rockwell hardnesses: **(a)** 93 HRB, **(b)** 40 HRC, and **(c)** 27 HRC.

11.38 The room-temperature tensile strengths of pure copper and pure silver are 209 MPa and 125 MPa, respectively.

(a) Make a schematic graph of the room-temperature tensile strength versus composition for all compositions between pure copper and pure silver.

(b) On this same graph schematically plot tensile strength versus composition at 600°C.

(c) Explain the shapes of these two curves, as well as any differences between them.

11.39 Compare precipitation hardening (Sections 11.10 and 11.11) and the hardening of steel by quenching and tempering (Sections 11.5, 11.6, and 11.8) with regard to

(a) The total heat treatment procedure.

(b) The microstructures that develop.

(c) How the mechanical properties change during the several heat treatment stages.

11.40 What is the principal difference between natural and artificial aging processes?

11.41* For each of the following pairs of polymers, plot and label schematic specific volume-versus-temperature curves on the same graph (i.e., make separate plots for parts a, b, and c):

(a) Spherulitic polypropylene, of 25% crystallinity, and having a weight-average molecular weight of 75,000 g/mol; spherulitic polystyrene, of 25% crystallinity, and having a weight-average molecular weight of 100,000 g/mol.

(b) Graft poly(styrene-butadiene) copolymer with 10% of available sites crosslinked; random poly(styrene-butadiene) copolymer with 15% of available sites crosslinked.

(c) Polyethylene having a density of 0.985 g/cm^3 and a number-average degree of polymerization of 2500; polyethylene having a density of 0.915 g/cm^3 and a degree of polymerization of 2000.

11.42* For each of the following pairs of polymers, do the following: (1) state whether or not it is possible to determine whether one polymer has a higher melting temperature than the other; (2) if it is possible, note which has the higher melting temperature and then cite reason(s) for your choice; and (3) if it is not possible to decide, then state why.

(a) Isotactic polystyrene that has a density of 1.12 g/cm^3 and a weight-average molecular weight of 150,000 g/mol; syndiotactic polystyrene that has a density of 1.10 g/cm^3 and a weight-average molecular weight of 125,000 g/mol.

(b) Linear polyethylene that has a number-average degree of polymerization of 5,000; linear and isotactic polypropylene that has a number-average degree of polymerization of 6,500.

(c) Branched and isotactic polystyrene that has a weight-average degree of polymerization of 4,000; linear and isotactic polypropylene that has a weight-average degree of polymerization of 7,500.

11.43* Make a schematic plot showing how the modulus of elasticity of an amorphous polymer depends on the glass transition temperature. Assume that molecular weight is held constant.

11.44 Name the following polymer(s) that would be suitable for the fabrication of cups to contain hot coffee: polyethylene, polypropylene, polyvinyl chloride, PET polyester, and polycarbonate. Why?

11.45 Of those polymers listed in Table 11.1, which polymer(s) would be best suited for use as ice cube trays? Why?

Design Problems

11.D1 Is it possible to produce an iron–carbon alloy of eutectoid composition that has a minimum hardness of 90 HRB and a minimum ductility of 35%RA? If so, describe the continuous cooling heat treatment to which the alloy would be subjected to achieve these properties. If it is not possible, explain why.

11.D2 Is it possible to produce an iron–carbon alloy that has a minimum tensile strength of 690 MPa (100,000 psi) and a minimum ductility of 40%RA? If so, what will be its composition and microstructure (coarse and fine pearlites and spheroidite are alternatives)? If this is not possible, explain why.

11.D3 It is desired to produce an iron–carbon alloy that has a minimum hardness of 175 HB and a minimum ductility of 52%RA. Is such an alloy possible? If so, what will be its composition and microstructure (coarse and fine pearlites and spheroidite are alternatives)? If this is not possible, explain why.

11.D4 (a) For a 1080 steel that has been water quenched, estimate the tempering time at 425°C (800°F) to achieve a hardness of 50 HRC.

(b) What will be the tempering time at 315°C (600°F) necessary to attain the same hardness?

11.D5 An alloy steel (4340) is to be used in an application requiring a minimum tensile strength of 1380 MPa (200,000 psi) and a minimum ductility of 43%RA. Oil quenching followed by tempering is to be used. Briefly describe the tempering heat treatment.

11.D6 Is it possible to produce an oil-quenched and tempered 4340 steel that has a minimum yield strength of 1400 MPa (203,000 psi) and a ductility of at least 42%RA? If this is possible, describe the tempering heat treatment. If it is not possible, explain why.

11.D7 Copper-rich copper–beryllium alloys are precipitation hardenable. After consulting

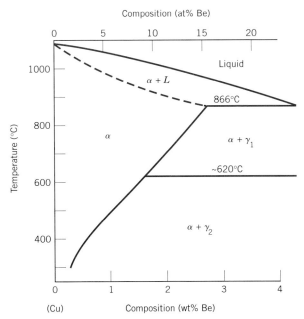

FIGURE 11.40 The copper-rich side of the copper–beryllium phase diagram. (Adapted from *Binary Alloy Phase Diagrams*, 2nd edition, Vol. 2, T. B. Massalski, Editor-in-Chief, 1990. Reprinted by permission of ASM International, Materials Park, OH.)

the portion of the phase diagram (Figure 11.40), do the following:

(a) Specify the range of compositions over which these alloys may be precipitation hardened.

(b) Briefly describe the heat-treatment procedures (in terms of temperatures) that would be used to precipitation harden an alloy having a composition of your choosing, yet lying within the range given for part a.

11.D8 A solution heat-treated 2014 aluminum alloy is to be precipitation hardened to have a minimum tensile strength of 450 MPa (65,250 psi) and a ductility of at least 15%EL. Specify a practical precipitation heat treatment in terms of temperature and time that would give these mechanical characteristics. Justify your answer.

11.D9 Is it possible to produce a precipitation-hardened 2014 aluminum alloy having a minimum tensile strength of 425 MPa (61,625 psi) and a ductility of at least 12%EL? If so, specify the precipitation heat treatment. If it is not possible, explain why.

Chapter 12 / Electrical Properties

It was noted in {Section 5.12} that an image is generated on a scanning electron micrograph as a beam of electrons scans the surface of the specimen being examined. The electrons in this beam cause some of the specimen surface atoms to emit x-rays; the energy of an x-ray photon depends on the particular atom from which it radiates. It is possible to selectively filter out all but the x-rays emitted from one kind of atom. When projected on a cathode ray tube, small white dots are produced indicating the locations of the particular atom type; thus, a "dot map" of the image is generated.

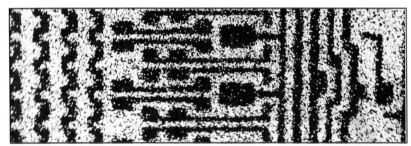

Top: Scanning electron micrograph of an integrated circuit.

Center: A silicon dot map for the integrated circuit above, showing regions where silicon atoms are concentrated. Doped silicon is the semiconducting material from which integrated circuit elements are made.

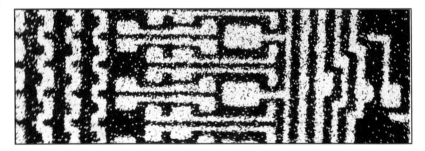

Bottom: An aluminum dot map. Metallic aluminum is an electrical conductor and, as such, wires the circuit elements together. Approximately 200×.

Why Study the *Electrical Properties* of Materials?

Consideration of the electrical properties of materials is often important when materials selection and processing decisions are being made during the design of a component or structure. {For example, we discuss in Sections 20.12 through 20.17 materials that are used in the several components of one type of integrated circuit package.} The electrical behaviors of the various materials are diverse. Some need to be highly electrically conductive (e.g., connecting wires), whereas electrical insulativity is required of others (e.g., the protective package encapsulation).

365

After careful study of this chapter you should be able to do the following:

1. Describe the four possible electron band structures for solid materials.
2. Briefly describe electron excitation events that produce free electrons/holes in (a) metals, (b) semiconductors (intrinsic and extrinsic), and (c) insulators.
3. Calculate the electrical conductivities of metals, semiconductors (intrinsic and extrinsic), and insulators given their charge carrier density(s) and mobility(s).
4. Distinguish between *intrinsic* and *extrinsic* semiconducting materials.
5. Note the manner in which electrical conductivity changes with increasing temperature for (a) metals, (b) semiconductors, and (c) insulating materials.
{6. For a *p–n* junction, explain the rectification process in terms of electron and hole motions.}
{7. Calculate the capacitance of a parallel-plate capacitor.}
{8. Define dielectric constant in terms of permittivities.}
{9. Briefly explain how the charge storing capacity of a capacitor may be increased by the insertion and polarization of a dielectric material between its plates.}
{10. Name and describe the three types of polarization.}

12.1 INTRODUCTION

The prime objective of this chapter is to explore the electrical properties of materials, that is, their responses to an applied electric field. We begin with the phenomenon of electrical conduction: the parameters by which it is expressed, the mechanism of conduction by electrons, and how the electron energy band structure of a material influences its ability to conduct. These principles are extended to metals, semiconductors, and insulators. Particular attention is given to the characteristics of semiconductors, and then to semiconducting devices. Also treated are the dielectric characteristics of insulating materials. The final sections are devoted to the peculiar phenomena of ferroelectricity and piezoelectricity.

ELECTRICAL CONDUCTION

12.2 OHM'S LAW

One of the most important electrical characteristics of a solid material is the ease with which it transmits an electric current. **Ohm's law** relates the current I—or time rate of charge passage—to the applied voltage V as follows:

$$V = IR \tag{12.1}$$

where R is the resistance of the material through which the current is passing. The units for V, I, and R are, respectively, volts (J/C), amperes (C/s), and ohms (V/A). The value of R is influenced by specimen configuration, and for many materials is independent of current. The **resistivity** ρ is independent of specimen geometry but related to R through the expression

$$\rho = \frac{RA}{l} \tag{12.2}$$

where l is the distance between the two points at which the voltage is measured, and A is the cross-sectional area perpendicular to the direction of the current. The units for ρ are ohm-meters (Ω-m). From the expression for Ohm's law and

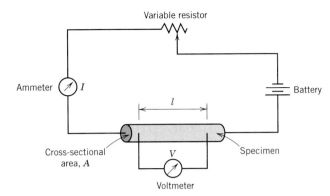

Equation 12.2,

$$\rho = \frac{VA}{Il}$$

(12.3)

Figure 12.1 is a schematic diagram of an experimental arrangement for measuring electrical resistivity.

12.3 ELECTRICAL CONDUCTIVITY

Sometimes, **electrical conductivity** σ is used to specify the electrical character of a material. It is simply the reciprocal of the resistivity, or

$$\sigma = \frac{1}{\rho}$$

(12.4)

and is indicative of the ease with which a material is capable of conducting an electric current. The units for σ are reciprocal ohm-meters [$(\Omega\text{-m})^{-1}$, or mho/m]. The following discussions on electrical properties use both resistivity and conductivity.

In addition to Equation 12.1, Ohm's law may be expressed as

$$J = \sigma \mathscr{E}$$

(12.5)

in which J is the current density, the current per unit of specimen area I/A, and $\mathscr{E}$ is the electric field intensity, or the voltage difference between two points divided by the distance separating them, that is,

$$\mathscr{E} = \frac{V}{l}$$

(12.6)

The demonstration of the equivalence of the two Ohm's law expressions (Equations 12.1 and 12.5) is left as a homework exercise.

Solid materials exhibit an amazing range of electrical conductivities, extending over 27 orders of magnitude; probably no other physical property experiences this breadth of variation. In fact, one way of classifying solid materials is according to the ease with which they conduct an electric current; within this classification scheme there are three groupings: *conductors, semiconductors,* and *insulators.* **Metals** are good conductors, typically having conductivities on the order of 10^7 $(\Omega\text{-m})^{-1}$. At the other extreme are materials with very low conductivities, ranging between 10^{-10} and 10^{-20} $(\Omega\text{-m})^{-1}$; these are electrical **insulators.** Materials with intermediate conductivities, generally from 10^{-6} to 10^4 $(\Omega\text{-m})^{-1}$, are termed **semiconductors.**

12.4 ELECTRONIC AND IONIC CONDUCTION

An electric current results from the motion of electrically charged particles in response to forces that act on them from an externally applied electric field. Positively charged particles are accelerated in the field direction, negatively charged particles in the direction opposite. Within most solid materials a current arises from the flow of electrons, which is termed *electronic conduction.* In addition, for ionic materials a net motion of charged ions is possible that produces a current; such is termed **ionic conduction.** The present discussion deals with electronic conduction; ionic conduction is treated briefly in Section 12.15.

12.5 ENERGY BAND STRUCTURES IN SOLIDS

In all conductors, semiconductors, and many insulating materials, only electronic conduction exists, and the magnitude of the electrical conductivity is strongly dependent on the number of electrons available to participate in the conduction process. However, not all electrons in every atom will accelerate in the presence of an electric field. The number of electrons available for electrical conduction in a particular material is related to the arrangement of electron states or levels with respect to energy, and then the manner in which these states are occupied by electrons. A thorough exploration of these topics is complicated and involves principles of quantum mechanics that are beyond the scope of this book; the ensuing development omits some concepts and simplifies others.

Concepts relating to electron energy states, their occupancy, and the resulting electron configuration for isolated atoms were discussed in Section 2.3. By way of review, for each individual atom there exist discrete energy levels that may be occupied by electrons, arranged into shells and subshells. Shells are designated by integers (1, 2, 3, etc.), and subshells by letters (*s*, *p*, *d*, and *f*). For each of *s*, *p*, *d*, and *f* subshells, there exist, respectively, one, three, five, and seven states. The electrons in most atoms fill just the states having the lowest energies, two electrons of opposite spin per state, in accordance with the Pauli exclusion principle. The electron configuration of an isolated atom represents the arrangement of the electrons within the allowed states.

Let us now make an extrapolation of some of these concepts to solid materials. A solid may be thought of as consisting of a large number, say, N, of atoms initially separated from one another, which are subsequently brought together and bonded to form the ordered atomic arrangement found in the crystalline material. At relatively large separation distances, each atom is independent of all the others and will have the atomic energy levels and electron configuration as if isolated. However, as the atoms come within close proximity of one another, electrons are acted upon, or perturbed, by the electrons and nuclei of adjacent atoms. This influence is such that each distinct atomic state may split into a series of closely spaced electron states in the solid, to form what is termed an **electron energy band.** The extent of splitting depends on interatomic separation (Figure 12.2) and begins with the outermost electron shells, since they are the first to be perturbed as the atoms coalesce. Within each band, the energy states are discrete, yet the difference between

FIGURE 12.2
Schematic plot of
electron energy versus
interatomic separation
for an aggregate of 12
atoms ($N = 12$). Upon
close approach, each of
the $1s$ and $2s$ atomic
states splits to form an
electron energy band
consisting of 12 states.

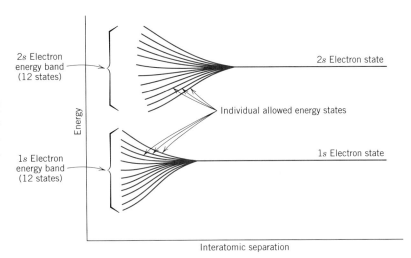

adjacent states is exceedingly small. At the equilibrium spacing, band formation may not occur for the electron subshells nearest the nucleus, as illustrated in Figure 12.3*b*. Furthermore, gaps may exist between adjacent bands, as also indicated in the figure; normally, energies lying within these band gaps are not available for electron occupancy. The conventional way of representing electron band structures in solids is shown in Figure 12.3*a*.

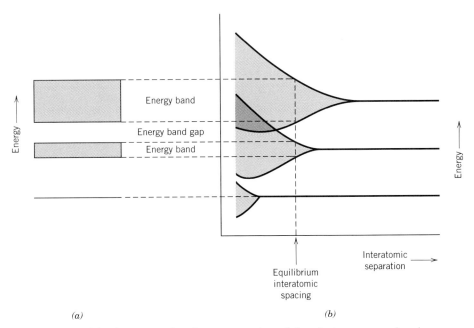

FIGURE 12.3 (*a*) The conventional representation of the electron energy band structure for a solid material at the equilibrium interatomic separation.
(*b*) Electron energy versus interatomic separation for an aggregate of atoms, illustrating how the energy band structure at the equilibrium separation in (*a*) is generated. (From Z. D. Jastrzebski, *The Nature and Properties of Engineering Materials,* 3rd edition. Copyright © 1987 by John Wiley & Sons, Inc. Reprinted by permission of John Wiley & Sons, Inc.)

The number of states within each band will equal the total of all states contributed by the N atoms. For example, an s band will consist of N states, and a p band of $3N$ states. With regard to occupancy, each energy state may accommodate two electrons, which must have oppositely directed spins. Furthermore, bands will contain the electrons that resided in the corresponding levels of the isolated atoms; for example, a $4s$ energy band in the solid will contain those isolated atom's $4s$ electrons. Of course, there will be empty bands and, possibly, bands that are only partially filled.

The electrical properties of a solid material are a consequence of its electron band structure, that is, the arrangement of the outermost electron bands and the way in which they are filled with electrons.

Four different types of band structures are possible at 0 K. In the first (Figure 12.4a), one outermost band is only partially filled with electrons. The energy corresponding to the highest filled state at 0 K is called the **Fermi energy** E_f, as indicated. This energy band structure is typified by some metals, in particular those that have a single s valence electron (e.g., copper). Each copper atom has one $4s$ electron; however, for a solid comprised of N atoms, the $4s$ band is capable of accommodating $2N$ electrons. Thus only half the available electron positions within this $4s$ band are filled.

For the second band structure, also found in metals (Figure 12.4b), there is an overlap of an empty band and a filled band. Magnesium has this band structure. Each isolated Mg atom has two $3s$ electrons. However, when a solid is formed, the $3s$ and $3p$ bands overlap. In this instance and at 0 K, the Fermi energy is taken as that energy below which, for N atoms, N states are filled, two electrons per state.

The final two band structures are similar; one band (the **valence band**) that is completely filled with electrons is separated from an empty **conduction band;** and an **energy band gap** lies between them. For very pure materials, electrons may not have energies within this gap. The difference between the two band structures lies in the magnitude of the energy gap; for materials that are insulators, the band gap is relatively wide (Figure 12.4c), whereas for semiconductors it is narrow (Figure

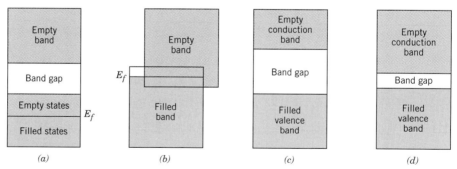

FIGURE 12.4 The various possible electron band structures in solids at 0 K. (a) The electron band structure found in metals such as copper, in which there are available electron states above and adjacent to filled states, in the same band. (b) The electron band structure of metals such as magnesium, wherein there is an overlap of filled and empty outer bands. (c) The electron band structure characteristic of insulators; the filled valence band is separated from the empty conduction band by a relatively large band gap (>2 eV). (d) The electron band structure found in the semiconductors, which is the same as for insulators except that the band gap is relatively narrow (<2 eV).

12.4*d*). The Fermi energy for these two band structures lies within the band gap—near its center.

12.6 CONDUCTION IN TERMS OF BAND AND ATOMIC BONDING MODELS

At this point in the discussion, it is vital that another concept be understood, namely, that only electrons with energies greater than the Fermi energy may be acted on and accelerated in the presence of an electric field. These are the electrons that participate in the conduction process, which are termed **free electrons.** Another charged electronic entity called a **hole** is found in semiconductors and insulators. Holes have energies less than E_f and also participate in electronic conduction. As the ensuing discussion reveals, the electrical conductivity is a direct function of the numbers of free electrons and holes. In addition, the distinction between conductors and nonconductors (insulators and semiconductors) lies in the numbers of these free electron and hole charge carriers.

METALS

For an electron to become free, it must be excited or promoted into one of the empty and available energy states above E_f. For metals having either of the band structures shown in Figures 12.4*a* and 12.4*b*, there are vacant energy states adjacent to the highest filled state at E_f. Thus, very little energy is required to promote electrons into the low-lying empty states, as shown in Figure 12.5. Generally, the energy provided by an electric field is sufficient to excite large numbers of electrons into these conducting states.

For the metallic bonding model discussed in Section 2.6, it was assumed that all the valence electrons have freedom of motion and form an "electron gas," which is uniformly distributed throughout the lattice of ion cores. Even though these electrons are not locally bound to any particular atom, they, nevertheless, must experience some excitation to become conducting electrons that are truly free. Thus, although only a fraction are excited, this still gives rise to a relatively large number of free electrons and, consequently, a high conductivity.

INSULATORS AND SEMICONDUCTORS

For insulators and semiconductors, empty states adjacent to the top of the filled valence band are not available. To become free, therefore, electrons must be pro-

FIGURE 12.5 For a metal, occupancy of electron states (*a*) before and (*b*) after an electron excitation.

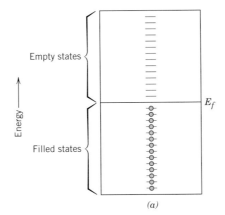

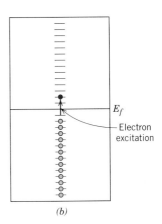

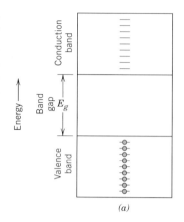 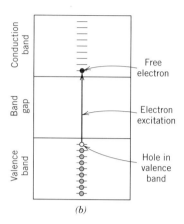

moted across the energy band gap and into empty states at the bottom of the conduction band. This is possible only by supplying to an electron the difference in energy between these two states, which is approximately equal to the band gap energy E_g. This excitation process is demonstrated in Figure 12.6. For many materials this band gap is several electron volts wide. Most often the excitation energy is from a nonelectrical source such as heat or light, usually the former.

The number of electrons excited thermally (by heat energy) into the conduction band depends on the energy band gap width as well as temperature. At a given temperature, the larger the E_g, the lower the probability that a valence electron will be promoted into an energy state within the conduction band; this results in fewer conduction electrons. In other words, the larger the band gap, the lower the electrical conductivity at a given temperature. Thus, the distinction between semiconductors and insulators lies in the width of the band gap; for semiconductors it is narrow, whereas for insulating materials it is relatively wide.

Increasing the temperature of either a semiconductor or an insulator results in an increase in the thermal energy that is available for electron excitation. Thus, more electrons are promoted into the conduction band, which gives rise to an enhanced conductivity.

The conductivity of insulators and semiconductors may also be viewed from the perspective of atomic bonding models discussed in Section 2.6. For electrically insulating materials, interatomic bonding is ionic or strongly covalent. Thus, the valence electrons are tightly bound to or shared with the individual atoms. In other words, these electrons are highly localized and are not in any sense free to wander throughout the crystal. The bonding in semiconductors is covalent (or predominantly covalent) and relatively weak, which means that the valence electrons are not as strongly bound to the atoms. Consequently, these electrons are more easily removed by thermal excitation than they are for insulators.

12.7 ELECTRON MOBILITY

When an electric field is applied, a force is brought to bear on the free electrons; as a consequence, they all experience an acceleration in a direction opposite to that of the field, by virtue of their negative charge. According to quantum mechanics, there is no interaction between an accelerating electron and atoms in a perfect crystal lattice. Under such circumstances all the free electrons should accelerate as long as the electric field is applied, which would give rise to a continuously increasing

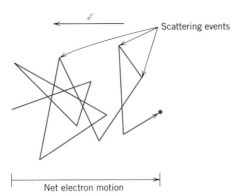

FIGURE 12.7 Schematic diagram showing the path of an electron that is deflected by scattering events.

electric current with time. However, we know that a current reaches a constant value the instant that a field is applied, indicating that there exist what might be termed "frictional forces," which counter this acceleration from the external field. These frictional forces result from the scattering of electrons by imperfections in the crystal lattice, including impurity atoms, vacancies, interstitial atoms, dislocations, and even the thermal vibrations of the atoms themselves. Each scattering event causes an electron to lose kinetic energy and to change its direction of motion, as represented schematically in Figure 12.7. There is, however, some net electron motion in the direction opposite to the field, and this flow of charge is the electric current.

The scattering phenomenon is manifested as a resistance to the passage of an electric current. Several parameters are used to describe the extent of this scattering, these include the *drift velocity* and the **mobility** of an electron. The drift velocity v_d represents the average electron velocity in the direction of the force imposed by the applied field. It is directly proportional to the electric field as follows:

$$v_d = \mu_e \mathscr{E} \tag{12.7}$$

The constant of proportionality μ_e is called the electron mobility, which is an indication of the frequency of scattering events; its units are square meters per volt-second (m²/V-s).

The conductivity σ of most materials may be expressed as

$$\sigma = n|e|\mu_e \tag{12.8}$$

where n is the number of free or conducting electrons per unit volume (e.g., per cubic meter), and $|e|$ is the absolute magnitude of the electrical charge on an electron (1.6×10^{-19} C). Thus, the electrical conductivity is proportional to both the number of free electrons and the electron mobility.

12.8 ELECTRICAL RESISTIVITY OF METALS

As mentioned previously, most metals are extremely good conductors of electricity; room-temperature conductivities for several of the more common metals are contained in Table 12.1. (Table B.9 in Appendix B lists the electrical resistivities of a large number of metals and alloys.) Again, metals have high conductivities because

Table 12.1   Room-Temperature Electrical Conductivities for Nine Common Metals and Alloys

Metal	Electrical Conductivity $[(\Omega\text{-}m)^{-1}]$
Silver	6.8×10^7
Copper	6.0×10^7
Gold	4.3×10^7
Aluminum	3.8×10^7
Iron	1.0×10^7
Brass (70 Cu–30 Zn)	1.6×10^7
Platinum	0.94×10^7
Plain carbon steel	0.6×10^7
Stainless steel	0.2×10^7

of the large numbers of free electrons that have been excited into empty states above the Fermi energy. Thus n has a large value in the conductivity expression, Equation 12.8.

At this point it is convenient to discuss conduction in metals in terms of the resistivity, the reciprocal of conductivity; the reason for this switch should become apparent in the ensuing discussion.

Since crystalline defects serve as scattering centers for conduction electrons in metals, increasing their number raises the resistivity (or lowers the conductivity). The concentration of these imperfections depends on temperature, composition, and the degree of cold work of a metal specimen. In fact, it has been observed experimentally that the total resistivity of a metal is the sum of the contributions from thermal vibrations, impurities, and plastic deformation; that is, the scattering mechanisms act independently of one another. This may be represented in mathematical form as follows:

$$\rho_{\text{total}} = \rho_t + \rho_i + \rho_d \tag{12.9}$$

in which ρ_t, ρ_i, and ρ_d represent the individual thermal, impurity, and deformation resistivity contributions, respectively. Equation 12.9 is sometimes known as **Matthiessen's rule.** The influence of each ρ variable on the total resistivity is demonstrated in Figure 12.8, as a plot of resistivity versus temperature for copper and several copper–nickel alloys in annealed and deformed states. The additive nature of the individual resistivity contributions is demonstrated at $-100°C$.

INFLUENCE OF TEMPERATURE

For the pure metal and all the copper–nickel alloys shown in Figure 12.8, the resistivity rises linearly with temperature above about $-200°C$. Thus,

$$\rho_t = \rho_0 + aT \tag{12.10}$$

where ρ_0 and a are constants for each particular metal. This dependence of the thermal resistivity component on temperature is due to the increase with temperature in thermal vibrations and other lattice irregularities (e.g., vacancies), which serve as electron-scattering centers.

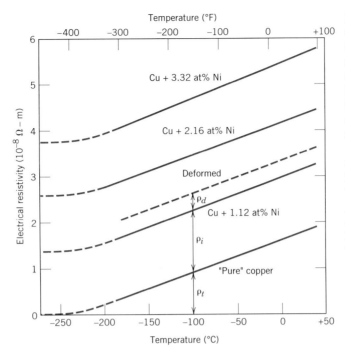

FIGURE 12.8 The electrical resistivity versus temperature for copper and three copper–nickel alloys, one of which has been deformed. Thermal, impurity, and deformation contributions to the resistivity are indicated at $-100°C$. [Adapted from J. O. Linde, *Ann. Physik,* **5,** 219 (1932); and C. A. Wert and R. M. Thomson, *Physics of Solids,* 2nd edition, McGraw-Hill Book Company, New York, 1970.]

INFLUENCE OF IMPURITIES

For additions of a single impurity that forms a solid solution, the impurity resistivity ρ_i is related to the impurity concentration c_i in terms of the atom fraction (at%/100) as follows:

$$\rho_i = Ac_i(1 - c_i) \tag{12.11}$$

where A is a composition-independent constant that is a function of both the impurity and host metals. The influence of nickel impurity additions on the room-temperature resistivity of copper is demonstrated in Figure 12.9, up to 50 wt% Ni; over this composition range nickel is completely soluble in copper (Figure 10.2). Again, nickel atoms in copper act as scattering centers, and increasing the concentration of nickel in copper results in an enhancement of resistivity.

For a two-phase alloy consisting of α and β phases, a rule-of-mixtures expression may be utilized to approximate the resistivity as follows:

$$\rho_i = \rho_\alpha V_\alpha + \rho_\beta V_\beta \tag{12.12}$$

where the V's and ρ's represent volume fractions and individual resistivities for the respective phases.

INFLUENCE OF PLASTIC DEFORMATION

Plastic deformation also raises the electrical resistivity as a result of increased numbers of electron-scattering dislocations. The effect of deformation on resistivity is also represented in Figure 12.8.

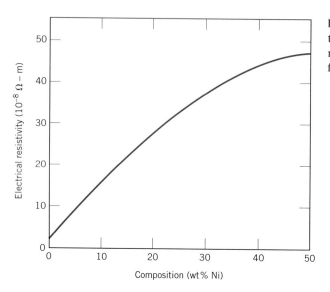

FIGURE 12.9 Room-temperature electrical resistivity versus composition for copper–nickel alloys.

12.9 ELECTRICAL CHARACTERISTICS OF COMMERCIAL ALLOYS

Electrical and other properties of copper render it the most widely used metallic conductor. Oxygen-free high-conductivity (OFHC) copper, having extremely low oxygen and other impurity contents, is produced for many electrical applications. Aluminum, having a conductivity only about one half that of copper, is also frequently used as an electrical conductor. Silver has a higher conductivity than either copper or aluminum; however, its use is restricted on the basis of cost.

On occasion, it is necessary to improve the mechanical strength of a metal alloy without impairing significantly its electrical conductivity. Both solid solution alloying and cold working (Section 8.11) improve strength at the expense of conductivity, and thus, a tradeoff must be made for these two properties. Most often, strength is enhanced by introducing a second phase that does not have so adverse an effect on conductivity. For example, copper–beryllium alloys are precipitation hardened (Section 11.11); but even so, the conductivity is reduced by about a factor of 5 over high-purity copper.

For some applications, such as furnace heating elements, a high electrical resistivity is desirable. The energy loss by electrons that are scattered is dissipated as heat energy. Such materials must have not only a high resistivity, but also a resistance to oxidation at elevated temperatures and, of course, a high melting temperature. Nichrome, a nickel–chromium alloy, is commonly employed in heating elements.

SEMICONDUCTIVITY

The electrical conductivity of the semiconducting materials is not as high as that of the metals; nevertheless, they have some unique electrical characteristics that render them especially useful. The electrical properties of these materials are extremely sensitive to the presence of even minute concentrations of impurities. **Intrinsic semiconductors** are those in which the electrical behavior is based on the electronic structure inherent to the pure material. When the electrical characteristics are dictated by impurity atoms, the semiconductor is said to be **extrinsic.**

12.10 INTRINSIC SEMICONDUCTION

Intrinsic semiconductors are characterized by the electron band structure shown in Figure 12.4*d*: at 0 K, a completely filled valence band, separated from an empty conduction band by a relatively narrow forbidden band gap, generally less than 2 eV. The two elemental semiconductors are silicon (Si) and germanium (Ge), having band gap energies of approximately 1.1 and 0.7 eV, respectively. Both are found in Group IVA of the periodic table (Figure 2.6) and are covalently bonded.[1] In addition, a host of compound semiconducting materials also display intrinsic behavior. One such group is formed between elements of Groups IIIA and VA, for example, gallium arsenide (GaAs) and indium antimonide (InSb); these are frequently called III–V compounds. The compounds composed of elements of Groups IIB and VIA also display semiconducting behavior; these include cadmium sulfide (CdS) and zinc telluride (ZnTe). As the two elements forming these compounds become more widely separated with respect to their relative positions in the periodic table (i.e., the electronegativities become more dissimilar, Figure 2.7), the atomic bonding becomes more ionic and the magnitude of the band gap energy increases—the materials tend to become more insulative. Table 12.2 gives the band gaps for some compound semiconductors.

CONCEPT OF A HOLE

In intrinsic semiconductors, for every electron excited into the conduction band there is left behind a missing electron in one of the covalent bonds, or in the band scheme, a vacant electron state in the valence band, as shown in Figure 12.6*b*. Under the influence of an electric field, the position of this missing electron within

Table 12.2 Band Gap Energies, Electron and Hole Mobilities, and Intrinsic Electrical Conductivities at Room Temperature for Semiconducting Materials

Material	Band Gap (eV)	Electrical Conductivity $[(\Omega\text{-}m)^{-1}]$	Electron Mobility $(m^2/V\text{-}s)$	Hole Mobility $(m^2/V\text{-}s)$
		Elemental		
Si	1.11	4×10^{-4}	0.14	0.05
Ge	0.67	2.2	0.38	0.18
		III–V Compounds		
GaP	2.25	—	0.05	0.002
GaAs	1.42	10^{-6}	0.85	0.45
InSb	0.17	2×10^4	7.7	0.07
		II–VI Compounds		
CdS	2.40	—	0.03	—
ZnTe	2.26	—	0.03	0.01

---

[1] The valence bands in silicon and germanium correspond to sp^3 hybrid energy levels for the isolated atom; these hybridized valence bands are completely filled at 0 K.

the crystalline lattice may be thought of as moving by the motion of other valence electrons that repeatedly fill in the incomplete bond (Figure 12.10). This process is expedited by treating a missing electron from the valence band as a positively charged particle called a *hole*. A hole is considered to have a charge that is of the same magnitude as that for an electron, but of opposite sign ($+1.6 \times 10^{-19}$ C).

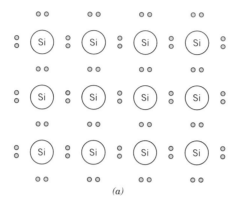

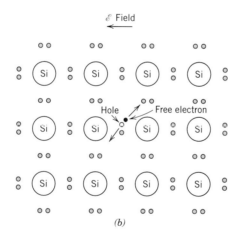

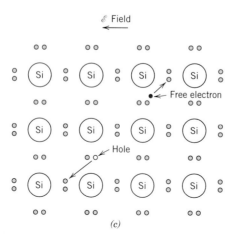

FIGURE 12.10 Electron bonding model of electrical conduction in intrinsic silicon: (*a*) before excitation; (*b*) and (*c*) after excitation (the subsequent free-electron and hole motions in response to an external electric field).

Thus, in the presence of an electric field, excited electrons and holes move in opposite directions. Furthermore, in semiconductors both electrons and holes are scattered by lattice imperfections.

INTRINSIC CONDUCTIVITY

Since there are two types of charge carrier (free electrons and holes) in an intrinsic semiconductor, the expression for electrical conduction, Equation 12.8, must be modified to include a term to account for the contribution of the hole current. Therefore, we write

$$\sigma = n|e|\mu_e + p|e|\mu_h \tag{12.13}$$

where p is the number of holes per cubic meter and μ_h is the hole mobility. The magnitude of μ_h is always less than μ_e for semiconductors. For intrinsic semiconductors, every electron promoted across the band gap leaves behind a hole in the valence band, thus,

$$n = p \tag{12.14}$$

and

$$\sigma = n|e|(\mu_e + \mu_h) = p|e|(\mu_e + \mu_h) \tag{12.15}$$

The room-temperature intrinsic conductivities and electron and hole mobilities for several semiconducting materials are also presented in Table 12.2.

EXAMPLE PROBLEM 12.1

For intrinsic silicon, the room-temperature electrical conductivity is 4×10^{-4} $(\Omega\text{-m})^{-1}$; the electron and hole mobilities are, respectively, 0.14 and 0.048 m^2/V-s. Compute the electron and hole concentrations at room temperature.

SOLUTION

Since the material is intrinsic, electron and hole concentrations will be the same, and therefore, from Equation 12.15,

$$n = p = \frac{\sigma}{|e|(\mu_e + \mu_h)}$$

$$= \frac{4 \times 10^{-4}\,(\Omega\text{-m})^{-1}}{(1.6 \times 10^{-19}\,\text{C})(0.14 + 0.048\,\text{m}^2/\text{V-s})}$$

$$= 1.33 \times 10^{16}\,\text{m}^{-3}$$

12.11 EXTRINSIC SEMICONDUCTION

Virtually all commercial semiconductors are extrinsic; that is, the electrical behavior is determined by impurities, which, when present in even minute concentrations, introduce excess electrons or holes. For example, an impurity concentration of one atom in 10^{12} is sufficient to render silicon extrinsic at room temperature.

n-TYPE EXTRINSIC SEMICONDUCTION

To illustrate how extrinsic semiconduction is accomplished, consider again the elemental semiconductor silicon. An Si atom has four electrons, each of which is covalently bonded with one of four adjacent Si atoms. Now, suppose that an impurity atom with a valence of 5 is added as a substitutional impurity; possibilities would include atoms from the Group VA column of the periodic table (e.g., P, As, and Sb). Only four of five valence electrons of these impurity atoms can participate in the bonding because there are only four possible bonds with neighboring atoms.

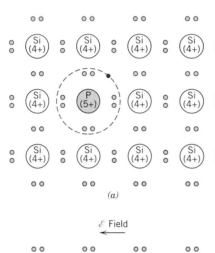

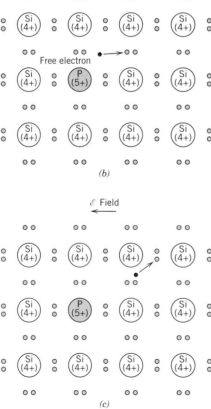

FIGURE 12.11 Extrinsic *n*-type semiconduction model (electron bonding). (*a*) An impurity atom such as phosphorus, having five valence electrons, may substitute for a silicon atom. This results in an extra bonding electron, which is bound to the impurity atom and orbits it. (*b*) Excitation to form a free electron. (*c*) The motion of this free electron in response to an electric field.

FIGURE 12.12 (*a*) Electron energy band scheme for a donor impurity level located within the band gap and just below the bottom of the conduction band. (*b*) Excitation from a donor state in which a free electron is generated in the conduction band.

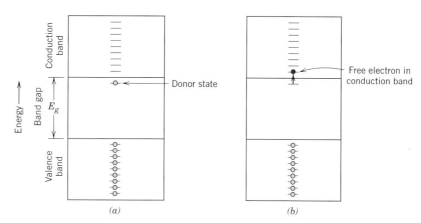

The extra nonbonding electron is loosely bound to the region around the impurity atom by a weak electrostatic attraction, as illustrated in Figure 12.11*a*. The binding energy of this electron is relatively small (on the order of 0.01 eV); thus, it is easily removed from the impurity atom, in which case it becomes a free or conducting electron (Figures 12.11*b* and 12.11*c*).

The energy state of such an electron may be viewed from the perspective of the electron band model scheme. For each of the loosely bound electrons, there exists a single energy level, or energy state, which is located within the forbidden band gap just below the bottom of the conduction band (Figure 12.12*a*). The electron binding energy corresponds to the energy required to excite the electron from one of these impurity states to a state within the conduction band. Each excitation event (Figure 12.12*b*), supplies or donates a single electron to the conduction band; an impurity of this type is aptly termed a *donor*. Since each donor electron is excited from an impurity level, no corresponding hole is created within the valence band.

At room temperature, the thermal energy available is sufficient to excite large numbers of electrons from **donor states;** in addition, some intrinsic valence–conduction band transitions occur, as in Figure 12.6*b*, but to a negligible degree. Thus, the number of electrons in the conduction band far exceeds the number of holes in the valence band (or $n \gg p$), and the first term on the right-hand side of Equation 12.13 overwhelms the second; that is,

$$\sigma \cong n|e|\mu_e \tag{12.16}$$

A material of this type is said to be an *n-type* extrinsic semiconductor. The electrons are *majority carriers* by virtue of their density or concentration; holes, on the other hand, are the *minority charge carriers*. For *n*-type semiconductors, the Fermi level is shifted upward in the band gap, to within the vicinity of the donor state; its exact position is a function of both temperature and donor concentration.

p-TYPE EXTRINSIC SEMICONDUCTION

An opposite effect is produced by the addition to silicon or germanium of trivalent substitutional impurities such as aluminum, boron, and gallium from Group IIIA of the periodic table. One of the covalent bonds around each of these atoms is deficient in an electron; such a deficiency may be viewed as a hole that is weakly bound to the impurity atom. This hole may be liberated from the impurity atom by the transfer of an electron from an adjacent bond as illustrated in Figure 12.13.

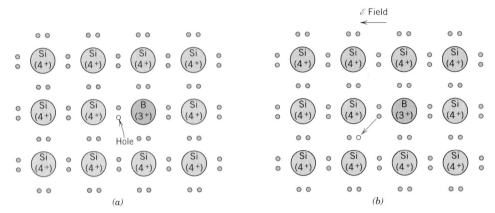

FIGURE 12.13   Extrinsic *p*-type semiconduction model (electron bonding). (*a*) An impurity atom such as boron, having three valence electrons, may substitute for a silicon atom. This results in a deficiency of one valence electron, or a hole associated with the impurity atom. (*b*) The motion of this hole in response to an electric field.

In essence, the electron and the hole exchange positions. A moving hole is considered to be in an excited state and participates in the conduction process, in a manner analogous to an excited donor electron, as described above.

Extrinsic excitations, in which holes are generated, may also be represented using the band model. Each impurity atom of this type introduces an energy level within the band gap, above yet very close to the top of the valence band (Figure 12.14*a*). A hole is imagined to be created in the valence band by the thermal excitation of an electron from the valence band into this impurity electron state, as demonstrated in Figure 12.14*b*. With such a transition, only one carrier is produced—a hole in the valence band; a free electron is *not* created in either the impurity level or the conduction band. An impurity of this type is called an *acceptor,* because it is capable of accepting an electron from the valence band, leaving behind a hole. It follows that the energy level within the band gap introduced by this type of impurity is called an **acceptor state.**

For this type of extrinsic conduction, holes are present in much higher concentrations than electrons (i.e., $p \gg n$), and under these circumstances a material is

FIGURE 12.14
(*a*) Energy band scheme for an acceptor impurity level located within the band gap and just above the top of the valence band. (*b*) Excitation of an electron into the acceptor level, leaving behind a hole in the valence band.

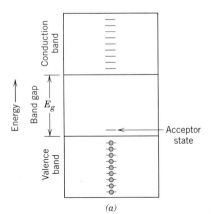

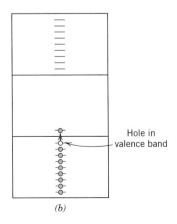

termed *p-type* because positively charged particles are primarily responsible for electrical conduction. Of course, holes are the majority carriers, and electrons are present in minority concentrations. This gives rise to a predominance of the second term on the right-hand side of Equation 12.13, or

$$\sigma \cong p|e|\mu_h \qquad (12.17)$$

For *p*-type semiconductors, the Fermi level is positioned within the band gap and near to the acceptor level.

Extrinsic semiconductors (both *n*- and *p*-type) are produced from materials that are initially of extremely high purity, commonly having total impurity contents on the order of 10^{-7} at%. Controlled concentrations of specific donors or acceptors are then intentionally added, using various techniques. Such an alloying process in semiconducting materials is termed **doping.**

In extrinsic semiconductors, large numbers of charge carriers (either electrons or holes, depending on the impurity type) are created at room temperature, by the available thermal energy. As a consequence, relatively high room-temperature electrical conductivities are obtained in extrinsic semiconductors. Most of these materials are designed for use in electronic devices to be operated at ambient conditions.

EXAMPLE PROBLEM 12.2

Phosphorus is added to high-purity silicon to give a concentration of 10^{23} m^{-3} of charge carriers at room temperature.

(a) Is this material *n*-type or *p*-type?

(b) Calculate the room-temperature conductivity of this material, assuming that electron and hole mobilities are the same as for the intrinsic material.

SOLUTION

(a) Phosphorus is a Group VA element (Figure 2.6) and, therefore, will act as a donor in silicon. Thus, the 10^{23} m^{-3} charge carriers will be virtually all electrons. This electron concentration is greater than that for the intrinsic case (1.33×10^{16} m^{-3}, Example Problem 12.1); hence, this material is extrinsically *n*-type.

(b) In this case the conductivity may be determined using Equation 12.16, as follows:

$$\sigma = n|e|\mu_e = (10^{23} \text{ m}^{-3})(1.6 \times 10^{-19} \text{ C})(0.14 \text{ m}^2/\text{V-s})$$
$$= 2240 \ (\Omega\text{-m})^{-1}$$

12.12 THE TEMPERATURE VARIATION OF CONDUCTIVITY AND CARRIER CONCENTRATION

Figure 12.15 plots the logarithm of the electrical conductivity as a function of the logarithm of absolute temperature for intrinsic silicon, and also for silicon that has been doped with 0.0013 and 0.0052 at% boron; again, boron acts as an acceptor in silicon. Worth noting from this figure is that the electrical conductivity in the intrinsic

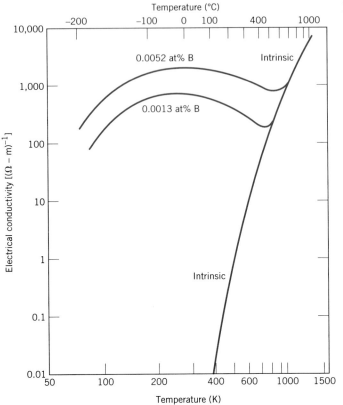

FIGURE 12.15 The temperature dependence of the electrical conductivity (log–log scales) for intrinsic silicon and boron-doped silicon at two doping levels. [Adapted from G. L. Pearson and J. Bardeen, *Phys. Rev.*, **75**, 865 (1949).]

specimen increases dramatically with rising temperature. The numbers of both electrons and holes increase with temperature because more thermal energy is available to excite electrons from the valence to the conduction band. Thus, both the values of n and p in the intrinsic conductivity expression, Equation 12.15, are enhanced. The magnitudes of electron and hole mobilities decrease slightly with temperature as a result of more effective electron and hole scattering by the thermal vibrations. However, these reductions in μ_e and μ_h by no means offset the increase in n and p, and the net effect of a rise in temperature is to produce a conductivity increase.

Mathematically, the dependence of intrinsic conductivity σ on the absolute temperature T is approximately

$$\ln \sigma \cong C - \frac{E_g}{2kT} \tag{12.18}$$

where C represents a temperature-independent constant and E_g and k are the band gap energy and Boltzmann's constant, respectively. Since the increase of n and p with rising temperature is so much greater than the decrease in μ_e and μ_h, the dependence of carrier concentration on temperature for intrinsic behavior is virtually the same as for the conductivity, or

$$\ln n = \ln p \cong C' - \frac{E_g}{2kT} \tag{12.19}$$

FIGURE 12.16
The logarithm of carrier (electron and hole) concentration as a function of the reciprocal of the absolute temperature for intrinsic silicon and two boron-doped silicon materials. (Adapted from G. L. Pearson and J. Bardeen, *Phys. Rev.*, **75**, 865, 1949.)

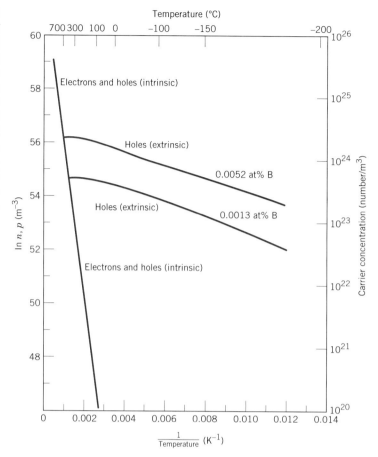

The parameter C' is a constant that is independent of temperature, yet is different from C in Equation 12.18.

In light of Equation 12.19, another method of representing the temperature dependence of the electrical behavior of semiconductors is as the natural logarithm of electron and hole concentrations versus the reciprocal of the absolute temperature. Figure 12.16 is such a plot using data taken from Figure 12.15; and, as may be noted (Figure 12.16), a straight line segment results for the intrinsic material; such a plot expedites the determination of the band gap energy. According to Equation 12.19, the slope of this line segment is equal to $-E_g/2k$, or E_g may be determined as follows:

$$E_g = -2k \left(\frac{\Delta \ln p}{\Delta(1/T)} \right)$$

$$= -2k \left(\frac{\Delta \ln n}{\Delta(1/T)} \right)$$

(12.20)

This is indicated in the schematic plot of Figure 12.17.

Another important feature of the behavior shown in Figures 12.15 and 12.16 is that at temperatures below about 800 K (527°C), the boron-doped materials are extrinsically *p*-type; that is, virtually all the carrier holes result from extrinsic

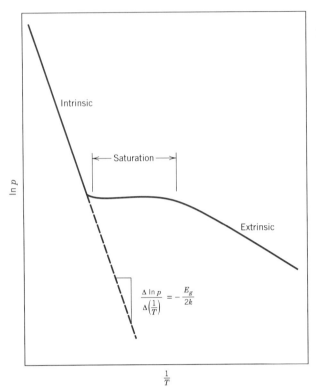

ln p

Intrinsic

Saturation

Extrinsic

$$\dfrac{\Delta \ln p}{\Delta \left(\dfrac{1}{T}\right)} = -\dfrac{E_g}{2k}$$

$\dfrac{1}{T}$

FIGURE 12.17 Schematic plot of the natural logarithm of hole concentration as a function of the reciprocal of absolute temperature for a p-type semiconductor that exhibits extrinsic, saturation, and intrinsic behavior.

excitations—electron transitions from the valence band into the boron acceptor level, which leave behind valence band holes (Figure 12.14). The available thermal energies at these temperatures are sufficient to promote significant numbers of these excitations, yet insufficient to stimulate many electrons from the valence band across the band gap. Thus, the extrinsic conductivity far exceeds that of the intrinsic material. For example, at 400 K (127°C) the conductivities for intrinsic silicon and extrinsic 0.0013 at% boron-doped material are approximately 10^{-2} and 600 $(\Omega\text{-m})^{-1}$, respectively (Figure 12.15). This comparison indicates the sensitivity of conductivity to even extremely small concentrations of some impurity elements.

Furthermore, the extrinsic conductivity is also sensitive to temperature, as indicated in Figure 12.15, for both boron-doped materials. Beginning at about 75 K (−200°C), the conductivity first increases with temperature, reaches a maximum, and then decreases slightly prior to becoming intrinsic. Or, in terms of carrier (i.e., hole) concentration, Figure 12.16, ln p first increases linearly with decreasing $1/T$ (or increasing temperature). Large numbers of extrinsic excitations are possible even at these relatively low temperatures inasmuch as the acceptor level lies just above the top of the valence band. With further temperature increase ($1/T$ decrease), the hole concentration eventually becomes independent of temperature, Figure 12.16. At this point virtually all of the boron atoms have accepted electrons from the valence band, or are said to be *saturated;* this is appropriately termed the *saturation region* (Figure 12.17). (Donor impurities become *exhausted* instead of saturated.) The number of holes in this region is approximately equal to the number of dopant impurity (i.e., boron) atoms.

The decrease of conductivity with increasing temperature within the saturation region for the two extrinsic curves in Figure 12.15 may be explained by the reduction in hole mobility with rising temperature. From the extrinsic conductivity expression, Equation 12.17, both e and p are independent of temperature in this region, and the only temperature dependence comes from the mobility.

Also worth noting from Figures 12.15 and 12.16 is that at about 800 K (527°C), the conductivity of both boron-doped materials becomes intrinsic. At the outset of intrinsic behavior, the number of intrinsic valence band-to-conduction band transitions becomes greater than the number of holes that are extrinsically generated.

A couple of final comments relate to the influence of boron acceptor content on the electrical behavior of silicon. First, the extrinsic and saturation conductivities and hole concentrations are greater for the material with the higher boron content (Figures 12.15 and 12.16), a result not unexpected, since more B atoms are present from which holes may be produced. Also, the intrinsic outset temperature becomes elevated as the dopant content increases.

EXAMPLE PROBLEM 12.3

If the room-temperature [25°C (298 K)] electrical conductivity of intrinsic germanium is 2.2 $(\Omega\text{-m})^{-1}$, estimate its conductivity at 150°C (423 K).

SOLUTION

This problem is solved by employment of Equation 12.18. First, we determine the value of the constant C using the room-temperature data, after which the value at 150°C may be computed. From Table 12.2, the value of E_g for germanium is 0.67 eV, and, therefore,

$$C = \ln \sigma + \frac{E_g}{2kT}$$

$$= \ln(2.2) + \frac{0.67 \text{ eV}}{(2)(8.62 \times 10^{-5} \text{ eV/K})(298 \text{ K})} = 13.83$$

Now, at 150°C (423 K),

$$\ln \sigma = C - \frac{E_g}{2kT}$$

$$= 13.83 - \frac{0.67 \text{ eV}}{(2)(8.62 \times 10^{-5} \text{ eV/K})(423 \text{ K})} = 4.64$$

or

$$\sigma = 103.8 \ (\Omega\text{-m})^{-1}$$

DESIGN EXAMPLE 12.1

The room-temperature electrical conductivity of intrinsic silicon is 4×10^{-4} $(\Omega\text{-m})^{-1}$ (Table 12.2). An extrinsic n-type silicon material is desired having a room-temperature conductivity of 150 $(\Omega\text{-m})^{-1}$. Specify a donor impurity type that may be used as well as its concentration in atom percent to yield these electrical characteristics. Assume that the electron and hole mobilities are the same as for the intrinsic material, and that at room temperature the donor impurities are exhausted.

SOLUTION

First of all, those elements which, when added to silicon render it *n*-type, lie one group to the right of silicon in the periodic table; these include the group VA elements (Figure 2.6)—i.e., nitrogen, phosphorus, arsenic, and antimony.

Since this material is extrinsic and *n*-type, $n \gg p$, the electrical conductivity is a function of the free electron concentration according to Equation 12.16. Furthermore, the design stipulates that the donor impurity atoms are exhausted; therefore, the number of free electrons is about equal to the number of donor impurities, N_d. That is

$$n \sim N_d$$

We now solve Equation 12.16 for n using the stipulated conductivity $[150 \ (\Omega\text{-m})^{-1}]$ and the electron mobility value provided in Table 12.2 ($0.14 \ \text{m}^2/\text{V-s}$). Thus

$$n = N_d = \frac{\sigma}{|e|\mu_e}$$

$$= \frac{150 \ (\Omega\text{-m})^{-1}}{(1.6 \times 10^{-19} \ \text{C})(0.14 \ \text{m}^2/\text{V-s})}$$

$$= 6.7 \times 10^{21} \ \text{m}^{-3}$$

It next becomes necessary to calculate the concentration of donor impurities in atom percent. This computation first requires the determination of the number of silicon atoms per cubic meter, N_{Si}, using Equation 5.2, which is as follows:

$$N_{Si} = \frac{N_A \rho_{Si}}{A_{Si}}$$

$$= \frac{(6.023 \times 10^{23} \ \text{atoms/mol})(2.33 \ \text{g/cm}^3)(10^6 \ \text{cm}^3/\text{m}^3)}{28.09 \ \text{g/mol}}$$

$$= 5 \times 10^{28} \ \text{m}^{-3}$$

The concentration of donor impurities in atom percent (C_d') is just the ratio of N_d and $N_d + N_{Si}$ multiplied by 100 as

$$C_d' = \frac{N_d}{N_d + N_{Si}} \times 100$$

$$= \frac{6.7 \times 10^{21} \ \text{m}^{-3}}{(6.7 \times 10^{21} \ \text{m}^{-3}) + (5 \times 10^{28} \ \text{m}^{-3})} \times 100 = 1.34 \times 10^{-5}$$

Thus, a silicon material having a room-temperature *n*-type electrical conductivity of $150 \ (\Omega\text{-m})^{-1}$ must contain 1.34×10^{-5} at% nitrogen, phosphorus, arsenic, or antimony.

12.13 THE HALL EFFECT (CD-ROM)

12.14 SEMICONDUCTOR DEVICES (CD-ROM)

ELECTRICAL CONDUCTION IN IONIC
CERAMICS AND IN POLYMERS

Most polymers and ionic ceramics are insulating materials at room temperature and, therefore, have electron energy band structures similar to that represented in Figure 12.4c; a filled valence band is separated from an empty conduction band by a relatively large band gap, usually greater than 2 eV. Thus, at normal temperatures only very few electrons may be excited across the band gap by the available thermal energy, which accounts for the very small values of conductivity; Table 12.3 gives the room-temperature electrical conductivity of several of these materials. (The electrical resistivities of a large number of ceramic and polymeric materials are provided in Table B.9, Appendix B.) Of course, many materials are utilized on the basis of their ability to insulate, and thus a high electrical resistivity is desirable. With rising temperature insulating materials experience an increase in electrical conductivity, which may ultimately be greater than that for semiconductors.

12.15 CONDUCTION IN IONIC MATERIALS

Both cations and anions in ionic materials possess an electric charge and, as a consequence, are capable of migration or diffusion when an electric field is present. Thus an electric current will result from the net movement of these charged ions, which will be present in addition to that due to any electron motion. Of course, anion and cation migrations will be in opposite directions. The total conductivity of an ionic material σ_{total} is thus equal to the sum of both electronic and ionic contributions, as follows:

$$\sigma_{total} = \sigma_{electronic} + \sigma_{ionic} \tag{12.25}$$

Table 12.3   Typical Room-Temperature Electrical Conductivities for 13 Nonmetallic Materials

Material	Electrical Conductivity $[(\Omega\text{-}m)^{-1}]$
Graphite	3×10^4–2×10^5
Ceramics	
Concrete (dry)	10^{-9}
Soda–lime glass	10^{-10}–10^{-11}
Porcelain	10^{-10}–10^{-12}
Borosilicate glass	$\sim 10^{-13}$
Aluminum oxide	$< 10^{-13}$
Fused silica	$< 10^{-18}$
Polymers	
Phenol-formaldehyde	10^{-9}–10^{-10}
Polymethyl methacrylate	$< 10^{-12}$
Nylon 6,6	10^{-12}–10^{-13}
Polystyrene	$< 10^{-14}$
Polyethylene	10^{-15}–10^{-17}
Polytetrafluoroethylene	$< 10^{-17}$

Either contribution may predominate depending on the material, its purity, and, of course, temperature.

A mobility μ_i may be associated with each of the ionic species as follows:

$$\mu_i = \frac{n_i e D_i}{kT} \tag{12.26}$$

where n_i and D_i represent, respectively, the valence and diffusion coefficient of a particular ion; e, k, and T denote the same parameters as explained earlier in the chapter. Thus, the ionic contribution to the total conductivity increases with increasing temperature, as does the electronic component. However, in spite of the two conductivity contributions, most ionic materials remain insulative, even at elevated temperatures.

12.16 ELECTRICAL PROPERTIES OF POLYMERS

Most polymeric materials are poor conductors of electricity (Table 12.3) because of the unavailability of large numbers of free electrons to participate in the conduction process. The mechanism of electrical conduction in these materials is not well understood, but it is felt that conduction in polymers of high purity is electronic.

CONDUCTING POLYMERS

Within the past several years, polymeric materials have been synthesized that have electrical conductivities on par with metallic conductors; they are appropriately termed *conducting polymers*. Conductivities as high as 1.5×10^7 $(\Omega\text{-m})^{-1}$ have been achieved in these materials; on a volume basis, this value corresponds to one fourth of the conductivity of copper, or twice its conductivity on the basis of weight.

This phenomenon is observed in a dozen or so polymers, including polyacetylene, polyparaphenylene, polypyrrole, and polyaniline that have been doped with appropriate impurities. As is the case with semiconductors, these polymers may be made either *n*-type (i.e., free-electron dominant) or *p*-type (i.e., hole dominant) depending on the dopant. However, unlike semiconductors, the dopant atoms or molecules do not substitute for or replace any of the polymer atoms.

High-purity polymers have electron band structures characteristic of electrical insulators (Figure 12.4c). The mechanism by which large numbers of free electrons and holes are generated in these conducting polymers is complex and not well understood. In very simple terms, it appears that the dopant atoms lead to the formation of new energy bands that overlap the valence and conduction bands of the intrinsic polymer, giving rise to a partially filled band, and the production at room temperature of a high concentration of free electrons or holes. Orienting the polymer chains, either mechanically (Section 8.17) or magnetically, during synthesis results in a highly anisotropic material having a maximum conductivity along the direction of orientation.

These conducting polymers have the potential to be used in a host of applications inasmuch as they have low densities, are highly flexible, and are easy to produce. Rechargeable batteries are currently being manufactured that employ polymer electrodes; in many respects these are superior to their metallic counterpart batteries. Other possible applications include wiring in aircraft and aerospace components, antistatic coatings for clothing, electromagnetic screening materials, and electronic devices (e.g., transistors and diodes).

DIELECTRIC BEHAVIOR

12.17 CAPACITANCE (CD-ROM)

12.18 FIELD VECTORS AND POLARIZATION (CD-ROM)

12.19 TYPES OF POLARIZATION (CD-ROM)

12.20 FREQUENCY DEPENDENCE OF THE DIELECTRIC CONSTANT (CD-ROM)

12.21 DIELECTRIC STRENGTH (CD-ROM)

12.22 DIELECTRIC MATERIALS (CD-ROM)

OTHER ELECTRICAL CHARACTERISTICS OF MATERIALS

12.23 FERROELECTRICITY (CD-ROM)

12.24 PIEZOELECTRICITY (CD-ROM)

SUMMARY

The ease with which a material is capable of transmitting an electric current is expressed in terms of electrical conductivity or its reciprocal, resistivity. On the basis of its conductivity, a solid material may be classified as a metal, a semiconductor, or an insulator.

For most materials, an electric current results from the motion of free electrons, which are accelerated in response to an applied electric field. The number of these free electrons depends on the electron energy band structure of the material. An electron band is just a series of electron states that are closely spaced with respect to energy, and one such band may exist for each electron subshell found in the isolated atom. By "electron energy band structure" is meant the manner in which the outermost bands are arranged relative to one another and then filled with electrons. A distinctive band structure type exists for metals, for semiconductors, and for insulators. An electron becomes free by being excited from a filled state in one band, to an available empty state above the Fermi energy. Relatively small energies are required for electron excitations in metals, giving rise to large numbers of free electrons. Larger energies are required for electron excitations in semiconductors and insulators, which accounts for their lower free electron concentrations and smaller conductivity values.

Free electrons being acted on by an electric field are scattered by imperfections in the crystal lattice. The magnitude of electron mobility is indicative of the frequency of these scattering events. In many materials, the electrical conductivity is proportional to the product of the electron concentration and the mobility.

For metallic materials, electrical resistivity increases with temperature, impurity content, and plastic deformation. The contribution of each to the total resistivity is additive.

Semiconductors may be either elements (Si and Ge) or covalently bonded compounds. With these materials, in addition to free electrons, holes (missing electrons in the valence band) may also participate in the conduction process. On the basis of electrical behavior, semiconductors are classified as either intrinsic or extrinsic. For intrinsic behavior, the electrical properties are inherent to the pure material, and electron and hole concentrations are equal; electrical behavior is dictated by impurities for extrinsic semiconductors. Extrinsic semiconductors may be either n- or p-type depending on whether electrons or holes, respectively, are the predominant charge carriers. Donor impurities introduce excess electrons; acceptor impurities, excess holes.

The electrical conductivity of semiconducting materials is particularly sensitive to impurity type and content, as well as to temperature. The addition of even minute concentrations of some impurities enhances the conductivity drastically. Furthermore, with rising temperature, intrinsic conductivity experiences an exponential increase. Extrinsic conductivity may also increase with temperature.

{A number of semiconducting devices employ the unique electrical characteristics of these materials to perform specific electronic functions. Included are the p–n rectifying junction, and junction and MOSFET transistors. Transistors are used for amplification of electrical signals, as well as for switching devices in computer circuitries.}

{Dielectric materials are electrically insulative, yet susceptible to polarization in the presence of an electric field. This polarization phenomenon accounts for the ability of the dielectrics to increase the charge storing capability of capacitors, the efficiency of which is expressed in terms of a dielectric constant. Polarization results from the inducement by, or orientation with the electric field of atomic or molecular dipoles; a dipole is said to exist when there is a net spatial separation of positively and negatively charged entities. Possible polarization types include electronic, ionic, and orientation; not all types need be present in a particular dielectric. For alternating electric fields, whether a specific polarization type contributes to the total polarization and dielectric constant depends on frequency; each polarization mechanism ceases to function when the applied field frequency exceeds its relaxation frequency.}

{This chapter concluded with brief discussions of two other electrical phenomena. Ferroelectric materials are those that may exhibit polarization spontaneously, that is, in the absence of any external electric field. Finally, piezoelectricity is the phenomenon whereby polarization is induced in a material by the imposition of external forces.}

IMPORTANT TERMS AND CONCEPTS

Acceptor state
Capacitance
Conduction band
Conductivity, electrical
Dielectric
Dielectric constant
Dielectric displacement
Dielectric strength
Diode
Dipole, electric
Donor state
Doping
Electrical resistance
Electron energy band
Energy band gap

Extrinsic semiconductor
Fermi energy
Ferroelectric
Forward bias
Free electron
Hall effect
Hole
Insulator
Integrated circuit
Intrinsic semiconductor
Ionic conduction
Junction transistor
Matthiessen's rule
Metal
Mobility

MOSFET
Ohm's law
Permittivity
Piezoelectric
Polarization
Polarization, electronic
Polarization, ionic
Polarization, orientation
Rectifying junction
Relaxation frequency
Resistivity, electrical
Reverse bias
Semiconductor
Valence band

REFERENCES

Azaroff, L. V. and J. J. Brophy, *Electronic Processes in Materials,* McGraw-Hill Book Company, New York, 1963. Reprinted by TechBooks, Marietta, OH, 1990. Chapters 6–12.

Bube, R. H., *Electrons in Solids,* 3rd edition, Academic Press, San Diego, 1992.

Bylander, E. G., *Materials for Semiconductor Functions,* Hayden Book Company, New York, 1971. Good fundamental treatment of the physics of semiconductors and various semiconducting devices.

Chaudhari, P., "Electronic and Magnetic Materials," *Scientific American,* Vol. 255, No. 4, October 1986, pp. 136–144.

Ehrenreich, H., "The Electrical Properties of Materials," *Scientific American,* Vol. 217, No. 3, September 1967, pp. 194–204.

Hummel, R. E., *Electronic Properties of Materials,* 2nd edition, Springer-Verlag New York, Inc., New York, 1994.

Kingery, W. D., H. K. Bowen, and D. R. Uhlmann, *Introduction to Ceramics,* 2nd edition, John Wiley & Sons, New York, 1976. Chapters 17 and 18.

Kittel, C., *Introduction to Solid State Physics,* 7th edition, John Wiley & Sons, Inc., New York, 1995. An advanced treatment.

Kwok, H. L., *Electronic Materials,* Brooks/Cole Publishing Company, Pacific Grove, CA, 1997.

Livingston, J., *Electronic Properties,* John Wiley & Sons, New York, 1999.

Meindl, J. D., "Microelectronic Circuit Elements," *Scientific American,* Vol. 237, No. 3, September 1977, pp. 70–81.

Navon, D. H., *Semiconductor Microdevices and Materials,* Oxford University, Oxford, 1995.

Noyce, R. N., "Microelectronics," *Scientific American,* Vol. 237, No. 3, September 1977, pp. 62–69.

Oldham, W. G., "The Fabrication of Microelectronic Circuits," *Scientific American,* Vol. 237, No. 3, September 1977, pp. 110–128.

Rose, R. M., L. A. Shepard, and J. Wulff, *The Structure and Properties of Materials,* Vol. IV,

Electronic Properties, John Wiley & Sons, New York, 1966. Chapters 1, 2, 4–8, and 12.

Warnes, L. A. A., *Electronic Materials,* Chapman and Hall, New York, 1990.

Wert, C. A. and R. M. Thomson, *Physics of Solids,* 2nd edition, McGraw-Hill Book Company, New York, 1970. Chapters 9 and 11–19.

QUESTIONS AND PROBLEMS

Note: To solve those problems having an asterisk (*) by their numbers, consultation of supplementary topics [appearing only on the CD-ROM (and not in print)] will probably be necessary.

12.1 **(a)** Compute the electrical conductivity of a 5.1-mm (0.2-in.) diameter cylindrical silicon specimen 51 mm (2 in.) long in which a current of 0.1 A passes in an axial direction. A voltage of 12.5 V is measured across two probes that are separated by 38 mm (1.5 in.). **(b)** Compute the resistance over the entire 51 mm (2 in.) of the specimen.

12.2 A copper wire 100 m long must experience a voltage drop of less than 1.5 V when a current of 2.5 A passes through it. Using the data in Table 12.1, compute the minimum diameter of the wire.

12.3 An aluminum wire 4 mm in diameter is to offer a resistance of no more than 2.5 Ω. Using the data in Table 12.1, compute the maximum wire length.

12.4 Demonstrate that the two Ohm's law expressions, Equations 12.1 and 12.5, are equivalent.

12.5 **(a)** Using the data in Table 12.1, compute the resistance of a copper wire 3 mm (0.12 in.) in diameter and 2 m (78.7 in.) long. **(b)** What would be the current flow if the potential drop across the ends of the wire is 0.05 V? **(c)** What is the current density? **(d)** What is the magnitude of the electric field across the ends of the wire?

12.6 What is the distinction between electronic and ionic conduction?

12.7 How does the electron structure of an isolated atom differ from that of a solid material?

12.8 In terms of electron energy band structure, discuss reasons for the difference in electrical conductivity between metals, semiconductors, and insulators.

12.9 If a metallic material is cooled through its melting temperature at an extremely rapid rate, it will form a noncrystalline solid (i.e., a metallic glass). Will the electrical conductivity of the noncrystalline metal be greater or less than its crystalline counterpart? Why?

12.10 Briefly tell what is meant by the drift velocity and mobility of a free electron.

12.11 **(a)** Calculate the drift velocity of electrons in germanium at room temperature and when the magnitude of the electric field is 1000 V/m. **(b)** Under these circumstances, how long does it take an electron to traverse a 25 mm (1 in.) length of crystal?

12.12 An *n*-type semiconductor is known to have an electron concentration of 3×10^{18} m^{-3}. If the electron drift velocity is 100 m/s in an electric field of 500 V/m, calculate the conductivity of this material.

12.13 At room temperature the electrical conductivity and the electron mobility for copper are 6.0×10^7 (Ω-m)$^{-1}$ and 0.0030 m^2/V-s, respectively. **(a)** Compute the number of free electrons per cubic meter for copper at room temperature. **(b)** What is the number of free electrons per copper atom? Assume a density of 8.9 g/cm^3.

12.14 **(a)** Calculate the number of free electrons per cubic meter for gold assuming that there

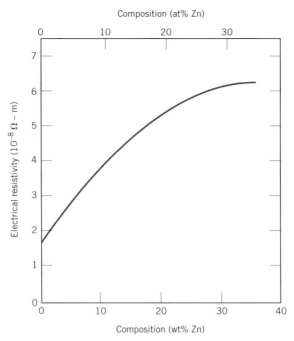

Composition (at% Zn)

Electrical resistivity (10^{-8} Ω-m)

Composition (wt% Zn)

FIGURE 12.35 Room-temperature electrical resistivity versus composition for copper–zinc alloys. (Adapted from *Metals Handbook: Properties and Selection: Nonferrous Alloys and Pure Metals*, Vol. 2, 9th edition, H. Baker, Managing Editor, American Society for Metals, 1979, p. 315.)

are 1.5 free electrons per gold atom. The electrical conductivity and density for Au are 4.3×10^{7} (Ω-m)$^{-1}$ and 19.32 g/cm^{3}, respectively. **(b)** Now compute the electron mobility for Au.

12.15 From Figure 12.35, estimate the value of A in Equation 12.11 for zinc as an impurity in copper–zinc alloys.

12.16 **(a)** Using the data in Figure 12.8, determine the values of ρ_0 and a from Equation 12.10 for pure copper. Take the temperature T to be in degrees Celsius. **(b)** Determine the value of A in Equation 12.11 for nickel as an impurity in copper, using the data in Figure 12.8. **(c)** Using the results of parts a and b, estimate the electrical resistivity of copper containing 1.75 at% Ni at 100°C.

12.17 Determine the electrical conductivity of a Cu-Ni alloy that has a yield strength of 125 MPa (18,000 psi). You will find Figure 8.16*b* helpful.

12.18 Tin bronze has a composition of 92 wt% Cu and 8 wt% Sn, and consists of two phases at room temperature: an α phase, which is copper containing a very small amount of tin in solid solution, and an ϵ phase, which

consists of approximately 37 wt% Sn. Compute the room temperature conductivity of this alloy given the following data:

Phase	Electrical Resistivity (Ω-m)	Density (g/cm³)
α	1.88×10^{-8}	8.94
ϵ	5.32×10^{-7}	8.25

12.19 The room-temperature electrical resistivities of pure lead and pure tin are 2.06×10^{-7} and 1.11×10^{-7} Ω-m, respectively.

(a) Make a schematic graph of the room-temperature electrical resistivity versus composition for all compositions between pure lead and pure tin.

(b) On this same graph schematically plot electrical resistivity versus composition at 150°C.

(c) Explain the shapes of these two curves, as well as any differences between them.

12.20 A cylindrical metal wire 2 mm (0.08 in.) in diameter is required to carry a current of 10 A with a minimum of 0.03 V drop per foot (300 mm) of wire. Which of the metals and alloys listed in Table 12.1 are possible candidates?

12.21 **(a)** Compute the number of free electrons and holes that exist in intrinsic germanium at room temperature, using the data in Table 12.2. **(b)** Now calculate the number of free electrons per atom for germanium and silicon (Example Problem 12.1). **(c)** Explain the difference. You will need the densities for Ge and Si, which are 5.32 and 2.33 g/cm^3, respectively.

12.22 For intrinsic semiconductors, both electron and hole concentrations depend on temperature as follows:

$$n, p \propto \exp\left(-\frac{E_g}{2kT}\right) \qquad (12.38)$$

or, taking natural logarithms,

$$\ln n, \ln p \propto -\frac{E_g}{2kT}$$

Thus, a plot of the intrinsic $\ln n$ (or $\ln p$) versus $1/T$ (K)$^{-1}$ should be linear and yield a slope of $-E_g/2k$. Using this information and Figure 12.16, determine the band gap energy for silicon. Compare this value with the one given in Table 12.2.

12.23 Define the following terms as they pertain to semiconducting materials: intrinsic, extrinsic, compound, elemental. Now provide an example of each.

12.24 Is it possible for compound semiconductors to exhibit intrinsic behavior? Explain your answer.

12.25 For each of the following pairs of semiconductors, decide which will have the smaller band gap energy E_g and then cite the reason for your choice: **(a)** ZnS and CdSe, **(b)** Si and C (diamond), **(c)** Al$_2$O$_3$ and ZnTe, **(d)** InSb and ZnSe, and **(e)** GaAs and AlP.

12.26 **(a)** In your own words, explain how donor impurities in semiconductors give rise to free electrons in numbers in excess of those generated by valence band–conduction band excitations. **(b)** Also explain how acceptor impurities give rise to holes in numbers in excess of those generated by valence band–conduction band excitations.

12.27 **(a)** Explain why no hole is generated by the electron excitation involving a donor impurity atom. **(b)** Explain why no free electron is generated by the electron excitation involving an acceptor impurity atom.

12.28 Will each of the following elements act as a donor or an acceptor when added to the indicated semiconducting material? Assume that the impurity elements are substitutional.

Impurity	Semiconductor
N	Si
B	Ge
Zn	GaAs
S	InSb
In	CdS
As	ZnTe

12.29 **(a)** At approximately what position is the Fermi energy for an intrinsic semiconductor? **(b)** At approximately what position is the Fermi energy for an n-type semiconductor? **(c)** Make a schematic plot of Fermi energy versus temperature for an n-type semiconductor up to a temperature at which it becomes intrinsic. Also note on this plot energy positions corresponding to the top of the valence band and the bottom of the conduction band.

12.30 **(a)** The room-temperature electrical conductivity of a silicon specimen is 10^3 (Ω-m)$^{-1}$. The hole concentration is known to be 1.0×10^{23} m^{-3}. Using the electron and hole mobilities for silicon in Table 12.2, compute the electron concentration. **(b)** On the basis of the result in part a, is the specimen intrinsic, n-type extrinsic, or p-type extrinsic? Why?

12.31 Using the data in Table 12.2, compute the electron and hole concentrations for intrinsic InSb at room temperature.

12.32 Germanium to which 5×10^{22} m^{-3} Sb atoms have been added is an extrinsic semiconductor at room temperature, and virtually all the Sb atoms may be thought of as being ionized (i.e., one charge carrier exists for each Sb atom). **(a)** Is this material n-type or p-type? **(b)** Calculate the electrical conductivity of this material, assuming electron and hole mobilities of 0.1 and 0.05 m²/V-s, respectively.

12.33 The following electrical characteristics have been determined for both intrinsic and *p*-type extrinsic indium phosphide (InP) at room temperature:

	$\sigma(\Omega\text{-}m)^{-1}$	$n\ (m^{-3})$	$p\ (m^{-3})$
Intrinsic	2.5×10^{-6}	3.0×10^{13}	3.0×10^{13}
Extrinsic (*n*-type)	3.6×10^{-5}	4.5×10^{14}	2.0×10^{12}

Calculate electron and hole mobilities.

12.34 Compare the temperature dependence of the conductivity for metals and intrinsic semiconductors. Briefly explain the difference in behavior.

12.35 Using the data in Table 12.2, estimate the electrical conductivity of intrinsic GaAs at 150°C (423 K).

12.36 Briefly explain the presence of the factor 2 in the denominator of the second term on the right-hand side of Equation 12.19.

12.37 Using the data in Table 12.2, estimate the temperature at which the electrical conductivity of intrinsic GaAs is $4 \times 10^{-4}\ (\Omega\text{-}m)^{-1}$.

12.38 The intrinsic electrical conductivities of a semiconductor at 20 and 100°C (293 and 373 K) are 1.0 and 500 $(\Omega\text{-}m)^{-1}$, respectively. Determine the approximate band gap energy for this material.

12.39 Below, the intrinsic electrical conductivity of a semiconductor at two temperatures is tabulated:

$T\ (K)$	$\sigma\ (\Omega\text{-}m)^{-1}$
450	0.12
550	2.25

(a) Determine the band gap energy (in eV) for this material.

(b) Estimate the electrical conductivity at 300 K (27°C).

12.40 At room temperature, the temperature dependence of electron and hole mobilities for intrinsic germanium is found to be proportional to $T^{-3/2}$ for T in Kelvins; thus, a more accurate form of Equation 12.18 is as follows:

$$\sigma = C'' T^{-3/2} \exp\left(-\frac{E_g}{2kT}\right)$$
(12.39a)

or, in logarithmic form

$$\ln \sigma = \ln C'' - \frac{3}{2}\ln T - \frac{E_g}{2kT}$$
(12.39b)

where C'' is a temperature-independent constant.

(a) Calculate the intrinsic electrical conductivity for intrinsic germanium at 150°C, and compare this value with that obtained in Example Problem 12.3, which uses Equation 12.18.

(b) Now compute the number of free electrons and holes for intrinsic germanium at 150°C assuming this $T^{-3/2}$ dependence of electron and hole mobilities.

12.41 Estimate the temperature at which GaAs has an electrical conductivity of 3.7×10^{-3} $(\Omega\text{-}m)^{-1}$ assuming the temperature dependence for σ of Equation 12.39a. The data shown in Table 12.2 might prove helpful.

12.42 The slope of the extrinsic portions of the curves in Figure 12.16 is related to the position of the acceptor level in the band gap (Figure 12.14). Write an expression for the dependence of p on the position of this level.

12.43 We noted in Section 5.3 (Figure 5.4) that in FeO (wüstite), the iron ions can exist in both Fe^{2+} and Fe^{3+} states. The number of each of these ion types depends on temperature and the ambient oxygen pressure. Furthermore, it was also noted that in order to retain electroneutrality, one Fe^{2+} vacancy will be created for every two Fe^{3+} ions that are formed; consequently, in order to reflect the existence of these vacancies the formula for wüstite is often represented as $Fe_{(1-x)}O$ where x is some small fraction less than unity.

In this nonstoichiometric $Fe_{(1-x)}O$ material, conduction is electronic, and, in fact, it behaves as a *p*-type semiconductor. That is, the Fe^{3+} ions act as electron acceptors, and it is relatively easy to excite an electron from the valence band into an Fe^{3+} acceptor state, with the formation of a hole. Determine the electrical conductivity of a specimen of wüstite that has a hole mobility of 1.0×10^{-5} m²/V-s and for which the value of *x* is 0.060. Assume that the acceptor states are saturated (i.e., one hole exists for every Fe^{3+} ion). Wüstite has the sodium chloride crystal structure with a unit cell edge length of 0.437 nm.

12.44* Some hypothetical metal is known to have an electrical resistivity of 4×10^{-8} (Ω-m). Through a specimen of this metal that is 25 mm thick is passed a current of 30 A; when a magnetic field of 0.75 tesla is simultaneously imposed in a direction perpendicular to that of the current, a Hall voltage of -1.26×10^{-7} V is measured. Compute **(a)** the electron mobility for this metal, and **(b)** the number of free electrons per cubic meter.

12.45* Some metal alloy is known to have electrical conductivity and electron mobility values of 1.5×10^{7} (Ω-m)$^{-1}$ and 0.0020 m²/V-s, respectively. Through a specimen of this alloy that is 35 mm thick is passed a current of 45 A. What magnetic field would need to be imposed to yield a Hall voltage of -1.0×10^{-7} V?

12.46* Briefly describe electron and hole motions in a *p–n* junction for forward and reverse biases; then explain how these lead to rectification.

12.47* How is the energy in the reaction described by Equation 12.24 dissipated?

12.48* What are the two functions that a transistor may perform in an electronic circuit?

12.49* Would you expect increasing temperature to influence the operation of *p–n* junction rectifiers and transistors? Explain.

12.50* Cite the differences in operation and application for junction transistors and MOSFETs.

12.51 At temperatures between 775°C (1048 K) and 1100°C (1373 K), the activation energy and preexponential for the diffusion coef-

ficient of Fe^{2+} in FeO are 102,000 J/mol and 7.3×10^{-8} m²/s, respectively. Compute the mobility for an Fe^{2+} ion at 1000°C (1273 K).

12.52* A parallel-plate capacitor using a dielectric material having an ϵ_r of 2.5 has a plate spacing of 1 mm (0.04 in.). If another material having a dielectric constant of 4.0 is used and the capacitance is to be unchanged, what must be the new spacing between the plates?

12.53* A parallel-plate capacitor with dimensions of 100 mm by 25 mm and a plate separation of 3 mm must have a minimum capacitance of 38 pF (3.8×10^{-11} F) when an ac potential of 500 V is applied at a frequency of 1 MHz. Which of those materials listed in Table 12.4 are possible candidates? Why?

12.54* Consider a parallel-plate capacitor having an area of 2500 mm² and a plate separation of 2 mm, and with a material of dielectric constant 4.0 positioned between the plates. **(a)** What is the capacitance of this capacitor? **(b)** Compute the electric field that must be applied for a charge of 8.0×10^{-9} C to be stored on each plate.

12.55* In your own words, explain the mechanism by which charge storing capacity is increased by the insertion of a dielectric material within the plates of a capacitor.

12.56* For NaCl, the ionic radii for Na$^+$ and Cl$^-$ ions are 0.102 and 0.181 nm, respectively. If an externally applied electric field produces a 5% expansion of the lattice, compute the dipole moment for each Na$^+$–Cl$^-$ pair. Assume that this material is completely unpolarized in the absence of an electric field.

12.57* The polarization *P* of a dielectric material positioned within a parallel-plate capacitor is to be 1.0×10^{-6} C/m².

(a) What must be the dielectric constant if an electric field of 5×10^{4} V/m is applied?

(b) What will be the dielectric displacement *D*?

12.58* A charge of 3.5×10^{-11} C is to be stored on each plate of a parallel-plate capacitor having an area of 160 mm² (0.25 in.²) and a plate separation of 3.5 mm (0.14 in.).

(a) What voltage is required if a material having a dielectric constant of 5.0 is positioned within the plates?

(b) What voltage would be required if a vacuum is used?

(c) What are the capacitances for parts a and b?

(d) Compute the dielectric displacement for part a.

(e) Compute the polarization for part a.

12.59* **(a)** For each of the three types of polarization, briefly describe the mechanism by which dipoles are induced and/or oriented by the action of an applied electric field. **(b)** For solid lead titanate ($PbTiO_3$), gaseous neon, diamond, solid KCl, and liquid NH_3 what kind(s) of polarization is (are) possible? Why?

12.60* The dielectric constant for a soda–lime glass measured at very high frequencies (on the order of 10^{15} Hz) is approximately 2.3. What fraction of the dielectric constant at relatively low frequencies (1 MHz) is attributed to ionic polarization? Neglect any orientation polarization contributions.

12.61* **(a)** Compute the magnitude of the dipole moment associated with each unit cell of $BaTiO_3$, as illustrated in Figure 12.33.

(b) Compute the maximum polarization that is possible for this material.

12.62* Briefly explain why the ferroelectric behavior of $BaTiO_3$ ceases above its ferroelectric Curie temperature.

12.63* Would you expect the physical dimensions of a piezoelectric material such as $BaTiO_3$ to change when it is subjected to an electric field? Why or why not?

Design Problems

12.D1 A 95 wt% Pt-5 wt% Ni alloy is known to have an electrical resistivity of 2.35×10^{-7} Ω-m at room temperature (25°C). Calculate the composition of a platinum-nickel alloy that gives a room-temperature resistivity of 1.75×10^{-7} Ω-m. The room-temperature resistivity of pure platinum may be determined from the data in Table 12.1; assume that platinum and nickel form a solid solution.

12.D2 Using information contained in Figures 12.8 and 12.35, determine the electrical conductivity of an 80 wt% Cu-20 wt% Zn alloy at -150°C (-240°F).

12.D3 Is it possible to alloy copper with nickel to achieve a minimum tensile strength of 375 MPa (54,400 psi) and yet maintain an electrical conductivity of 2.5×10^6 $(\Omega\text{-m})^{-1}$? If not, why? If so, what concentration of nickel is required? You may want to consult Figure 8.16a.

12.D4 Specify an acceptor impurity type and concentration (in weight percent) that will produce a p-type silicon material having a room temperature electrical conductivity of 50 $(\Omega\text{-m})^{-1}$. Use intrinsic electron and hole mobilities, and assume that the acceptor impurities are saturated.

12.D5 One integrated circuit design calls for diffusing boron into very high purity silicon at an elevated temperature. It is necessary that at a distance 0.2 μm from the surface of the silicon wafer, the room-temperature electrical conductivity be 1.2×10^4 $(\Omega\text{-m})^{-1}$. The concentration of B at the surface of the Si is maintained at a constant level of 1.0×10^{25} m^{-3}; furthermore, it is assumed that the concentration of B in the original Si material is negligible, and that at room temperature the boron atoms are saturated. Specify the temperature at which this diffusion heat treatment is to take place if the treatment time is to be one hour. The diffusion coefficient for the diffusion of B in Si is a function of temperature as

$$D(\text{m}^2/\text{s}) = 2.4 \times 10^{-4} \exp\left(-\frac{347 \text{ kJ/mol}}{RT}\right)$$

12.D6 Problem 12.43 noted that FeO (wüstite) may behave as a semiconductor by virtue of the transformation of Fe^{2+} to Fe^{3+} and the creation of Fe^{2+} vacancies; the maintenance of electroneutrality requires that for every two Fe^{3+} ions, one vacancy is formed. The existence of these vacancies is reflected in the chemical formula of this nonstoichiometric wüstite as $Fe_{(1-x)}O$, where x is a small number having a value less than unity. The degree of nonstoichiometry (i.e., the value of x) may be varied by changing temperature and oxygen partial pressure. Compute the value of x that is required to produce an $Fe_{(1-x)}O$ material having a p-type electrical conductivity of 2000 $(\Omega\text{-m})^{-1}$; assume that the hole mobility is 1.0×10^{-5} m^2/V-s, and that the acceptor states are saturated.

12.D7 The base semiconducting material used in virtually all of our modern integrated circuits is silicon. However, silicon has some limitations and restrictions. Write an essay comparing the properties and applications (and/or potential applications) of silicon and gallium arsenide.

Chapter 13 / Types and Applications of Materials

Why Study *Types and Applications of Materials?*

Engineers are often involved in materials selection
decisions, which necessitates that they have some fa-
miliarity with the general characteristics of a wide
variety of materials. In addition, access to data
bases containing property values for a large number
of materials may be required. {For example, in Section 20.2 we discuss a materials selection process
that is applied to a cylindrical shaft that is stressed
in torsion.}

Learning Objectives

After careful study of this chapter you should be able to do the following:

1. Name four different types of steels and, for each, cite compositional differences, distinctive properties, and typical uses.
2. Name the four cast iron types and, for each, describe its microstructure and note its general mechanical characteristics.
3. Name seven different types of nonferrous alloys and, for each, cite its distinctive physical and mechanical characteristics; in addition, list at least three typical applications.
4. Describe the process that is used to produce glass–ceramics.

5. Name the two types of clay products and give two examples of each.
6. Cite three important requirements that normally must be met by refractory ceramics and abrasive ceramics.
7. Describe the mechanism by which cement hardens when water is added.
8. Cite the seven different polymer application types and, for each, note its general characteristics.

13.1 INTRODUCTION

Many times a materials problem is really one of selecting that material which has the right combination of characteristics for a specific application. Therefore, the persons who are involved in the decision making should have some knowledge of the available options. This extremely abbreviated presentation provides an overview of some of the types of metal alloys, ceramics, and polymeric materials, their general properties, and their limitations.

TYPES OF METAL ALLOYS

Metal alloys, by virtue of composition, are often grouped into two classes—ferrous and nonferrous. Ferrous alloys, those in which iron is the principal constituent, include steels and cast irons. These alloys and their characteristics are the first topics of discussion of this section. The nonferrous ones—all the alloys that are not iron based—are treated next.

13.2 FERROUS ALLOYS

Ferrous alloys—those of which iron is the prime constituent—are produced in larger quantities than any other metal type. They are especially important as engineering construction materials. Their widespread use is accounted for by three factors: (1) iron-containing compounds exist in abundant quantities within the earth's crust; (2) metallic iron and steel alloys may be produced using relatively economical extraction, refining, alloying, and fabrication techniques; and (3) ferrous alloys are extremely versatile, in that they may be tailored to have a wide range of mechanical and physical properties. The principal disadvantage of many ferrous alloys is their susceptibility to corrosion. These sections discuss compositions, microstructures, and properties of a number of different classes of steels and cast irons. A taxonomic classification scheme for the various ferrous alloys is presented in Figure 13.1.

STEELS

Steels are iron–carbon alloys that may contain appreciable concentrations of other alloying elements; there are thousands of alloys that have different compositions and/or heat treatments. The mechanical properties are sensitive to the content of

402

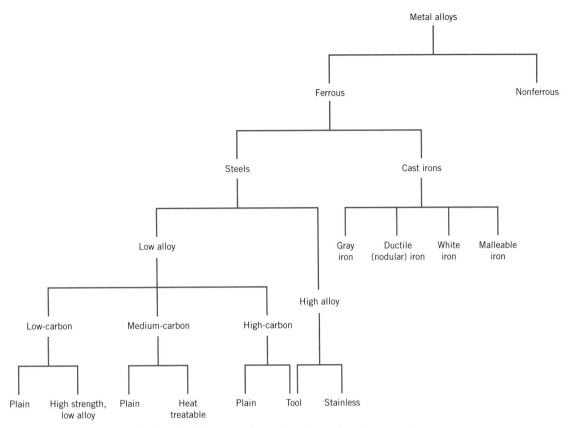

FIGURE 13.1   Classification scheme for the various ferrous alloys.

carbon, which is normally less than 1.0 wt%. Some of the more common steels are classified according to carbon concentration, namely, into low-, medium-, and high-carbon types. Subclasses also exist within each group according to the concentration of other alloying elements. **Plain carbon steels** contain only residual concentrations of impurities other than carbon and a little manganese. For **alloy steels,** more alloying elements are intentionally added in specific concentrations.

Low-Carbon Steels

Of all the different steels, those produced in the greatest quantities fall within the low-carbon classification. These generally contain less than about 0.25 wt% C and are unresponsive to heat treatments intended to form martensite; strengthening is accomplished by cold work. Microstructures consist of ferrite and pearlite constituents. As a consequence, these alloys are relatively soft and weak, but have outstanding ductility and toughness; in addition, they are machinable, weldable, and, of all steels, are the least expensive to produce. Typical applications include automobile body components, structural shapes (I-beams, channel and angle iron), and sheets that are used in pipelines, buildings, bridges, and tin cans. Tables 13.1a and 13.1b, respectively, present the compositions and mechanical properties of several plain low-carbon steels. They typically have a yield strength of 275 MPa (40,000 psi), tensile strengths between 415 and 550 MPa (60,000 and 80,000 psi), and a ductility of 25%EL.

Table 13.1a   Compositions of Five Plain Low-Carbon Steels and Three High-Strength, Low-Alloy Steels

Designation[a]		*Composition (wt%)*[b]		
AISI/SAE or ASTM Number	*UNS Number*	*C*	*Mn*	*Other*
Plain Low-Carbon Steels				
1010	G10100	0.10	0.45	
1020	G10200	0.20	0.45	
A36	K02600	0.29	1.00	0.20 Cu (min)
A516 Grade 70	K02700	0.31	1.00	0.25 Si
High-Strength, Low-Alloy Steels				
A440	K12810	0.28	1.35	0.30 Si (max), 0.20 Cu (min)
A633 Grade E	K12002	0.22	1.35	0.30 Si, 0.08 V, 0.02 N, 0.03 Nb
A656 Grade 1	K11804	0.18	1.60	0.60 Si, 0.1 V, 0.20 Al, 0.015 N

[a] The codes used by the American Iron and Steel Institute (AISI), the Society of Automotive Engineers (SAE), and the American Society for Testing and Materials (ASTM), and in the Uniform Numbering System (UNS) are explained in the text.

[b] Also a maximum of 0.04 wt% P, 0.05 wt% S, and 0.30 wt% Si (unless indicated otherwise).

Source: Adapted from *Metals Handbook: Properties and Selection: Irons and Steels,* Vol. 1, 9th edition, B. Bardes (Editor), American Society for Metals, 1978, pp. 185, 407.

Table 13.1b   Mechanical Characteristics of Hot-Rolled Material and Typical Applications for Various Plain Low-Carbon and High-Strength, Low-Alloy Steels

AISI/SAE or ASTM Number	*Tensile Strength [MPa (ksi)]*	*Yield Strength [MPa (ksi)]*	*Ductility [%EL in 50 mm (2 in.)]*	*Typical Applications*
Plain Low-Carbon Steels				
1010	325 (47)	180 (26)	28	Automobile panels, nails, and wire
1020	380 (55)	205 (30)	25	Pipe; structural and sheet steel
A36	400 (58)	220 (32)	23	Structural (bridges and buildings)
A516 Grade 70	485 (70)	260 (38)	21	Low-temperature pressure vessels
High-Strength, Low-Alloy Steels				
A440	435 (63)	290 (42)	21	Structures that are bolted or riveted
A633 Grade E	520 (75)	380 (55)	23	Structures used at low ambient temperatures
A656 Grade 1	655 (95)	552 (80)	15	Truck frames and railway cars

Another group of low-carbon alloys are the **high-strength, low-alloy (HSLA) steels.** They contain other alloying elements such as copper, vanadium, nickel, and molybdenum in combined concentrations as high as 10 wt%, and possess higher strengths than the plain low-carbon steels. Most may be strengthened by heat treatment, giving tensile strengths in excess of 480 MPa (70,000 psi); in addition, they are ductile, formable, and machinable. Several are listed in Table 13.1. In normal atmospheres, the HSLA steels are more resistant to corrosion than the plain carbon steels, which they have replaced in many applications where structural strength is critical (e.g., bridges, towers, support columns in high-rise buildings, and pressure vessels).

Medium-Carbon Steels

The medium-carbon steels have carbon concentrations between about 0.25 and 0.60 wt%. These alloys may be heat treated by austenitizing, quenching, and then tempering to improve their mechanical properties. They are most often utilized in the tempered condition, having microstructures of tempered martensite. The plain medium-carbon steels have low hardenabilities {(Section 14.6)} and can be successfully heat treated only in very thin sections and with very rapid quenching rates. Additions of chromium, nickel, and molybdenum improve the capacity of these alloys to be heat treated, giving rise to a variety of strength–ductility combinations. These heat-treated alloys are stronger than the low-carbon steels, but at a sacrifice of ductility and toughness. Applications include railway wheels and tracks, gears, crankshafts, and other machine parts and high-strength structural components calling for a combination of high strength, wear resistance, and toughness.

The compositions of several of these alloyed medium-carbon steels are presented in Table 13.2a. Some comment is in order regarding the designation schemes that are also included. The Society of Automotive Engineers (SAE), the American Iron and Steel Institute (AISI), and the American Society for Testing and Materials (ASTM) are responsible for the classification and specification of steels as well as other alloys. The AISI/SAE designation for these steels is a four-digit number: the first two digits indicate the alloy content; the last two, the carbon concentration. For plain carbon steels, the first two digits are 1 and 0; alloy steels are designated by other initial two-digit combinations (e.g., 13, 41, 43). The third and fourth digits represent the weight percent carbon multiplied by 100. For example, a 1060 steel is a plain carbon steel containing 0.60 wt% C.

A unified numbering system (UNS) is used for uniformly indexing both ferrous and nonferrous alloys. Each UNS number consists of a single-letter prefix followed by a five-digit number. The letter is indicative of the family of metals to which an alloy belongs. The UNS designation for these alloys begins with a G, followed by the AISI/SAE number; the fifth digit is a zero. Table 13.2b contains the mechanical characteristics and typical applications of several of these steels, which have been quenched and tempered.

High-Carbon Steels

The high-carbon steels, normally having carbon contents between 0.60 and 1.4 wt%, are the hardest, strongest, and yet least ductile of the carbon steels. They are almost always used in a hardened and tempered condition and, as such, are especially wear resistant and capable of holding a sharp cutting edge. The tool and die steels are high-carbon alloys, usually containing chromium, vanadium, tungsten, and molybdenum. These alloying elements combine with carbon to form very hard and wear-resistant

Table 13.2a   AISI/SAE and UNS Designation Systems and Composition Ranges for Plain Carbon Steel and Various Low-Alloy Steels

AISI/SAE Designation[a]	UNS Designation	Composition Ranges (wt% of Alloying Elements in Addition to C)[b]			
		Ni	Cr	Mo	Other
10xx, Plain carbon	G10xx0				
11xx, Free machining	G11xx0				0.08–0.33S
12xx, Free machining	G12xx0				0.10–0.35S, 0.04–0.12P
13xx	G13xx0				1.60–1.90Mn
40xx	G40xx0			0.20–0.30	
41xx	G41xx0		0.80–1.10	0.15–0.25	
43xx	G43xx0	1.65–2.00	0.40–0.90	0.20–0.30	
46xx	G46xx0	0.70–2.00		0.15–0.30	
48xx	G48xx0	3.25–3.75		0.20–0.30	
51xx	G51xx0		0.70–1.10		
61xx	G61xx0		0.50–1.10		0.10–0.15V
86xx	G86xx0	0.40–0.70	0.40–0.60	0.15–0.25	
92xx	G92xx0				1.80–2.20Si

[a] The carbon concentration, in weight percent times 100, is inserted in the place of "xx" for each specific steel.

[b] Except for 13xx alloys, manganese concentration is less than 1.00 wt%.
Except for 12xx alloys, phosphorus concentration is less than 0.35 wt%.
Except for 11xx and 12xx alloys, sulfur concentration is less than 0.04 wt%.
Except for 92xx alloys, silicon concentration varies between 0.15 and 0.35 wt%.

Table 13.2b   Typical Applications and Mechanical Property Ranges for Oil-Quenched and Tempered Plain Carbon and Alloy Steels

AISI Number	UNS Number	Tensile Strength [MPa (ksi)]	Yield Strength [MPa (ksi)]	Ductility [%EL in 50 mm (2 in.)]	Typical Applications
Plain Low-Carbon Steels					
1040	G10400	605–780 (88–113)	430–585 (62–85)	33–19	Crankshafts, bolts
1080[a]	G10800	800–1310 (116–190)	480–980 (70–142)	24–13	Chisels, hammers
1095[a]	G10950	760–1280 (110–186)	510–830 (74–120)	26–10	Knives, hacksaw blades
Alloy Steels					
4063	G40630	786–2380 (114–345)	710–1770 (103–257)	24–4	Springs, hand tools
4340	G43400	980–1960 (142–284)	895–1570 (130–228)	21–11	Bushings, aircraft tubing
6150	G61500	815–2170 (118–315)	745–1860 (108–270)	22–7	Shafts, pistons, gears

[a] Classified as high-carbon steels.

Table 13.3 Designations, Compositions, and Applications for Six Tool Steels

AISI Number	UNS Number	Composition (wt%)[a]						Typical Applications
		C	*Cr*	*Ni*	*Mo*	*W*	*V*	
M1	T11301	0.85	3.75	0.30 max	8.70	1.75	1.20	Drills, saws; lathe and planer tools
A2	T30102	1.00	5.15	0.30 max	1.15	—	0.35	Punches, embossing dies
D2	T30402	1.50	12	0.30 max	0.95	—	1.10 max	Cutlery, drawing dies
O1	T31501	0.95	0.50	0.30 max	—	0.50	0.30 max	Shear blades, cutting tools
S1	T41901	0.50	1.40	0.30 max	0.50 max	2.25	0.25	Pipe cutters, concrete drills
W1	T72301	1.10	0.15 max	0.20 max	0.10 max	0.15 max	0.10 max	Blacksmith tools, wood-working tools

[a] The balance of the composition is iron. Manganese concentrations range between 0.10 and 1.4 wt%, depending on alloy; silicon concentrations between 0.20 and 1.2 wt% depending on alloy.

Source: Adapted from *ASM Handbook,* Vol. 1, *Properties and Selection: Irons, Steels, and High-Performance Alloys,* 1990. Reprinted by permission of ASM International, Materials Park, OH.

carbide compounds (e.g., $Cr_{23}C_6$, V_4C_3, and WC). Some tool steel compositions and their applications are listed in Table 13.3. These steels are utilized as cutting tools and dies for forming and shaping materials, as well as in knives, razors, hacksaw blades, springs, and high-strength wire.

Stainless Steels

The **stainless steels** are highly resistant to corrosion (rusting) in a variety of environments, especially the ambient atmosphere. Their predominant alloying element is chromium; a concentration of at least 11 wt% Cr is required. Corrosion resistance may also be enhanced by nickel and molybdenum additions.

Stainless steels are divided into three classes on the basis of the predominant phase constituent of the microstructure—martensitic, ferritic, or austenitic. Table 13.4 lists several stainless steels, by class, along with composition, typical mechanical properties, and applications. A wide range of mechanical properties combined with excellent resistance to corrosion make stainless steels very versatile in their applicability.

Martensitic stainless steels are capable of being heat treated in such a way that martensite is the prime microconstituent. Additions of alloying elements in significant concentrations produce dramatic alterations in the iron–iron carbide phase diagram (Figure 10.26). For austenitic stainless steels, the austenite (or γ) phase field is extended to room temperature. Ferritic stainless steels are composed of the α ferrite (BCC) phase. Austenitic and ferritic stainless steels are hardened and strengthened by cold work because they are not heat treatable. The austenitic stainless steels are the most corrosion resistant because of the high chromium contents and also the nickel additions; and they are produced in the largest quantities. Both martensitic and ferritic stainless steels are magnetic; the austenitic stainlesses are not.

Table 13.4 Designations, Compositions, Mechanical Properties, and Typical Applications for Austenitic, Ferritic, Martensitic, and Precipitation-Hardenable Stainless Steels

AISI Number	UNS Number	Composition (wt%)[a]	Condition[b]	Tensile Strength [MPa (ksi)]	Yield Strength [MPa (ksi)]	Ductility [%EL in 50 mm (2 in.)]	Typical Applications
				Ferritic			
409	S40900	0.08 C, 11.0 Cr, 1.0 Mn, 0.50 Ni, 0.75 Ti	Annealed	380 (55)	205 (30)	20	Automotive exhaust components, tanks for agricultural sprays
446	S44600	0.20 C, 25 Cr, 1.5 Mn	Annealed	515 (75)	275 (40)	20	Valves (high temperature), glass molds, combustion chambers
				Austenitic			
304	S30400	0.08 C, 19 Cr, 9 Ni, 2.0 Mn	Annealed	515 (75)	205 (30)	40	Chemical and food processing equipment, cryogenic vessels
316L	S31603	0.03 C, 17 Cr, 12 Ni, 2.5 Mo, 2.0 Mn	Annealed	485 (70)	170 (25)	40	Welding construction
				Martensitic			
410	S41000	0.15 C, 12.5 Cr, 1.0 Mn	Annealed Q & T	485 (70) 825 (120)	275 (40) 620 (90)	20 12	Rifle barrels, cutlery, jet engine parts
440A	S44002	0.70 C, 17 Cr, 0.75 Mo, 1.0 Mn	Annealed Q & T	725 (105) 1790 (260)	415 (60) 1650 (240)	20 5	Cutlery, bearings, surgical tools
				Precipitation Hardenable			
17-7PH	S17700	0.09 C, 17 Cr, 7 Ni, 1.0 Al, 1.0 Mn	Precipitation hardened	1450 (210)	1310 (190)	1–6	Springs, knives, pressure vessels

[a] The balance of the composition is iron.

[b] Q & T denotes quenched and tempered.

Source: Adapted from *ASM Handbook,* Vol. 1, *Properties and Selection: Irons, Steels, and High-Performance Alloys,* 1990. Reprinted by permission of ASM International, Materials Park, OH.

Some stainless steels are frequently used at elevated temperatures and in severe environments because they resist oxidation and maintain their mechanical integrity under such conditions; the upper temperature limit in oxidizing atmospheres is about 1000°C (1800°F). Equipment employing these steels includes gas turbines, high-temperature steam boilers, heat-treating furnaces, aircraft, missiles, and nuclear power generating units. Also included in Table 13.4 is one ultrahigh-strength stainless steel (17-7PH), which is unusually strong and corrosion resistant. Strengthening is accomplished by precipitation-hardening heat treatments (Section 11.10).

CAST IRONS

Generically, **cast irons** are a class of ferrous alloys with carbon contents above 2.14 wt%; in practice, however, most cast irons contain between 3.0 and 4.5 wt% C and, in addition, other alloying elements. A reexamination of the iron–iron carbide phase diagram (Figure 10.26) reveals that alloys within this composition range become completely liquid at temperatures between approximately 1150 and 1300°C (2100 and 2350°F), which is considerably lower than for steels. Thus, they are easily melted and amenable to casting. Furthermore, some cast irons are very brittle, and casting is the most convenient fabrication technique.

Cementite (Fe_3C) is a metastable compound, and under some circumstances it can be made to dissociate or decompose to form α ferrite and graphite, according to the reaction

$$Fe_3C \rightarrow 3Fe\ (\alpha) + C\ (graphite) \tag{13.1}$$

Thus, the true equilibrium diagram for iron and carbon is not that presented in Figure 10.26, but rather as shown in Figure 13.2. The two diagrams are virtually identical on the iron-rich side (e.g., eutectic and eutectoid temperatures for the Fe–Fe_3C system are 1147 and 727°C, respectively, as compared to 1153 and 740°C for Fe–C); however, Figure 13.2 extends to 100 wt% carbon such that graphite is the carbon-rich phase, instead of cementite at 6.7 wt% C (Figure 10.26).

This tendency to form graphite is regulated by the composition and rate of cooling. Graphite formation is promoted by the presence of silicon in concentrations greater than about 1 wt%. Also, slower cooling rates during solidification favor graphitization (the formation of graphite). For most cast irons, the carbon exists

FIGURE 13.2 The true equilibrium iron–carbon phase diagram with graphite instead of cementite as a stable phase. (Adapted from *Binary Alloy Phase Diagrams*, T. B. Massalski, Editor-in-Chief, 1990. Reprinted by permission of ASM International, Materials Park, OH.)

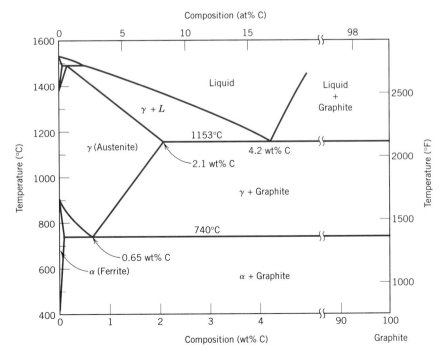

as graphite, and both microstructure and mechanical behavior depend on composition and heat treatment. The most common cast iron types are gray, nodular, white, and malleable.

Gray Iron

The carbon and silicon contents of **gray cast irons** vary between 2.5 and 4.0 wt% and 1.0 and 3.0 wt%, respectively. For most of these cast irons, the graphite exists in the form of flakes (similar to corn flakes), which are normally surrounded by an α ferrite or pearlite matrix; the microstructure of a typical gray iron is shown in Figure 13.3a. Because of these graphite flakes, a fractured surface takes on a gray appearance, hence its name.

Mechanically, gray iron is comparatively weak and brittle in tension as a consequence of its microstructure; the tips of the graphite flakes are sharp and pointed, and may serve as points of stress concentration when an external tensile stress is applied. Strength and ductility are much higher under compressive loads. Typical mechanical properties and compositions of several of the common gray cast irons are listed in Table 13.5. Gray irons do have some desirable characteristics and, in fact, are utilized extensively. They are very effective in damping vibrational energy; this is represented in Figure 13.4, which compares the relative damping capacities of steel and gray iron. Base structures for machines and heavy equipment that are exposed to vibrations are frequently constructed of this material. In addition, gray irons exhibit a high resistance to wear. Furthermore, in the molten state they have a high fluidity at casting temperature, which permits casting pieces having intricate shapes; also, casting shrinkage is low. Finally, and perhaps most important, gray cast irons are among the least expensive of all metallic materials.

Gray irons having microstructures different from that shown in Figure 13.3a may be generated by adjustment of composition and/or by using an appropriate treatment. For example, lowering the silicon content or increasing the cooling rate may prevent the complete dissociation of cementite to form graphite (Equation 13.1). Under these circumstances the microstructure consists of graphite flakes embedded in a pearlite matrix. Figure 13.5 compares schematically the several cast iron microstructures obtained by varying the composition and heat treatment.

Ductile (or Nodular) Iron

Adding a small amount of magnesium and/or cerium to the gray iron before casting produces a distinctly different microstructure and set of mechanical properties. Graphite still forms, but as nodules or spherelike particles instead of flakes. The resulting alloy is called **nodular** or **ductile iron,** and a typical microstructure is shown in Figure 13.3b. The matrix phase surrounding these particles is either pearlite or ferrite, depending on heat treatment (Figure 13.5); it is normally pearlite for an as-cast piece. However, a heat treatment for several hours at about 700°C (1300°F) will yield a ferrite matrix as in this photomicrograph. Castings are stronger and much more ductile than gray iron, as a comparison of their mechanical properties in Table 13.5 shows. In fact, ductile iron has mechanical characteristics approaching those of steel. For example, ferritic ductile irons have tensile strengths ranging between 380 and 480 MPa (55,000 and 70,000 psi), and ductilities (as percent elongation) from 10 to 20%. Typical applications for this material include valves, pump bodies, crankshafts, gears, and other automotive and machine components.

White Iron and Malleable Iron

For low-silicon cast irons (containing less than 1.0 wt% Si) and rapid cooling rates, most of the carbon exists as cementite instead of graphite, as indicated in Figure

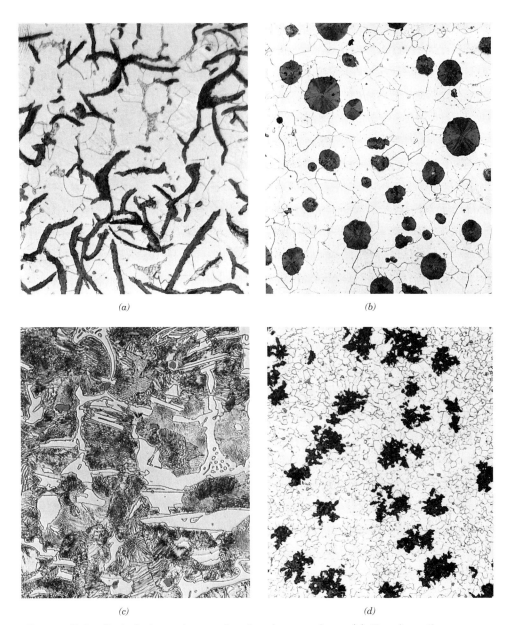

FIGURE 13.3 Optical photomicrographs of various cast irons. (a) Gray iron: the dark graphite flakes are embedded in an α-ferrite matrix. 500×. (Courtesy of C. H. Brady, National Bureau of Standards, Washington, DC.) (b) Nodular (ductile) iron: the dark graphite nodules are surrounded by an α-ferrite matrix. 200×. (Courtesy of C. H. Brady and L. C. Smith, National Bureau of Standards, Washington, DC.) (c) White iron: the light cementite regions are surrounded by pearlite, which has the ferrite–cementite layered structure. 400×. (Courtesy of Amcast Industrial Corporation.) (d) Malleable iron: dark graphite rosettes (temper carbon) in an α-ferrite matrix. 150×. (Reprinted with permission of the Iron Castings Society, Des Plaines, IL.)

Table 13.5 Designations, Minimum Mechanical Properties, Approximate Compositions, and Typical Applications for Various Gray, Nodular, and Malleable Cast Irons

Grade	UNS Number	Composition (wt%)[a]	Matrix Structure	Tensile Strength MPa (ksi)	Yield Strength MPa (ksi)	Ductility [%EL in 50 mm (2 in.)]	Typical Applications
Gray Iron							
SAE G1800	F10004	3.40–3.7 C, 2.55 Si, 0.7 Mn	Ferrite + Pearlite	124 (18)	—	—	Miscellaneous soft iron castings in which strength is not a primary consideration
SAE G2500	F10005	3.2–3.5 C, 2.20 Si, 0.8 Mn	Ferrite + Pearlite	173 (25)	—	—	Small cylinder blocks, cylinder heads, pistons, clutch plates, transmission cases
SAE G4000	F10008	3.0–3.3 C, 2.0 Si, 0.8 Mn	Pearlite	276 (40)	—	—	Diesel engine castings, liners, cylinders, and pistons
Ductile (Nodular) Iron							
ASTM A536 60-40-18	F32800	3.5–3.8 C, 2.0–2.8 Si, 0.05 Mg, <0.20 Ni, <0.10 Mo	Ferrite	414 (60)	276 (40)	18	Pressure-containing parts such as valve and pump bodies
100-70-03	F34800		Pearlite	689 (100)	483 (70)	3	High-strength gears and machine components
120-90-02	F36200		Tempered martensite	827 (120)	621 (90)	2	Pinions, gears, rollers, slides
Malleable Iron							
32510	F22200	2.3–2.7 C, 1.0–1.75 Si, <0.55 Mn	Ferrite	345 (50)	224 (32)	10	General engineering service at normal and elevated temperatures
45006	—	2.4–2.7 C, 1.25–1.55 Si, <0.55 Mn	Ferrite + Pearlite	448 (65)	310 (45)	6	

[a] The balance of the composition is iron.

Source: Adapted from *ASM Handbook*, Vol. 1, *Properties and Selection: Irons, Steels, and High-Performance Alloys*, 1990. Reprinted by permission of ASM International, Materials Park, OH.

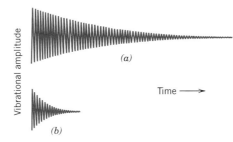

FIGURE 13.4 Comparison of the relative vibrational damping capacities of (a) steel and (b) gray cast iron. (From *Metals Engineering Quarterly*, February 1961. Copyright 1961 American Society for Metals.)

13.5. A fracture surface of this alloy has a white appearance, and thus it is termed **white cast iron.** An optical photomicrograph showing the microstructure of white iron is presented in Figure 13.3c. Thick sections may have only a surface layer of white iron that was "chilled" during the casting process; gray iron forms at interior regions, which cool more slowly. As a consequence of large amounts of the cementite

FIGURE 13.5 From the iron–carbon phase diagram, composition ranges for commercial cast irons. Also shown are microstructures that result from a variety of heat treatments. G_f, flake graphite; G_r, graphite rosettes; G_n, graphite nodules; P, pearlite; α, ferrite. (Adapted from W. G. Moffatt, G. W. Pearsall, and J. Wulff, *The Structure and Properties of Materials,* Vol. 1, *Structure,* p. 195. Copyright © 1964 by John Wiley & Sons, New York. Reprinted by permission of John Wiley & Sons, Inc.)

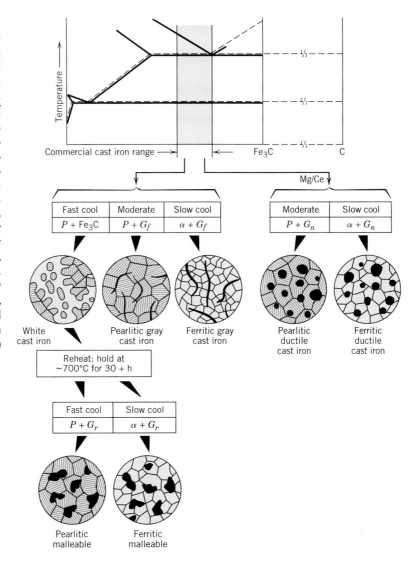

phase, white iron is extremely hard but also very brittle, to the point of being virtually unmachinable. Its use is limited to applications that necessitate a very hard and wear-resistant surface, and without a high degree of ductility—for example, as rollers in rolling mills. Generally, white iron is used as an intermediary in the production of yet another cast iron, **malleable iron.**

Heating white iron at temperatures between 800 and 900°C (1470 and 1650°F) for a prolonged time period and in a neutral atmosphere (to prevent oxidation) causes a decomposition of the cementite, forming graphite, which exists in the form of clusters or rosettes surrounded by a ferrite or pearlite matrix, depending on cooling rate, as indicated in Figure 13.5. A photomicrograph of a ferritic malleable iron is presented in Figure 13.3d. The microstructure is similar to that for nodular iron (Figure 13.3b), which accounts for relatively high strength and appreciable ductility or malleability. Some typical mechanical characteristics are also listed in Table 13.5. Representative applications include connecting rods, transmission gears, and differential cases for the automotive industry, and also flanges, pipe fittings, and valve parts for railroad, marine, and other heavy-duty services.

13.3 NONFERROUS ALLOYS

Steel and other ferrous alloys are consumed in exceedingly large quantities because they have such a wide range of mechanical properties, may be fabricated with relative ease, and are economical to produce. However, they have some distinct limitations, chiefly: (1) a relatively high density, (2) a comparatively low electrical conductivity, and (3) an inherent susceptibility to corrosion in some common environments. Thus, for many applications it is advantageous or even necessary to utilize other alloys having more suitable property combinations. Alloy systems are classified either according to the base metal or according to some specific characteristic that a group of alloys share. This section discusses the following metal and alloy systems: copper, aluminum, magnesium, and titanium alloys, the refractory metals, the superalloys, the noble metals, and miscellaneous alloys, including those that have nickel, lead, tin, zirconium, and zinc as base metals.

On occasion, a distinction is made between cast and wrought alloys. Alloys that are so brittle that forming or shaping by appreciable deformation is not possible ordinarily are cast; these are classified as *cast alloys*. On the other hand, those that are amenable to mechanical deformation are termed **wrought alloys.**

In addition, the heat treatability of an alloy system is mentioned frequently. "Heat treatable" designates an alloy whose mechanical strength is improved by precipitation hardening or a martensitic transformation (normally the former), both of which involve specific heat-treating procedures.

COPPER AND ITS ALLOYS

Copper and copper-based alloys, possessing a desirable combination of physical properties, have been utilized in quite a variety of applications since antiquity. Unalloyed copper is so soft and ductile that it is difficult to machine; also, it has an almost unlimited capacity to be cold worked. Furthermore, it is highly resistant to corrosion in diverse environments including the ambient atmosphere, seawater, and some industrial chemicals. The mechanical and corrosion-resistance properties of copper may be improved by alloying. Most copper alloys cannot be hardened or strengthened by heat-treating procedures; consequently, cold working and/or solid-solution alloying must be utilized to improve these mechanical properties.

The most common copper alloys are the **brasses** for which zinc, as a substitutional impurity, is the predominant alloying element. As may be observed for the copper–zinc phase diagram (Figure 10.17), the α phase is stable for concentrations up to approximately 35 wt% Zn. This phase has an FCC crystal structure, and α brasses are relatively soft, ductile, and easily cold worked. Brass alloys having a higher zinc content contain both α and β' phases at room temperature. The β' phase has an ordered BCC crystal structure and is harder and stronger than the α phase; consequently, $\alpha + \beta'$ alloys are generally hot worked.

Some of the common brasses are yellow, naval, and cartridge brass, muntz metal, and gilding metal. The compositions, properties, and typical uses of several of these alloys are listed in Table 13.6. Some of the common uses for brass alloys include costume jewelry, cartridge casings, automotive radiators, musical instruments, electronic packaging, and coins.

Table 13.6 Compositions, Mechanical Properties, and Typical Applications for Eight Copper Alloys

Alloy Name	UNS Number	Composition (wt%)[a]	Condition	Tensile Strength [MPa (ksi)]	Yield Strength [MPa (ksi)]	Ductility [%EL in 50 mm (2 in.)]	Typical Applications
				Mechanical Properties			
Wrought Alloys							
Electrolytic tough pitch	C11000	0.04 O	Annealed	220 (32)	69 (10)	45	Electrical wire, rivets, screening, gaskets, pans, nails, roofing
Beryllium copper	C17200	1.9 Be, 0.20 Co	Precipitation hardened	1140–1310 (165–190)	690–860 (100–125)	4–10	Springs, bellows, firing pins, bushings, valves, diaphragms
Cartridge brass	C26000	30 Zn	Annealed	300 (44)	75 (11)	68	Automotive radiator cores, ammunition components, lamp fixtures, flashlight shells, kickplates
			Cold-worked (H04 hard)	525 (76)	435 (63)	8	
Phosphor bronze, 5% A	C51000	5 Sn, 0.2 P	Annealed	325 (47)	130 (19)	64	Bellows, clutch disks, diaphragms, fuse clips, springs, welding rods
			Cold-worked (H04 hard)	560 (81)	515 (75)	10	
Copper-nickel, 30%	C71500	30 Ni	Annealed	380 (55)	125 (18)	36	Condenser and heat-exchanger components, saltwater piping
			Cold-worked (H02 hard)	515 (75)	485 (70)	15	
Cast Alloys							
Leaded yellow brass	C85400	29 Zn, 3 Pb, 1 Sn	As cast	234 (34)	83 (12)	35	Furniture hardware, radiator fittings, light fixtures, battery clamps
Tin bronze	C90500	10 Sn, 2 Zn	As cast	310 (45)	152 (22)	25	Bearings, bushings, piston rings, steam fittings, gears
Aluminum bronze	C95400	4 Fe, 11 Al	As cast	586 (85)	241 (35)	18	Bearings, gears, worms, bushings, valve seats and guards, pickling hooks

[a] The balance of the composition is copper.

Source: Adapted from *ASM Handbook,* Vol. 2, *Properties and Selection: Nonferrous Alloys and Special-Purpose Materials,* 1990. Reprinted by permission of ASM International, Materials Park, OH.

The **bronzes** are alloys of copper and several other elements, including tin, aluminum, silicon, and nickel. These alloys are somewhat stronger than the brasses, yet they still have a high degree of corrosion resistance. Table 13.6 contains several of the bronze alloys, their compositions, properties, and applications. Generally they are utilized when, in addition to corrosion resistance, good tensile properties are required.

The most common precipitation hardenable copper alloys are the beryllium coppers. They possess a remarkable combination of properties: tensile strengths as high as 1400 MPa (200,000 psi), excellent electrical and corrosion properties, and wear resistance when properly lubricated; they may be cast, hot worked, or cold worked. High strengths are attained by precipitation-hardening heat treatments. These alloys are costly because of the beryllium additions, which range between 1.0 and 2.5 wt%. Applications include jet aircraft landing gear bearings and bushings, springs, and surgical and dental instruments. One of these alloys (C17200) is included in Table 13.6.

ALUMINUM AND ITS ALLOYS

Aluminum and its alloys are characterized by a relatively low density (2.7 g/cm^3 as compared to 7.9 g/cm^3 for steel), high electrical and thermal conductivities, and a resistance to corrosion in some common environments, including the ambient atmosphere. Many of these alloys are easily formed by virtue of high ductility; this is evidenced by the thin aluminum foil sheet into which the relatively pure material may be rolled. Since aluminum has an FCC crystal structure, its ductility is retained even at very low temperatures. The chief limitation of aluminum is its low melting temperature [660°C (1220°F)], which restricts the maximum temperature at which it can be used.

The mechanical strength of aluminum may be enhanced by cold work and by alloying; however, both processes tend to diminish resistance to corrosion. Principal alloying elements include copper, magnesium, silicon, manganese, and zinc. Non-heat-treatable alloys consist of a single phase, for which an increase in strength is achieved by solid solution strengthening. Others are rendered heat treatable (capable of being precipitation hardened) as a result of alloying. In several of these alloys precipitation hardening is due to the precipitation of two elements other than aluminum, to form an intermetallic compound such as $MgZn_2$.

Generally, aluminum alloys are classified as either cast or wrought. Composition for both types is designated by a four-digit number that indicates the principal impurities, and in some cases, the purity level. For cast alloys, a decimal point is located between the last two digits. After these digits is a hyphen and the basic **temper designation**—a letter and possibly a one- to three-digit number, which indicates the mechanical and/or heat treatment to which the alloy has been subjected. For example, F, H, and O represent, respectively, the as-fabricated, strain-hardened, and annealed states; T3 means that the alloy was solution heat treated, cold worked, and then naturally aged (age hardened). A solution heat treatment followed by artificial aging is indicated by T6. The compositions, properties, and applications of several wrought and cast alloys are contained in Table 13.7. Some of the more common applications of aluminum alloys include aircraft structural parts, beverage cans, bus bodies, and automotive parts (engine blocks, pistons, and manifolds).

Recent attention has been given to alloys of aluminum and other low-density metals (e.g., Mg and Ti) as engineering materials for transportation, to effect reductions in fuel consumption. An important characteristic of these materials is **specific**

Table 13.7 Compositions, Mechanical Properties, and Typical Applications for Several Common Aluminum Alloys

Aluminum Association Number	UNS Number	Composition (wt%)[a]	Condition (Temper Designation)	Mechanical Properties		Ductility [%EL in 50 mm (2 in.)]	Typical Applications/ Characteristics
				Tensile Strength [MPa (ksi)]	Yield Strength [MPa (ksi)]		
Wrought, Nonheat-Treatable Alloys							
1100	A91100	0.12 Cu	Annealed (O)	90 (13)	35 (5)	35–45	Food/chemical handling & storage equipment, heat exchangers, light reflectors
3003	A93003	0.12 Cu, 1.2 Mn, 0.1 Zn	Annealed (O)	110 (16)	40 (6)	30–40	Cooking utensils, pressure vessels and piping
5052	A95052	2.5 Mg, 0.25 Cr	Strain hardened (H32)	230 (33)	195 (28)	12–18	Aircraft fuel & oil lines, fuel tanks, appliances, rivets, and wire
Wrought, Heat-Treatable Alloys							
2024	A92024	4.4 Cu, 1.5 Mg, 0.6 Mn	Heat treated (T4)	470 (68)	325 (47)	20	Aircraft structures, rivets, truck wheels, screw machine products
6061	A96061	1.0 Mg, 0.6 Si, 0.30 Cu, 0.20 Cr	Heat treated (T4)	240 (35)	145 (21)	22–25	Trucks, canoes, railroad cars, furniture, pipelines
7075	A97075	5.6 Zn, 2.5 Mg, 1.6 Cu, 0.23 Cr	Heat treated (T6)	570 (83)	505 (73)	11	Aircraft structural parts and other highly stressed applications
Cast, Heat-Treatable Alloys							
295.0	A02950	4.5 Cu, 1.1 Si	Heat treated (T4)	221 (32)	110 (16)	8.5	Flywheel and rear-axle housings, bus and aircraft wheels, crankcases
356.0	A03560	7.0 Si, 0.3 Mg	Heat treated (T6)	228 (33)	164 (24)	3.5	Aircraft pump parts, automotive transmission cases, water-cooled cylinder blocks
Aluminum–Lithium Alloys							
2090	—	2.7 Cu, 0.25 Mg, 2.25 Li, 0.12 Zr	Heat treated, cold worked (T83)	455 (66)	455 (66)	5	Aircraft structures and cryogenic tankage structures
8090	—	1.3 Cu, 0.95 Mg, 2.0 Li, 0.1 Zr	Heat treated, cold worked (T651)	465 (67)	360 (52)	—	Aircraft structures that must be highly damage tolerant

[a] The balance of the composition is aluminum.

Source: Adapted from *ASM Handbook,* Vol. 2, *Properties and Selection: Nonferrous Alloys and Special-Purpose Materials,* 1990. Reprinted by permission of ASM International, Materials Park, OH.

strength, which is quantified by the tensile strength–specific gravity ratio. Even though an alloy of one of these metals may have a tensile strength that is inferior to a more dense material (such as steel), on a weight basis it will be able to sustain a larger load.

A generation of new aluminum-lithium alloys have been developed recently for use by the aircraft and aerospace industries. These materials have relatively low densities (between about 2.5 and 2.6 g/cm^3), high specific moduli (elastic modulus-specific gravity ratios), and excellent fatigue and low-temperature toughness properties. Furthermore, some of them may be precipitation hardened. However, these materials are more costly to manufacture than the conventional aluminum alloys because special processing techniques are required as a result of lithium's chemical reactivity.

MAGNESIUM AND ITS ALLOYS

Perhaps the most outstanding characteristic of magnesium is its density, 1.7 g/cm^3, which is the lowest of all the structural metals; therefore, its alloys are used where light weight is an important consideration (e.g., in aircraft components). Magnesium has an HCP crystal structure, is relatively soft, and has a low elastic modulus: 45 GPa (6.5×10^6 psi). At room temperature magnesium and its alloys are difficult to deform; in fact, only small degrees of cold work may be imposed without annealing. Consequently, most fabrication is by casting or hot working at temperatures between 200 and 350°C (400 and 650°F). Magnesium, like aluminum, has a moderately low melting temperature [651°C (1204°F)]. Chemically, magnesium alloys are relatively unstable and especially susceptible to corrosion in marine environments. On the other hand, corrosion or oxidation resistance is reasonably good in the normal atmosphere; it is believed that this behavior is due to impurities rather than being an inherent characteristic of Mg alloys. Fine magnesium powder ignites easily when heated in air; consequently, care should be exercised when handling it in this state.

These alloys are also classified as either cast or wrought, and some of them are heat treatable. Aluminum, zinc, manganese, and some of the rare earths are the major alloying elements. A composition–temper designation scheme similar to that for aluminum alloys is also used. Table 13.8 lists several common magnesium alloys, their compositions, properties, and applications. These alloys are used in aircraft and missile applications, as well as in luggage. Furthermore, in the last several years the demand for magnesium alloys has increased dramatically in a host of different industries. For many applications, magnesium alloys have replaced engineering plastics that have comparable densities inasmuch as the magnesium materials are stiffer, more recyclable, and less costly to produce. For example, magnesium is now employed in a variety of hand-held devices (e.g., chain saws, power tools, hedge clippers), in automobiles (e.g., steering wheels and columns, seat frames, transmission cases), and in audio-video-computer-communications equipment (e.g., laptop computers, camcorders, TV sets, cellular telephones).

TITANIUM AND ITS ALLOYS

Titanium and its alloys are relatively new engineering materials that possess an extraordinary combination of properties. The pure metal has a relatively low density (4.5 g/cm^3), a high melting point [1668°C (3035°F)], and an elastic modulus of 107 GPa (15.5×10^6 psi). Titanium alloys are extremely strong; room temperature tensile strengths as high as 1400 MPa (200,000 psi) are attainable, yielding remarkable specific strengths. Furthermore, the alloys are highly ductile and easily forged and machined.

Table 13.8 Compositions, Mechanical Properties, and Typical Applications for Six Common Magnesium Alloys

ASTM Number	UNS Number	Composition (wt%)[a]	Condition	Tensile Strength [MPa (ksi)]	Yield Strength [MPa (ksi)]	Ductility [%EL in 50 mm (2 in.)]	Typical Applications
			Wrought Alloys				
AZ31B	M11311	3.0 Al, 1.0 Zn, 0.2 Mn	As extruded	262 (38)	200 (29)	15	Structures and tubing, cathodic protection
HK31A	M13310	3.0 Th, 0.6 Zr	Strain hardened, partially annealed	255 (37)	200 (29)	9	High strength to 315°C (600°F)
ZK60A	M16600	5.5 Zn, 0.45 Zr	Artificially aged	350 (51)	285 (41)	11	Forgings of maximum strength for aircraft
			Cast Alloys				
AZ91D	M11916	9.0 Al, 0.15 Mn, 0.7 Zn	As cast	230 (33)	150 (22)	3	Die-cast parts for automobiles, luggage, and electronic devices
AM60A	M10600	6.0 Al, 0.13 Mn	As cast	220 (32)	130 (19)	6	Automotive wheels
AS41A	M10410	4.3 Al, 1.0 Si, 0.35 Mn	As cast	210 (31)	140 (20)	6	Die castings requiring good creep resistance

[a] The balance of the composition is magnesium.

Source: Adapted from *ASM Handbook,* Vol. 2, *Properties and Selection: Nonferrous Alloys and Special-Purpose Materials,* 1990. Reprinted by permission of ASM International, Materials Park, OH.

The major limitation of titanium is its chemical reactivity with other materials at elevated temperatures. This property has necessitated the development of non-conventional refining, melting, and casting techniques; consequently, titanium alloys are quite expensive. In spite of this high temperature reactivity, the corrosion resistance of titanium alloys at normal temperatures is unusually high; they are virtually immune to air, marine, and a variety of industrial environments. Table 13.9 presents several titanium alloys along with their typical properties and applications. They are commonly utilized in airplane structures, space vehicles, surgical implants, and in the petroleum and chemical industries.

THE REFRACTORY METALS

Metals that have extremely high melting temperatures are classified as the refractory metals. Included in this group are niobium (Nb), molybdenum (Mo), tungsten (W), and tantalum (Ta). Melting temperatures range between 2468°C (4474°F) for niobium and 3410°C (6170°F), the highest melting temperature of any metal, for tungsten. Interatomic bonding in these metals is extremely strong, which accounts for the melting temperatures, and, in addition, large elastic moduli and high strengths and hardnesses, at ambient as well as elevated temperatures. The applications of these metals are varied. For example, tantalum and molybdenum are alloyed with stainless steel to improve its corrosion resistance. Molybdenum alloys are utilized for extrusion dies and structural parts in space vehicles; incandescent light filaments,

Table 13.9 Compositions, Mechanical Properties, and Typical Applications for Several Common Titanium Alloys

Alloy Type	Common Name (UNS Number)	Composition (wt%)	Condition	Average Mechanical Properties			Typical Applications
				Tensile Strength [MPa (ksi)]	Yield Strength [MPa (ksi)]	Ductility [%EL in 50 mm (2 in.)]	
Commercially pure	Unalloyed (R50500)	99.1 Ti	Annealed	484 (70)	414 (60)	25	Jet engine shrouds, cases and airframe skins, corrosion-resistant equipment for marine and chemical processing industries
α	Ti-5Al-2.5Sn (R54520)	5 Al, 2.5 Sn, balance Ti	Annealed	826 (120)	784 (114)	16	Gas turbine engine casings and rings; chemical processing equipment requiring strength to temperatures of 480°C (900°F)
Near α	Ti-8Al-1Mo-1V (R54810)	8 Al, 1 Mo, 1 V, balance Ti	Annealed (duplex)	950 (138)	890 (129)	15	Forgings for jet engine components (compressor disks, plates, and hubs)
α-β	Ti-6Al-4V (R56400)	6 Al, 4 V, balance Ti	Annealed	947 (137)	877 (127)	14	High-strength prosthetic implants, chemical-processing equipment, airframe structural components
α-β	Ti-6Al-6V-2Sn (R56620)	6 Al, 2 Sn, 6 V, 0.75 Cu, balance Ti	Annealed	1050 (153)	985 (143)	14	Rocket engine case airframe applications and high-strength airframe structures
β	Ti-10V-2Fe-3Al	10 V, 2 Fe, 3 Al, balance Ti	Solution + aging	1223 (178)	1150 (167)	10	Best combination of high strength and toughness of any commercial titanium alloy; used for applications requiring uniformity of tensile properties at surface and center locations; high-strength airframe components

Source: Adapted from *ASM Handbook, Vol. 2, Properties and Selection: Nonferrous Alloys and Special-Purpose Materials*, 1990. Reprinted by permission of ASM International, Materials Park, OH.

x-ray tubes, and welding electrodes employ tungsten alloys. Tantalum is immune to chemical attack by virtually all environments at temperatures below 150°C, and is frequently used in applications requiring such a corrosion-resistant material.

THE SUPERALLOYS

The superalloys have superlative combinations of properties. Most are used in aircraft turbine components, which must withstand exposure to severely oxidizing environments and high temperatures for reasonable time periods. Mechanical integrity under these conditions is critical; in this regard, density is an important consideration because centrifugal stresses are diminished in rotating members when the density is reduced. These materials are classified according to the predominant metal in the alloy, which may be cobalt, nickel, or iron. Other alloying elements include the refractory metals (Nb, Mo, W, Ta), chromium, and titanium. In addition to turbine applications, these alloys are utilized in nuclear reactors and petrochemical equipment.

THE NOBLE METALS

The noble or precious metals are a group of eight elements that have some physical characteristics in common. They are expensive (precious) and are superior or notable (noble) in properties—i.e., characteristically soft, ductile, and oxidation resistant. The noble metals are silver, gold, platinum, palladium, rhodium, ruthenium, iridium, and osmium; the first three are most common and are used extensively in jewelry. Silver and gold may be strengthened by solid-solution alloying with copper; sterling silver is a silver–copper alloy containing approximately 7.5 wt% Cu. Alloys of both silver and gold are employed as dental restoration materials; also, some integrated circuit electrical contacts are of gold. Platinum is used for chemical laboratory equipment, as a catalyst (especially in the manufacture of gasoline), and in thermocouples to measure elevated temperatures.

MISCELLANEOUS NONFERROUS ALLOYS

The discussion above covers the vast majority of nonferrous alloys; however, a number of others are found in a variety of engineering applications, and a brief exposure of these is worthwhile.

Nickel and its alloys are highly resistant to corrosion in many environments, especially those that are basic (alkaline). Nickel is often coated or plated on some metals that are susceptible to corrosion as a protective measure. Monel, a nickel-based alloy containing approximately 65 wt% Ni and 28 wt% Cu (the balance iron), has very high strength and is extremely corrosion resistant; it is used in pumps, valves, and other components that are in contact with some acid and petroleum solutions. As already mentioned, nickel is one of the principal alloying elements in stainless steels, and one of the major constituents in the superalloys.

Lead, tin, and their alloys find some use as engineering materials. Both are mechanically soft and weak, have low melting temperatures, are quite resistant to many corrosion environments, and have recrystallization temperatures below room temperature. Many common solders are lead–tin alloys, which have low melting temperatures. Applications for lead and its alloys include x-ray shields and storage batteries. The primary use of tin is as a very thin coating on the inside of plain carbon steel cans (tin cans) that are used for food containers; this coating inhibits chemical reactions between the steel and the food products.

Unalloyed zinc also is a relatively soft metal having a low melting temperature and a subambient recrystallization temperature. Chemically, it is reactive in a number of common environments and, therefore, susceptible to corrosion. Galvanized steel is just plain carbon steel that has been coated with a thin zinc layer; the zinc preferentially corrodes and protects the steel {(Section 16.9).} Typical applications of galvanized steel are familiar (sheet metal, fences, screen, screws, etc.). Common applications of zinc alloys include padlocks, automotive parts (door handles and grilles), and office equipment.

Although zirconium is relatively abundant in the earth's crust, it was not until quite recent times that commercial refining techniques were developed. Zirconium and its alloys are ductile and have other mechanical characteristics that are comparable to those of titanium alloys and the austenitic stainless steels. However, the primary asset of these alloys is their resistance to corrosion in a host of corrosive media, including superheated water. Furthermore, zirconium is transparent to thermal neutrons, so that its alloys have been used as cladding for uranium fuel in water-cooled nuclear reactors. In terms of cost, these alloys are also often the materials of choice for heat exchangers, reactor vessels, and piping systems for the chemical-processing and nuclear industries. They are also used in incendiary ordnance and in sealing devices for vacuum tubes.

 In Appendix B is tabulated a wide variety of properties (e.g., density, elastic modulus, yield and tensile strengths, electrical conductivity, coefficient of thermal expansion, etc.) for a large number of metals and alloys.

TYPES OF CERAMICS

The preceding discussions of the properties of materials have demonstrated that there is a significant disparity between the physical characteristics of metals and ceramics. Consequently, these materials are utilized in totally different kinds of applications and, in this regard, tend to complement each other, and also the polymers. Most ceramic materials fall into an application–classification scheme that includes the following groups: glasses, structural clay products, whitewares, refractories, abrasives, cements, {and the newly developed advanced ceramics.} Figure 13.6 presents a taxonomy of these several types; some discussion is devoted to each. We have also chosen to discuss the characteristics and applications of diamond and graphite in this section.

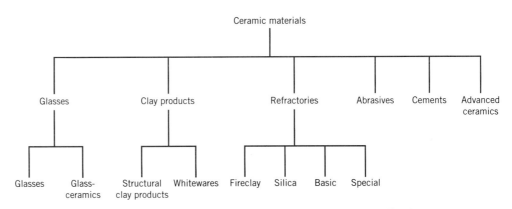

FIGURE 13.6 Classification of ceramic materials on the basis of application.

Table 13.10   Compositions and Characteristics of Some of the Common Commercial Glasses

Glass Type	Composition (wt%)						Characteristics and Applications
	SiO_2	Na_2O	CaO	Al_2O_3	B_2O_3	Other	
Fused silica	>99.5						High melting temperature, very low coefficient of expansion (shock resistant)
96% Silica (Vycor)	96			4			Thermally shock and chemically resistant—laboratory ware
Borosilicate (Pyrex)	81	3.5		2.5	13		Thermally shock and chemically resistant—ovenware
Container (soda–lime)	74	16	5	1		4MgO	Low melting temperature, easily worked, also durable
Fiberglass	55		16	15	10	4MgO	Easily drawn into fibers—glass–resin composites
Optical flint	54	1				37PbO, 8K$_2$O	High density and high index of refraction—optical lenses
Glass–ceramic (Pyroceram)	43.5	14		30	5.5	6.5TiO$_2$, 0.5As$_2$O$_3$	Easily fabricated; strong; resists thermal shock—ovenware

13.4 GLASSES

The glasses are a familiar group of ceramics; containers, windows, lenses, and fiberglass represent typical applications. As already mentioned, they are noncrystalline silicates containing other oxides, notably CaO, Na_2O, K_2O, and Al_2O_3, which influence the glass properties. A typical soda–lime glass consists of approximately 70 wt% SiO_2, the balance being mainly Na_2O (soda) and CaO (lime). The compositions of several common glass materials are contained in Table 13.10. Possibly the two prime assets of these materials are their optical transparency and the relative ease with which they may be fabricated.

13.5 GLASS–CERAMICS

Most inorganic glasses can be made to transform from a noncrystalline state to one that is crystalline by the proper high-temperature heat treatment. This process is called **devitrification,** and the product is a fine-grained polycrystalline material which is often called a **glass–ceramic.** A nucleating agent (frequently titanium dioxide) must be added to induce the crystallization or devitrification process. Desirable characteristics of glass–ceramics include a low coefficient of thermal expansion, such that the glass–ceramic ware will not experience thermal shock; in addition, relatively high mechanical strengths and thermal conductivities are achieved. Some glass–ceramics may be made optically transparent; others are opaque. Possibly the most attractive attribute of this class of materials is the ease with which they may be fabricated; conventional glass-forming techniques may be used conveniently in the mass production of nearly pore-free ware.

Glass–ceramics are manufactured commercially under the trade names of Pyroceram, Corningware, Cercor, and Vision. The most common uses for these materials are as ovenware and tableware, primarily because of their strength, excellent resistance to thermal shock, and their high thermal conductivity. They also serve as electrical insulators and as substrates for printed circuit boards, and are utilized for architectural cladding, and for heat exchangers and regenerators. A typical glass–ceramic is also included in Table 13.10, and the microstructure of a commercial material is shown in the chapter-opening photograph for this chapter.

13.6 CLAY PRODUCTS

One of the most widely used ceramic raw materials is clay. This inexpensive ingredient, found naturally in great abundance, often is used as mined without any upgrading of quality. Another reason for its popularity lies in the ease with which clay products may be formed; when mixed in the proper proportions, clay and water form a plastic mass that is very amenable to shaping. The formed piece is dried to remove some of the moisture, after which it is fired at an elevated temperature to improve its mechanical strength.

Most of the clay-based products fall within two broad classifications: the **structural clay products** and the **whitewares.** Structural clay products include building bricks, tiles, and sewer pipes—applications in which structural integrity is important. The whiteware ceramics become white after the high-temperature **firing.** Included in this group are porcelain, pottery, tableware, china, and plumbing fixtures (sanitary ware). In addition to clay, many of these products also contain other ingredients, each of which has some role to play in the processing and characteristics of the finished piece {(Section 14.8).}

13.7 REFRACTORIES

Another important class of ceramics that are utilized in large tonnages is the **refractory ceramics.** The salient properties of these materials include the capacity to withstand high temperatures without melting or decomposing, and the capacity to remain unreactive and inert when exposed to severe environments. In addition, the ability to provide thermal insulation is often an important consideration. Refractory materials are marketed in a variety of forms, but bricks are the most common. Typical applications include furnace linings for metal refining, glass manufacturing, metallurgical heat treatment, and power generation.

Of course, the performance of a refractory ceramic, to a large degree, depends on its composition. On this basis, there are several classifications, namely, fireclay, silica, basic, and special refractories. Compositions for a number of commercial refractories are listed in Table 13.11. For many commercial materials, the raw ingredients consist of both large (or grog) particles and fine particles, which may

Table 13.11   Compositions of Five Common Ceramic Refractory Materials

Refractory Type	Composition (wt%)							Apparent Porosity (%)
	Al_2O_3	SiO_2	MgO	Cr_2O_3	Fe_2O_3	CaO	TiO_2	
Fireclay	25–45	70–50	0–1		0–1	0–1	1–2	10–25
High-alumina fireclay	90–50	10–45	0–1		0–1	0–1	1–4	18–25
Silica	0.2	96.3	0.6			2.2		25
Periclase	1.0	3.0	90.0	0.3	3.0	2.5		22
Periclase–chrome ore	9.0	5.0	73.0	8.2	2.0	2.2		21

Source: From W. D. Kingery, H. K. Bowen, and D. R. Uhlmann, *Introduction to Ceramics,* 2nd edition. Copyright © 1976 by John Wiley & Sons, New York. Reprinted by permission of John Wiley & Sons, Inc.

have different compositions. Upon firing, the fine particles normally are involved in the formation of a bonding phase, which is responsible for the increased strength of the brick; this phase may be predominantly either glassy or crystalline. The service temperature is normally below that at which the refractory piece was fired.

Porosity is one microstructural variable that must be controlled to produce a suitable refractory brick. Strength, load-bearing capacity, and resistance to attack by corrosive materials all increase with porosity reduction. At the same time, thermal insulation characteristics and resistance to thermal shock are diminished. Of course, the optimum porosity depends on the conditions of service.

FIRECLAY REFRACTORIES (CD-ROM)

SILICA REFRACTORIES (CD-ROM)

BASIC REFRACTORIES (CD-ROM)

SPECIAL REFRACTORIES (CD-ROM)

13.8 ABRASIVES

Abrasive ceramics are used to wear, grind, or cut away other material, which necessarily is softer. Therefore, the prime requisite for this group of materials is hardness or wear resistance; in addition, a high degree of toughness is essential to ensure that the abrasive particles do not easily fracture. Furthermore, high temperatures may be produced from abrasive frictional forces, so some refractoriness is also desirable.

Diamonds, both natural and synthetic, are utilized as abrasives; however, they are relatively expensive. The more common ceramic abrasives include silicon carbide, tungsten carbide (WC), aluminum oxide (or corundum), and silica sand.

Abrasives are used in several forms—bonded to grinding wheels, as coated abrasives, and as loose grains. In the first case, the abrasive particles are bonded to a wheel by means of a glassy ceramic or an organic resin. The surface structure should contain some porosity; a continual flow of air currents or liquid coolants within the pores that surround the refractory grains will prevent excessive heating. Figure 13.7 shows the microstructure of a bonded abrasive, revealing abrasive grains, the bonding phase, and pores.

Coated abrasives are those in which an abrasive powder is coated on some type of paper or cloth material; sandpaper is probably the most familiar example. Wood, metals, ceramics, and plastics are all frequently ground and polished using this form of abrasive.

Grinding, lapping, and polishing wheels often employ loose abrasive grains that are delivered in some type of oil- or water-based vehicle. Diamonds, corundum, silicon carbide, and rouge (an iron oxide) are used in loose form over a variety of grain size ranges.

13.9 CEMENTS

Several familiar ceramic materials are classified as inorganic **cements:** cement, plaster of paris, and lime, which, as a group, are produced in extremely large quantities. The characteristic feature of these materials is that when mixed with water, they form a paste that subsequently sets and hardens. This trait is especially useful in

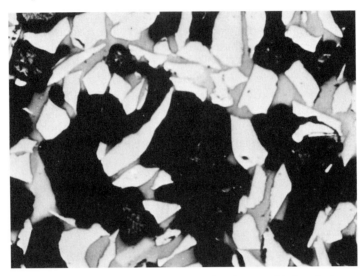

FIGURE 13.7 Photomicrograph of an aluminum oxide bonded ceramic abrasive. The light regions are the Al_2O_3 abrasive grains; the gray and dark areas are the bonding phase and porosity, respectively. 100×. (From W. D. Kingery, H. K. Bowen, and D. R. Uhlmann, *Introduction to Ceramics,* 2nd edition, p. 568. Copyright © 1976 by John Wiley & Sons. Reprinted by permission of John Wiley & Sons, Inc.)

that solid and rigid structures having just about any shape may be expeditiously formed. Also, some of these materials act as a bonding phase that chemically binds particulate aggregates into a single cohesive structure. Under these circumstances, the role of the cement is similar to that of the glassy bonding phase that forms when clay products and some refractory bricks are fired. One important difference, however, is that the cementitious bond develops at room temperature.

Of this group of materials, portland cement is consumed in the largest tonnages. It is produced by grinding and intimately mixing clay and lime-bearing minerals in the proper proportions, and then heating the mixture to about 1400°C (2550°F) in a rotary kiln; this process, sometimes called **calcination,** produces physical and chemical changes in the raw materials. The resulting "clinker" product is then ground into a very fine powder to which is added a small amount of gypsum ($CaSO_4$–$2H_2O$) to retard the setting process. This product is portland cement. The properties of portland cement, including setting time and final strength, to a large degree depend on its composition.

Several different constituents are found in portland cement, the principal ones being tricalcium silicate ($3CaO$–SiO_2) and dicalcium silicate ($2CaO$–SiO_2). The setting and hardening of this material result from relatively complicated hydration reactions that occur between the various cement constituents and the water that is added. For example, one hydration reaction involving dicalcium silicate is as follows:

$$2CaO\text{–}SiO_2 + xH_2O = 2CaO\text{–}SiO_2\text{–}xH_2O \qquad (13.2)$$

where x is variable and depends on how much water is available. These hydrated products are in the form of complex gels or crystalline substances that form the cementitious bond. Hydration reactions begin just as soon as water is added to the cement. These are first manifested as setting (i.e., the stiffening of the once-plastic paste), which takes place soon after mixing, usually within several hours. Hardening of the mass follows as a result of further hydration, a relatively slow process that may continue for as long as several years. It should be emphasized that the process by which cement hardens is not one of drying, but rather, of hydration in which water actually participates in a chemical bonding reaction.

Portland cement is termed a hydraulic cement because its hardness develops by chemical reactions with water. It is used primarily in mortar and concrete to bind, into a cohesive mass, aggregates of inert particles (sand and/or gravel); these are considered to be composite materials {(see Section 15.2).} Other cement materials, such as lime, are nonhydraulic; that is, compounds other than water (e.g., CO_2) are involved in the hardening reaction.

13.10 ADVANCED CERAMICS (CD-ROM)

13.11 DIAMOND AND GRAPHITE

DIAMOND

The physical properties of diamond make it an extremely attractive material. It is extremely hard (the hardest known material) and has a very low electrical conductivity; these characteristics are due to its crystal structure and the strong interatomic covalent bonds. Furthermore, it has an unusually high thermal conductivity for a nonmetallic material, is optically transparent in the visible and infrared regions of the electromagnetic spectrum, and has a high index of refraction. Relatively large diamond single crystals are used as gem stones. Industrially, diamonds are utilized to grind or cut other softer materials (Section 13.8). Techniques to produce synthetic diamonds have been developed, beginning in the mid-1950s, that have been refined to the degree that today a large proportion of the industrial-quality materials are man-made, in addition to some of those of gem quality.

Over the last several years, diamond in the form of thin films has been produced. Film growth techniques involve vapor-phase chemical reactions followed by the film deposition. Maximum film thicknesses are on the order of a millimeter. Furthermore, none of the films yet produced has the long-range crystalline regularity of natural diamond. The diamond is polycrystalline and may consist of very small and/or relatively large grains; in addition, amorphous carbon and graphite may be present. A scanning electron micrograph of the surface of a diamond thin film is shown in Figure 13.8. The mechanical, electrical, and optical properties of diamond films approach those of the bulk diamond material. These desirable properties have been and will continue to be exploited so as to create new and better products. For example, the surfaces of drills, dies, bearings, knives, and other tools have been coated with diamond films to increase surface hardness; some lenses and radomes have been made stronger while remaining transparent by the application of diamond coatings; coatings have also been applied to loudspeaker tweeters and to high-precision micrometers. Potential applications for these films include application to the surface of machine components such as gears and bearings, to optical recording heads and disks, and as substrates for semiconductor devices.

GRAPHITE

The structure of graphite is represented in Figure 3.17; in addition, the discussion of graphite in Section 3.9 noted that the electron bonding between the layers of hexagonally arranged carbon atoms is of the van der Waals type. As a consequence

FIGURE 13.8 Scanning electron micrograph of a diamond thin film in which is shown numerous multifaceted microcrystals. 1000×. (Photograph courtesy of the Norton Company.)

of these weak interplanar bonds, interplanar cleavage is facile, which gives rise to the excellent lubricative properties of graphite. Also, the electrical conductivity is relatively high in crystallographic directions parallel to the hexagonal sheets.

Other desirable properties of graphite include the following: high strength and good chemical stability at elevated temperatures and in nonoxidizing atmospheres, high thermal conductivity, low coefficient of thermal expansion and high resistance to thermal shock, high adsorption of gases, and good machinability. Graphite is commonly used as heating elements for electric furnaces, as electrodes for arc welding, in metallurgical crucibles, in casting molds for metal alloys and ceramics, for high-temperature refractories and insulations, in rocket nozzles, in chemical reactor vessels, for electrical contacts, brushes and resistors, as electrodes in batteries, and in air purification devices.

TYPES OF POLYMERS

There are many different polymeric materials that are familiar to us and find a wide variety of applications. These include plastics, elastomers (or rubbers), fibers, coatings, adhesives, foams, and films. Depending on its properties, a particular polymer may be used in two or more of these application categories. For example, a plastic, if crosslinked and utilized above its glass transition temperature, may make a satisfactory elastomer. Or, a fiber material may be used as a plastic if it is not drawn into filaments. This portion of the chapter includes a brief discussion of each of these types of polymer.

13.12 PLASTICS

Possibly the largest number of different polymeric materials come under the plastic classification. Polyethylene, polypropylene, polyvinyl chloride, polystyrene, and the fluorocarbons, epoxies, phenolics, and polyesters may all be classified as **plastics.**

They have a wide variety of combinations of properties. Some plastics are very rigid and brittle; others are flexible, exhibiting both elastic and plastic deformations when stressed, and sometimes experiencing considerable deformation before fracture.

Polymers falling within this classification may have any degree of crystallinity, and all molecular structures and configurations (linear, branched, isotactic, etc.) are possible. Plastic materials may be either thermoplastic or thermosetting; in fact, this is the manner in which they are usually subclassified. The trade names, characteristics, and typical applications for a number of plastics are given in Table 13.12.

Table 13.12 Trade Names, Characteristics, and Typical Applications for a Number of Plastic Materials

Material Type	Trade Names	Major Application Characteristics	Typical Applications
		Thermoplastics	
Acrylonitrile-butadiene-styrene (ABS)	Abson Cycolac Kralastic Lustran Novodur Tybrene	Outstanding strength and toughness, resistant to heat distortion; good electrical properties; flammable and soluble in some organic solvents	Refrigerator linings, lawn and garden equipment, toys, highway safety devices
Acrylics (polymethyl methacrylate)	Acrylite Diakon Lucite Plexiglas	Outstanding light transmission and resistance to weathering; only fair mechanical properties	Lenses, transparent aircraft enclosures, drafting equipment, outdoor signs
Fluorocarbons (PTFE or TFE)	Teflon Fluon Halar Halon Hostaflon TF	Chemically inert in almost all environments, excellent electrical properties; low coefficient of friction; may be used to 260°C (500°F); relatively weak and poor cold-flow properties	Anticorrosive seals, chemical pipes and valves, bearings, antiadhesive coatings, high-temperature electronic parts
Polyamides (nylons)	Nylon Durethan Herox Nomex Ultramid Zytel	Good mechanical strength, abrasion resistance, and toughness; low coefficient of friction; absorbs water and some other liquids	Bearings, gears, cams, bushings, handles, and jacketing for wires and cables
Polycarbonates	Baylon Iupilon Lexan Makrolon Merlon Nuclon	Dimensionally stable; low water absorption; transparent; very good impact resistance and ductility; chemical resistance not outstanding	Safety helmets, lenses, light globes, base for photographic film
Polyethylene	Alathon Alkathene Ethron Fortiflex Hi-fax Petrothene Rigidex Zendel	Chemically resistant, and electrically insulating; tough and relatively low coefficient of friction; low strength and poor resistance to weathering	Flexible bottles, toys, tumblers, battery parts, ice trays, film wrapping materials

Table 13.12 (*Continued*)

Material Type	Trade Names	Major Application Characteristics	Typical Applications
Polypropylene	Bexphane Herculon Meraklon Moplen Poly-pro Pro-fax Propathene	Resistant to heat distortion; excellent electrical properties and fatigue strength; chemically inert; relatively inexpensive; poor resistance to UV light	Sterilizable bottles, packaging film, TV cabinets, luggage
Polystyrene	Carinex Celatron Hostyren Lustrex Styron Vestyron	Excellent electrical properties and optical clarity; good thermal and dimensional stability; relatively inexpensive	Wall tile, battery cases, toys, indoor lighting panels, appliance housings
Vinyls	Darvic Exon Geon Pee Vee Cee Pliovic Saran Tygon	Good low-cost, general-purpose materials; ordinarily rigid, but may be made flexible with plasticizers; often copolymerized; susceptible to heat distortion	Floor coverings, pipe, electrical wire insulation, garden hose, phonograph records
Polyester (PET or PETE)	Celanar Crastin Dacron Hylar Melinex Mylar Terylem	One of the toughest of plastic films; excellent fatigue and tear strength, and resistance to humidity, acids, greases, oils, and solvents	Magnetic recording tapes, clothing, automotive tire cords, beverage containers
Thermosetting Polymers			
Epoxies	Araldite Epikote Epon Epi-rez Lekutherm Nepoxide	Excellent combination of mechanical properties and corrosion resistance; dimensionally stable; good adhesion; relatively inexpensive; good electrical properties	Electrical moldings, sinks, adhesives, protective coatings, used with fiberglass laminates
Phenolics	Bakelite Amberol Arofene Durite Resinox	Excellent thermal stability to over 150°C (300°F); may be compounded with a large number of resins, fillers, etc.; inexpensive	Motor housings, telephones, auto distributors, electrical fixtures
Polyesters	Aropol Baygal Derakane Laguval Laminac Selectron	Excellent electrical properties and low cost; can be formulated for room- or high-temperature use; often fiber reinforced	Helmets, fiberglass boats, auto body components, chairs, fans

Source: Adapted from C. A. Harper, Editor, *Handbook of Plastics and Elastomers.* Copyright © 1975 by McGraw-Hill Book Company. Reproduced with permission.

Several plastics exhibit especially outstanding properties. For applications in which optical transparency is critical, polystyrene and polymethyl methacrylate are especially well suited; however, it is imperative that the material be highly amorphous or, if semicrystalline, have very small crystallites. The fluorocarbons have a low coefficient of friction and are extremely resistant to attack by a host of chemicals, even at relatively high temperatures. They are utilized as coatings on nonstick cookware, in bearings and bushings, and for high-temperature electronic components.

13.13 ELASTOMERS

The characteristics of and deformation mechanism for elastomers were treated previously (Section 8.19). The present discussion, therefore, focuses on the types of elastomeric materials.

Table 13.13 lists properties and applications of common elastomers; these properties are typical and, of course, depend on the degree of vulcanization and on whether any reinforcement is used. Natural rubber is still utilized to a large degree because it has an outstanding combination of desirable properties. However, the

Table 13.13 Tabulation of Important Characteristics and Typical Applications for Five Commercial Elastomers

Chemical Type	Trade (Common) Name	Elongation (%)	Useful Temperature Range [°C (°F)]	Major Application Characteristics	Typical Applications
Natural poly-isoprene	Natural Rubber (NR)	500–760	−60 to 120 (−75 to 250)	Excellent physical properties; good resistance to cutting, gouging, and abrasion; low heat, ozone, and oil resistance; good electrical properties	Pneumatic tires and tubes; heels and soles; gaskets
Styrene–butadiene copolymer	GRS, Buna S (SBR)	450–500	−60 to 120 (−75 to 250)	Good physical properties; excellent abrasion resistance; not oil, ozone, or weather resistant; electrical properties good, but not outstanding	Same as natural rubber
Acrylonitrile–butadiene copolymer	Buna A, Nitrile (NBR)	400–600	−50 to 150 (−60 to 300)	Excellent resistance to vegetable, animal, and petroleum oils; poor low-temperature properties; electrical properties not outstanding	Gasoline, chemical, and oil hose; seals and O-rings; heels and soles
Chloroprene	Neoprene (CR)	100–800	−50 to 105 (−60 to 225)	Excellent ozone, heat, and weathering resistance; good oil resistance; excellent flame resistance; not as good in electrical applications as natural rubber	Wire and cable; chem. tank linings; belts, hoses, seals, and gaskets
Polysiloxane	Silicone (VMQ)	100–800	−115 to 315 (−175 to 600)	Excellent resistance to high and low temperatures; low strength; excellent electrical properties	High- and low-temperature insulation; seals, diaphragms; tubing for food and medical uses

Sources: Adapted from: C. A. Harper, Editor, *Handbook of Plastics and Elastomers.* Copyright © 1975 by McGraw-Hill Book Company, reproduced with permission; and Materials Engineering's *Materials Selector,* copyright Penton/IPC.

most important synthetic elastomer is SBR, which is used predominantly in automobile tires, reinforced with carbon black. NBR, which is highly resistant to degradation and swelling, is another common synthetic elastomer.

For many applications (e.g., automobile tires), the mechanical properties of even vulcanized rubbers are not satisfactory in terms of tensile strength, abrasion and tear resistance, and stiffness. These characteristics may be further improved by additives such as carbon black {(Section 15.2).}

Finally, some mention should be made of the silicone rubbers. For these materials, the backbone carbon chain is replaced by a chain that alternates silicon and oxygen atoms:

$$-\underset{\underset{R'}{|}}{\overset{\overset{R}{|}}{Si}}-O-$$

where R and R' represent side-bonded atoms such as hydrogen or groups of atoms such as CH_3. For example, polydimethylsiloxane has the mer structure

$$-\underset{\underset{CH_3}{|}}{\overset{\overset{CH_3}{|}}{Si}}-O-$$

Of course, as elastomers, these materials are crosslinked.

The silicone elastomers possess a high degree of flexibility at low temperatures [to $-90°C$ ($-130°F$)] and yet are stable to temperatures as high as $250°C$ ($480°F$). In addition, they are resistant to weathering and lubricating oils. A further attractive characteristic is that some silicone rubbers vulcanize at room temperature (RTV rubbers).

13.14 FIBERS

The **fiber** polymers are capable of being drawn into long filaments having at least a 100:1 length-to-diameter ratio. Most commercial fiber polymers are utilized in the textile industry, being woven or knit into cloth or fabric. In addition, the aramid fibers are employed in composite materials, {Section 15.8.} To be useful as a textile material, a fiber polymer must have a host of rather restrictive physical and chemical properties. While in use, fibers may be subjected to a variety of mechanical deformations—stretching, twisting, shearing, and abrasion. Consequently, they must have a high tensile strength (over a relatively wide temperature range) and a high modulus of elasticity, as well as abrasion resistance. These properties are governed by the chemistry of the polymer chains and also by the fiber drawing process.

The molecular weight of fiber materials should be relatively high. Also, since the tensile strength increases with degree of crystallinity, the structure and configuration of the chains should allow the production of a highly crystalline polymer; that translates into a requirement for linear and unbranched chains that are symmetrical and have regularly repeating mer units.

Convenience in washing and maintaining clothing depends primarily on the thermal properties of the fiber polymer, that is, its melting and glass transition temperatures. Furthermore, fiber polymers must exhibit chemical stability to a rather extensive variety of environments, including acids, bases, bleaches, dry cleaning solvents, and sunlight. In addition, they must be relatively nonflammable and amenable to drying.

13.15 MISCELLANEOUS APPLICATIONS

COATINGS

Coatings are frequently applied to the surface of materials to serve one or more of the following functions: (1) to protect the item from the environment that may produce corrosive or deteriorative reactions; (2) to improve the item's appearance; and (3) to provide electrical insulation. Many of the ingredients in coating materials are polymers, the majority of which are organic in origin. These organic coatings fall into several different classifications, as follows: paint, varnish, enamel, lacquer, and shellac.

ADHESIVES

An **adhesive** is a substance used to join together the surfaces of two solid materials (termed "adherends") to produce a joint with a high shear strength. The bonding forces between the adhesive and adherend surfaces are thought to be electrostatic, similar to the secondary bonding forces between the molecular chains in thermoplastic polymers. Even though the inherent strength of the adhesive may be much less than that of the adherend materials, nevertheless, a strong joint may be produced if the adhesive layer is thin and continuous. If a good joint is formed, the adherend material may fracture or rupture before the adhesive.

Polymeric materials that fall within the classifications of thermoplastics, thermosetting resins, elastomeric compounds, and natural adhesives (animal glue, casein, starch, and rosin) may serve adhesive functions. Polymer adhesives may be used to join a large variety of material combinations: metal–metal, metal–plastic, metal–ceramic, and so on. The primary drawback is the service temperature limitation. Organic polymers maintain their mechanical integrity only at relatively low temperatures, and strength decreases rapidly with increasing temperature.

FILMS

Within relatively recent times, polymeric materials have found widespread use in the form of thin *films*. Films having thicknesses between 0.025 and 0.125 mm (0.001 and 0.005 in.) are fabricated and used extensively as bags for packaging food products and other merchandise, as textile products, and a host of other uses. Important characteristics of the materials produced and used as films include low density, a high degree of flexibility, high tensile and tear strengths, resistance to attack by moisture and other chemicals, and low permeability to some gases, especially water vapor. Some of the polymers that meet these criteria and are manufactured in film form are polyethylene, polypropylene, cellophane, and cellulose acetate.

FOAMS

Foams are plastic materials that contain a relatively high volume percent of small pores. Both thermoplastic and thermosetting materials are used as foams; these include polyurethane, rubber, polystyrene, and polyvinyl chloride. Foams are commonly used as cushions in automobiles and furniture as well as in packaging and thermal insulation. The foaming process is carried out by incorporating into the batch of material a blowing agent that upon heating, decomposes with the liberation of a gas. Gas bubbles are generated throughout the now-fluid mass, which remain as pores upon cooling and give rise to a spongelike structure. The same effect is produced by bubbling an inert gas through a material while it is in a molten state.

13.16 ADVANCED POLYMERIC MATERIALS (CD-ROM)

S U M M A R Y

With regard to composition, metals and alloys are classified as either ferrous or nonferrous. Ferrous alloys (steels and cast irons) are those in which iron is the prime constituent. Most steels contain less than 1.0 wt% C, and, in addition, other alloying elements, which render them susceptible to heat treatment (and an enhancement of mechanical properties) and/or more corrosion resistant. Plain low-carbon steels and high-strength low-alloy, medium-carbon, tool, and stainless steels are the most common types.

Cast irons contain a higher carbon content, normally between 3.0 and 4.5 wt% C, and other alloying elements, notably silicon. For these materials, most of the carbon exists in graphite form rather than combined with iron as cementite. Gray, ductile (or nodular), and malleable irons are the three most widely used cast irons; the latter two are reasonably ductile.

All other alloys fall within the nonferrous category, which is further subdivided according to base metal or some distinctive characteristic that is shared by a group of alloys. The compositions, typical properties, and applications of copper, aluminum, magnesium, titanium, nickel, lead, tin, zirconium, and zinc alloys, as well as the refractory metals, the superalloys, and the noble metals were discussed.

Also discussed in this chapter were various types of ceramic materials. The familiar glass materials are noncrystalline silicates that contain other oxides; the most desirable trait of these materials is their optical transparency. Glass–ceramics are initially fabricated as a glass, then crystallized or devitrified.

Clay is the principal component of the whitewares and structural clay products. Other ingredients may be added, such as feldspar and quartz, which influence the changes that occur during firing.

The materials that are employed at elevated temperatures and often in reactive environments are the refractory ceramics; on occasion, their ability to thermally insulate is also utilized. On the basis of composition and application, the four main subdivisions are fireclay, silica, basic, and special.

The abrasive ceramics, being hard and tough, are utilized to cut, grind, and polish other softer materials. Diamond, silicon carbide, tungsten carbide, corundum, and silica sand are the most common examples. The abrasives may be employed in the form of loose grains, bonded to an abrasive wheel, or coated on paper or a fabric.

When mixed with water, inorganic cements form a paste that is capable of assuming just about any desired shape. Subsequent setting or hardening is a result of chemical reactions involving the cement particles and occurs at the ambient temperature. For hydraulic cements, of which portland cement is the most common, the chemical reaction is one of hydration.

{Many of our modern technologies utilize and will continue to utilize advanced ceramics because of their unique mechanical, chemical, electrical, magnetic, and optical properties and property combinations. Advanced materials characterization, processing, and reliability techniques need to be developed to make these materials cost effective.}

The properties and some applications for diamond and graphite were presented. Diamond is a gemstone and, because of its hardness, is used to cut and grind softer materials. Furthermore, it is now being produced and utilized in thin films. The layered structure of graphite gives rise to its excellent lubricative properties and a relatively high electrical conductivity. Graphite is also known for its high strength and chemical stability at elevated temperatures and in nonoxidizing atmospheres.

The various types and applications of polymeric materials were also discussed. Plastic materials are perhaps the most widely used group of polymers, which include the following: polyethylene, polypropylene, polyvinyl chloride, polystyrene, and the fluorocarbons, epoxies, phenolics, and polyesters.

Another polymer classification includes the elastomeric materials that may experience very large elastic deformations. Most of these materials are copolymers, whereas the silicone elastomers are really inorganic materials.

Many polymeric materials may be spun into fibers, which are used primarily in textiles. Mechanical, thermal, and chemical characteristics of these materials are especially critical.

Other miscellaneous applications that employ polymers include coatings, adhesives, films, and foams.

{This chapter concluded with discussions of three advanced polymeric materials—ultrahigh molecular weight polyethylene, liquid crystal polymers, and thermoplastic elastomers. These materials have unusual properties and are used in a host of high-technology applications.}

IMPORTANT TERMS AND CONCEPTS

Abrasive (ceramic)	Fiber	Refractory (ceramic)
Adhesive	Foam	Specific strength
Alloy steel	Glass–ceramic	Stainless steel
Brass	Gray cast iron	Structural clay product
Bronze	High-strength, low-alloy	Temper designation
Calcination	(HSLA) steel	Thermoplastic elastomer
Cast iron	Liquid crystal polymer	Ultrahigh molecular weight
Cement	Malleable iron	polyethylene
Devitrification	Nonferrous alloy	White cast iron
Ductile (nodular) iron	Plain carbon steel	Whiteware
Ferrous alloy	Plastic	Wrought alloy

REFERENCES

ASM Handbook, Vol. 1, *Properties and Selection: Irons, Steels, and High-Performance Alloys,* ASM International, Materials Park, OH, 1990.

ASM Handbook, Vol. 2, *Properties and Selection: Nonferrous Alloys and Special-Purpose Mate-rials,* ASM International, Materials Park, OH, 1991.

Billmeyer, F. W., Jr., *Textbook of Polymer Science,* 3rd edition, Wiley-Interscience, New York, 1984.

Bowen, H. K., "Advanced Ceramics." *Scientific American,* Vol. 255, No. 4, October 1986, pp. 168–176.

Brick, R. M., A. W. Pense, and R. B. Gordon, *Structure and Properties of Engineering Materials,* 4th edition, McGraw-Hill Book Company, New York, 1977.

Coes, L., Jr., *Abrasives,* Springer-Verlag, New York, 1971.

Engineered Materials Handbook, Vol. 2, *Engineering Plastics,* ASM International, Metals Park, OH, 1988.

Engineered Materials Handbook, Vol. 4, *Ceramics and Glasses,* ASM International, Materials Park, OH, 1991.

Frick, J. (Editor), *Woldman's Engineering Alloys,* 8th edition, ASM International, Materials Park, OH, 1994.

Harper, C. A. (Editor), *Handbook of Plastics, Elastomers and Composites,* 3rd edition, McGraw-Hill Book Company, New York, 1996.

Lea, F. M., *The Chemistry of Cement and Concrete,* Chemical Publishing Company, New York, 1971.

Metals and Alloys in the Unified Numbering System, 7th edition, Society of Automotive Engineers, and American Society for Testing and Materials, Warrendale, PA, 1996.

Walton, C. F. and T. F. Opar (Editors), *Iron Castings Handbook,* Iron Castings Society, Des Plaines, IL, 1981.

Worldwide Guide to Equivalent Irons and Steels, 3rd edition, ASM International, Materials Park, OH, 1993.

Worldwide Guide to Equivalent Nonferrous Metals and Alloys, 3rd edition, ASM International, Materials Park, OH, 1996.

QUESTIONS AND PROBLEMS

Note: To solve those problems having an asterisk (*) by their numbers, consultation of supplementary topics [appearing only on the CD-ROM (and not in print)] will probably be necessary.

13.1 **(a)** List the four classifications of steels. **(b)** For each, briefly describe the properties and typical applications.

13.2 **(a)** Cite three reasons why ferrous alloys are used so extensively. **(b)** Cite three characteristics of ferrous alloys that limit their utilization.

13.3 Briefly explain why ferritic and austenitic stainless steels are not heat treatable.

13.4 What is the function of alloying elements in tool steels?

13.5 Compute the volume percent of graphite V_{Gr} in a 3.5 wt% C cast iron, assuming that all the carbon exists as the graphite phase. Assume densities of 7.9 and 2.3 g/cm^3 for ferrite and graphite, respectively.

13.6 On the basis of microstructure, briefly explain why gray iron is brittle and weak in tension.

13.7 Compare gray and malleable cast irons with respect to **(a)** composition and heat treatment, **(b)** microstructure, and **(c)** mechanical characteristics.

13.8 It is possible to produce cast irons that consist of a martensite matrix in which graphite is embedded in either flake, nodule, or rosette form. Briefly describe the treatment necessary to produce each of these three microstructures.

13.9 Compare white and nodular cast irons with respect to **(a)** composition and heat treatment, **(b)** microstructure, and **(c)** mechanical characteristics.

13.10 Is it possible to produce malleable cast iron in pieces having large cross-sectional dimensions? Why or why not?

13.11 What is the principal difference between wrought and cast alloys?

13.12 What is the main difference between a brass and a bronze?

13.13 Why must rivets of a 2017 aluminum alloy be refrigerated before they are used?

13.14 Explain why, under some circumstances, it is not advisable to weld a structure that is fabricated with a 3003 aluminum alloy.

13.15 What is the chief difference between heat-treatable and nonheat-treatable alloys?

13.16 Give the distinctive features, limitations, and applications of the following alloy groups: titanium alloys, refractory metals, superalloys, and noble metals.

13.17 Cite the two desirable characteristics of glasses.

13.18 **(a)** What is devitrification?

(b) Cite two properties that may be improved by devitrification and two that may be impaired.

13.19* Briefly explain why glass–ceramics are generally not transparent. You may want to consult Chapter 19.

13.20 For refractory ceramic materials, cite three characteristics that improve with and two characteristics that are adversely affected by increasing porosity.

13.21* Find the maximum temperature to which the following two magnesia–alumina refractory materials may be heated before a liquid phase will appear.

(a) A spinel-bonded alumina material of composition 95 wt% Al_2O_3-5 wt% MgO.

(b) A magnesia–alumina spinel of composition 65 wt% Al_2O_3-35 wt% MgO. Consult Figure 10.22.

13.22* Upon consideration of the SiO_2–Al_2O_3 phase diagram, Figure 10.24, for each pair of the following list of compositions, which would you judge to be the more desirable refractory? Justify your choices.

(a) 20 wt% Al_2O_3-80 wt% SiO_2 and 25 wt% Al_2O_3-75 wt% SiO_2.

(b) 70 wt% Al_2O_3-30 wt% SiO_2 and 80 wt% Al_2O_3-20 wt% SiO_2.

13.23* Compute the mass fractions of liquid in the following refractory materials at 1600°C (2910°F):

(a) 6 wt% Al_2O_3-94 wt% SiO_2.

(b) 10 wt% Al_2O_3-90 wt% SiO_2.

(c) 30 wt% Al_2O_3-70 wt% SiO_2.

(d) 80 wt% Al_2O_3-20 wt% SiO_2.

13.24* **(a)** For the SiO_2–Al_2O_3 system, what is the maximum temperature that is possible without the formation of a liquid phase? At what composition or over what range of compositions will this maximum temperature be achieved?

(b) For the MgO–Al_2O_3 system, what is the maximum temperature that is possible without the formation of a liquid phase? At what composition or over what range of compositions will this maximum temperature be achieved?

13.25 Compare the manner in which the aggregate particles become bonded together in clay-based mixtures during firing and in cements during setting.

13.26 Explain why it is important to grind cement into a fine powder.

13.27 During the winter months, the temperature in some parts of Alaska may go as low as −55°C (−65°F). Of the elastomers natural isoprene, styrene-butadiene, acrylonitrile-butadiene, chloroprene, and polysiloxane, which would be suitable for automobile tires under these conditions? Why?

13.28 Briefly explain the difference in molecular chemistry between silicone polymers and other polymeric materials.

13.29 Silicone polymers may be prepared to exist as liquids at room temperature. Cite differences in molecular structure between them and the silicone elastomers.

13.30 List two important characteristics for polymers that are to be used in fiber applications.

13.31 Cite five important characteristics for polymers that are to be used in thin-film applications.

Design Problems

13.D1 Of the following alloys, pick the one(s) that may be strengthened by heat treatment, cold work, or both: R50500 titanium, AZ31B magnesium, 6061 aluminum, C51000 phosphor bronze, lead, 6150 steel, 304 stainless steel, and C17200 beryllium copper.

13.D2 A structural member 100 mm (4 in.) long must be able to support a load of 50,000 N (11,250 lb$_f$) without experiencing any plastic deformation. Given the following data for brass, steel, aluminum, and titanium, rank them from least to greatest weight in accordance with these criteria.

Alloy	Yield Strength MPa (ksi)	Density (g/cm³)
Brass	415 (60)	8.5
Steel	860 (125)	7.9
Aluminum	310 (45)	2.7
Titanium	550 (80)	4.5

13.D3 Discuss whether it would be advisable to hot work or cold work the following metals and alloys on the basis of melting temperature, oxidation resistance, yield strength, and degree of brittleness: tin, tungsten, aluminum alloys, magnesium alloys, and a 4140 steel.

13.D4 Below is a list of metals and alloys:

Plain carbon steel	Magnesium
Brass	Zinc
Gray cast iron	Tool steel
Platinum	Aluminum
Stainless steel	Tungsten
Titanium alloy	

Select from this list the one metal or alloy that is best suited for each of the following applications, and cite at least one reason for your choice:

(a) The base for a milling machine.

(b) The walls of a steam boiler.

(c) High-speed aircraft.

(d) Drill bit.

(e) Cryogenic (i.e., very low temperature) container.

(f) As a pyrotechnic (i.e., in flares and fireworks).

(g) High-temperature furnace elements to be used in oxidizing atmospheres.

13.D5 (a) List at least three important characteristics required of metal alloys that are used for coins. **(b)** Write an essay in which you cite which metal alloys are employed in the coinage of your country, and then provide the rationale for their use.

13.D6 Some of our modern kitchen cookware is made of ceramic materials.

(a) List at least three important characteristics required of a material to be used for this application.

(b) Make a comparison of three ceramic materials as to their relative properties, and, in addition, to cost.

(c) On the basis of this comparison, select that material most suitable for the cookware.

13.D7 (a) List several advantages and disadvantages of using transparent polymeric materials for eyeglass lenses.

(b) Cite four properties (in addition to being transparent) that are important for this application.

(c) Note three polymers that may be candidates for eyeglass lenses, and then tabulate values of the properties noted in part b for these three materials.

13.D8 Write an essay on polymeric materials that are used in the packaging of food products and drinks. Include a list of the general requisite characteristics of materials that are used for these applications. Now cite a specific material that is utilized for each of three different container types and the rationale for each choice.

Appendix A / The International System of Units (SI)

Units in the *International System of Units* fall into two classifications: base and derived. Base units are fundamental and not reducible. Table A.1 lists the base units of interest in the discipline of materials science and engineering.

Derived units are expressed in terms of the base units, using mathematical signs for multiplication and division. For example, the SI units for density are kilogram per cubic meter (kg/m^3). For some derived units, special names and symbols exist; for example, N is used to denote the newton, the unit of force, which is equivalent to 1 kg-m/s^2. Table A.2 contains a number of the important derived units.

It is sometimes necessary, or convenient, to form names and symbols that are decimal multiples or submultiples of SI units. Only one prefix is used when a multiple of an SI unit is formed, which should be in the numerator. These prefixes and their approved symbols are given in Table A.3. Symbols for all units used in this book, SI or otherwise, are contained inside the front cover.

Table A.1   The SI Base Units

Quantity	*Name*	*Symbol*
Length	meter, metre	m
Mass	kilogram	kg
Time	second	s
Electric current	ampere	A
Thermodynamic temperature	kelvin	K
Amount of substance	mole	mol

Table A.2 Some of the SI Derived Units

Quantity	Name	Formula	Special Symbol[a]
Area	square meter	m^2	—
Volume	cubic meter	m^3	—
Velocity	meter per second	m/s	—
Density	kilogram per cubic meter	kg/m^3	—
Concentration	moles per cubic meter	mol/m^3	—
Force	newton	$kg\text{-}m/s^2$	N
Energy	joule	$kg\text{-}m^2/s^2$, N-m	J
Stress	pascal	$kg/m\text{-}s^2$, N/m^2	Pa
Strain	—	m/m	—
Power, radiant flux	watt	$kg\text{-}m^2/s^3$, J/s	W
Viscosity	pascal-second	kg/m-s	Pa-s
Frequency (of a periodic phenomenon)	hertz	s^{-1}	Hz
Electric charge	coulomb	A-s	C
Electric potential	volt	$kg\text{-}m^2/s^2\text{-}C$	V
Capacitance	farad	$s^2\text{-}C/kg\text{-}m^2$	F
Electric resistance	ohm	$kg\text{-}m^2/s\text{-}C^2$	Ω
Magnetic flux	weber	$kg\text{-}m^2/s\text{-}C$	Wb
Magnetic flux density	tesla	kg/s-C, Wb/m^2	(T)

[a] Special symbols in parentheses are approved in the SI but not used in this text; here, the name is used.

Table A.3 SI Multiple and Submultiple Prefixes

Factor by Which Multiplied	Prefix	Symbol
10^9	giga	G
10^6	mega	M
10^3	kilo	k
10^{-2}	centi[a]	c
10^{-3}	milli	m
10^{-6}	micro	μ
10^{-9}	nano	n
10^{-12}	pico	p

[a] Avoided when possible.

Appendix B / Properties of Selected Engineering Materials

This appendix represents a compilation of important properties for approximately one hundred common engineering materials. Each table contains data values of one particular property for this chosen set of materials; also included is a tabulation of the compositions of the various metal alloys that are considered (Table B.10). Data are tabulated by material type (viz., metals and metal alloys; graphite, ceramics, and semiconducting materials; polymers; fiber materials; and composites); within each classification, the materials are listed alphabetically.

A couple of comments are appropriate relative to the content of these tables. First, data entries are expressed either as ranges of values or as single values that are typically measured. Also, on occasion, "(min)" is associated with an entry; this means that the value cited is a minimum one.

Table B.1   Room-Temperature Density Values for Various Engineering Materials

	Density	
Material	g/cm^3	$lb_m/in.^3$
METALS AND METAL ALLOYS		
Plain Carbon and Low Alloy Steels		
Steel alloy A36	7.85	0.283
Steel alloy 1020	7.85	0.283
Steel alloy 1040	7.85	0.283
Steel alloy 4140	7.85	0.283
Steel alloy 4340	7.85	0.283
Stainless Steels		
Stainless alloy 304	8.00	0.289
Stainless alloy 316	8.00	0.289

Table B.1 (*Continued*)

Material	Density	
	g/cm³	*lb_m/in.³*
Stainless alloy 405	7.80	0.282
Stainless alloy 440A	7.80	0.282
Stainless alloy 17-7PH	7.65	0.276
Cast Irons		
Gray irons		
• Grade G1800	7.30	0.264
• Grade G3000	7.30	0.264
• Grade G4000	7.30	0.264
Ductile irons		
• Grade 60-40-18	7.10	0.256
• Grade 80-55-06	7.10	0.256
• Grade 120-90-02	7.10	0.256
Aluminum Alloys		
Alloy 1100	2.71	0.0978
Alloy 2024	2.77	0.100
Alloy 6061	2.70	0.0975
Alloy 7075	2.80	0.101
Alloy 356.0	2.69	0.0971
Copper Alloys		
C11000 (electrolytic tough pitch)	8.89	0.321
C17200 (beryllium–copper)	8.25	0.298
C26000 (cartridge brass)	8.53	0.308
C36000 (free-cutting brass)	8.50	0.307
C71500 (copper–nickel, 30%)	8.94	0.323
C93200 (bearing bronze)	8.93	0.322
Magnesium Alloys		
Alloy AZ31B	1.77	0.0639
Alloy AZ91D	1.81	0.0653
Titanium Alloys		
Commercially pure (ASTM grade 1)	4.51	0.163
Alloy Ti-5Al-2.5Sn	4.48	0.162
Alloy Ti-6Al-4V	4.43	0.160
Precious Metals		
Gold (commercially pure)	19.32	0.697
Platinum (commercially pure)	21.45	0.774
Silver (commercially pure)	10.49	0.379
Refractory Metals		
Molybdenum (commercially pure)	10.22	0.369
Tantalum (commercially pure)	16.6	0.599
Tungsten (commercially pure)	19.3	0.697
Miscellaneous Nonferrous Alloys		
Nickel 200	8.89	0.321
Inconel 625	8.44	0.305
Monel 400	8.80	0.318
Haynes alloy 25	9.13	0.330
Invar	8.05	0.291

Table B.1 (*Continued*)

Material	Density	
	g/cm^3	$lb_m/in.^3$
Super invar	8.10	0.292
Kovar	8.36	0.302
Chemical lead	11.34	0.409
Antimonial lead (6%)	10.88	0.393
Tin (commercially pure)	7.17	0.259
Lead–Tin solder (60Sn-40Pb)	8.52	0.308
Zinc (commercially pure)	7.14	0.258
Zirconium, reactor grade 702	6.51	0.235

GRAPHITE, CERAMICS, AND SEMICONDUCTING MATERIALS

Material	g/cm^3	$lb_m/in.^3$
Aluminum oxide		
• 99.9% pure	3.98	0.144
• 96%	3.72	0.134
• 90%	3.60	0.130
Concrete	2.4	0.087
Diamond		
• Natural	3.51	0.127
• Synthetic	3.20–3.52	0.116–0.127
Gallium arsenide	5.32	0.192
Glass, borosilicate (Pyrex)	2.23	0.0805
Glass, soda–lime	2.5	0.0903
Glass ceramic (Pyroceram)	2.60	0.0939
Graphite		
• Extruded	1.71	0.0616
• Isostatically molded	1.78	0.0643
Silica, fused	2.2	0.079
Silicon	2.33	0.0841
Silicon carbide		
• Hot pressed	3.3	0.119
• Sintered	3.2	0.116
Silicon nitride		
• Hot pressed	3.3	0.119
• Reaction bonded	2.7	0.0975
• Sintered	3.3	0.119
Zirconia, 3 mol% Y_2O_3, sintered	6.0	0.217

POLYMERS

Material	g/cm^3	$lb_m/in.^3$
Elastomers		
• Butadiene-acrylonitrile (nitrile)	0.98	0.0354
• Styrene-butadiene (SBR)	0.94	0.0339
• Silicone	1.1–1.6	0.040–0.058
Epoxy	1.11–1.40	0.0401–0.0505
Nylon 6,6	1.14	0.0412
Phenolic	1.28	0.0462
Polybutylene terephthalate (PBT)	1.34	0.0484
Polycarbonate (PC)	1.20	0.0433
Polyester (thermoset)	1.04–1.46	0.038–0.053
Polyetheretherketone (PEEK)	1.31	0.0473
Polyethylene		
• Low density (LDPE)	0.925	0.0334
• High density (HDPE)	0.959	0.0346

Table B.1 (*Continued*)

Material	Density g/cm³	Density $lb_m/in.^3$
• Ultrahigh molecular weight (UHMWPE)	0.94	0.0339
Polyethylene terephthalate (PET)	1.35	0.0487
Polymethyl methacrylate (PMMA)	1.19	0.0430
Polypropylene (PP)	0.905	0.0327
Polystyrene (PS)	1.05	0.0379
Polytetrafluoroethylene (PTFE)	2.17	0.0783
Polyvinyl chloride (PVC)	1.30–1.58	0.047–0.057
FIBER MATERIALS		
Aramid (Kevlar 49)	1.44	0.0520
Carbon (PAN precursor)		
• Standard modulus	1.78	0.0643
• Intermediate modulus	1.78	0.0643
• High modulus	1.81	0.0653
E Glass	2.58	0.0931
COMPOSITE MATERIALS		
Aramid fibers-epoxy matrix ($V_f = 0.60$)	1.4	0.050
High modulus carbon fibers-epoxy matrix ($V_f = 0.60$)	1.7	0.061
E glass fibers-epoxy matrix ($V_f = 0.60$)	2.1	0.075
Wood		
• Douglas fir (12% moisture)	0.46–0.50	0.017–0.018
• Red oak (12% moisture)	0.61–0.67	0.022–0.024

Sources: *ASM Handbooks, Volumes 1 and 2, Engineered Materials Handbook, Volume 4, Metals Handbook: Properties and Selection: Nonferrous Alloys and Pure Metals,* Vol. 2, 9th edition, and *Advanced Materials & Processes,* Vol. 146, No. 4, ASM International Materials Park, OH; *Modern Plastics Encyclopedia '96,* The McGraw-Hill Companies, New York, NY; R. F. Floral and S. T. Peters, "Composite Structures and Technologies," tutorial notes, 1989; and manufacturers' technical data sheets.

Table B.2 Room-Temperature Modulus of Elasticity Values for Various Engineering Materials

Material	Modulus of Elasticity GPa	Modulus of Elasticity 10^6 psi
METALS AND METAL ALLOYS		
Plain Carbon and Low Alloy Steels		
Steel alloy A36	207	30
Steel alloy 1020	207	30
Steel alloy 1040	207	30
Steel alloy 4140	207	30
Steel alloy 4340	207	30

Table B.2 (*Continued*)

Material	Modulus of Elasticity	
	GPa	*10^6 psi*
Stainless Steels		
Stainless alloy 304	193	28
Stainless alloy 316	193	28
Stainless alloy 405	200	29
Stainless alloy 440A	200	29
Stainless alloy 17-7PH	204	29.5
Cast Irons		
Gray irons		
• Grade G1800	66–97[a]	9.6–14[a]
• Grade G3000	90–113[a]	13.0–16.4[a]
• Grade G4000	110–138[a]	16–20[a]
Ductile irons		
• Grade 60-40-18	169	24.5
• Grade 80-55-06	168	24.4
• Grade 120-90-02	164	23.8
Aluminum Alloys		
Alloy 1100	69	10
Alloy 2024	72.4	10.5
Alloy 6061	69	10
Alloy 7075	71	10.3
Alloy 356.0	72.4	10.5
Copper Alloys		
C11000 (electrolytic tough pitch)	115	16.7
C17200 (beryllium–copper)	128	18.6
C26000 (cartridge brass)	110	16
C36000 (free-cutting brass)	97	14
C71500 (copper–nickel, 30%)	150	21.8
C93200 (bearing bronze)	100	14.5
Magnesium Alloys		
Alloy AZ31B	45	6.5
Alloy AZ91D	45	6.5
Titanium Alloys		
Commercially pure (ASTM grade 1)	103	14.9
Alloy Ti-5Al-2.5Sn	110	16
Alloy Ti-6Al-4V	114	16.5
Precious Metals		
Gold (commercially pure)	77	11.2
Platinum (commercially pure)	171	24.8
Silver (commercially pure)	74	10.7
Refractory Metals		
Molybdenum (commercially pure)	320	46.4
Tantalum (commercially pure)	185	27
Tungsten (commercially pure)	400	58
Miscellaneous Nonferrous Alloys		
Nickel 200	204	29.6
Inconel 625	207	30
Monel 400	180	26
Haynes alloy 25	236	34.2

Table B.2 (*Continued*)

Material	Modulus of Elasticity	
	GPa	*10⁶ psi*
Invar	141	20.5
Super invar	144	21
Kovar	207	30
Chemical lead	13.5	2
Tin (commercially pure)	44.3	6.4
Lead–Tin solder (60Sn-40Pb)	30	4.4
Zinc (commercially pure)	104.5	15.2
Zirconium, reactor grade 702	99.3	14.4

GRAPHITE, CERAMICS, AND SEMICONDUCTING MATERIALS

Aluminum oxide		
• 99.9% pure	380	55
• 96%	303	44
• 90%	275	40
Concrete	25.4–36.6[a]	3.7–5.3[a]
Diamond		
• Natural	700–1200	102–174
• Synthetic	800–925	116–134
Gallium arsenide, single crystal		
• In the ⟨100⟩ direction	85	12.3
• In the ⟨110⟩ direction	122	17.7
• In the ⟨111⟩ direction	142	20.6
Glass, borosilicate (Pyrex)	70	10.1
Glass, soda–lime	69	10
Glass ceramic (Pyroceram)	120	17.4
Graphite		
• Extruded	11	1.6
• Isostatically molded	11.7	1.7
Silica, fused	73	10.6
Silicon, single crystal		
• In the ⟨100⟩ direction	129	18.7
• In the ⟨110⟩ direction	168	24.4
• In the ⟨111⟩ direction	187	27.1
Silicon carbide		
• Hot pressed	207–483	30–70
• Sintered	207–483	30–70
Silicon nitride		
• Hot pressed	304	44.1
• Reaction bonded	304	44.1
• Sintered	304	44.1
Zirconia, 3 mol% Y_2O_3	205	30

POLYMERS

Elastomers		
• Butadiene-acrylonitrile (nitrile)	0.0034[b]	0.00049[b]
• Styrene-butadiene (SBR)	0.002–0.010[b]	0.0003–0.0015[b]
Epoxy	2.41	0.35
Nylon 6,6	1.59–3.79	0.230–0.550
Phenolic	2.76–4.83	0.40–0.70
Polybutylene terephthalate (PBT)	1.93–3.00	0.280–0.435
Polycarbonate (PC)	2.38	0.345
Polyester (thermoset)	2.06–4.41	0.30–0.64

Table B.2 (*Continued*)

Material	Modulus of Elasticity	
	GPa	10⁶ psi
Polyetheretherketone (PEEK)	1.10	0.16
Polyethylene		
• Low density (LDPE)	0.172–0.282	0.025–0.041
• High density (HDPE)	1.08	0.157
• Ultrahigh molecular weight (UHMWPE)	0.69	0.100
Polyethylene terephthalate (PET)	2.76–4.14	0.40–0.60
Polymethyl methacrylate (PMMA)	2.24–3.24	0.325–0.470
Polypropylene (PP)	1.14–1.55	0.165–0.225
Polystyrene (PS)	2.28–3.28	0.330–0.475
Polytetrafluoroethylene (PTFE)	0.40–0.55	0.058–0.080
Polyvinyl chloride (PVC)	2.41–4.14	0.35–0.60
FIBER MATERIALS		
Aramid (Kevlar 49)	131	19
Carbon (PAN precursor)		
• Standard modulus	230	33.4
• Intermediate modulus	285	41.3
• High modulus	400	58
E Glass	72.5	10.5
COMPOSITE MATERIALS		
Aramid fibers-epoxy matrix ($V_f = 0.60$)		
Longitudinal	76	11
Transverse	5.5	0.8
High modulus carbon fibers-epoxy matrix ($V_f = 0.60$)		
Longitudinal	220	32
Transverse	6.9	1.0
E glass fibers-epoxy matrix ($V_f = 0.60$)		
Longitudinal	45	6.5
Transverse	12	1.8
Wood		
• Douglas fir (12% moisture)		
Parallel to grain	10.8–13.6[c]	1.57–1.97[c]
Perpendicular to grain	0.54–0.68[c]	0.078–0.10[c]
• Red oak (12% moisture)		
Parallel to grain	11.0–14.1[c]	1.60–2.04[c]
Perpendicular to grain	0.55–0.71[c]	0.08–0.10[c]

[a] Secant modulus taken at 25% of ultimate strength.

[b] Modulus taken at 100% elongation.

[c] Measured in bending.

Sources: *ASM Handbooks, Volumes 1 and 2, Engineered Materials Handbooks, Volumes 1 and 4, Metals Handbook: Properties and Selection: Nonferrous Alloys and Pure Metals,* Vol. 2, 9th edition, and *Advanced Materials & Processes,* Vol. 146, No. 4, ASM International, Materials Park, OH; *Modern Plastics Encyclopedia '96,* The McGraw-Hill Companies, New York, NY; R. F. Floral and S. T. Peters, "Composite Structures and Technologies," tutorial notes, 1989; and manufacturers' technical data sheets.

Table B.3 Room-Temperature Poisson's Ratio Values for Various Engineering Materials

Material	Poisson's Ratio	Material	Poisson's Ratio
METALS AND METAL ALLOYS		**Refractory Metals**	
Plain Carbon and Low Alloy Steels		Molybdenum (commercially pure)	0.32
Steel alloy A36	0.30	Tantalum (commercially pure)	0.35
Steel alloy 1020	0.30	Tungsten (commercially pure)	0.28
Steel alloy 1040	0.30	**Miscellaneous Nonferrous Alloys**	
Steel alloy 4140	0.30	Nickel 200	0.31
Steel alloy 4340	0.30	Inconel 625	0.31
Stainless Steels		Monel 400	0.32
Stainless alloy 304	0.30	Chemical lead	0.44
Stainless alloy 316	0.30	Tin (commercially pure)	0.33
Stainless alloy 405	0.30	Zinc (commercially pure)	0.25
Stainless alloy 440A	0.30	Zirconium, reactor grade 702	0.35
Stainless alloy 17-7PH	0.30	**GRAPHITE, CERAMICS, AND SEMICONDUCTING MATERIALS**	
Cast Irons		Aluminum oxide	
Gray irons		• 99.9% pure	0.22
• Grade G1800	0.26	• 96%	0.21
• Grade G3000	0.26	• 90%	0.22
• Grade G4000	0.26	Concrete	0.20
Ductile irons		Diamond	
• Grade 60-40-18	0.29	• Natural	0.10–0.30
• Grade 80-55-06	0.31	• Synthetic	0.20
• Grade 120-90-02	0.28	Gallium arsenide	
Aluminum Alloys		• ⟨100⟩ orientation	0.30
Alloy 1100	0.33	Glass, borosilicate (Pyrex)	0.20
Alloy 2024	0.33	Glass, soda–lime	0.23
Alloy 6061	0.33	Glass ceramic (Pyroceram)	0.25
Alloy 7075	0.33	Silica, fused	0.17
Alloy 356.0	0.33	Silicon	
Copper Alloys		• ⟨100⟩ orientation	0.28
C11000 (electrolytic tough pitch)	0.33	• ⟨111⟩ orientation	0.36
C17200 (beryllium–copper)	0.30	Silicon carbide	
C26000 (cartridge brass)	0.35	• Hot pressed	0.17
C36000 (free-cutting brass)	0.34	• Sintered	0.16
C71500 (copper–nickel, 30%)	0.34	Silicon nitride	
C93200 (bearing bronze)	0.34	• Hot pressed	0.30
Magnesium Alloys		• Reaction bonded	0.22
Alloy AZ31B	0.35	• Sintered	0.28
Alloy AZ91D	0.35	Zirconia, 3 mol% Y_2O_3	0.31
Titanium Alloys		**POLYMERS**	
Commercially pure (ASTM grade 1)	0.34	Nylon 6,6	0.39
Alloy Ti-5Al-2.5Sn	0.34	Polycarbonate (PC)	0.36
Alloy Ti-6Al-4V	0.34	Polystyrene (PS)	0.33
Precious Metals		Polytetrafluoroethylene (PTFE)	0.46
Gold (commercially pure)	0.42	Polyvinyl chloride (PVC)	0.38
Platinum (commercially pure)	0.39	**FIBER MATERIALS**	
Silver (commercially pure)	0.37	E Glass	0.22

Table B.3 (*Continued*)

Material	Poisson's Ratio	Material	Poisson's Ratio
COMPOSITE MATERIALS		E glass fibers-epoxy matrix ($V_f = 0.6$)	0.19
Aramid fibers-epoxy matrix ($V_f = 0.6$)	0.34		
High modulus carbon fibers-epoxy matrix ($V_f = 0.6$)	0.25		

Sources: *ASM Handbooks, Volumes 1 and 2,* and *Engineered Materials Handbooks, Volumes 1 and 4,* ASM International, Materials Park, OH; R. F. Floral and S. T. Peters, "Composite Structures and Technologies," tutorial notes, 1989; and manufacturers' technical data sheets.

Table B.4 Typical Room-Temperature Yield Strength, Tensile Strength, and Ductility (Percent Elongation) Values for Various Engineering Materials

Material/Condition	Yield Strength (MPa [ksi])	Tensile Strength (MPa [ksi])	Percent Elongation
METALS AND METAL ALLOYS			
Plain Carbon and Low Alloy Steels			
Steel alloy A36			
• Hot rolled	220–250 (32–36)	400–500 (58–72.5)	23
Steel alloy 1020			
• Hot rolled	210 (30) (min)	380 (55) (min)	25 (min)
• Cold drawn	350 (51) (min)	420 (61) (min)	15 (min)
• Annealed (@ 870°C)	295 (42.8)	395 (57.3)	36.5
• Normalized (@ 925°C)	345 (50.3)	440 (64)	38.5
Steel alloy 1040			
• Hot rolled	290 (42) (min)	520 (76) (min)	18 (min)
• Cold drawn	490 (71) (min)	590 (85) (min)	12 (min)
• Annealed (@ 785°C)	355 (51.3)	520 (75.3)	30.2
• Normalized (@ 900°C)	375 (54.3)	590 (85)	28.0
Steel alloy 4140			
• Annealed (@ 815°C)	417 (60.5)	655 (95)	25.7
• Normalized (@ 870°C)	655 (95)	1020 (148)	17.7
• Oil-quenched and tempered (@ 315°C)	1570 (228)	1720 (250)	11.5
Steel alloy 4340			
• Annealed (@ 810°C)	472 (68.5)	745 (108)	22
• Normalized (@ 870°C)	862 (125)	1280 (185.5)	12.2
• Oil-quenched and tempered (@ 315°C)	1620 (235)	1760 (255)	12
Stainless Steels			
Stainless alloy 304			
• Hot finished and annealed	205 (30) (min)	515 (75) (min)	40 (min)
• Cold worked ($\frac{1}{4}$ hard)	515 (75) (min)	860 (125) (min)	10 (min)
Stainless alloy 316			
• Hot finished and annealed	205 (30) (min)	515 (75) (min)	40 (min)
• Cold drawn and annealed	310 (45) (min)	620 (90) (min)	30 (min)
Stainless alloy 405			
• Annealed	170 (25)	415 (60)	20

Table B.4 (*Continued*)

Material/Condition	Yield Strength (MPa [ksi])	Tensile Strength (MPa [ksi])	Percent Elongation
Stainless alloy 440A			
• Annealed	415 (60)	725 (105)	20
• Tempered @ 315°C	1650 (240)	1790 (260)	5
Stainless alloy 17-7PH			
• Cold rolled	1210 (175) (min)	1380 (200) (min)	1 (min)
• Precipitation hardened @ 510°C	1310 (190) (min)	1450 (210) (min)	3.5 (min)
Cast Irons			
Gray irons			
• Grade G1800 (as cast)	—	124 (18) (min)	—
• Grade G3000 (as cast)	—	207 (30) (min)	—
• Grade G4000 (as cast)	—	276 (40) (min)	—
Ductile irons			
• Grade 60-40-18 (annealed)	276 (40) (min)	414 (60) (min)	18 (min)
• Grade 80-55-06 (as cast)	379 (55) (min)	552 (80) (min)	6 (min)
• Grade 120-90-02 (oil quenched and tempered)	621 (90) (min)	827 (120) (min)	2 (min)
Aluminum Alloys			
Alloy 1100			
• Annealed (O temper)	34 (5)	90 (13)	40
• Strain hardened (H14 temper)	117 (17)	124 (18)	15
Alloy 2024			
• Annealed (O temper)	75 (11)	185 (27)	20
• Heat treated and aged (T3 temper)	345 (50)	485 (70)	18
• Heat treated and aged (T351 temper)	325 (47)	470 (68)	20
Alloy 6061			
• Annealed (O temper)	55 (8)	124 (18)	30
• Heat treated and aged (T6 and T651 tempers)	276 (40)	310 (45)	17
Alloy 7075			
• Annealed (O temper)	103 (15)	228 (33)	17
• Heat treated and aged (T6 temper)	505 (73)	572 (83)	11
Alloy 356.0			
• As cast	124 (18)	164 (24)	6
• Heat treated and aged (T6 temper)	164 (24)	228 (33)	3.5
Copper Alloys			
C11000 (electrolytic tough pitch)			
• Hot rolled	69 (10)	220 (32)	50
• Cold worked (H04 temper)	310 (45)	345 (50)	12
C17200 (beryllium–copper)			
• Solution heat treated	195–380 (28–55)	415–540 (60–78)	35–60
• Solution heat treated, aged @ 330°C	965–1205 (140–175)	1140–1310 (165–190)	4–10
C26000 (cartridge brass)			
• Annealed	75–150 (11–22)	300–365 (43.5–53.0)	54–68
• Cold worked (H04 temper)	435 (63)	525 (76)	8
C36000 (free-cutting brass)			
• Annealed	125 (18)	340 (49)	53
• Cold worked (H02 temper)	310 (45)	400 (58)	25
C71500 (copper–nickel, 30%)			
• Hot rolled	140 (20)	380 (55)	45
• Cold worked (H80 temper)	545 (79)	580 (84)	3

Table B.4 (*Continued*)

Material/Condition	Yield Strength (MPa [ksi])	Tensile Strength (MPa [ksi])	Percent Elongation
C93200 (bearing bronze)			
• Sand cast	125 (18)	240 (35)	20
Magnesium Alloys			
Alloy AZ31B			
• Rolled	220 (32)	290 (42)	15
• Extruded	200 (29)	262 (38)	15
Alloy AZ91D			
• As cast	97–150 (14–22)	165–230 (24–33)	3
Titanium Alloys			
Commercially pure (ASTM grade 1)			
• Annealed	170 (25) (min)	240 (35) (min)	30
Alloy Ti-5Al-2.5Sn			
• Annealed	760 (110) (min)	790 (115) (min)	16
Alloy Ti-6Al-4V			
• Annealed	830 (120) (min)	900 (130) (min)	14
• Solution heat treated and aged	1103 (160)	1172 (170)	10
Precious Metals			
Gold (commercially pure)			
• Annealed	nil	130 (19)	45
• Cold worked (60% reduction)	205 (30)	220 (32)	4
Platinum (commercially pure)			
• Annealed	<13.8 (2)	125–165 (18–24)	30–40
• Cold worked (50%)	—	205–240 (30–35)	1–3
Silver (commercially pure)			
• Annealed	—	170 (24.6)	44
• Cold worked (50%)	—	296 (43)	3.5
Refractory Metals			
Molybdenum (commercially pure)	500 (72.5)	630 (91)	25
Tantalum (commercially pure)	165 (24)	205 (30)	40
Tungsten (commercially pure)	760 (110)	960 (139)	2
Miscellaneous Nonferrous Alloys			
Nickel 200 (annealed)	148 (21.5)	462 (67)	47
Inconel 625 (annealed)	517 (75)	930 (135)	42.5
Monel 400 (annealed)	240 (35)	550 (80)	40
Haynes alloy 25	445 (65)	970 (141)	62
Invar (annealed)	276 (40)	517 (75)	30
Super invar (annealed)	276 (40)	483 (70)	30
Kovar (annealed)	276 (40)	517 (75)	30
Chemical lead	6–8 (0.9–1.2)	16–19 (2.3–2.7)	30–60
Antimonial lead (6%) (chill cast)	—	47.2 (6.8)	24
Tin (commercially pure)	11 (1.6)	—	57
Lead–Tin solder (60Sn-40Pb)	—	52.5 (7.6)	30–60
Zinc (commercially pure)			
• Hot rolled (anisotropic)	—	134–159 (19.4–23.0)	50–65
• Cold rolled (anisotropic)	—	145–186 (21–27)	40–50
Zirconium, reactor grade 702			
• Cold worked and annealed	207 (30) (min)	379 (55) (min)	16 (min)

Table B.4 (*Continued*)

Material/Condition	Yield Strength (MPa [ksi])	Tensile Strength (MPa [ksi])	Percent Elongation
GRAPHITE, CERAMICS, AND SEMICONDUCTING MATERIALS[a]			
Aluminum oxide			
• 99.9% pure	—	282–551 (41–80)	—
• 96%	—	358 (52)	—
• 90%	—	337 (49)	—
Concrete[b]	—	37.3–41.3 (5.4–6.0)	—
Diamond			
• Natural	—	1050 (152)	—
• Synthetic	—	800–1400 (116–203)	—
Gallium arsenide			
• {100} orientation, polished surface	—	66 (9.6)[c]	—
• {100} orientation, as-cut surface	—	57 (8.3)[c]	—
Glass, borosilicate (Pyrex)	—	69 (10)	—
Glass, soda–lime	—	69 (10)	—
Glass ceramic (Pyroceram)	—	123–370 (18–54)	—
Graphite			
• Extruded (with the grain direction)	—	13.8–34.5 (2.0–5.0)	—
• Isostatically molded	—	31–69 (4.5–10)	—
Silica, fused	—	104 (15)	—
Silicon			
• {100} orientation, as-cut surface	—	130 (18.9)	—
• {100} orientation, laser scribed	—	81.8 (11.9)	—
Silicon carbide			
• Hot pressed	—	230–825 (33–120)	—
• Sintered	—	96–520 (14–75)	—
Silicon nitride			
• Hot pressed	—	700–1000 (100–150)	—
• Reaction bonded	—	250–345 (36–50)	—
• Sintered	—	414–650 (60–94)	—
Zirconia, 3 mol% Y_2O_3 (sintered)	—	800–1500 (116–218)	—
POLYMERS			
Elastomers			
• Butadiene-acrylonitrile (nitrile)	—	6.9–24.1 (1.0–3.5)	400–600
• Styrene-butadiene (SBR)	—	12.4–20.7 (1.8–3.0)	450–500
• Silicone	—	10.3 (1.5)	100–800
Epoxy	—	27.6–90.0 (4.0–13)	3–6
Nylon 6,6			
• Dry, as molded	55.1–82.8 (8–12)	94.5 (13.7)	15–80
• 50% relative humidity	44.8–58.6 (6.5–8.5)	75.9 (11)	150–300
Phenolic	—	34.5–62.1 (5.0–9.0)	1.5–2.0
Polybutylene terephthalate (PBT)	56.6–60.0 (8.2–8.7)	56.6–60.0 (8.2–8.7)	50–300
Polycarbonate (PC)	62.1 (9)	62.8–72.4 (9.1–10.5)	110–150
Polyester (thermoset)	—	41.4–89.7 (6.0–13.0)	<2.6
Polyetheretherketone (PEEK)	91 (13.2)	70.3–103 (10.2–15.0)	30–150
Polyethylene			
• Low density (LDPE)	9.0–14.5 (1.3–2.1)	8.3–31.4 (1.2–4.55)	100–650
• High density (HDPE)	26.2–33.1 (3.8–4.8)	22.1–31.0 (3.2–4.5)	10–1200
• Ultrahigh molecular weight (UHMWPE)	21.4–27.6 (3.1–4.0)	38.6–48.3 (5.6–7.0)	350–525
Polyethylene terephthalate (PET)	59.3 (8.6)	48.3–72.4 (7.0–10.5)	30–300

Table B.4 (*Continued*)

Material/Condition	Yield Strength (MPa [ksi])	Tensile Strength (MPa [ksi])	Percent Elongation
Polymethyl methacrylate (PMMA)	53.8–73.1 (7.8–10.6)	48.3–72.4 (7.0–10.5)	2.0–5.5
Polypropylene (PP)	31.0–37.2 (4.5–5.4)	31.0–41.4 (4.5–6.0)	100–600
Polystyrene (PS)	—	35.9–51.7 (5.2–7.5)	1.2–2.5
Polytetrafluoroethylene (PTFE)	—	20.7–34.5 (3.0–5.0)	200–400
Polyvinyl chloride (PVC)	40.7–44.8 (5.9–6.5)	40.7–51.7 (5.9–7.5)	40–80
FIBER MATERIALS			
Aramid (Kevlar 49)	—	3600–4100 (525–600)	2.8
Carbon (PAN precursor)			
• Standard modulus (longitudinal)	—	3800–4200 (550–610)	2
• Intermediate modulus (longitudinal)	—	4650–6350 (675–920)	1.8
• High modulus (longitudinal)	—	2500–4500 (360–650)	0.6
E Glass	—	3450 (500)	4.3
COMPOSITE MATERIALS			
Aramid fibers-epoxy matrix (aligned, $V_f = 0.6$)			
• Longitudinal direction	—	1380 (200)	1.8
• Transverse direction	—	30 (4.3)	0.5
High modulus carbon fibers-epoxy matrix (aligned, $V_f = 0.6$)			
• Longitudinal direction	—	760 (110)	0.3
• Transverse direction	—	28 (4)	0.4
E glass fibers-epoxy matrix (aligned, $V_f = 0.6$)			
• Longitudinal direction	—	1020 (150)	2.3
• Transverse direction	—	40 (5.8)	0.4
Wood			
• Douglas fir (12% moisture)			
Parallel to grain	—	108 (15.6)	—
Perpendicular to grain	—	2.4 (0.35)	—
• Red oak (12% moisture)			
Parallel to grain	—	112 (16.3)	—
Perpendicular to grain	—	7.2 (1.05)	—

[a] The strength of graphite, ceramics, and semiconducting materials is taken as flexural strength.

[b] The strength of concrete is measured in compression.

[c] Flexural strength value at 50% fracture probability.

Sources: *ASM Handbooks, Volumes 1 and 2, Engineered Materials Handbooks, Volumes 1 and 4, Metals Handbook: Properties and Selection: Nonferrous Alloys and Pure Metals,* Vol. 2, 9th edition, *Advanced Materials & Processes,* Vol. 146, No. 4, and *Materials & Processing Databook (1985),* ASM International, Materials Park, OH; *Modern Plastics Encyclopedia '96,* The McGraw-Hill Companies, New York, NY; R. F. Floral and S. T. Peters, "Composite Structures and Technologies," tutorial notes, 1989; and manufacturers' technical data sheets.

Table B.5   Room-Temperature Plane Strain Fracture Toughness and Strength Values for Various Engineering Materials

| Material | Fracture Toughness | | Strength[a] |
	$MPa\sqrt{m}$	$ksi\sqrt{in.}$	(MPa)
METALS AND METAL ALLOYS			
Plain Carbon and Low Alloy Steels			
Steel alloy 1040	54.0	49.0	260
Steel alloy 4140			
• Tempered @ 370°C	55–65	50–59	1375–1585
• Tempered @ 482°C	75–93	68.3–84.6	1100–1200
Steel alloy 4340			
• Tempered @ 260°C	50.0	45.8	1640
• Tempered @ 425°C	87.4	80.0	1420
Stainless Steels			
Stainless alloy 17-7PH			
• Precipitation hardened @ 510°C	76	69	1310
Aluminum Alloys			
Alloy 2024-T3	44	40	345
Alloy 7075-T651	24	22	495
Magnesium Alloys			
Alloy AZ31B			
• Extruded	28.0	25.5	200
Titanium Alloys			
Alloy Ti-5Al-2.5Sn			
• Air cooled	71.4	65.0	876
Alloy Ti-6Al-4V			
• Equiaxed grains	44–66	40–60	910
GRAPHITE, CERAMICS, AND SEMICONDUCTING MATERIALS			
Aluminum oxide			
• 99.9% pure	4.2–5.9	3.8–5.4	282–551
• 96%	3.85–3.95	3.5–3.6	358
Concrete	0.2–1.4	0.18–1.27	—
Diamond			
• Natural	3.4	3.1	1050
• Synthetic	6.0–10.7	5.5–9.7	800–1400
Gallium arsenide			
• In the {100} orientation	0.43	0.39	66
• In the {110} orientation	0.31	0.28	—
• In the {111} orientation	0.45	0.41	—
Glass, borosilicate (Pyrex)	0.77	0.70	69
Glass, soda-lime	0.75	0.68	69
Glass ceramic (Pyroceram)	1.6–2.1	1.5–1.9	123–370
Silica, fused	0.79	0.72	104
Silicon			
• In the {100} orientation	0.95	0.86	—
• In the {110} orientation	0.90	0.82	—
• In the {111} orientation	0.82	0.75	—
Silicon carbide			
• Hot pressed	4.8–6.1	4.4–5.6	230–825
• Sintered	4.8	4.4	96–520

Table B.5 (*Continued*)

Material	Fracture Toughness		Strength[a] (MPa)
	MPa√m	ksi√in.	
Silicon nitride			
• Hot pressed	4.1–6.0	3.7–5.5	700–1000
• Reaction bonded	3.6	3.3	250–345
• Sintered	5.3	4.8	414–650
Zirconia, 3 mol% Y_2O_3	7.0–12.0	6.4–10.9	800–1500
POLYMERS			
Epoxy	0.6	0.55	—
Nylon 6,6	2.5–3.0	2.3–2.7	44.8–58.6
Polycarbonate (PC)	2.2	2.0	62.1
Polyester (thermoset)	0.6	0.55	—
Polyethylene terephthalate (PET)	5.0	4.6	59.3
Polymethyl methacrylate (PMMA)	0.7–1.6	0.6–1.5	53.8–73.1
Polypropylene (PP)	3.0–4.5	2.7–4.1	31.0–37.2
Polystyrene (PS)	0.7–1.1	0.6–1.0	—
Polyvinyl chloride (PVC)	2.0–4.0	1.8–3.6	40.7–44.8

[a] For metal alloys and polymers, strength is taken as yield strength; for ceramic materials, flexural strength is used.

Sources: *ASM Handbooks, Volumes 1 and 19, Engineered Materials Handbooks, Volumes 2* and *4,* and *Advanced Materials & Processes,* Vol. 137, No. 6, ASM International, Materials Park, OH.

Table B.6   Room-Temperature Linear Coefficient of Thermal Expansion Values for Various Engineering Materials

Material	Coefficient of Thermal Expansion	
	10^{-6} (°C)$^{-1}$	10^{-6} (°F)$^{-1}$
METALS AND METAL ALLOYS		
Plain Carbon and Low Alloy Steels		
Steel alloy A36	11.7	6.5
Steel alloy 1020	11.7	6.5
Steel alloy 1040	11.3	6.3
Steel alloy 4140	12.3	6.8
Steel alloy 4340	12.3	6.8
Stainless Steels		
Stainless alloy 304	17.2	9.6
Stainless alloy 316	15.9	8.8
Stainless alloy 405	10.8	6.0
Stainless alloy 440A	10.2	5.7
Stainless alloy 17-7PH	11.0	6.1
Cast Irons		
Gray irons		
• Grade G1800	11.4	6.3
• Grade G3000	11.4	6.3
• Grade G4000	11.4	6.3

Table B.6 (*Continued*)

Material	Coefficient of Thermal Expansion	
	10^{-6} $(°C)^{-1}$	10^{-6} $(°F)^{-1}$
Ductile irons		
• Grade 60-40-18	11.2	6.2
• Grade 80-55-06	10.6	5.9
Aluminum Alloys		
Alloy 1100	23.6	13.1
Alloy 2024	22.9	12.7
Alloy 6061	23.6	13.1
Alloy 7075	23.4	13.0
Alloy 356.0	21.5	11.9
Copper Alloys		
C11000 (electrolytic tough pitch)	17.0	9.4
C17200 (beryllium–copper)	16.7	9.3
C26000 (cartridge brass)	19.9	11.1
C36000 (free-cutting brass)	20.5	11.4
C71500 (copper–nickel, 30%)	16.2	9.0
C93200 (bearing bronze)	18.0	10.0
Magnesium Alloys		
Alloy AZ31B	26.0	14.4
Alloy AZ91D	26.0	14.4
Titanium Alloys		
Commercially pure (ASTM grade 1)	8.6	4.8
Alloy Ti-5Al-2.5Sn	9.4	5.2
Alloy Ti-6Al-4V	8.6	4.8
Precious Metals		
Gold (commerically pure)	14.2	7.9
Platinum (commercially pure)	9.1	5.1
Silver (commercially pure)	19.7	10.9
Refractory Metals		
Molybdenum (commercially pure)	4.9	2.7
Tantalum (commercially pure)	6.5	3.6
Tungsten (commercially pure)	4.5	2.5
Miscellaneous Nonferrous Alloys		
Nickel 200	13.3	7.4
Inconel 625	12.8	7.1
Monel 400	13.9	7.7
Haynes alloy 25	12.3	6.8
Invar	1.6	0.9
Super invar	0.72	0.40
Kovar	5.1	2.8
Chemical lead	29.3	16.3
Antimonial lead (6%)	27.2	15.1
Tin (commercially pure)	23.8	13.2
Lead–Tin solder (60Sn-40Pb)	24.0	13.3
Zinc (commercially pure)	23.0–32.5	12.7–18.1
Zirconium, reactor grade 702	5.9	3.3

Table B.6 (*Continued*)

Material	Coefficient of Thermal Expansion	
	10^{-6} (°C)$^{-1}$	10^{-6} (°F)$^{-1}$
GRAPHITE, CERAMICS, AND SEMICONDUCTING MATERIALS		
Aluminum oxide		
• 99.9% pure	7.4	4.1
• 96%	7.4	4.1
• 90%	7.0	3.9
Concrete	10.0–13.6	5.6–7.6
Diamond (natural)	0.11–1.23	0.06–0.68
Gallium arsenide	5.9	3.3
Glass, borosilicate (Pyrex)	3.3	1.8
Glass, soda–lime	9.0	5.0
Glass ceramic (Pyroceram)	6.5	3.6
Graphite		
• Extruded	2.0–2.7	1.1–1.5
• Isostatically molded	2.2–6.0	1.2–3.3
Silica, fused	0.4	0.22
Silicon	2.5	1.4
Silicon carbide		
• Hot pressed	4.6	2.6
• Sintered	4.1	2.3
Silicon nitride		
• Hot pressed	2.7	1.5
• Reaction bonded	3.1	1.7
• Sintered	3.1	1.7
Zirconia, 3 mol% Y_2O_3	9.6	5.3
POLYMERS		
Elastomers		
• Butadiene-acrylonitrile (nitrile)	235	130
• Styrene-butadiene (SBR)	220	125
• Silicone	270	150
Epoxy	81–117	45–65
Nylon 6,6	144	80
Phenolic	122	68
Polybutylene terephthalate (PBT)	108–171	60–95
Polycarbonate (PC)	122	68
Polyester (thermoset)	100–180	55–100
Polyetheretherketone (PEEK)	72–85	40–47
Polyethylene		
• Low density (LDPE)	180–400	100–220
• High density (HDPE)	106–198	59–110
• Ultrahigh molecular weight (UHMWPE)	234–360	130–200
Polyethylene terephthalate (PET)	117	65
Polymethyl methacrylate (PMMA)	90–162	50–90
Polypropylene (PP)	146–180	81–100
Polystyrene (PS)	90–150	50–83
Polytetrafluoroethylene (PTFE)	126–216	70–120
Polyvinyl chloride (PVC)	90–180	50–100

Table B.6 (*Continued*)

Material	Coefficient of Thermal Expansion	
	10^{-6} (°C)$^{-1}$	10^{-6} (°F)$^{-1}$
FIBER MATERIALS		
Aramid (Kevlar 49)		
• Longitudinal direction	−2.0	−1.1
• Transverse direction	60	33
Carbon (PAN precursor)		
• Standard modulus		
Longitudinal direction	−0.6	−0.3
Transverse direction	10.0	5.6
• Intermediate modulus		
Longitudinal direction	−0.6	−0.3
• High modulus		
Longitudinal direction	−0.5	−0.28
Transverse direction	7.0	3.9
E Glass	5.0	2.8
COMPOSITE MATERIALS		
Aramid fibers-epoxy matrix ($V_f = 0.6$)		
• Longitudinal direction	−4.0	−2.2
• Transverse direction	70	40
High modulus carbon fibers-epoxy matrix ($V_f = 0.6$)		
• Longitudinal direction	−0.5	−0.3
• Transverse direction	32	18
E glass fibers-epoxy matrix ($V_f = 0.6$)		
• Longitudinal direction	6.6	3.7
• Transverse direction	30	16.7
Wood		
• Douglas fir (12% moisture)		
Parallel to grain	3.8–5.1	2.2–2.8
Perpendicular to grain	25.4–33.8	14.1–18.8
• Red oak (12% moisture)		
Parallel to grain	4.6–5.9	2.6–3.3
Perpendicular to grain	30.6–39.1	17.0–21.7

Sources: *ASM Handbooks, Volumes 1* and *2, Engineered Materials Handbooks, Volumes 1* and *4, Metals Handbook: Properties and Selection: Nonferrous Alloys and Pure Metals,* Vol. 2, 9th edition, and *Advanced Materials & Processes,* Vol. 146, No. 4, ASM International, Materials Park, OH; *Modern Plastics Encyclopedia '96,* The McGraw-Hill Companies, New York, NY; R. F. Floral and S. T. Peters, "Composite Structures and Technologies," tutorial notes, 1989; and manufacturers' technical data sheets.

Table B.7   Room-Temperature Thermal Conductivity Values for Various Engineering Materials

Material	Thermal Conductivity	
	W/m-K	Btu/ft-h-°F
METALS AND METAL ALLOYS		
Plain Carbon and Low Alloy Steels		
Steel alloy A36	51.9	30
Steel alloy 1020	51.9	30
Steel alloy 1040	51.9	30
Stainless Steels		
Stainless alloy 304 (annealed)	16.2	9.4
Stainless alloy 316 (annealed)	16.2	9.4
Stainless alloy 405 (annealed)	27.0	15.6
Stainless alloy 440A (annealed)	24.2	14.0
Stainless alloy 17-7PH (annealed)	16.4	9.5
Cast Irons		
Gray irons		
• Grade G1800	46.0	26.6
• Grade G3000	46.0	26.6
• Grade G4000	46.0	26.6
Ductile irons		
• Grade 60-40-18	36.0	20.8
• Grade 80-55-06	36.0	20.8
• Grade 120-90-02	36.0	20.8
Aluminum Alloys		
Alloy 1100 (annealed)	222	128
Alloy 2024 (annealed)	190	110
Alloy 6061 (annealed)	180	104
Alloy 7075-T6	130	75
Alloy 356.0-T6	151	87
Copper Alloys		
C11000 (electrolytic tough pitch)	388	224
C17200 (beryllium–copper)	105–130	60–75
C26000 (cartridge brass)	120	70
C36000 (free-cutting brass)	115	67
C71500 (copper–nickel, 30%)	29	16.8
C93200 (bearing bronze)	59	34
Magnesium Alloys		
Alloy AZ31B	96[a]	55[a]
Alloy AZ91D	72[a]	43[a]
Titanium Alloys		
Commercially pure (ASTM grade 1)	16	9.2
Alloy Ti-5Al-2.5Sn	7.6	4.4
Alloy Ti-6Al-4V	6.7	3.9
Precious Metals		
Gold (commercially pure)	315	182
Platinum (commercially pure)	71[b]	41[b]
Silver (commercially pure)	428	247

Table B.7 (*Continued*)

Material	Thermal Conductivity	
	W/m-K	*Btu/ft-h-°F*
Refractory Metals		
Molybdenum (commercially pure)	142	82
Tantalum (commercially pure)	54.4	31.4
Tungsten (commercially pure)	155	89.4
Miscellaneous Nonferrous Alloys		
Nickel 200	70	40.5
Inconel 625	9.8	5.7
Monel 400	21.8	12.6
Haynes alloy 25	9.8	5.7
Invar	10	5.8
Super invar	10	5.8
Kovar	17	9.8
Chemical lead	35	20.2
Antimonial lead (6%)	29	16.8
Tin (commercially pure)	60.7	35.1
Lead–Tin solder (60Sn-40Pb)	50	28.9
Zinc (commercially pure)	108	62
Zirconium, reactor grade 702	22	12.7
GRAPHITE, CERAMICS, AND SEMICONDUCTING MATERIALS		
Aluminum oxide		
• 99.9% pure	39	22.5
• 96%	35	20
• 90%	16	9.2
Concrete	1.25–1.75	0.72–1.0
Diamond		
• Natural	1450–4650	840–2700
• Synthetic	3150	1820
Gallium arsenide	45.5	26.3
Glass, borosilicate (Pyrex)	1.4	0.81
Glass, soda–lime	1.7	1.0
Glass ceramic (Pyroceram)	3.3	1.9
Graphite		
• Extruded	130–190	75–110
• Isostatically molded	104–130	60–75
Silica, fused	1.4	0.81
Silicon	141	82
Silicon carbide		
• Hot pressed	80	46.2
• Sintered	71	41
Silicon nitride		
• Hot pressed	29	17
• Reaction bonded	10	6
• Sintered	33	19.1
Zirconia, 3 mol% Y_2O_3	2.0–3.3	1.2–1.9
POLYMERS		
Elastomers		
• Butadiene-acrylonitrile (nitrile)	0.25	0.14
• Styrene-butadiene (SBR)	0.25	0.14
• Silicone	0.23	0.13

Table B.7 (*Continued*)

Material	Thermal Conductivity	
	W/m-K	Btu/ft-h-°F
Epoxy	0.19	0.11
Nylon 6,6	0.24	0.14
Phenolic	0.15	0.087
Polybutylene terephthalate (PBT)	0.18–0.29	0.10–0.17
Polycarbonate (PC)	0.20	0.12
Polyester (thermoset)	0.17	0.10
Polyethylene		
• Low density (LDPE)	0.33	0.19
• High density (HDPE)	0.48	0.28
• Ultrahigh molecular weight (UHMWPE)	0.33	0.19
Polyethylene terephthalate (PET)	0.15	0.087
Polymethyl methacrylate (PMMA)	0.17–0.25	0.10–0.15
Polypropylene (PP)	0.12	0.069
Polystyrene (PS)	0.13	0.075
Polytetrafluoroethylene (PTFE)	0.25	0.14
Polyvinyl chloride (PVC)	0.15–0.21	0.08–0.12
FIBER MATERIALS		
Carbon (PAN precursor), longitudinal		
• Standard modulus	11	6.4
• Intermediate modulus	15	8.7
• High modulus	70	40
E Glass	1.3	0.75
COMPOSITE MATERIALS		
Wood		
• Douglas fir (12% moisture) Perpendicular to grain	0.14	0.08
• Red oak (12% moisture) Perpendicular to grain	0.18	0.11

[a] At 100°C.
[b] At 0°C.

Sources: *ASM Handbooks, Volumes 1 and 2, Engineered Materials Handbooks, Volumes 1 and 4, Metals Handbook: Properties and Selection: Nonferrous Alloys and Pure Metals,* Vol. 2, 9th edition, and *Advanced Materials & Processes,* Vol. 146, No. 4, ASM International, Materials Park, OH; *Modern Plastics Encyclopedia '96* and *Modern Plastics Encyclopedia 1977–1978,* The McGraw-Hill Companies, New York, NY; and manufacturers' technical data sheets.

Table B.8 Room-Temperature Specific Heat Values for Various Engineering Materials

Material	Specific Heat	
	$J/kg\text{-}K$	$10^{-2}Btu/lb_m-°F$

METALS AND METAL ALLOYS
Plain Carbon and Low Alloy Steels

Steel alloy A36	486[a]	11.6[a]
Steel alloy 1020	486[a]	11.6[a]
Steel alloy 1040	486[a]	11.6[a]

Stainless Steels

Stainless alloy 304	500	12.0
Stainless alloy 316	500	12.0
Stainless alloy 405	460	11.0
Stainless alloy 440A	460	11.0
Stainless alloy 17-7PH	460	11.0

Cast Irons

Gray irons		
• Grade G1800	544	13
• Grade G3000	544	13
• Grade G4000	544	13
Ductile irons		
• Grade 60-40-18	544	13
• Grade 80-55-06	544	13
• Grade 120-90-02	544	13

Aluminum Alloys

Alloy 1100	904	21.6
Alloy 2024	875	20.9
Alloy 6061	896	21.4
Alloy 7075	960[b]	23.0[b]
Alloy 356.0	963[b]	23.0[b]

Copper Alloys

C11000 (electrolytic tough pitch)	385	9.2
C17200 (beryllium–copper)	420	10.0
C26000 (cartridge brass)	375	9.0
C36000 (free-cutting brass)	380	9.1
C71500 (copper–nickel, 30%)	380	9.1
C93200 (bearing bronze)	376	9.0

Magnesium Alloys

Alloy AZ31B	1024	24.5
Alloy AZ91D	1050	25.1

Titanium Alloys

Commercially pure (ASTM grade 1)	528[c]	12.6[c]
Alloy Ti-5Al-2.5Sn	470[c]	11.2[c]
Alloy Ti-6Al-4V	610[c]	14.6[c]

Precious Metals

Gold (commercially pure)	130	3.1
Platinum (commercially pure)	132[d]	3.2[d]
Silver (commercially pure)	235	5.6

Table B.8 (*Continued*)

Material	Specific Heat	
	J/kg-K	$10^{-2} Btu/lb_m-°F$
Refractory Metals		
Molybdenum (commercially pure)	276	6.6
Tantalum (commercially pure)	139	3.3
Tungsten (commercially pure)	138	3.3
Miscellaneous Nonferrous Alloys		
Nickel 200	456	10.9
Inconel 625	410	9.8
Monel 400	427	10.2
Haynes alloy 25	377	9.0
Invar	500	12.0
Super invar	500	12.0
Kovar	460	11.0
Chemical lead	129	3.1
Antimonial lead (6%)	135	3.2
Tin (commercially pure)	222	5.3
Lead–Tin solder (60Sn-40Pb)	150	3.6
Zinc (commercially pure)	395	9.4
Zirconium, reactor grade 702	285	6.8
GRAPHITE, CERAMICS, AND SEMICONDUCTING MATERIALS		
Aluminum oxide		
• 99.9% pure	775	18.5
• 96%	775	18.5
• 90%	775	18.5
Concrete	850–1150	20.3–27.5
Diamond (natural)	520	12.4
Gallium arsenide	350	8.4
Glass, borosilicate (Pyrex)	850	20.3
Glass, soda–lime	840	20.0
Glass ceramic (Pyroceram)	975	23.3
Graphite		
• Extruded	830	19.8
• Isostatically molded	830	19.8
Silica, fused	740	17.7
Silicon	700	16.7
Silicon carbide		
• Hot pressed	670	16.0
• Sintered	590	14.1
Silicon nitride		
• Hot pressed	750	17.9
• Reaction bonded	870	20.7
• Sintered	1100	26.3
Zirconia, 3 mol% Y_2O_3	481	11.5
POLYMERS		
Epoxy	1050	25
Nylon 6,6	1670	40
Phenolic	1590–1760	38–42
Polybutylene terephthalate (PBT)	1170–2300	28–55
Polycarbonate (PC)	840	20

Table B.8 (*Continued*)

Material	Specific Heat	
	J/kg-K	*$10^{-2}Btu/lb_m-°F$*
Polyester (thermoset)	710–920	17–22
Polyethylene		
• Low density (LDPE)	2300	55
• High density (HDPE)	1850	44.2
Polyethylene terephthalate (PET)	1170	28
Polymethyl methacrylate (PMMA)	1460	35
Polypropylene (PP)	1925	46
Polystyrene (PS)	1170	28
Polytetrafluoroethylene (PTFE)	1050	25
Polyvinyl chloride (PVC)	1050–1460	25–35
FIBER MATERIALS		
Aramid (Kevlar 49)	1300	31
E Glass	810	19.3
COMPOSITE MATERIALS		
Wood		
• Douglas fir (12% moisture)	2900	69.3
• Red oak (12% moisture)	2900	69.3

[a] At temperatures between 50°C and 100°C.
[b] At 100°C.
[c] At 50°C.
[d] At 0°C.

Sources: *ASM Handbooks, Volumes 1 and 2, Engineered Materials Handbooks, Volumes 1, 2, and 4, Metals Handbook: Properties and Selection: Nonferrous Alloys and Pure Metals,* Vol. 2, 9th edition, and *Advanced Materials & Processes,* Vol. 146, No. 4, ASM International, Materials Park, OH; *Modern Plastics Encyclopedia 1977–1978,* The McGraw-Hill Companies, New York, NY; and manufacturers' technical data sheets.

Table B.9   Room-Temperature Electrical Resistivity Values for Various Engineering Materials

Material	Electrical Resistivity, Ω-m
METALS AND METAL ALLOYS	
Plain Carbon and Low Alloy Steels	
Steel alloy A36[a]	1.60×10^{-7}
Steel alloy 1020 (annealed)[a]	1.60×10^{-7}
Steel alloy 1040 (annealed)[a]	1.60×10^{-7}
Steel alloy 4140 (quenched and tempered)	2.20×10^{-7}
Steel alloy 4340 (quenched and tempered)	2.48×10^{-7}
Stainless Steels	
Stainless alloy 304 (annealed)	7.2×10^{-7}
Stainless alloy 316 (annealed)	7.4×10^{-7}
Stainless alloy 405 (annealed)	6.0×10^{-7}
Stainless alloy 440A (annealed)	6.0×10^{-7}
Stainless alloy 17-7PH (annealed)	8.3×10^{-7}

Table B.9 (*Continued*)

Material	Electrical Resistivity, Ω-m
Cast Irons	
Gray irons	
• Grade G1800	15.0×10^{-7}
• Grade G3000	9.5×10^{-7}
• Grade G4000	8.5×10^{-7}
Ductile irons	
• Grade 60-40-18	5.5×10^{-7}
• Grade 80-55-06	6.2×10^{-7}
• Grade 120-90-02	6.2×10^{-7}
Aluminum Alloys	
Alloy 1100 (annealed)	2.9×10^{-8}
Alloy 2024 (annealed)	3.4×10^{-8}
Alloy 6061 (annealed)	3.7×10^{-8}
Alloy 7075 (T6 treatment)	5.22×10^{-8}
Alloy 356.0 (T6 treatment)	4.42×10^{-8}
Copper Alloys	
C11000 (electrolytic tough pitch, annealed)	1.72×10^{-8}
C172000 (beryllium–copper)	5.7×10^{-8}–1.15×10^{-7}
C26000 (cartridge brass)	6.2×10^{-8}
C36000 (free-cutting brass)	6.6×10^{-8}
C71500 (copper–nickel, 30%)	37.5×10^{-8}
C93200 (bearing bronze)	14.4×10^{-8}
Magnesium Alloys	
Alloy AZ31B	9.2×10^{-8}
Alloy AZ91D	17.0×10^{-8}
Titanium Alloys	
Commercially pure (ASTM grade 1)	4.2×10^{-7}–5.2×10^{-7}
Alloy Ti-5Al-2.5Sn	15.7×10^{-7}
Alloy Ti-6Al-4V	17.1×10^{-7}
Precious Metals	
Gold (commercially pure)	2.35×10^{-8}
Platinum (commercially pure)	10.60×10^{-8}
Silver (commercially pure)	1.47×10^{-8}
Refractory Metals	
Molybdenum (commercially pure)	5.2×10^{-8}
Tantalum (commercially pure)	13.5×10^{-8}
Tungsten (commercially pure)	5.3×10^{-8}
Miscellaneous Nonferrous Alloys	
Nickel 200	0.95×10^{-7}
Inconel 625	12.90×10^{-7}
Monel 400	5.47×10^{-7}
Haynes alloy 25	8.9×10^{-7}
Invar	8.2×10^{-7}
Super invar	8.0×10^{-7}
Kovar	4.9×10^{-7}
Chemical lead	2.06×10^{-7}
Antimonial lead (6%)	2.53×10^{-7}
Tin (commercially pure)	1.11×10^{-7}
Lead–Tin solder (60Sn-40Pb)	1.50×10^{-7}

Table B.9 (*Continued*)

Material	Electrical Resistivity, Ω-m
Zinc (commercially pure)	62.0×10^{-7}
Zirconium, reactor grade 702	3.97×10^{-7}

GRAPHITE, CERAMICS, AND SEMICONDUCTING MATERIALS

Aluminum oxide	
• 99.9% pure	$>10^{13}$
• 96%	$>10^{12}$
• 90%	$>10^{12}$
Concrete (dry)	10^{9}
Diamond	
• Natural	$10-10^{14}$
• Synthetic	1.5×10^{-2}
Gallium arsenide (intrinsic)	10^{6}
Glass, borosilicate (Pyrex)	$\sim 10^{13}$
Glass, soda–lime	$10^{10}-10^{11}$
Glass ceramic (Pyroceram)	2×10^{14}
Graphite	
• Extruded (with grain direction)	$7 \times 10^{-6}-20 \times 10^{-6}$
• Isostatically molded	$10 \times 10^{-6}-18 \times 10^{-6}$
Silica, fused	$>10^{18}$
Silicon (intrinsic)	2500
Silicon carbide	
• Hot pressed	$1.0-10^{9}$
• Sintered	$1.0-10^{9}$
Silicon nitride	
• Hot isostatic pressed	$>10^{12}$
• Reaction bonded	$>10^{12}$
• Sintered	$>10^{12}$
Zirconia, 3 mol% Y_2O_3	10^{10}

POLYMERS

Elastomers	
• Butadiene-acrylonitrile (nitrile)	3.5×10^{8}
• Styrene-butadiene (SBR)	6×10^{11}
• Silicone	10^{13}
Epoxy	$10^{10}-10^{13}$
Nylon 6,6	$10^{12}-10^{13}$
Phenolic	$10^{9}-10^{10}$
Polybutylene terephthalate (PBT)	4×10^{14}
Polycarbonate (PC)	2×10^{14}
Polyester (thermoset)	10^{13}
Polyetheretherketone (PEEK)	6×10^{14}
Polyethylene	
• Low density (LDPE)	$10^{15}-5 \times 10^{16}$
• High density (HDPE)	$10^{15}-5 \times 10^{16}$
• Ultrahigh molecular weight (UHMWPE)	$>5 \times 10^{14}$
Polyethylene terephthalate (PET)	10^{12}
Polymethyl methacrylate (PMMA)	$>10^{12}$
Polypropylene (PP)	$>10^{14}$
Polystyrene (PS)	$>10^{14}$
Polytetrafluoroethylene (PTFE)	10^{17}
Polyvinyl chloride (PVC)	$>10^{14}$

Table B.9 (*Continued*)

Material	Electrical Resistivity, Ω-m
FIBER MATERIALS	
Carbon (PAN precursor)	
• Standard modulus	17×10^{-6}
• Intermediate modulus	15×10^{-6}
• High modulus	9.5×10^{-6}
E Glass	4×10^{14}
COMPOSITE MATERIALS	
Wood	
• Douglas fir (oven dry)	
Parallel to grain	10^{14}–10^{16}
Perpendicular to grain	10^{14}–10^{16}
• Red oak (oven dry)	
Parallel to grain	10^{14}–10^{16}
Perpendicular to grain	10^{14}–10^{16}

[a] At 0°C.

Sources: *ASM Handbooks, Volumes 1 and 2, Engineered Materials Handbooks, Volumes 1, 2, and 4, Metals Handbook: Properties and Selection: Nonferrous Alloys and Pure Metals*, Vol. 2, 9th edition, and *Advanced Materials & Processes*, Vol. 146, No. 4, ASM International, Materials Park, OH; *Modern Plastics Encyclopedia 1977–1978*, The McGraw-Hill Companies, New York, NY; and manufacturers' technical data sheets.

Table B.10 **Compositions of Metal Alloys for Which Data Are Included in Tables B.1 through B.9**

Alloy (UNS Designation)	Composition (wt%)
PLAIN-CARBON AND LOW-ALLOY STEELS	
A 36 (ASTM A 36)	98.0 Fe (min), 0.29 C, 1.0 Mn, 0.28 Si
1020 (G10200)	99.1 Fe (min), 0.20 C, 0.45 Mn
1040 (G10400)	98.6 Fe (min), 0.40 C, 0.75 Mn
4140 (G41400)	96.8 Fe (min), 0.40 C, 0.90 Cr, 0.20 Mo, 0.9 Mn
4340 (G43400)	95.2 Fe (min), 0.40 C, 1.8 Ni, 0.80 Cr, 0.25 Mo, 0.7 Mn
STAINLESS STEELS	
304 (S30400)	66.4 Fe (min), 0.08 C, 19.0 Cr, 9.25 Ni, 2.0 Mn
316 (S31600)	61.9 Fe (min), 0.08 C, 17.0 Cr, 12.0 Ni, 2.5 Mo, 2.0 Mn
405 (S40500)	83.1 Fe (min), 0.08 C, 13.0 Cr, 0.20 Al, 1.0 Mn
440A (S44002)	78.4 Fe (min), 0.70 C, 17.0 Cr, 0.75 Mo. 1.0 Mn
17-7PH (S17700)	70.6 Fe (min), 0.09 C, 17.0 Cr, 7.1 Ni, 1.1 Al, 1.0 Mn
CAST IRONS	
Grade G1800 (F10004)	Fe (bal), 3.4–3.7 C, 2.8–2.3 Si, 0.65 Mn, 0.15 P, 0.15 S
Grade G3000 (F10006)	Fe (bal), 3.1–3.4 C, 2.3–1.9 Si, 0.75 Mn, 0.10 P, 0.15 S
Grade G4000 (F10008)	Fe (bal), 3.0–3.3 C, 2.1–1.8 Si, 0.85 Mn, 0.07 P, 0.15 S
Grade 60-40-18 (F32800)	Fe (bal), 3.4–4.0 C, 2.0–2.8 Si, 0–1.0 Ni, 0.05 Mg
Grade 80-55-06 (F33800)	Fe (bal), 3.3–3.8 C, 2.0–3.0 Si, 0–1.0 Ni, 0.05 Mg
Grade 120-90-02 (F36200)	Fe (bal), 3.4–3.8 C, 2.0–2.8 Si, 0–2.5 Ni, 0–1.0 Mo, 0.05 Mg

Table B.10 (*Continued*)

Alloy (UNS Designation)	Composition (wt%)
ALUMINUM ALLOYS	
1100 (A91100)	99.00 Al (min), 0.20 Cu (max)
2024 (A92024)	90.75 Al (min), 4.4 Cu, 0.6 Mn, 1.5 Mg
6061 (A96061)	95.85 Al (min), 1.0 Mg, 0.6 Si, 0.30 Cu, 0.20 Cr
7075 (A97075)	87.2 Al (min), 5.6 Zn, 2.5 Mg, 1.6 Cu, 0.23 Cr
356.0 (A03560)	90.1 Al (min), 7.0 Si, 0.3 Mg
COPPER ALLOYS	
(C11000)	99.90 Cu (min), 0.04 O (max)
(C17200)	96.7 Cu (min), 1.9 Be, 0.20 Co
(C26000)	Zn (bal), 70 Cu, 0.07 Pb, 0.05 Fe (max)
(C36000)	60.0 Cu (min), 35.5 Zn, 3.0 Pb
(C71500)	63.75 Cu (min), 30.0 Ni
(C93200)	81.0 Cu (min), 7.0 Sn, 7.0 Pb, 3.0 Zn
MAGNESIUM ALLOYS	
AZ31B (M11311)	94.4 Mg (min), 3.0 Al, 0.20 Mn (min), 1.0 Zn, 0.1 Si (max)
AZ91D (M11916)	89.0 Mg (min), 9.0 Al, 0.13 Mn (min), 0.7 Zn, 0.1 Si (max)
TITANIUM ALLOYS	
Commercial, grade 1 (R50250)	99.5 Ti (min)
Ti-5Al-2.5Sn (R54520)	90.2 Ti (min), 5.0 Al, 2.5 Sn
Ti-6Al-4V (R56400)	87.7 Ti (min), 6.0 Al, 4.0 V
MISCELLANEOUS ALLOYS	
Nickel 200	99.0 Ni (min)
Inconel 625	58.0 Ni (min), 21.5 Cr, 9.0 Mo, 5.0 Fe, 3.65 Nb + Ta, 1.0 Co
Monel 400	63.0 Ni (min), 31.0 Cu, 2.5 Fe, 0.2 Mn, 0.3 C, 0.5 Si
Haynes alloy 25	49.4 Co (min), 20 Cr, 15 W, 10 Ni, 3 Fe (max), 0.10 C, 1.5 Mn
Invar (K93601)	64 Fe, 36 Ni
Super invar	63 Fe, 32 Ni, 5 Co
Kovar	54 Fe, 29 Ni, 17 Co
Chemical lead (L51120)	99.90 Pb (min)
Antimonial lead, 6% (L53105)	94 Pb, 6 Sb
Tin (commercially pure) (ASTM B339A)	98.85 Pb (min)
Lead–Tin solder (60Sn-40Pb) (ASTM B32 grade 60)	60 Sn, 40 Pb
Zinc (commercially pure) (Z21210)	99.9 Zn (min), 0.10 Pb (max)
Zirconium, reactor grade 702 (R60702)	99.2 Zr + Hf (min), 4.5 Hf (max), 0.2 Fe + Cr

Sources: *ASM Handbooks, Volumes 1 and 2,* ASM International, Materials Park, OH.

Appendix C / Costs and Relative Costs for Selected Engineering Materials

This appendix contains price information for the same set of materials for which the properties are included in Appendix B. The collection of valid cost data for materials is an extremely difficult task, which explains the dearth of materials-pricing information in the literature. One reason for this is that there are three pricing tiers: manufacturer, distributor, and retail. Under most circumstances, we have cited distributor prices. For some materials (e.g., specialized ceramics such as silicon carbide and silicon nitride), it was necessary to use manufacturer's prices. In addition, there may be significant variation in the cost for a specific material. There are several reasons for this. First, each vendor has its own pricing scheme. Furthermore, cost will depend on quantity of material purchased and, in addition, how it was processed or treated. We have endeavored to collect data for relatively large orders—i.e., quantities on the order of 900 kg (2000 lb$_m$) for materials that are ordinarily sold in bulk lots—and, also, for common shapes/treatments. When possible, we obtained price quotes from at least three distributors/manufacturers.

This pricing information was collected between May and August of 1998. Cost data are in U.S. dollars per kilogram; in addition, these data are expressed as both price ranges and single-price values. The absence of a price range (i.e., when a single value is cited) means that either the variation is small, or that, on the basis of limited data, it is not possible to identify a range of prices. Furthermore, inasmuch as material prices change over time, it was decided to use a relative cost index; this index represents the per-unit mass cost (or average per-unit mass cost) of a material divided by the average per-unit mass cost of a common engineering material—A36 plain carbon steel. Although the price of a specific material will vary over time, the price ratio between that material and another will, most likely, change more slowly.

Material/Condition	Cost ($US/kg)	Relative Cost
PLAIN CARBON AND LOW ALLOY STEELS		
Steel alloy A36		
• Plate, hot rolled	0.50–0.90	1.00
• Angle bar, hot rolled	1.15	1.6
Steel alloy 1020		
• Plate, hot rolled	0.50–0.60	0.8
• Plate, cold rolled	0.85–1.45	1.6
Steel alloy 1040		
• Plate, hot rolled	0.75–0.85	1.1
• Plate, cold rolled	1.30	1.9

Material/Condition	Cost ($US/kg)	Relative Cost
Steel alloy 4140		
• Bar, normalized	1.75–1.95	2.6
• H grade (round), normalized	2.85–3.05	4.2
Steel alloy 4340		
• Bar, annealed	2.45	3.5
• Bar, normalized	3.30	4.7
STAINLESS STEELS		
Stainless alloy 304		
• Plate, hot finished and annealed	2.15–3.50	4.0
Stainless alloy 316		
• Plate, hot finished and annealed	3.00–4.40	5.3
• Round, cold drawn and annealed	6.20	8.9
Stainless alloy 440A		
• Plate, annealed	4.40–5.00	6.7
Stainless alloy 17-7PH		
• Plate, cold rolled	6.85–10.00	12.0
CAST IRONS		
Gray irons (all grades)		
• High production	1.20–1.50	1.9
• Low production	3.30	4.7
Ductile irons (all grades)		
• High production	1.45–1.85	2.4
• Low production	3.30–5.00	5.9
ALUMINUM ALLOYS		
Alloy 1100		
• Sheet, annealed	7.25–10.00	12.3
Alloy 2024		
• Sheet, T3 temper	8.80–11.00	14.1
• Bar, T351 temper	11.35	16.2
Alloy 6061		
• Sheet, T6 temper	4.40–6.20	7.6
• Bar, T651 temper	6.10	8.7
Alloy 7075		
• Sheet, T6 temper	9.00–9.70	13.4
Alloy 356		
• As cast, high production	4.40–6.60	7.9
• As cast, custom pieces	11.00	15.7
• T6 temper, custom pieces	11.65	16.6
COPPER ALLOYS		
Alloy C11000 (electrolytic tough pitch), sheet	4.00–7.00	7.9
Alloy C17200 (beryllium–copper), sheet	25.00–47.00	51.4
Alloy C26000 (cartridge brass), sheet	3.50–4.85	6.0
Alloy C36000 (free-cutting brass), sheet, rod	3.20–4.00	5.1
Alloy C71500 (copper–nickel, 30%), sheet	8.50–9.50	12.9
Alloy C93200 (bearing bronze)		
• Bar	4.50–6.50	7.9
• As cast, custom piece	12.20	17.4

Material/Condition	Cost ($US/kg)	Relative Cost
MAGNESIUM ALLOYS		
Alloy AZ31B		
• Sheet (rolled)	11.00	15.7
• Extruded	8.80	12.6
Alloy AZ91D (as cast)	3.80	5.4
TITANIUM ALLOYS		
Commercially pure		
• ASTM grade 1, annealed	28.00–65.00	66.4
Alloy Ti-5Al-2.5V	90.00–130.00	157
Alloy Ti-6Al-4V	55.00–130.00	132
PRECIOUS METALS		
Gold, bullion	9,500–10,250	14,100
Platinum, bullion	11,400–14,400	18,400
Silver, bullion	170–210	271
REFRACTORY METALS		
Molybdenum		
• Commercially pure, sheet and rod	85.00–115.00	143
Tantalum		
• Commercially pure, sheet and rod	390–440	593
Tungsten		
• Commercially pure, sheet	77.50	111
• Commercially pure, rod ($1/2$–$3/8$ in. dia.)	97.00–135.00	166
MISCELLANEOUS NONFERROUS ALLOYS		
Nickel 200	19.00–25.00	31.4
Inconel 625	20.00–29.00	35.0
Monel 400	15.50–16.50	22.9
Haynes alloy 25	85.50–103.50	135
Invar	17.25–19.75	26.4
Super invar	22.00–33.00	39.3
Kovar	30.75–39.75	50.4
Chemical lead		
• Ingot	1.20	1.7
• Plate	1.55–1.95	2.5
Antimonial lead (6%)		
• Ingot	1.50	2.1
• Plate	2.00–2.70	3.4
Tin, commercial purity (99.9+%), ingot	6.85–8.85	11.2
Solder (60Sn-40Pb), bar	5.50–7.50	9.3
Zinc, commercial purity		
• Ingot	1.20	1.7
• Anode	1.65–2.45	2.9
Zirconium, reactor grade 702 (plate)	44.00–48.50	66.1
GRAPHITE, CERAMICS, AND SEMICONDUCTING MATERIALS		
Aluminum oxide		
• Calcined powder, 99.8% pure, particle size between 0.4 and 5 μm	1.40–1.60	2.1
• Ball grinding media, 99% pure, $1/4$ in. dia.	28.65	41
• Ball grinding media, 96% pure, $1/4$ in. dia.	29.75	42.5
• Ball grinding media, 90% pure, $1/4$ in. dia.	15.20	21.7

Material/Condition	Cost ($US/kg)	Relative Cost
Concrete, mixed	0.04	0.06
Diamond		
• Natural, $^1/_3$ carat, industrial grade	36,000–90,000	90,000
• Synthetic, 30–40 mesh, industrial grade	18,750	27,000
• Natural, powder, 45 μm, polishing abrasive	5,000	7,100
Gallium arsenide		
• Mechanical grade, 75 mm dia. wafers, ~625 μm thick	1650–2700	3100
• Prime grade, 75 mm dia. wafers, ~625 μm thick	8500–10,000	13,200
Glass, borosilicate (Pyrex), plate	8.50–17.00	18.2
Glass, soda–lime, plate	1.75–2.35	2.9
Glass ceramic (Pyroceram), plate	12.25–19.25	22.5
Graphite		
• Powder, synthetic, 99% pure, particle size ~10 μm	5.00	7.1
• Powder, synthetic, 99.7% pure, particle size ~10 μm	7.50	10.7
• Extruded, high purity, fine (<0.75 mm) particle size	6.00–7.00	9.3
• Isostatically pressed parts, high purity, ~20 μm particle size	15.00–25.50	29
Silica, fused, plate	315–395	500
Silicon		
• Test grade, undoped, 100 mm dia. wafers, ~425 μm thick	900–2000	2070
• Prime grade, undoped, 100 mm dia. wafers, ~425 μm thick	2075–2525	3300
Silicon carbide		
• α-phase sinterable powder, particle size between 1 and 10 μm	22.00–58.00	57.1
• α-phase, polishing abrasive	4.50–21.50	18.6
• β-phase sinterable powder, particle size between 1 and 10 μm	40.00–100.00	100
• β-phase, polishing abrasive, 1200 to 400 mesh	8.00–22.00	21.4
• α-phase ball grinding media, $^1/_4$ in. dia., sintered	250.00	360
Silicon nitride		
• Sinterable powder, submicron particle size	100.00	143
• Balls, unfinished, 0.25 in. to 0.50 in. in diameter, hot isostatic pressed	875–1100	1400
• Balls, finished ground, 0.25 in. to 0.50 in. diameter, hot isostatic pressed	2000–4000	4300
Zirconia, partially stabilized (3 mol% Y_2O_3)		
• Sinterable powder, submicron particle size	45.00–50.00	68
• Sinterable powder, particle size greater than a micron	22.00–33.00	39.3
• Ball grinding media, 15 mm dia., sintered	125.00–175.00	215

Material/Condition	Cost ($US/kg)	Relative Cost
POLYMERS		
Butadiene-acrylonitrile (nitrile) rubber		
• Raw and unprocessed	2.90	4.1
• Extruded sheet ($1/4$ to $1/8$ in. thick)	9.90–10.50	14.6
• Calendered sheet ($1/4$ to $1/8$ in. thick)	8.40	12.0
Styrene-butadiene (SBR) rubber		
• Raw and unprocessed	1.20	1.7
• Extruded sheet ($1/4$ to $1/8$ in. thick)	7.60–12.20	14.1
• Calendered sheet ($1/4$ to $1/8$ in. thick)	6.80	9.7
Silicone rubber		
• Raw and unprocessed	5.50	7.9
• Extruded sheet ($1/4$ to $1/8$ in. thick)	12.60–26.20	27.7
• Calendered sheet ($1/4$ to $1/8$ in. thick)	31.50–38.50	50.0
Epoxy resin, raw form	3.00–4.00	5.0
Nylon 6,6		
• Raw form	4.40–6.00	7.4
• Extruded	9.40	13.4
Phenolic resin, raw form	6.50–12.00	13.2
Polybutylene terephthalate (PBT)		
• Raw form	4.00	5.7
• Sheet	9.75	13.9
Polycarbonate (PC)		
• Raw form	4.85–5.30	7.3
• Sheet	7.00–10.00	12.1
Polyester (thermoset), raw form	1.50–4.40	4.2
Polyetheretherketone (PEEK), raw form	90.00–110.00	143
Polyethylene		
• Low density (LDPE), raw form	1.20–1.35	1.8
• High density (HDPE), raw form	1.00–1.70	1.9
• Ultrahigh molecular weight (UHMWPE), raw form	3.00–8.50	8.2
Polyethylene terephthalate (PET)		
• Raw form	1.90–2.10	2.9
• Sheet	3.30–7.70	7.9
Polymethyl methacrylate (PMMA)		
• Raw form	2.40	3.4
• Calendered sheet	4.20	6.0
• Cell cast	5.85	8.4
Polypropylene (PP), raw form	0.85–1.65	1.8
Polystyrene (PS), raw form	1.00–1.10	1.5
Polytetrafluoroethylene (PTFE)		
• Raw form	20.00–26.50	33.2
• Sheet	38.00	54
Polyvinyl chloride (PVC), raw form	1.40–2.80	3.0
FIBER MATERIALS		
Aramid (Kevlar 49), continuous	31.00	44.3
Carbon (PAN precursor), continuous		
• Standard modulus	31.50–41.50	52.1
• Intermediate modulus	70.00–105.00	125
• High modulus	175.00–225.00	285
E glass, continuous	1.90–3.30	3.7

Material/Condition	Cost ($US/kg)	Relative Cost
COMPOSITE MATERIALS		
Aramid (Kevlar 49) continuous-fiber, epoxy prepreg	55.00–62.00	84
Carbon continuous-fiber, epoxy prepreg		
• Standard modulus	40.00–60.00	71
• Intermediate modulus	100.00–130.00	164
• High modulus	200.00–275.00	340
E-glass continuous-fiber, epoxy prepreg	22.00	31.4
Woods		
Douglas fir	0.54–0.60	0.8
Red oak	2.55–3.35	4.2

Chemical Name	Mer Chemical Structure
Epoxy (diglycidyl ether of bisphenol A, DGEPA)	
Melamine-formaldehyde (melamine)	
Phenol-formaldehyde (phenolic)	
Polyacrylonitrile (PAN)	
Polyamide-imide (PAI)	

Chemical Name	Mer Chemical Structure
Polybutadiene	
Polybutylene terephthalate (PBT)	
Polycarbonate (PC)	
Polychloroprene	
Polychlorotrifluoroethylene	
Polydimethyl siloxane (silicone rubber)	
Polyetheretherketone (PEEK)	
Polyethylene (PE)	
Polyethylene terephthalate (PET)	

Chemical Name	*Mer Chemical Structure*
Polyhexamethylene adipamide (nylon 6,6)	
Polyimide	
Polyisobutylene	
cis-Polyisoprene (natural rubber)	
Polymethyl methacrylate (PMMA)	
Polyphenylene oxide (PPO)	
Polyphenylene sulfide (PPS)	
Polyparaphenylene terephthalamide (aramid)	

Chemical Name	Mer Chemical Structure
Polypropylene (PP)	
Polystyrene (PS)	
Polytetrafluoroethylene (PTFE)	
Polyvinyl acetate (PVAc)	
Polyvinyl alcohol (PVA)	
Polyvinyl chloride (PVC)	
Polyvinyl fluoride (PVF)	
Polyvinylidene chloride (PVDC)	
Polyvinylidene fluoride (PVDF)	

Polymer	Glass Transition Temperature [°C (°F)]	Melting Temperature [°C (°F)]
Aramid	375 (705)	~640 (~1185)
Polyimide (thermoplastic)	280–330 (535–625)	a
Polyamide-imide	277–289 (530–550)	a
Polycarbonate	150 (300)	265 (510)
Polyetheretherketone	143 (290)	334 (635)
Polyacrylonitrile	104 (220)	317 (600)
Polystyrene		
• Atactic	100 (212)	a
• Isotactic	100 (212)	240 (465)
Polybutylene terephthalate	—	220–267 (428–513)
Polyvinyl chloride	87 (190)	212 (415)
Polyphenylene sulfide	85 (185)	285 (545)
Polyethylene terephthalate	69 (155)	265 (510)
Nylon 6,6	57 (135)	265 (509)
Polymethyl methacrylate		
• Syndiotactic	3 (35)	105 (220)
• Isotactic	3 (35)	45 (115)
Polypropylene		
• Isotactic	−10 (15)	175 (347)
• Atactic	−18 (0)	175 (347)
Polyvinylidene chloride	−17 (1)	198 (390)
• Atactic	−18 (0)	175 (347)
Polyvinyl fluoride	−20 (−5)	200 (390)
Polyvinylidene fluoride	−35 (−30)	—
Polychloroprene (chloroprene rubber or neoprene)	−50 (−60)	80 (175)
Polyisobutylene	−70 (−95)	128 (260)
cis-Polyisoprene	−73 (−100)	28 (80)
Polybutadiene		
• Syndiotactic	−90 (−130)	154 (310)
• Isotactic	−90 (−130)	120 (250)
High density polyethylene	−90 (−130)	137 (279)
Polytetrafluoroethylene	−97 (−140)	327 (620)
Low density polyethylene	−110 (−165)	115 (240)
Polydimethylsiloxane (silicone rubber)	−123 (−190)	−54 (−65)

a These polymers normally exist at least 95% noncrystalline.

Glossary

A

Abrasive. A hard and wear-resistant material (commonly a ceramic) that is used to wear, grind, or cut away other material.

Absorption. The optical phenomenon whereby the energy of a photon of light is assimilated within a substance, normally by electronic polarization or by an electron excitation event.

Acceptor level. For a semiconductor or insulator, an energy level lying within yet near the bottom of the energy band gap, which may accept electrons from the valence band, leaving behind holes. The level is normally introduced by an impurity atom.

Activation energy (Q). The energy required to initiate a reaction, such as diffusion.

Activation polarization. The condition wherein the rate of an electrochemical reaction is controlled by the one slowest step in a sequence of steps that occur in series.

Addition (or chain reaction) polymerization. The process by which bifunctional monomer units are attached one at a time, in chainlike fashion, to form a linear polymer macromolecule.

Adhesive. A substance that bonds together the surfaces of two other materials (termed adherends).

Age hardening. See **Precipitation hardening.**

Allotropy. The possibility of existence of two or more different crystal structures for a substance (generally an elemental solid).

Alloy. A metallic substance that is composed of two or more elements.

Alloy steel. A ferrous (or iron-based) alloy that contains appreciable concentrations of alloying elements (other than C and residual amounts of Mn, Si, S, and P). These alloying elements are usually added to improve mechanical and corrosion resistance properties.

Alternating copolymer. A copolymer in which two different mer units alternate positions along the molecular chain.

Amorphous. Having a noncrystalline structure.

Anelastic deformation. Time-dependent elastic (nonpermanent) deformation.

Anion. A negatively charged, nonmetallic ion.

Anisotropic. Exhibiting different values of a property in different crystallographic directions.

Annealing. A generic term used to denote a heat treatment wherein the microstructure and, consequently, the properties of a material are altered. "Annealing" frequently refers to a heat treatment whereby a previously cold-worked metal is softened by allowing it to recrystallize.

Annealing point (glass). That temperature at which residual stresses in a glass are eliminated within about 15 min; this corresponds to a glass viscosity of about 10^{12} Pa-s (10^{13} P).

Anode. The electrode in an electrochemical cell or galvanic couple that experiences oxidation, or gives up electrons.

Antiferromagnetism. A phenomenon observed in some materials (e.g., MnO); complete magnetic moment cancellation occurs as a result of antiparallel coupling of adjacent atoms or ions. The macroscopic solid possesses no net magnetic moment.

Artificial aging. For precipitation hardening, aging above room temperature.

Atactic. A type of polymer chain configuration wherein side groups are randomly positioned on one side of the chain or the other.

Athermal transformation. A reaction that is not thermally activated, and usually diffusionless, as with the martensitic transformation. Normally, the transformation takes place with great speed (i.e., is independent of time), and the extent of reaction depends on temperature.

Atomic mass unit (amu). A measure of atomic mass; one twelfth of the mass of an atom of C^{12}.

Atomic number (Z). For a chemical element, the number of protons within the atomic nucleus.

Atomic packing factor (APF). The fraction of the volume of a unit cell that is occupied by "hard sphere" atoms or ions.

Atomic vibration. The vibration of an atom about its normal position in a substance.

Atomic weight (A). The weighted average of the atomic masses of an atom's naturally occurring isotopes. It may be expressed in terms of atomic mass units (on an atomic basis), or the mass per mole of atoms.

Atom percent (at%). Concentration specification on the basis of the number of moles (or atoms) of a particular element relative to the total number of moles (or atoms) of all elements within an alloy.

Austenite. Face-centered cubic iron; also iron and steel alloys that have the FCC crystal structure.

Austenitizing. Forming austenite by heating a ferrous alloy above its upper critical temperature—to within the austenite phase region from the phase diagram.

B

Bainite. An austenitic transformation product found in some steels and cast irons. It forms at temperatures between those at which pearlite and martensite transformations occur. The microstructure consists of α-ferrite and a fine dispersion of cementite.

Band gap energy (E_g). For semiconductors and insulators, the energies that lie between the valence and conduction bands; for intrinsic materials, electrons are forbidden to have energies within this range.

Bifunctional. Designating monomer units that have two active bonding positions.

Block copolymer. A linear copolymer in which identical mer units are clustered in blocks along the molecular chain.

Body-centered cubic (BCC). A common crystal structure found in some elemental metals. Within the cubic unit cell, atoms are located at corner and cell center positions.

Bohr atomic model. An early atomic model, in which electrons are assumed to revolve around the nucleus in discrete orbitals.

Bohr magneton (μ_B). The most fundamental magnetic moment, of magnitude 9.27×10^{-24} A-m^2.

Boltzmann's constant (k). A thermal energy constant having the value of 1.38×10^{-23} J/atom-K (8.62×10^{-5} eV/atom-K). See also **Gas constant.**

Bonding energy. The energy required to separate two atoms that are chemically bonded to each other. It may be expressed on a per-atom basis, or per mole of atoms.

Bragg's law. A relationship {(Equation 3.10)} which stipulates the condition for diffraction by a set of crystallographic planes.

Branched polymer. A polymer having a molecular structure of secondary chains that extend from the primary main chains.

Brass. A copper-rich copper–zinc alloy.

Brazing. A metal joining technique that uses a molten filler metal alloy having a melting temperature greater than about 425°C (800°F).

Brittle fracture. Fracture that occurs by rapid crack propagation and without appreciable macroscopic deformation.

Bronze. A copper-rich copper–tin alloy; aluminum, silicon, and nickel bronzes are also possible.

Burgers vector (b). A vector that denotes the magnitude and direction of lattice distortion associated with a dislocation.

C

Calcination. A high-temperature reaction whereby one solid material dissociates to form a gas and another solid. It is one step in the production of cement.

Capacitance (C). The charge-storing ability of a capacitor, defined as the magnitude of charge stored on either plate divided by the applied voltage.

Carbon-carbon composite. A composite that is composed of continuous fibers of carbon that are imbedded in a carbon matrix. The matrix was originally a polymer resin that was subsequently pyrolyzed to form carbon.

Carburizing. The process by which the surface carbon concentration of a ferrous alloy is increased by diffusion from the surrounding environment.

Case hardening. Hardening of the outer surface (or "case") of a steel component by a carburizing or nitriding process; used to improve wear and fatigue resistance.

Cast iron. Generically, a ferrous alloy, the carbon content of which is greater than the maximum solubility in austenite at the eutectic temperature. Most commercial cast irons contain between 3.0 and 4.5 wt% C, and between 1 and 3 wt% Si.

Cathode. The electrode in an electrochemical cell or galvanic couple at which a reduction reaction occurs; thus the electrode that receives electrons from an external circuit.

Cathodic protection. A means of corrosion prevention whereby electrons are supplied to the structure to be protected from an external source such as another more reactive metal or a dc power supply.

Cation. A positively charged metallic ion.

Cement. A substance (often a ceramic) that by chemical reaction binds particulate aggregates into a cohesive structure. With hydraulic cements the chemical reaction is one of hydration, involving water.

Cementite. Iron carbide (Fe$_3$C).

Ceramic. A compound of metallic and nonmetallic elements, for which the interatomic bonding is predominantly ionic.

Ceramic-matrix composite (CMC). A composite for which both matrix and dispersed phases are ceramic materials. The dispersed phase is normally added to improve fracture toughness.

Cermet. A composite material consisting of a combination of ceramic and metallic materials. The most common cermets are the cemented carbides, composed of an extremely hard ceramic (e.g., WC, TiC), bonded together by a ductile metal such as cobalt or nickel.

Chain-folded model. For crystalline polymers, a model that describes the structure of platelet crystallites. Molecular alignment is accomplished by chain folding that occurs at the crystallite faces.

Charpy test. One of two tests (see also **Izod test**) that may be used to measure the impact energy or notch toughness of a standard notched specimen. An impact blow is imparted to the specimen by means of a weighted pendulum.

Cis. For polymers, a prefix denoting a type of molecular structure. For some unsaturated carbon chain atoms within a mer unit, a side atom or group may be situated on one side of the chain or directly opposite at a 180° rotation position. In a cis structure, two such side groups within the same mer reside on the same side (e.g., *cis*-isoprene).

Coarse pearlite. Pearlite for which the alternating ferrite and cementite layers are relatively thick.

Coercivity (or coercive field, H_c). The applied magnetic field necessary to reduce to zero the magnetic flux density of a magnetized ferromagnetic or ferrimagnetic material.

Cold working. The plastic deformation of a metal at a temperature below that at which it recrystallizes.

Color. Visual perception that is stimulated by the combination of wavelengths of light that are transmitted to the eye.

Colorant. An additive that imparts a specific color to a polymer.

Component. A chemical constituent (element or compound) of an alloy, which may be used to specify its composition.

Composition (C_i). The relative content of a particular element or constituent (i) within an alloy, usually expressed in weight percent or atom percent.

Concentration. See **Composition**.

Concentration gradient (dC/dx). The slope of the concentration profile at a specific position.

Concentration polarization. The condition wherein the rate of an electrochemical reaction is limited by the rate of diffusion in the solution.

Concentration profile. The curve that results when the concentration of a chemical species is plotted versus position in a material.

Concrete. A composite material consisting of aggregate particles bound together in a solid body by a cement.

Condensation (or step reaction) polymerization. The formation of polymer macromolecules by an intermolecular reaction involving at least two monomer species, usually with the production of a by-product of low molecular weight, such as water.

Conduction band. For electrical insulators and semiconductors, the lowest lying electron energy band that is empty of electrons at 0 K. Conduction electrons are those that have been excited to states within this band.

Conductivity, electrical (σ). The proportionality constant between current density and applied electric field; also a measure of the ease with which a material is capable of conducting an electric current.

Congruent transformation. A transformation of one phase to another of the same composition.

Continuous cooling transformation (*CCT*) diagram. A plot of temperature versus the logarithm of time for a steel alloy of definite composition. Used to indicate when transformations occur as the initially austenitized material is continuously cooled at a specified rate; in addition, the final microstructure and mechanical characteristics may be predicted.

Coordination number. The number of atomic or ionic nearest neighbors.

Copolymer. A polymer that consists of two or more dissimilar mer units in combination along its molecular chains.

Corrosion. Deteriorative loss of a metal as a result of dissolution environmental reactions.

Corrosion fatigue. A type of failure that results from the simultaneous action of a cyclic stress and chemical attack.

Corrosion penetration rate (CPR). Thickness loss of material per unit of time as a result of corrosion; usually expressed in terms of mils per year or millimeters per year.

Coulombic force. A force between charged particles such as ions; the force is attractive when the particles are of opposite charge.

Covalent bond. A primary interatomic bond that is formed by the sharing of electrons between neighboring atoms.

Creep. The time-dependent permanent deformation that occurs under stress; for most materials it is important only at elevated temperatures.

Crevice corrosion. A form of corrosion that occurs within narrow crevices and under deposits of dirt or corrosion products (i.e., in regions of localized depletion of oxygen in the solution).

Critical resolved shear stress (τ_{crss}). That shear stress, resolved within a slip plane and direction, which is required to initiate slip.

Crosslinked polymer. A polymer in which adjacent linear molecular chains are joined at various positions by covalent bonds.

Crystalline. The state of a solid material characterized by a periodic and repeating three-dimensional array of atoms, ions, or molecules.

Crystallinity. For polymers, the state wherein a periodic and repeating atomic arrangement is achieved by molecular chain alignment.

Crystallite. A region within a crystalline polymer in which all the molecular chains are ordered and aligned.

Crystal structure. For crystalline materials, the manner in which atoms or ions are arrayed in space. It is defined in terms of the unit cell geometry and the atom positions within the unit cell.

Crystal system. A scheme by which crystal structures are classified according to unit cell geometry. This geometry is specified in terms of the relationships between edge lengths and interaxial angles. There are seven different crystal systems.

Curie temperature (T_c). That temperature above which a ferromagnetic or ferrimagnetic material becomes paramagnetic.

D

Defect structure. Relating to the kinds and concentrations of vacancies and interstitials in a ceramic compound.

Degradation. A term used to denote the deteriorative processes that occur with polymeric materials. These processes include swelling, dissolution, and chain scission.

Degree of polymerization. The average number of mer units per polymer chain molecule.

Design stress (σ_d). Product of the calculated stress level (on the basis of estimated maximum load) and a design factor (which has a value greater than unity). Used to protect against unanticipated failure.

Devitrification. The process in which a glass (noncrystalline or vitreous solid) transforms to a crystalline solid.

Diamagnetism. A weak form of induced or nonpermanent magnetism for which the magnetic susceptibility is negative.

Dielectric. Any material that is electrically insulating.

Dielectric constant (ϵ_r). The ratio of the permittivity of a medium to that of a vacuum. Often called the relative dielectric constant or relative permittivity.

Dielectric displacement (D). The magnitude of charge per unit area of capacitor plate.

Dielectric (breakdown) strength. The magnitude of an electric field necessary to cause significant current passage through a dielectric material.

Diffraction (x-ray). Constructive interference of x-ray beams that are scattered by atoms of a crystal.

Diffusion. Mass transport by atomic motion.

Diffusion coefficient (D). The constant of proportionality between the diffusion flux and the concentration gradient in Fick's first law. Its magnitude is indicative of the rate of atomic diffusion.

Diffusion flux (J). The quantity of mass diffusing through and perpendicular to a unit cross-sectional area of material per unit time.

Diode. An electronic device that rectifies an electrical current—i.e., allows current flow in one direction only.

Dipole (electric). A pair of equal yet opposite electrical charges that are separated by a small distance.

Dislocation. A linear crystalline defect around which there is atomic misalignment. Plastic deformation corresponds to the motion of dislocations in response to an applied shear stress. Edge, screw, and mixed dislocations are possible.

Dislocation density. The total dislocation length per unit volume of material; alternately, the number of dislocations that intersect a unit area of a random surface section.

Dislocation line. The line that extends along the end of the extra half-plane of atoms for an edge dislocation, and along the center of the spiral of a screw dislocation.

Dispersed phase. For composites and some two-phase alloys, the discontinuous phase that is surrounded by the matrix phase.

Dispersion strengthening. A means of strengthening materials wherein very small particles (usually less than 0.1 μm) of a hard yet inert phase are uniformly dispersed within a load-bearing matrix phase.

Domain. A volume region of a ferromagnetic or ferrimagnetic material in which all atomic or ionic magnetic moments are aligned in the same direction.

Donor level. For a semiconductor or insulator, an energy level lying within yet near the top of the energy band gap, and from which electrons may be excited into the conduction band. It is normally introduced by an impurity atom.

Doping. The intentional alloying of semiconducting materials with controlled concentrations of donor or acceptor impurities.

Drawing (metals). A forming technique used to fabricate metal wire and tubing. Deformation is accomplished by pulling the material through a die by means of a tensile force applied on the exit side.

Drawing (polymers). A deformation technique wherein polymer fibers are strengthened by elongation.

Driving force. The impetus behind a reaction, such as diffusion, grain growth, or a phase transformation. Usually attendant to the reaction is a reduction in some type of energy (e.g., free energy).

Ductile fracture. A mode of fracture that is attended by extensive gross plastic deformation.

Ductile iron. A cast iron that is alloyed with silicon and a small concentration of magnesium and/or cerium and in which the free graphite exists in nodular form. Sometimes called nodular iron.

Ductile-to-brittle transition. The transition from ductile to brittle behavior with a decrease in temperature exhibited by BCC alloys; the temperature range over which the transition occurs is determined by Charpy and Izod impact tests.

Ductility. A measure of a material's ability to undergo appreciable plastic deformation before fracture; it may be expressed as percent elongation (%EL) or percent reduction in area (%RA) from a tensile test.

E

Edge dislocation. A linear crystalline defect associated with the lattice distortion produced in the vicinity of the end of an extra half-plane of atoms within a crystal. The Burgers vector is perpendicular to the dislocation line.

Elastic deformation. Deformation that is nonpermanent, that is, totally recovered upon release of an applied stress.

Elastic recovery. Nonpermanent deformation that is recovered or regained upon the release of a mechanical stress.

Elastomer. A polymeric material that may experience large and reversible elastic deformations.

Electrical conductivity. See **Conductivity, electrical.**

Electric dipole. See **Dipole (electric).**

Electric field ($\mathscr{E}$). The gradient of voltage.

Electroluminescence. The emission of visible light by a p–n junction across which a forward-biased voltage is applied.

Electrolyte. A solution through which an electric current may be carried by the motion of ions.

Electromotive force (emf) series. A ranking of metallic elements according to their standard electrochemical cell potentials.

Electron configuration. For an atom, the manner in which possible electron states are filled with electrons.

Electronegative. For an atom, having a tendency to accept valence electrons. Also, a term used to describe nonmetallic elements.

Electron energy band. A series of electron energy states that are very closely spaced with respect to energy.

Electroneutrality. The state of having exactly the same numbers of positive and negative electrical charges (ionic and electronic), that is, of being electrically neutral.

Electron state (level). One of a set of discrete, quantized energies that are allowed for electrons. In the atomic case each state is specified by four quantum numbers.

Electron volt (eV). A convenient unit of energy for atomic and subatomic systems. It is equivalent to the energy acquired by an electron when it falls through an electric potential of 1 volt.

Electropositive. For an atom, having a tendency to release valence electrons. Also, a term used to describe metallic elements.

Endurance limit. See **Fatigue limit.**

Energy band gap. See **Band gap energy.**

Engineering strain. See **Strain, engineering.**

Engineering stress. See **Stress, engineering.**

Equilibrium (phase). The state of a system where the phase characteristics remain constant over indefinite time periods. At equilibrium the free energy is a minimum.

Erosion–corrosion. A form of corrosion that arises from the combined action of chemical attack and mechanical wear.

Eutectic phase. One of the two phases found in the eutectic structure.

Eutectic reaction. A reaction wherein, upon cooling, a liquid phase transforms isothermally and reversibly into two intimately mixed solid phases.

Eutectic structure. A two-phase microstructure resulting from the solidification of a liquid having the eutectic composition; the phases exist as lamellae that alternate with one another.

Eutectoid reaction. A reaction wherein, upon cooling, one solid phase transforms isothermally and reversibly into two new solid phases that are intimately mixed.

Excited state. An electron energy state, not normally occupied, to which an electron may be promoted (from a lower energy state) by the absorption of some type of energy (e.g., heat, radiative).

Extrinsic semiconductor. A semiconducting material for which the electrical behavior is determined by impurities.

Extrusion. A forming technique whereby a material is forced, by compression, through a die orifice.

F

Face-centered cubic (FCC). A crystal structure found in some of the common elemental metals. Within the cubic unit cell, atoms are located at all corner and face-centered positions.

Fatigue. Failure, at relatively low stress levels, of structures that are subjected to fluctuating and cyclic stresses.

Fatigue life (N_f). The total number of stress cycles that will cause a fatigue failure at some specified stress amplitude.

Fatigue limit. For fatigue, the maximum stress amplitude level below which a material can endure an essentially infinite number of stress cycles and not fail.

Fatigue strength. The maximum stress level that a material can sustain, without failing, for some specified number of cycles.

Fermi energy (E_f). For a metal, the energy corresponding to the highest filled electron state at 0 K.

Ferrimagnetism. Permanent and large magnetizations found in some ceramic materials. It results from antiparallel spin coupling and incomplete magnetic moment cancellation.

Ferrite (ceramic). Ceramic oxide materials composed of both divalent and trivalent cations (e.g., Fe^{2+} and Fe^{3+}), some of which are ferrimagnetic.

Ferrite (iron). Body-centered cubic iron; also iron and steel alloys that have the BCC crystal structure.

Ferroelectric. A dielectric material that may exhibit polarization in the absence of an electric field.

Ferromagnetism. Permanent and large magnetizations found in some metals (e.g., Fe, Ni, and Co), which result from the parallel alignment of neighboring magnetic moments.

Ferrous alloy. A metal alloy for which iron is the prime constituent.

Fiber. Any polymer, metal, or ceramic that has been drawn into a long and thin filament.

Fiber-reinforced composite. A composite in which the dispersed phase is in the form of a fiber (i.e., a filament that has a large length-to-diameter ratio).

Fiber reinforcement. Strengthening or reinforcement of a relatively weak material by embedding a strong fiber phase within the weak matrix material.

Fick's first law. The diffusion flux is proportional to the concentration gradient. This relationship is employed for steady-state diffusion situations.

Fick's second law. The time rate of change of concentration is proportional to the second derivative of concentration. This relationship is employed in nonsteady-state diffusion situations.

Filler. An inert foreign substance added to a polymer to improve or modify its properties.

Fine pearlite. Pearlite for which the alternating ferrite and cementite layers are relatively thin.

Firing. A high temperature heat treatment that increases the density and strength of a ceramic piece.

Flame retardant. A polymer additive that increases flammability resistance.

Flexural strength (σ_{fs}). Stress at fracture from a bend (or flexure) test.

Fluorescence. Luminescence that occurs for times much less than a second after an electron excitation event.

Foam. A polymer that has been made porous (or spongelike) by the incorporation of gas bubbles.

Forging. Mechanical forming of a metal by heating and hammering.

Forward bias. The conducting bias for a p–n junction rectifier such that electron flow is to the n side of the junction.

Fracture mechanics. A technique of fracture analysis used to determine the stress level at which preexisting cracks of known size will propagate, leading to fracture.

Fracture toughness (K_c). Critical value of the stress intensity factor for which crack extension occurs.

Free electron. An electron that has been excited into an energy state above the Fermi energy (or into the conduction band for semiconductors and insulators) and may participate in the electrical conduction process.

Free energy. A thermodynamic quantity that is a function of both the internal energy and entropy (or randomness) of a system. At equilibrium, the free energy is at a minimum.

Frenkel defect. In an ionic solid, a cation–vacancy and cation–interstitial pair.

Full annealing. For ferrous alloys, austenitizing, followed by cooling slowly to room temperature.

G

Galvanic corrosion. The preferential corrosion of the more chemically active of two metals that are electrically coupled and exposed to an electrolyte.

Galvanic series. A ranking of metals and alloys as to their relative electrochemical reactivity in seawater.

Gas constant (R). Boltzmann's constant per mole of atoms. $R = 8.31$ J/mol-K (1.987 cal/mol-K).

Gibbs phase rule. For a system at equilibrium, an equation {(Equation 10.16)} that expresses the relationship between the number of phases present and the number of externally controllable variables.

Glass–ceramic. A fine-grained crystalline ceramic material that was formed as a glass and subsequently devitrified (or crystallized).

Glass transition temperature (T_g). That temperature at which, upon cooling, a noncrystalline ceramic or polymer transforms from a supercooled liquid to a rigid glass.

Graft copolymer. A copolymer wherein homopolymer side branches of one mer type are grafted to homopolymer main chains of a different mer.

Grain. An individual crystal in a polycrystalline metal or ceramic.

Grain boundary. The interface separating two adjoining grains having different crystallographic orientations.

Grain growth. The increase in average grain size of a polycrystalline material; for most materials, an elevated-temperature heat treatment is necessary.

Grain size. The average grain diameter as determined from a random cross section.

Gray cast iron. A cast iron alloyed with silicon in which the graphite exists in the form of flakes. A fractured surface appears gray.

Green ceramic body. A ceramic piece, formed as a particulate aggregate, that has been dried but not fired.

Ground state. A normally filled electron energy state from which electron excitation may occur.

H

Hall effect. The phenomenon whereby a force is brought to bear on a moving electron or hole by a magnetic field that is applied perpendicular to the direction of motion. The force direction is perpendicular to both the magnetic field and the particle motion directions.

Hardenability. A measure of the depth to which a specific ferrous alloy may be hardened by the formation of martensite upon quenching from a temperature above the upper critical temperature.

Hard magnetic material. A ferrimagnetic or ferromagnetic material that has large coercive field and remanence values, normally used in permanent magnet applications.

Hardness. The measure of a material's resistance to deformation by surface indentation or by abrasion.

Heat capacity (C_p, C_v). The quantity of heat required to produce a unit temperature rise per mole of material.

Hexagonal close-packed (HCP). A crystal structure found for some metals. The HCP unit cell is of hexagonal geometry and is generated by the stacking of close-packed planes of atoms.

High polymer. A solid polymeric material having a molecular weight greater than about 10,000 g/mol.

High-strength, low-alloy (HSLA) steels. Relatively strong, low-carbon steels, with less than about 10 wt% total of alloying elements.

Hole (electron). For semiconductors and insulators, a vacant electron state in the valence band that behaves as a positive charge carrier in an electric field.

Homopolymer. A polymer having a chain structure in which all mer units are of the same type.

Hot working. Any metal forming operation that is performed above a metal's recrystallization temperature.

Hybrid composite. A composite that is fiber reinforced by two or more types of fibers (e.g., glass and carbon).

Hydrogen bond. A strong secondary interatomic bond that exists between a bound hydrogen atom (its unscreened proton) and the electrons of adjacent atoms.

Hydrogen embrittlement. The loss or reduction of ductility of a metal alloy (often steel) as a result of the diffusion of atomic hydrogen into the material.

Hydroplastic forming. The molding or shaping of clay-based ceramics that have been made plastic and pliable by adding water.

Hypereutectoid alloy. For an alloy system displaying a eutectoid, an alloy for which the concentration of solute is greater than the eutectoid composition.

Hypoeutectoid alloy. For an alloy system displaying a eutectoid, an alloy for which the concentration of solute is less than the eutectoid composition.

Hysteresis (magnetic). The irreversible magnetic flux density-versus-magnetic field strength (B-versus-H) behavior found for ferromagnetic and ferrimagnetic materials; a closed B–H loop is formed upon field reversal.

I

Impact energy (notch toughness). A measure of the energy absorbed during the fracture of a specimen of standard dimensions and geometry when subjected to very rapid (impact) loading. Charpy and Izod impact tests are used to

measure this parameter, which is important in assessing the ductile-to-brittle transition behavior of a material.

Imperfection. A deviation from perfection; normally applied to crystalline materials wherein there is a deviation from atomic/molecular order and/or continuity.

Index of refraction (n). The ratio of the velocity of light in a vacuum to the velocity in some medium.

Inhibitor. A chemical substance that, when added in relatively low concentrations, retards a chemical reaction.

Insulator (electrical). A nonmetallic material that has a filled valence band at 0 K and a relatively wide energy band gap. Consequently, the room-temperature electrical conductivity is very low, less than about 10^{-10} $(\Omega\text{-m})^{-1}$.

Integrated circuit. Thousands of electronic circuit elements (transistors, diodes, resistors, capacitors, etc.) incorporated on a very small silicon chip.

Interdiffusion. Diffusion of atoms of one metal into another metal.

Intergranular corrosion. Preferential corrosion along grain boundary regions of polycrystalline materials.

Intergranular fracture. Fracture of polycrystalline materials by crack propagation along grain boundaries.

Intermediate solid solution. A solid solution or phase having a composition range that does not extend to either of the pure components of the system.

Intermetallic compound. A compound of two metals that has a distinct chemical formula. On a phase diagram it appears as an intermediate phase that exists over a very narrow range of compositions.

Interstitial diffusion. A diffusion mechanism whereby atomic motion is from interstitial site to interstitial site.

Interstitial solid solution. A solid solution wherein relatively small

solute atoms occupy interstitial positions between the solvent or host atoms.

Intrinsic semiconductor. A semiconductor material for which the electrical behavior is characteristic of the pure material; that is, electrical conductivity depends only on temperature and the band gap energy.

Invariant point. A point on a binary phase diagram at which three phases are in equilibrium.

Ionic bond. A coulombic interatomic bond that exists between two adjacent and oppositely charged ions.

Isomerism. The phenomenon whereby two or more polymer molecules or mer units have the same composition but different structural arrangements and properties.

Isomorphous. Having the same structure. In the phase diagram sense, isomorphicity means having the same crystal structure or complete solid solubility for all compositions (see Figure 10.2a).

Isotactic. A type of polymer chain configuration wherein all side groups are positioned on the same side of the chain molecule.

Isothermal. At a constant temperature.

Isothermal transformation (T–T–T) diagram. A plot of temperature versus the logarithm of time for a steel alloy of definite composition. Used to determine when transformations begin and end for an isothermal (constant-temperature) heat treatment of a previously austenitized alloy.

Isotopes. Atoms of the same element that have different atomic masses.

Isotropic. Having identical values of a property in all crystallographic directions.

Izod test. One of two tests (see also **Charpy test**) that may be used to measure the impact energy of a standard notched specimen. An impact blow is imparted to the specimen by a weighted pendulum.

J

Jominy end-quench test. A standardized laboratory test that is used to assess the hardenability of ferrous alloys.

Junction transistor. A semiconducting device composed of appropriately biased n–p–n or p–n–p junctions, used to amplify an electrical signal.

K

Kinetics. The study of reaction rates and the factors that affect them.

L

Laminar composite. A series of two-dimensional sheets, each having a preferred high-strength direction, fastened one on top of the other at different orientations; strength in the plane of the laminate is highly isotropic.

Large-particle composite. A type of particle-reinforced composite wherein particle-matrix interactions cannot be treated on an atomic level; the particles reinforce the matrix phase.

Laser. Acronym for light amplification by stimulated emission of radiation—a source of light that is coherent.

Lattice. The regular geometrical arrangement of points in crystal space.

Lattice parameters. The combination of unit cell edge lengths and interaxial angles that defines the unit cell geometry.

Lattice strains. Slight displacements of atoms relative to their normal lattice positions, normally imposed by crystalline defects such as dislocations, and interstitial and impurity atoms.

Lever rule. Mathematical expression, such as Equation 10.1b or Equation 10.2b, whereby the relative phase amounts in a two-phase alloy at equilibrium may be computed.

Linear coefficient of thermal expansion. See **Thermal expansion coefficient, linear.**

Linear polymer. A polymer in which each molecule consists of bifunctional mer units joined end to end in a single chain.

Liquid crystal polymer (*LCP*). A group of polymeric materials having extended and rod-shaped molecules, which, structurally, do not fall within traditional liquid, amorphous, crystalline, or semicrystalline classifications. They are used in digital displays and a variety of applications in electronics and medical equipment industries.

Liquidus line. On a binary phase diagram, that line or boundary separating liquid and liquid + solid phase regions. For an alloy, the liquidus temperature is that temperature at which a solid phase first forms under conditions of equilibrium cooling.

Longitudinal direction. The lengthwise dimension. For a rod or fiber, in the direction of the long axis.

Lower critical temperature. For a steel alloy, the temperature below which, under equilibrium conditions, all austenite has transformed to ferrite and cementite phases.

Luminescence. The emission of visible light as a result of electron decay from an excited state.

M

Macromolecule. A huge molecule made up of thousands of atoms.

Magnetic field strength (*H*). The intensity of an externally applied magnetic field.

Magnetic flux density (*B*). The magnetic field produced in a substance by an external magnetic field.

Magnetic induction (*B*). See **Magnetic flux density.**

Magnetic susceptibility (χ_m). The proportionality constant between the magnetization M and the magnetic field strength H.

Magnetization (*M*). The total magnetic moment per unit volume of material. Also, a measure of the contribution to the magnetic flux by some material within an H field.

Malleable cast iron. White cast iron that has been heat treated to convert the cementite into graphite clusters; a relatively ductile cast iron.

Martensite. A metastable iron phase supersaturated in carbon that is the product of a diffusionless (athermal) transformation from austenite.

Matrix phase. The phase in a composite or two-phase alloy microstructure that is continuous or completely surrounds the other (or dispersed) phase.

Matthiessen's rule. The total electrical resistivity of a metal is equal to the sum of temperature-, impurity-, and cold work-dependent contributions.

Melting point (glass). The temperature at which the viscosity of a glass material is 10 Pa-s (100 P).

Mer. The group of atoms that constitutes a polymer chain repeat unit.

Metal. The electropositive elements and alloys based on these elements. The electron band structure of metals is characterized by a partially filled electron band.

Metallic bond. A primary interatomic bond involving the nondirectional sharing of nonlocalized valence electrons ("sea of electrons") that are mutually shared by all the atoms in the metallic solid.

Metal-matrix composite (*MMC*). A composite material which has a metal or metal alloy as the matrix phase. The dispersed phase may be particulates, fibers, or whiskers that normally are stiffer, stronger, and/or harder than the matrix.

Metastable. Nonequilibrium state that may persist for a very long time.

Microconstituent. An element of the microstructure that has an identifiable and characteristic structure. It may consist of more than one phase such as with pearlite.

Microscopy. The investigation of microstructural elements using some type of microscope.

Microstructure. The structural features of an alloy (e.g., grain and phase structure) that are subject to observation under a microscope.

Miller indices. A set of three integers (four for hexagonal) that designate crystallographic planes, as determined from reciprocals of fractional axial intercepts.

Mixed dislocation. A dislocation that has both edge and screw components.

Mobility (electron, μ_e, and hole, μ_h). The proportionality constant between the carrier drift velocity and applied electric field; also, a measure of the ease of charge carrier motion.

Modulus of elasticity (E). The ratio of stress to strain when deformation is totally elastic; also a measure of the stiffness of a material.

Molarity (M). Concentration in a liquid solution, in terms of the number of moles of a solute dissolved in 10^6 mm^3 (10^3 cm^3) of solution.

Molding (plastics). Shaping a plastic material by forcing it, under pressure and at an elevated temperature, into a mold cavity.

Mole. The quantity of a substance corresponding to 6.023×10^{23} atoms or molecules.

Molecular chemistry (polymer). With regard only to composition, not the structure of a mer.

Molecular structure (polymer). With regard to atomic arrangements within and interconnections between polymer molecules.

Molecular weight. The sum of the atomic weights of all the atoms in a molecule.

Molecule. A group of atoms that are bound together by primary interatomic bonds.

Monomer. A molecule consisting of a single mer.

MOSFET. Metal-oxide-silicon field effect transistor, an integrated circuit element.

N

n-Type semiconductor. A semiconductor for which the predominant charge carriers responsible for electrical conduction are electrons.

Normally, donor impurity atoms give rise to the excess electrons.

Natural aging. For precipitation hardening, aging at room temperature.

Network polymer. A polymer composed of trifunctional mer units that form three-dimensional molecules.

Nodular iron. See **Ductile iron.**

Noncrystalline. The solid state wherein there is no long-range atomic order. Sometimes the terms *amorphous*, *glassy*, and *vitreous* are used synonymously.

Nonferrous alloy. A metal alloy for which iron is *not* the prime constituent.

Nonsteady-state diffusion. The diffusion condition for which there is some net accumulation or depletion of diffusing species. The diffusion flux is dependent on time.

Normalizing. For ferrous alloys, austenitizing above the upper critical temperature, then cooling in air. The objective of this heat treatment is to enhance toughness by refining the grain size.

Nucleation. The initial stage in a phase transformation. It is evidenced by the formation of small particles (nuclei) of the new phase, which are capable of growing.

O

Octahedral position. The void space among close-packed, hard sphere atoms or ions for which there are six nearest neighbors. An octahedron (double pyramid) is circumscribed by lines constructed from centers of adjacent spheres.

Ohm's law. The applied voltage is equal to the product of the current and resistance; equivalently, the current density is equal to the product of the conductivity and electric field intensity.

Opaque. Being impervious to the transmission of light as a result of absorption, reflection, and/or scattering of incident light.

Overaging. During precipitation hardening, aging beyond the point

at which strength and hardness are at their maxima.

Oxidation. The removal of one or more electrons from an atom, ion, or molecule.

P

Paramagnetism. A relatively weak form of magnetism that results from the independent alignment of atomic dipoles (magnetic) with an applied magnetic field.

Particle-reinforced composite. A composite for which the dispersed phase is equiaxed.

Passivity. The loss of chemical reactivity, under particular environmental conditions, by some active metals and alloys.

Pauli exclusion principle. The postulate that for an individual atom, at most two electrons, which necessarily have opposite spins, can occupy the same state.

Pearlite. A two-phase microstructure found in some steels and cast irons; it results from the transformation of austenite of eutectoid composition and consists of alternating layers (or lamellae) of α-ferrite and cementite.

Periodic table. The arrangement of the chemical elements with increasing atomic number according to the periodic variation in electron structure. Nonmetallic elements are positioned at the far right-hand side of the table.

Peritectic reaction. A reaction wherein, upon cooling, a solid and a liquid phase transform isothermally and reversibly to a solid phase having a different composition.

Permeability (magnetic, μ). The proportionality constant between B and H fields. The value of the permeability of a vacuum (μ_0) is 1.257×10^{-6} H/m.

Permittivity (ϵ). The proportionality constant between the dielectric displacement D and the electric field $\mathscr{E}$. The value of the permittivity ϵ_0 for a vacuum is 8.85×10^{-12} F/m.

Phase. A homogeneous portion of a system that has uniform physical and chemical characteristics.

Phase diagram. A graphical representation of the relationships between environmental constraints (e.g., temperature and sometimes pressure), composition, and regions of phase stability, ordinarily under conditions of equilibrium.

Phase equilibrium. See **Equilibrium (phase)**.

Phase transformation. A change in the number and/or character of the phases that constitute the microstructure of an alloy.

Phonon. A single quantum of vibrational or elastic energy.

Phosphorescence. Luminescence that occurs at times greater than on the order of a second after an electron excitation event.

Photoconductivity. Electrical conductivity that results from photon-induced electron excitations in which light is absorbed.

Photomicrograph. The photograph made with a microscope, which records a microstructural image.

Photon. A quantum unit of electromagnetic energy.

Piezoelectric. A dielectric material in which polarization is induced by the application of external forces.

Pilling–Bedworth ratio (P–B ratio). The ratio of metal oxide volume to metal volume; used to predict whether or not a scale that forms will protect a metal from further oxidation.

Pitting. A form of very localized corrosion wherein small pits or holes form, usually in a vertical direction.

Plain carbon steel. A ferrous alloy in which carbon is the prime alloying element.

Planck's constant (h). A universal constant that has a value of 6.63×10^{-34} J-s. The energy of a photon of electromagnetic radiation is the product of h and the radiation frequency.

Plane strain. The condition, important in fracture mechanical analyses, wherein, for tensile loading, there is zero strain in a direction perpendicular to both the stress axis and the direction of crack propagation; this condition is found in thick plates, and the zero-strain direction is perpendicular to the plate surface.

Plane strain fracture toughness (K_{Ic}). The critical value of the stress intensity factor (i.e., at which crack propagation occurs) for the condition of plane strain.

Plastic. A solid material the primary ingredient of which is an organic polymer of high molecular weight; it may also contain additives such as fillers, plasticizers, flame retardants, and the like.

Plastic deformation. Deformation that is permanent or nonrecoverable after release of the applied load. It is accompanied by permanent atomic displacements.

Plasticizer. A low molecular weight polymer additive that enhances flexibility and workability and reduces stiffness and brittleness.

Point defect. A crystalline defect associated with one or, at most, several atomic sites.

Poisson's ratio (ν). For elastic deformation, the negative ratio of lateral and axial strains that result from an applied axial stress.

Polar molecule. A molecule in which there exists a permanent electric dipole moment by virtue of the asymmetrical distribution of positively and negatively charged regions.

Polarization (P). The total electric dipole moment per unit volume of dielectric material. Also, a measure of the contribution to the total dielectric displacement by a dielectric material.

Polarization (corrosion). The displacement of an electrode potential from its equilibrium value as a result of current flow.

Polarization (electronic). For an atom, the displacement of the center of the negatively charged electron cloud relative to the positive nucleus, which is induced by an electric field.

Polarization (ionic). Polarization as a result of the displacement of anions and cations in opposite directions.

Polarization (orientation). Polarization resulting from the alignment (by rotation) of permanent electric dipole moments with an applied electric field.

Polycrystalline. Referring to crystalline materials that are composed of more than one crystal or grain.

Polymer. A solid, nonmetallic (normally organic) compound of high molecular weight the structure of which is composed of small repeat (or mer) units.

Polymer-matrix composite (PMC). A composite material for which the matrix is a polymer resin, and having fibers (normally glass, carbon, or aramid) as the dispersed phase.

Polymorphism. The ability of a solid material to exist in more than one form or crystal structure.

Powder metallurgy (P/M). The fabrication of metal pieces having intricate and precise shapes by the compaction of metal powders, followed by a densification heat treatment.

Precipitation hardening. Hardening and strengthening of a metal alloy by extremely small and uniformly dispersed particles that precipitate from a supersaturated solid solution; sometimes also called *age hardening*.

Precipitation heat treatment. A heat treatment used to precipitate a new phase from a supersaturated solid solution. For precipitation hardening, it is termed *artificial aging*.

Prepreg. Continuous fiber reinforcement preimpregnated with a polymer resin that is then partially cured.

Prestressed concrete. Concrete into which compressive stresses have been introduced using steel wires or rods.

Primary bonds. Interatomic bonds that are relatively strong and for which bonding energies are relatively large. Primary bonding types are ionic, covalent, and metallic.

Primary phase. A phase that exists in addition to the eutectic structure.

Principle of combined action. The supposition, often valid, that new properties, better properties, better property combinations, and/or a higher level of properties can be fashioned by the judicious combination of two or more distinct materials.

Process annealing. Annealing of previously cold-worked products (commonly steel alloys in sheet or wire form) below the lower critical (eutectoid) temperature.

Proeutectoid cementite. Primary cementite that exists in addition to pearlite for hypereutectoid steels.

Proeutectoid ferrite. Primary ferrite that exists in addition to pearlite for hypoeutectoid steels.

Property. A material trait expressed in terms of the measured response to a specific imposed stimulus.

Proportional limit. The point on a stress–strain curve at which the straight line proportionality between stress and strain ceases.

p-Type semiconductor. A semiconductor for which the predominant charge carriers responsible for electrical conduction are holes. Normally, acceptor impurity atoms give rise to the excess holes.

Q

Quantum mechanics. A branch of physics that deals with atomic and subatomic systems; it allows only discrete values of energy that are separated from one another. By contrast, for classical mechanics, continuous energy values are permissible.

Quantum numbers. A set of four numbers, the values of which are used to label possible electron states. Three of the quantum numbers are integers, which also specify the size, shape, and spatial orientation of an electron's probability density; the fourth number designates spin orientation.

R

Random copolymer. A polymer in which two different mer units are randomly distributed along the molecular chain.

Recovery. The relief of some of the internal strain energy of a previously cold-worked metal, usually by heat treatment.

Recrystallization. The formation of a new set of strain-free grains within a previously cold-worked material; normally an annealing heat treatment is necessary.

Recrystallization temperature. For a particular alloy, the minimum temperature at which complete recrystallization will occur within approximately one hour.

Rectifying junction. A semiconductor $p–n$ junction that is conductive for a current flow in one direction and highly resistive for the opposite direction.

Reduction. The addition of one or more electrons to an atom, ion, or molecule.

Reflection. Deflection of a light beam at the interface between two media.

Refraction. Bending of a light beam upon passing from one medium into another; the velocity of light differs in the two media.

Refractory. A metal or ceramic that may be exposed to extremely high temperatures without deteriorating rapidly or without melting.

Reinforced concrete. Concrete that is reinforced (or strengthened in tension) by the incorporation of steel rods, wires, or mesh.

Relative magnetic permeability (μ_r). The ratio of the magnetic permeability of some medium to that of a vacuum.

Relaxation frequency. The reciprocal of the minimum reorientation time for an electric dipole within an alternating electric field.

Relaxation modulus [$E_r(t)$]. For viscoelastic polymers, the time-dependent modulus of elasticity. It is determined from stress relaxation measurements as the ratio of stress (taken at some time after the load application—normally 10 s) to strain.

Remanence (remanent induction, B_r). For a ferromagnetic or ferrimagnetic material, the magnitude of residual flux density that remains when a magnetic field is removed.

Residual stress. A stress that persists in a material that is free of external forces or temperature gradients.

Resilience. The capacity of a material to absorb energy when it is elastically deformed.

Resistivity (ρ). The reciprocal of electrical conductivity, and a measure of a material's resistance to the passage of electric current.

Resolved shear stress. An applied tensile or compressive stress resolved into a shear component along a specific plane and direction within that plane.

Reverse bias. The insulating bias for a $p–n$ junction rectifier; electrons flow into the p side of the junction.

Rolling. A metal-forming operation that reduces the thickness of sheet stock; also elongated shapes may be fashioned using grooved circular rolls.

Rule of mixtures. The properties of a multiphase alloy or composite material are a weighted average (usually on the basis of volume) of the properties of the individual constituents.

Rupture. Failure that is accompanied by significant plastic deformation; often associated with creep failure.

S

Sacrificial anode. An active metal or alloy that preferentially corrodes and protects another metal or alloy to which it is electrically coupled.

Safe stress (σ_w). A stress used for design purposes; for ductile metals, it is the yield strength divided by a factor of safety.

Sandwich panel. A type of structural composite consisting of two stiff and strong outer faces that are separated by a lightweight core material.

Saturated. A term describing a carbon atom that participates in only single covalent bonds with four other atoms.

Saturation magnetization, flux density (M_s, B_s). The maximum magnetization (or flux density) for a ferromagnetic or ferrimagnetic material.

Scanning electron microscope (SEM). A microscope that produces an image by using an electron beam that scans the surface of a specimen; an image is produced by reflected electron beams. Examination of surface and/or microstructural features at high magnifications is possible.

Scanning probe microscope (SPM). A microscope that does not produce an image using light radiation. Rather, a very small and sharp probe raster scans across the specimen surface; out-of-surface plane deflections in response to electronic or other interactions with the probe are monitored, from which a topographical map of the specimen surface (on a nanometer scale) is produced.

Schottky defect. In an ionic solid, a defect consisting of a cation–vacancy and anion–vacancy pair.

Scission. A polymer degradation process whereby molecular chain bonds are ruptured by chemical reactions or by exposure to radiation or heat.

Screw dislocation. A linear crystalline defect associated with the lattice distortion created when normally parallel planes are joined together to form a helical ramp. The Burgers vector is parallel to the dislocation line.

Secondary bonds. Interatomic and intermolecular bonds that are relatively weak and for which bonding energies are relatively small. Normally atomic or molecular dipoles are involved. Secondary bonding types are van der Waals and hydrogen.

Selective leaching. A form of corrosion wherein one element or constituent of an alloy is preferentially dissolved.

Self-diffusion. Atomic migration in pure metals.

Self-interstitial. A host atom or ion that is positioned on an interstitial lattice site.

Semiconductor. A nonmetallic material that has a filled valence band at 0 K and a relatively narrow energy band gap. The room temperature electrical conductivity ranges between about 10^{-6} and 10^4 (Ω-m)$^{-1}$.

Shear. A force applied so as to cause or tend to cause two adjacent parts of the same body to slide relative to each other, in a direction parallel to their plane of contact.

Shear strain (γ). The tangent of the shear angle that results from an applied shear load.

Shear stress (τ). The instantaneous applied shear load divided by the original cross-sectional area across which it is applied.

Single crystal. A crystalline solid for which the periodic and repeated atomic pattern extends throughout its entirety without interruption.

Sintering. Particle coalescence of a powdered aggregate by diffusion that is accomplished by firing at an elevated temperature.

Slip. Plastic deformation as the result of dislocation motion; also, the shear displacement of two adjacent planes of atoms.

Slip casting. A forming technique used for some ceramic materials. A slip, or suspension of solid particles in water, is poured into a porous mold. A solid layer forms on the inside wall as water is absorbed by the mold, leaving a shell (or ultimately a solid piece) having the shape of the mold.

Slip system. The combination of a crystallographic plane and, within that plane, a crystallographic direction along which slip (i.e., dislocation motion) occurs.

Softening point (glass). The maximum temperature at which a glass piece may be handled without permanent deformation; this corresponds to a viscosity of approximately 4×10^6 Pa-s (4×10^7 P).

Soft magnetic material. A ferromagnetic or ferrimagnetic material having a small B versus H hysteresis loop, which may be magnetized and demagnetized with relative ease.

Soldering. A technique for joining metals using a filler metal alloy that has a melting temperature less than about 425°C (800°F). Lead–tin alloys are common solders.

Solid solution. A homogeneous crystalline phase that contains two or more chemical species. Both substitutional and interstitial solid solutions are possible.

Solid-solution strengthening. Hardening and strengthening of metals that result from alloying in which a solid solution is formed. The presence of impurity atoms restricts dislocation mobility.

Solidus line. On a phase diagram, the locus of points at which solidification is complete upon equilibrium cooling, or at which melting begins upon equilibrium heating.

Solubility limit. The maximum concentration of solute that may be added without forming a new phase.

Solute. One component or element of a solution present in a minor concentration. It is dissolved in the solvent.

Solution heat treatment. The process used to form a solid solution by dissolving precipitate particles. Often, the solid solution is supersaturated and metastable at ambient conditions as a result of rapid cooling from an elevated temperature.

Solvent. The component of a solution present in the greatest amount. It is the component that dissolves a solute.

Solvus line. The locus of points on a phase diagram representing the limit of solid solubility as a function of temperature.

Specific heat (c_p, c_v). The heat capacity per unit mass of material.

Specific modulus (specific stiffness). The ratio of elastic modulus to specific gravity for a material.

Specific strength. The ratio of tensile strength to specific gravity for a material.

Spheroidite. Microstructure found in steel alloys consisting of sphere-like cementite particles within an α-ferrite matrix. It is produced by an appropriate elevated-temperature heat treatment of pearlite, bainite, or martensite, and is relatively soft.

Spheroidizing. For steels, a heat treatment carried out at a temperature just below the eutectoid in which the spheroidite microstructure is produced.

Spherulite. An aggregate of ribbonlike polymer crystallites radiating from a common center, which crystallites are separated by amorphous regions.

Spinning. The process by which fibers are formed. A multitude of fibers are spun as molten material is forced through many small orifices.

Stabilizer. A polymer additive that counteracts deteriorative processes.

Stainless steel. A steel alloy that is highly resistant to corrosion in a variety of environments. The predominant alloying element is chromium, which must be present in a concentration of at least 11 wt%; other alloy additions, to include nickel and molybdenum, are also possible.

Standard half-cell. An electrochemical cell consisting of a pure metal immersed in a $1M$ aqueous solution of its ions, which is electri-cally coupled to the standard hydrogen electrode.

Steady-state diffusion. The diffusion condition for which there is no net accumulation or depletion of diffusing species. The diffusion flux is independent of time.

Stereoisomerism. Polymer isomerism in which side groups within mer units are bonded along the molecular chain in the same order, but in different spatial arrangements.

Stoichiometry. For ionic compounds, the state of having exactly the ratio of cations to anions specified by the chemical formula.

Strain, engineering (ϵ). The change in gauge length of a specimen (in the direction of an applied stress) divided by its original gauge length.

Strain hardening. The increase in hardness and strength of a ductile metal as it is plastically deformed below its recrystallization temperature.

Strain point (glass). The maximum temperature at which glass fractures without plastic deformation; this corresponds to a viscosity of about 3×10^{13} Pa-s (3×10^{14} P).

Strain, true. See **True strain.**

Stress concentration. The concentration or amplification of an applied stress at the tip of a notch or small crack.

Stress corrosion (cracking). A form of failure that results from the combined action of a tensile stress and a corrosion environment; it occurs at lower stress levels than are required when the corrosion environment is absent.

Stress, engineering (σ). The instantaneous load applied to a specimen divided by its cross-sectional area before any deformation.

Stress intensity factor (K). A factor used in fracture mechanics to specify the stress intensity at the tip of a crack.

Stress raiser. A small flaw (internal or surface) or a structural discontinuity at which an applied tensile stress will be amplified and from which cracks may propagate.

Stress relief. A heat treatment for the removal of residual stresses.

Stress, true. See **True stress.**

Structural clay products. Ceramic products made principally of clay and used in applications where structural integrity is important (e.g., bricks, tiles, pipes).

Structural composite. A composite the properties of which depend on the geometrical design of the structural elements. Laminar composites and sandwich panels are two subclasses.

Structure. The arrangement of the internal components of matter: electron structure (on a subatomic level), crystal structure (on an atomic level), and microstructure (on a microscopic level).

Substitutional solid solution. A solid solution wherein the solute atoms replace or substitute for the host atoms.

Superconductivity. A phenomenon observed in some materials: the disappearance of the electrical resistivity at temperatures approaching 0 K.

Supercooling. Cooling to below a phase transition temperature without the occurrence of the transformation.

Superheating. Heating to above a phase transition temperature without the occurrence of the transformation.

Syndiotactic. A type of polymer chain configuration in which side groups regularly alternate positions on opposite sides of the chain.

System. Two meanings are possible: (1) a specific body of material that is being considered, and (2) a series of possible alloys consisting of the same components.

T

Temper designation. A letter–digit code used to designate the mechanical and/or thermal treatment to which a metal alloy has been subjected.

Tempered martensite. The microstructural product resulting from a tempering heat treatment of a martensitic steel. The microstructure consists of extremely small and uniformly dispersed cementite particles embedded within a continuous α-ferrite matrix. Toughness and ductility are enhanced significantly by tempering.

Tempering (glass). See **Thermal tempering.**

Tensile strength (TS). The maximum engineering stress, in tension, that may be sustained without fracture. Often termed *ultimate (tensile) strength*.

Terminal solid solution. A solid solution that exists over a composition range extending to either composition extremity of a binary phase diagram.

Tetrahedral position. The void space among close-packed, hard sphere atoms or ions for which there are four nearest neighbors.

Thermal conductivity (k). For steady-state heat flow, the proportionality constant between the heat flux and the temperature gradient. Also, a parameter characterizing the ability of a material to conduct heat.

Thermal expansion coefficient, linear (α_l). The fractional change in length divided by the change in temperature.

Thermal fatigue. A type of fatigue failure wherein the cyclic stresses are introduced by fluctuating thermal stresses.

Thermal shock. The fracture of a brittle material as a result of stresses that are introduced by a rapid temperature change.

Thermal stress. A residual stress introduced within a body resulting from a change in temperature.

Thermal tempering. Increasing the strength of a glass piece by the introduction of residual compressive stresses within the outer surface using an appropriate heat treatment.

Thermally activated transformation. A reaction that depends on atomic thermal fluctuations; the atoms having energies greater than an activation energy will spontaneously react or transform. The rate of this type of transformation depends on temperature according to Equation 11.3.

Thermoplastic (polymer). A polymeric material that softens when heated and hardens upon cooling. While in the softened state, articles may be formed by molding or extrusion.

Thermoplastic elastomer (TPE). A copolymeric material that exhibits elastomeric behavior yet is thermoplastic in nature. At the ambient temperature, domains of one mer type form at molecular chain ends that act as physical crosslinks.

Thermosetting (polymer). A polymeric material that, once having cured (or hardened) by a chemical reaction, will not soften or melt when subsequently heated.

Tie line. A horizontal line constructed across a two-phase region of a binary phase diagram; its intersections with the phase boundaries on either end represent the equilibrium compositions of the respective phases at the temperature in question.

Time–temperature–transformation (T–T–T) diagram. See **Isothermal transformation diagram.**

Toughness. A measure of the amount of energy absorbed by a material as it fractures. Toughness is indicated by the total area under the material's tensile stress–strain curve.

Trans. For polymers, a prefix denoting a type of molecular structure. To some unsaturated carbon chain atoms within a mer unit, a single side atom or group may be situated on one side of the chain, or directly opposite at a 180° rotation position. In a trans structure, two such side groups within the same mer reside on opposite chain sides (e.g., *trans*-isoprene).

Transformation rate. The reciprocal of the time necessary for a reaction to proceed halfway to its completion.

Transgranular fracture. Fracture of polycrystalline materials by crack propagation through the grains.

Translucent. Having the property of transmitting light only diffusely; objects viewed through a translucent medium are not clearly distinguishable.

Transmission electron microscope (TEM). A microscope that produces an image by using electron beams that are transmitted (pass through) the specimen. Examination of internal features at high magnifications is possible.

Transparent. Having the property of transmitting light with relatively little absorption, reflection, and scattering, such that objects viewed through a transparent medium can be distinguished readily.

Transverse direction. A direction that crosses (usually perpendicularly) the longitudinal or lengthwise direction.

Trifunctional mer. Designating mer units that have three active bonding positions.

True strain (ϵ_T). The natural logarithm of the ratio of instantaneous gauge length to original gauge length of a specimen being deformed by a uniaxial force.

True stress (σ_T). The instantaneous applied load divided by the instantaneous cross-sectional area of a specimen.

U

Ultimate (tensile) strength. See **Tensile strength.**

Ultrahigh molecular weight polyethylene (UHMWPE). A polyethylene polymer that has an extremely high molecular weight (approximately 4×10^6 g/mol). Distinctive characteristics of this material include high impact and abrasion resistance, and a low coefficient of friction.

Unit cell. The basic structural unit of a crystal structure. It is generally defined in terms of atom (or ion) positions within a parallelepiped volume.

Unsaturated. A term describing carbon atoms that participate in double or triple covalent bonds and, therefore, do not bond to a maximum of four other atoms.

Upper critical temperature. For a steel alloy, the minimum temperature above which, under equilibrium conditions, only austenite is present.

V

Vacancy. A normally occupied lattice site from which an atom or ion is missing.

Vacancy diffusion. The diffusion mechanism wherein net atomic migration is from lattice site to an adjacent vacancy.

Valence band. For solid materials, the electron energy band that contains the valence electrons.

Valence electrons. The electrons in the outermost occupied electron shell, which participate in interatomic bonding.

van der Waals bond. A secondary interatomic bond between adjacent molecular dipoles, which may be permanent or induced.

Viscoelasticity. A type of deformation exhibiting the mechanical characteristics of viscous flow and elastic deformation.

Viscosity (η). The ratio of the magnitude of an applied shear stress to the velocity gradient that it produces; that is, a measure of a noncrystalline material's resistance to permanent deformation.

Vitrification. During firing of a ceramic body, the formation of a liquid phase that upon cooling becomes a glass-bonding matrix.

Vulcanization. Nonreversible chemical reaction involving sulfur or other suitable agent wherein crosslinks are formed between molecular chains in rubber materials. The rubber's modulus of elasticity and strength are enhanced.

W

Wave-mechanical model. Atomic model in which electrons are treated as being wavelike.

Weight percent (wt%). Concentration specification on the basis of weight (or mass) of a particular element relative to the total alloy weight (or mass).

Weld decay. Intergranular corrosion that occurs in some welded stainless steels at regions adjacent to the weld.

Welding. A technique for joining metals in which actual melting of the pieces to be joined occurs in the vicinity of the bond. A filler metal may be used to facilitate the process.

Whisker. A very thin, single crystal of high perfection that has an extremely large length-to-diameter ratio. Whiskers are used as the reinforcing phase in some composites.

White cast iron. A low-silicon and very brittle cast iron, in which the carbon is in combined form as cementite; a fractured surface appears white.

Whiteware. A clay-based ceramic product that becomes white after high-temperature firing; whitewares include porcelain, china, and plumbing sanitary ware.

Working point (glass). The temperature at which a glass is easily deformed, which corresponds to a viscosity of 10^3 Pa-s (10^4 P).

Wrought alloy. A metal alloy that is relatively ductile and amenable to hot working or cold working during fabrication.

Y

Yielding. The onset of plastic deformation.

Yield strength (σ_y). The stress required to produce a very slight yet specified amount of plastic strain; a strain offset of 0.002 is commonly used.

Young's modulus. See **Modulus of elasticity.**

Answers to Selected Problems

Chapter 2

2.3 **(a)** 1.66×10^{-24} g/amu;

(b) 2.73×10^{26} atoms/lb-mol

2.13

$$r_0 = \left(\frac{A}{nB}\right)^{1/(1-n)}$$

$$E_0 = -\frac{A}{\left(\dfrac{A}{nB}\right)^{1/(1-n)}} + \frac{B}{\left(\dfrac{A}{nB}\right)^{n/(1-n)}}$$

2.14 **(c)** $r_0 = 0.279$ nm; $E_0 = -4.57$ eV

2.19 63.2% for TiO_2; 1.0% for InSb

Chapter 3

3.3 $V_C = 6.62 \times 10^{-29}$ m^3

3.9 $R = 0.136$ nm

3.12 **(a)** $V_C = 1.40 \times 10^{-28}$ m^3;

(b) $a = 0.323$ nm, $c = 0.515$ nm

3.15 Metal B: face-centered cubic

3.17 **(a)** $n = 8.0$; **(b)** $\rho = 4.96$ g/cm^3

3.20 $V_C = 8.63 \times 10^{-2}$ nm^3

3.29 **(a)** Cesium chloride; **(c)** sodium chloride

3.31 APF = 0.73

3.35 **(a)** $a = 0.421$ nm; $a = 0.424$ nm

3.37 **(a)** ρ (calculated) = 4.11 g/cm^3;

(b) ρ (measured) = 4.10 g/cm^3

3.39 **(a)** $\rho = 4.20$ g/cm^3

3.41 Cesium chloride

3.43 APF = 0.84

3.45 APF = 0.68

3.50 **(a)** Direction 1: [012]; **(b)** Plane 1: (020)

3.52 Direction A: [$0\bar{1}\bar{1}$]; Direction C: [112]

3.53 Direction B: [$2\bar{3}2$]; Direction D: [$13\bar{6}$]

3.54 **(b)** [$\bar{1}\bar{1}0$], [$\bar{1}10$], and [$1\bar{1}0$]

3.56 Plane B: ($\bar{1}\bar{1}2$) or ($11\bar{2}$)

3.57 Plane A: ($32\bar{2}$)

3.58 Plane B: (221)

3.60 **(a)** ($1\bar{1}00$)

3.64 **(a)** (100) and ($0\bar{1}0$)

3.65 **(c)** [010]

3.66 **(a)** FCC; **(b)** tetrahedral; **(c)** one half

3.68 **(a)** octahedral; **(b)** all

3.70* [100]: LD = 0.71

3.71* [111]: LD = 1.0

3.72* (100): PD = 0.79

3.73* (110): PD = 0.83

3.79* $2\theta = 81.38°$

3.80* $d_{110} = 0.2862$ nm

3.82* **(a)** $d_{321} = 0.1523$ nm; **(b)** $R = 0.2468$ nm

3.84* $d_{110} = 0.2015$ nm; $a = 0.285$ nm

Chapter 4

4.4 $n_n = 23,700$

4.6 **(a)** $\overline{M} = 33,040$ g/mol; **(c)** $n_n = 785$

4.9 **(a)** $C_{Cl} = 20.3$ wt%

4.11 $L = 1254$ nm; $r = 15.4$ nm

4.16 8530 of both styrene and butadiene mers

4.18 Propylene

4.21 f(isoprene) = 0.88; f(isobutylene) = 0.12

4.28 $\rho = 0.998$ g/cm^3

4.30 **(a)** $\rho_a = 2.000$ g/cm^3, $\rho_c = 2.301$ g/cm^3;

(b) % crystallinity = 87.9%

Chapter 5

5.1 $N_v/N = 2.41 \times 10^{-5}$

5.3 $Q_v = 1.10$ eV/atom

5.4 6.02×10^{28} atoms/m^3

5.9 For FCC, $r = 0.41R$

5.10 **(a)** O^{2-} vacancy; one O^{2-} vacancy for every two Li^+ added

5.13 $C_{Pb} = 10.0$ wt%; $C_{Sn} = 90.0$ wt%

5.15 $C'_{Sn} = 72.5$ at%; $C'_{Pb} = 27.5$ at%

5.18* $C'_{Fe} = 94.2$ at%; $C'_{Si} = 5.8$ at%

5.23 $N_{Au} = 3.36 \times 10^{21}$ atoms/cm^3

5.27 $C_{Nb} = 35.2$ wt%

5.30 **(a)** FCC: $\mathbf{b} = \dfrac{a}{2}[110]$;

(b) Al: $|\mathbf{b}| = 0.2862$ nm

5.35 $d \cong 0.07$ mm

5.37 **(a)** $N = 8$

5.D1* $C_{Li} = 1.537$ wt%

Chapter 6

6.6 $M = 2.6 \times 10^{-3}$ kg/h

6.8 $D = 3.9 \times 10^{-11}$ m^2/s

6.11 $t = 19.7$ h

6.15 $t = 40$ h

6.18 $T = 1152$ K (879°C)

6.21 **(a)** $Q_d = 252.4$ kJ/mol, $D_0 = 2.2 \times 10^{-5}$ m^2/s

(b) $D = 5.3 \times 10^{-15}$ m^2/s

6.24 $T = 1044$ K (771°C)

6.29 $x = 1.6$ mm

6.D1 Not possible

Chapter 7

7.4 $l_0 = 250$ mm (10 in.)

7.7 **(a)** $F = 89,400$ N (20,000 lb$_f$)

(b) $l_i = 115.28$ mm (4.511 in.)

7.9 $\Delta l = 0.10$ mm (0.004 in.)

7.12

$$\left(\frac{dF}{dr}\right)_{r_0} = -\frac{2A}{\left(\dfrac{A}{nB}\right)^{3/(1-n)}} + \frac{(nB)(n+1)}{\left(\dfrac{A}{nB}\right)^{(n+2)/(1-n)}}$$

7.14 **(a)** $\Delta l = 0.50$ mm (0.02 in.);

(b) $\Delta d = -1.62 \times 10^{-2}$ mm (-6.2×10^{-4} in.); decrease

7.15 $F = 16,250$ N (3770 lb$_f$)

7.16 $\nu = 0.280$

7.18 $E = 170.5$ GPa (24.7×10^6 psi)

7.21 **(a)** $\Delta l = 0.10$ mm (4×10^{-3} in.);

(b) $\Delta d = -3.6 \times 10^{-3}$ mm (-1.4×10^{-4} in.)

7.24 Steel

7.27 **(a)** Both elastic and plastic;

(b) $\Delta l = 4.0$ mm (0.16 in.)

7.29 **(b)** $E = 62$ GPa (9×10^6 psi)

(c) $\sigma_y = 285$ MPa (41,500 psi)

(d) $TS = 370$ MPa (53,500 psi)

(e) %EL = 16%

(f) $U_r = 0.66 \times 10^6$ J/m^2 (95.7 in.-lb$_m$/in.^{3})

7.32 Figure 7.12: $U_r = 3.32 \times 10^5$ J/m^3 (47.6 in.-lb$_f$/in.^{3})

7.34 $\sigma_y = 381$ MPa (55,500 psi)

7.39 $\epsilon_T = 0.237$

7.41 $\sigma_T = 440$ MPa (63,700 psi)

7.43 Toughness = 3.65×10^9 J/m^3 (5.29×10^5 in.-lb$_m$/in.^{3})

7.45 $n = 0.134$

7.47 **(a)** ϵ (elastic) $\cong 0.0027$; ϵ (plastic) $\cong 0.0023$

(b) $l_i = 461.1$ mm (18.05 in.)

7.49 $R = 4$ mm

7.50 $F_f = 10,100$ N (2165 lb$_f$)

7.52* **(a)** $E_0 = 342$ GPa; **(b)** $E = 280$ GPa

7.54* **(b)** $P = 0.186$

7.59* $E_r(10) = 4.25$ MPa (616 psi)

7.65 **(a)** 125 HB (70 HRB)

7.70 Figure 7.12: $\sigma_w = 125$ MPa (18,000 psi)

7.D2 **(a)** $\Delta x = 2.5$ mm; **(b)** $\sigma = 10$ MPa

Chapter 8

8.10* $\cos \lambda \cos \phi = 0.408$

8.12* **(b)** $\tau_{crss} = 0.80$ MPa (114 psi)

8.13* $\tau_{crss} = 0.45$ MPa (65.1 psi)

8.20 $d = 1.48 \times 10^{-2}$ mm

8.21 $d = 6.9 \times 10^{-3}$ mm

8.25 $r_d = 8.25$ mm

8.27 $r_0 = 10.6$ mm (0.424 in.)

8.29 $\tau_{crss} = 20.2$ MPa (2920 psi)

8.35 **(b)** $t \cong 150$ min

8.36 **(b)** $d = 0.085$ mm

8.45* $TS = 44$ MPa

8.54 Fraction sites vulcanized = 0.180

8.56 Fraction of mer sites crosslinked = 0.47

8.D1 Is possible

8.D6 Cold work to between 21 and 23%CW [to $d_0' \cong 12.8$ mm (0.50 in.)], anneal, then cold work to give a final diameter of 11.3 mm (0.445 in.).

Chapter 9

9.3 $\sigma_m = 2404$ MPa (354,000 psi)

9.7 $\rho_t = 0.39$ nm

9.8 $\sigma_c = 16.2$ MPa

9.10* **(a)** $\sigma_x = 171$ MPa (25,000 psi), $\sigma_y = 247$ MPa (35,800 psi)

(d) $\sigma_x = 41.7$ MPa (6050 psi), $\sigma_y = 126$ MPa (18,300 psi)

9.12* **(a)** $\sigma_m = 170$ MPa (24,650 psi)

9.13* **(a)** $\sigma_m = 120$ MPa (17,400 psi)

9.15* Aluminum 2024-T3: $B \geq 40.6$ mm (1.6 in.); 4340 steel (tempered at 260°C): $B \geq 2.3$ mm (0.10 in.)

9.17 Fracture will not occur

9.19 $a = 24$ mm (0.95 in.)

9.21 Is not subject to detection since $a < 4.0$ mm

9.26 **(b)** -105°C; **(c)** -95°C

9.29 **(a)** $\sigma_{max} = 275$ MPa (40,000 psi), $\sigma_{min} = -175$ MPa ($-25,500$ psi);

(b) $R = -0.64$; **(c)** $\sigma_r = 450$ MPa (65,500 psi)

9.31 $N_f = 1 \times 10^5$ cycles

9.33 **(b)** $S = 250$ MPa; **(c)** $N_f \cong 2.2 \times 10^6$ cycles

9.34 **(a)** $\tau = 130$ MPa; **(c)** $\tau = 195$ MPa

9.36 **(a)** $t = 120$ min; **(c)** $t = 220$ h

9.47 $\Delta\epsilon/\Delta t = 7.0 \times 10^{-3}$ min^{-1}

9.48 $\Delta l = 7.1$ mm (0.29 in.)

9.50 $t_r = 36,000$ h

9.52* 427°C: $n = 5.3$

9.53* **(a)** $Q_c = 186,200$ J/mol

9.55* $\dot{\epsilon}_s = 0.118$ (h)$^{-1}$

9.D1* $K_{Ic} = 67.9$ MPa$\sqrt{m}$ (62.3 ksi$\sqrt{in.}$)

9.D3* $W \geq 4.4$ mm

9.D7* $\sigma_{max} = 178$ MPa

9.D9* $a_c = 0.25$ in.

9.D14* $T = 991$ K (718°C)

9.D16* For 5 years: $\sigma = 260$ MPa (37,500 psi)

Chapter 10

10.5 **(a)** $\epsilon + \eta$; $C_\epsilon = 87$ wt% Zn-13 wt% Cu, $C_\eta = 97$ wt% Zn-3 wt% Cu;

(c) Liquid; $C_L = 55$ wt% Ag-45 wt% Cu;

(e) $\beta + \gamma$; $C_\beta = 49$ wt% Zn-51 wt% Cu, $C_\gamma = 57$ wt% Zn-43 wt% Cu;

(g) α; $C_\alpha = 63.8$ wt% Ni-36.2 wt% Cu

10.7 **(a)** $W_\epsilon = 0.70$, $W_\eta = 0.30$;

(c) $W_L = 1.0$;

(e) $W_\beta = 0.50$, $W_\gamma = 0.50$;

(g) $W_\alpha = 1.0$

10.9 **(a)** $V_\epsilon = 0.70$, $V_\eta = 0.30$

10.11 **(a)** $T = 300$°C (570°F)

10.12 **(a)** $m_s = 5022$ g;

(b) $C_L = 64$ wt% sugar;

(c) $m_s = 2355$ g

10.13 **(a)** The pressure must be raised to approximately 570 atm

10.18 Is possible

10.21 **(a)** $T = 550$°C (1020°F);

(b) $C_\alpha = 22$ wt% Pb-78 wt% Mg;

(c) $T = 465$°C (870°F);

(d) $C_L = 66$ wt% Pb-34 wt% Mg

10.24 **(a)** $T \cong 230$°C (445°F);

(b) $C_\alpha = 15$ wt% Sn; $C_L = 42$ wt% Sn

10.25* **(a)** $C = 45.9$ wt% Al_2O_3–54.1 wt% SiO_2

10.26 $C_\alpha = 90$ wt% A-10 wt% B; $C_\beta = 20.2$ wt% A-79.8 wt% B

10.28   Not possible

10.32* Is possible

10.35* $C_0 = 82.4$ wt% Sn-17.6 wt% Pb

10.38*

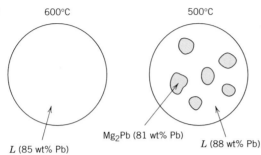

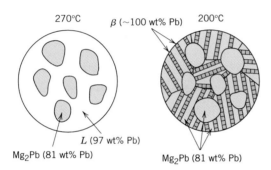

10.42   Eutectics: **(1)** 12 wt% Nd, 632°C, $L \rightarrow$ Al + $Al_{11}Nd_3$;

   (2) 97 wt% Nd, 635°C, $L \rightarrow AlNd_3$ + Nd;
   Congruent melting point: 73 wt% Nd, 1460°C, $L \rightarrow Al_2Nd$
   Peritectics: **(1)** 59 wt% Nd, 1235°C, L + $Al_2Nd \rightarrow Al_{11}Nd_3$;

   (2) 84 wt% Nd, 940°C, L + $Al_2Nd \rightarrow$ AlNd;

   (3) 91 wt% Nd, 795°C, L + AlNd $\rightarrow$ $AlNd_2$;

   (4) 94 wt% Nd, 675°C, L + $AlNd_2 \rightarrow$ $AlNd_3$
   No eutectoids are present.

10.45* **(a)** 8.1% of Mg^{2+} vacancies

10.47* For point B, $F = 2$

10.54 $C_0' = 0.42$ wt% C

10.57   **(a)** α-ferrite; **(b)** 2.26 kg of ferrite, 0.24 kg of Fe_3C;

   (c) 0.38 kg of proeutectoid ferrite, 2.12 kg of pearlite

10.59 $C_0 = 0.55$ wt% C

10.61 $C_0 = 0.61$ wt% C

10.64   Possible

10.67   Two answers are possible: $C_0 = 1.11$ wt% C and 0.72 wt% C

10.70   HB (alloy) = 128

10.73* **(a)** T (eutectoid) = 650°C (1200°F); **(b)** ferrite; **(c)** $W_{\alpha'} = 0.68$, $W_p = 0.32$

Chapter 11

11.2 $t = 305$ s

11.4 $r = 4.42 \times 10^{-3}$ min^{-1}

11.6 $y = 0.51$

11.7   **(c)** $t \cong 250$ days

11.10   **(b)** 265 HB (27 HRC)

11.14   **(a)** 50% coarse pearlite and 50% martensite; **(d)** 100% martensite; **(e)** 40% bainite and 60% martensite; **(g)** 100% fine pearlite

11.16   **(a)** martensite; **(c)** bainite; **(e)** ferrite, medium pearlite, bainite, and martensite; **(g)** proeutectoid ferrite, pearlite, and martensite

11.19* **(a)** martensite

11.24* **(a)** martensite; **(c)** martensite, proeutectoid ferrite, and bainite

11.34   **(b)** 180 HB (87 HRB); **(g)** 265 IIB (27 HRC)

11.36* **(c)** $TS = 915$ MPa (132,500 psi)

11.37   **(a)** Rapidly cool to about 675°C (1245°F), hold for at least 200 s, then cool to room temperature

11.D1   Not possible

11.D5   Temper at between 400 and 450°C (750 and 840°F) for 1 h

11.D8   For about 10 h at 149°C, or between about 35 and 400 h at 121°C

Chapter 12

12.2 $d = 1.88$ mm

12.5   **(a)** $R = 4.7 \times 10^{-3}$ Ω; **(b)** $I = 10.6$ A; **(c)** $J = 1.5 \times 10^6$ A/m^2; $\mathscr{E} = 2.5 \times 10^{-2}$ V/m

12.12 $\sigma = 0.096$ (Ω-m)$^{-1}$

12.13   **(a)** $n = 1.25 \times 10^{29}$ m^{-3};

   (b) 1.48 free electrons/atom

12.16 (a) $\rho_0 = 1.58 \times 10^{-8}$ Ω-m, $a = 6.5 \times 10^{-11}$ Ω-m/°C;

(b) $A = 1.12 \times 10^{-6}$ Ω-m;

(c) $\rho = 4.26 \times 10^{-8}$ Ω-m

12.18 $\sigma = 7.3 \times 10^6$ (Ω-m)$^{-1}$

12.21 (b) for Si, 2.7×10^{-13}; for Ge, 5.6×10^{-10}

12.30 (a) $n = 8.9 \times 10^{21}$ m^{-3}; (b) p-type extrinsic

12.33 $\mu_e = 0.50$ m^2/V-s; $\mu_h = 0.02$ m^2/V-s

12.38 $E_g = 1.46$ eV

12.39 (b) $\sigma = 3.8 \times 10^{-5}$ (Ω-m)$^{-1}$

12.40 (a) $\sigma = 61.4$ (Ω-m)$^{-1}$; (b) $n = p = 1.16 \times 10^{21}$ m^{-3}

12.45* $B_z = 0.58$ tesla

12.52* $l = 1.6$ mm

12.56* $p_i = 2.26 \times 10^{-30}$ C-m

12.58* (a) $V = 17.3$ V; (b) $V = 86.5$ V;

(e) $P = 1.75 \times 10^{-7}$ C/m^2

12.60* Fraction of ϵ_r due to $P_i = 0.67$

12.D2 $\sigma = 2.44 \times 10^7$(Ω-m)$^{-1}$

12.D3 Possible; 30 wt% $< C_{Ni} <$ 32.5 wt%

Chapter 13

13.5 $V_{Gr} = 11.1$ vol%

13.21* (a) $T = 2000$°C (3630°F)

13.23* (a) $W_L = 0.86$; (c) $W_L = 0.66$

13.24* (a) 1890°C (3435°F); between ~77 and 100 wt% Al$_2$O$_3$

Chapter 14

14.11 (a) 890–920°C (1635–1690°F)

14.12 (b) 790–815°C (1450–1500°F)

14.26 (b) $Q_{vis} = 362$ kJ/mol

14.43 (a) m(adipic acid) $= 117.7$ kg

(b) m(polyester) $= 153.2$ kg

14.D5 Maximum diameter $= 75$ mm (3 in.)

14.D7 Maximum diameter $= 70$ mm (2.75 in.)

Chapter 15

15.4 $k_{max} = 33.3$ W/m-K; $k_{min} = 29.7$ W/m-K

15.9 $\tau_c = 34.5$ MPa

15.12 Possible

15.14 $E_f = 70.4$ GPa (10.2×10^6 psi); $E_m = 2.79$ GPa (4.04×10^5 psi)

15.17 (a) $F_f/F_m = 23.4$;

(b) $F_f = 42,676$ N (9590 lb$_f$), $F_m = 1824$ N (410 lb$_f$)

(c) $\sigma_f = 445$ MPa (63,930 psi), $\sigma_m = 8.14$ MPa (1170 psi);

(d) $\epsilon = 3.4 \times 10^{-3}$

15.19 $\sigma_{cl}^* = 633$ MPa (91,700 psi)

15.21 $\sigma_{cd}^* = 1340$ MPa (194,400 psi)

15.28 $E_{cl} = 69.1$ GPa (10.0×10^6 psi)

15.D1 Carbon (PAN standard-modulus) and aramid

15.D2 Not possible

Chapter 16

16.5 (a) $\Delta V = +0.031$ V;

(b) Fe^{2+} + Cd $\longrightarrow$ Fe + Cd^{2+}

16.7 [Pb^{2+}] $= 2.5 \times 10^{-2}$ M

16.13 $t = 10$ yr

16.16 CPR $= 5.24$ mpy

16.20 (a) $r = 8.0 \times 10^{-14}$ mol/cm^2-s;

(b) $V_C = -0.019$ V

16.34 Sn: P-B ratio $= 1.33$; protective

16.36 (a) Parabolic kinetics; (b) $W = 1.51$ mg/cm^2

Chapter 17

17.2 $T_f = 49$°C (120°F)

17.4 (a) $c_v = 139$ J/kg-K; (b) $c_v = 925$ J/kg-K

17.9 $\Delta l = -9.2$ mm (-0.36 in.)

17.14 $T_f = 129.5$°C

17.16 (b) $dQ/dt = 9.3 \times 10^8$ J/h

17.24 k(upper) $= 26.4$ W/m-K

17.28 (a) $\sigma = 150$ MPa (21,800 psi); compression

17.29 $T_f = 39$°C (102°F)

17.30 $\Delta d = 0.0251$ mm

17.D1 $T_f = 42.2$°C (108°F)

17.D4 Glass ceramic: $\Delta T_f = 317$°C

Chapter 18

18.1 (a) $H = 10,000$ A-turns/m;

(b) $B_0 = 1.257 \times 10^{-2}$ tesla;

(c) $B \cong 1.257 \times 10^{-2}$ tesla;

(d) $M = 1.81$ A/m

18.6 **(a)** $\mu = 1.26 \times 10^{-6}$ H/m;
 (b) $\chi_m = 6 \times 10^{-3}$

18.8 **(a)** $M_s = 1.45 \times 10^6$ A/m

18.16 4.6 Bohr magnetons/Mn^{2+} ion

18.24 $M_s = 1.69 \times 10^6$ A/m

18.27 **(b)** $\mu_i = 3.0 \times 10^{-3}$ H/m, $\mu_{ri} = 2400$;
 (c) $\mu(max) \cong 9 \times 10^{-3}$ H/m

18.29 **(b)** **(i)** $\mu \cong 1.0 \times 10^{-2}$ H/m; **(iii)** $\chi_m \cong$ 7954

18.32 **(a)** 2.5 K: 1.33×10^4 A/m; **(b)** 5.96 K

Chapter 19

19.9 $v = 2.09 \times 10^8$ m/s

19.10 Silica: 0.53; soda-lime glass: 0.33

19.11 Fused silica: $\epsilon_r = 2.13$; polyethylene: $\epsilon_r = 2.28$

19.19 $I'_T/I'_0 = 0.81$

19.21 $l = 67.3$ mm

19.30 $\Delta E = 1.78$ eV

Chapter 20

20.D2 Stiffness: $P = \dfrac{\sqrt{G}}{\rho}$

20.D3 Stiffness $P = \dfrac{\sqrt{E}}{\rho}$; strength: $P = \dfrac{\sigma_y^{2/3}}{\rho}$

20.D6 **(a)** $F = 21.5$ N (4.8 lb_f);
 (b) $F = 53.6$ N (12.0 lb_f)

Index

Page numbers in *italics* refer to the glossary.

501

Unit Conversion Factors

Length

$1\text{ m} = 10^{10}\text{ Å}$ $\qquad$ $1\text{ Å} = 10^{-10}\text{ m}$

$1\text{ m} = 10^{9}\text{ nm}$ $\qquad$ $1\text{ nm} = 10^{-9}\text{ m}$

$1\text{ m} = 10^{6}\text{ }\mu\text{m}$ $\qquad$ $1\text{ }\mu\text{m} = 10^{-6}\text{ m}$

$1\text{ m} = 10^{3}\text{ mm}$ $\qquad$ $1\text{ mm} = 10^{-3}\text{ m}$

$1\text{ m} = 10^{2}\text{ cm}$ $\qquad$ $1\text{ cm} = 10^{-2}\text{ m}$

$1\text{ mm} = 0.0394\text{ in.}$ $\qquad$ $1\text{ in.} = 25.4\text{ mm}$

$1\text{ cm} = 0.394\text{ in.}$ $\qquad$ $1\text{ in.} = 2.54\text{ cm}$

$1\text{ m} = 3.28\text{ ft}$ $\qquad$ $1\text{ ft} = 0.3048\text{ m}$

Area

$1\text{ m}^2 = 10^{4}\text{ cm}^2$ $\qquad$ $1\text{ cm}^2 = 10^{-4}\text{ m}^2$

$1\text{ mm}^2 = 10^{-2}\text{ cm}^2$ $\qquad$ $1\text{ cm}^2 = 10^{2}\text{ mm}^2$

$1\text{ m}^2 = 10.76\text{ ft}^2$ $\qquad$ $1\text{ ft}^2 = 0.093\text{ m}^2$

$1\text{ cm}^2 = 0.1550\text{ in.}^2$ $\qquad$ $1\text{ in.}^2 = 6.452\text{ cm}^2$

Volume

$1\text{ m}^3 = 10^{6}\text{ cm}^3$ $\qquad$ $1\text{ cm}^3 = 10^{-6}\text{ m}^3$

$1\text{ mm}^3 = 10^{-3}\text{ cm}^3$ $\qquad$ $1\text{ cm}^3 = 10^{3}\text{ mm}^3$

$1\text{ m}^3 = 35.32\text{ ft}^3$ $\qquad$ $1\text{ ft}^3 = 0.0283\text{ m}^3$

$1\text{ cm}^3 = 0.0610\text{ in.}^3$ $\qquad$ $1\text{ in.}^3 = 16.39\text{ cm}^3$

Mass

$1\text{ Mg} = 10^{3}\text{ kg}$ $\qquad$ $1\text{ kg} = 10^{-3}\text{ Mg}$

$1\text{ kg} = 10^{3}\text{ g}$ $\qquad$ $1\text{ g} = 10^{-3}\text{ kg}$

$1\text{ kg} = 2.205\text{ lb}_{\text{m}}$ $\qquad$ $1\text{ lb}_{\text{m}} = 0.4536\text{ kg}$

$1\text{ g} = 2.205 \times 10^{-3}\text{ lb}_{\text{m}}$ $\qquad$ $1\text{ lb}_{\text{m}} = 453.6\text{ g}$

Density

$1\text{ kg/m}^3 = 10^{-3}\text{ g/cm}^3$ $\qquad$ $1\text{ g/cm}^3 = 10^{3}\text{ kg/m}^3$

$1\text{ Mg/m}^3 = 1\text{ g/cm}^3$ $\qquad$ $1\text{ g/cm}^3 = 1\text{ Mg/m}^3$

$1\text{ kg/m}^3 = 0.0624\text{ lb}_{\text{m}}/\text{ft}^3$ $\qquad$ $1\text{ lb}_{\text{m}}/\text{ft}^3 = 16.02\text{ kg/m}^3$

$1\text{ g/cm}^3 = 62.4\text{ lb}_{\text{m}}/\text{ft}^3$ $\qquad$ $1\text{ lb}_{\text{m}}/\text{ft}^3 = 1.602 \times 10^{-2}\text{ g/cm}^3$

$1\text{ g/cm}^3 = 0.0361\text{ lb}_{\text{m}}/\text{in.}^3$ $\qquad$ $1\text{ lb}_{\text{m}}/\text{in.}^3 = 27.7\text{ g/cm}^3$

Force

$1\text{ N} = 10^{5}\text{ dynes}$ $\qquad$ $1\text{ dyne} = 10^{-5}\text{ N}$

$1\text{ N} = 0.2248\text{ lb}_{\text{f}}$ $\qquad$ $1\text{ lb}_{\text{f}} = 4.448\text{ N}$

Stress

$1\text{ MPa} = 145\text{ psi}$ $\qquad$ $1\text{ psi} = 6.90 \times 10^{-3}\text{ MPa}$

$1\text{ MPa} = 0.102\text{ kg/mm}^2$ $\qquad$ $1\text{ kg/mm}^2 = 9.806\text{ MPa}$

$1\text{ Pa} = 10\text{ dynes/cm}^2$ $\qquad$ $1\text{ dyne/cm}^2 = 0.10\text{ Pa}$

$1\text{ kg/mm}^2 = 1422\text{ psi}$ $\qquad$ $1\text{ psi} = 7.03 \times 10^{-4}\text{ kg/mm}^2$

Fracture Toughness

$1\text{ psi }\sqrt{\text{in.}} = 1.099 \times 10^{-3}\text{ MPa }\sqrt{\text{m}}$ $\qquad$ $1\text{ MPa }\sqrt{\text{m}} = 910\text{ psi }\sqrt{\text{in.}}$

Energy

$1\text{ J} = 10^{7}\text{ ergs}$ $\qquad$ $1\text{ erg} = 10^{-7}\text{ J}$

$1\text{ J} = 6.24 \times 10^{18}\text{ eV}$ $\qquad$ $1\text{ eV} = 1.602 \times 10^{-19}\text{ J}$

$1\text{ J} = 0.239\text{ cal}$ $\qquad$ $1\text{ cal} = 4.184\text{ J}$

$1\text{ J} = 9.48 \times 10^{-4}\text{ Btu}$ $\qquad$ $1\text{ Btu} = 1054\text{ J}$

$1\text{ J} = 0.738\text{ ft-lb}_{\text{f}}$ $\qquad$ $1\text{ ft-lb}_{\text{f}} = 1.356\text{ J}$

$1\text{ eV} = 3.83 \times 10^{-20}\text{ cal}$ $\qquad$ $1\text{ cal} = 2.61 \times 10^{19}\text{ eV}$

$1\text{ cal} = 3.97 \times 10^{-3}\text{ Btu}$ $\qquad$ $1\text{ Btu} = 252.0\text{ cal}$